A. K. M
May 1991.

PHYSICS OF MATERIALS

PHYSICS OF MATERIALS

Richard J. Weiss
King's College, London

HEMISPHERE PUBLISHING CORPORATION
A member of the Taylor & Francis Group

New York Washington Philadelphia London

PHYSICS OF MATERIALS

1 2 3 4 5 6 7 8 9 0   B R B R   9 8 7 6 5 4 3 2 1 0

This book is revised and augmented from the previous edition, entitled *Solid State Physics for Metallurgists*.
Cover design by Renee Winfield.

A CIP catalogue record for this book is available from the British Library.

Library of Congress Cataloging-in-Publication Data

Weiss, Richard J. (Richard Jerome), date.
Physics of materials / Richard J. Weiss.
p. cm.

1. Solid state physics. 2. Materials. I. Title.
QC176.W38 1990
530.4′1—dc20 89-27222
CIP

ISBN 0-89116-968-7

To

A. W., M. W., D. P. W.

and C. W.

CONTENTS

Theory[*]

[*]The first four chapters of the book are primarily for reference and the student can set his own place for studying this material.

Experiment

FOREWORD

Engineers in general and metallurgists in particular are seeking a more basic understanding of the properties that make materials useful to mankind; consequently, they are evidencing deeper interest in the electronic structure of matter, which has become the domain of solid-state physics. Unfortunately, this branch of physics suffers from a tremendous gap between the quantum theory which deals fundamentally with the positions and momenta of electrons on the one hand, and the experimental observations which reflect the macroscopic properties on the other. For example, the electrical resistivity of a metal is relatively easy to measure but extremely difficult to calculate from first principles.

Pedagogically, one may enter solid-state physics from either the theoretical or experimental points of view. The former approach is the common one, perhaps because of the power and elegance of the Schrödinger equation, but the subject soon becomes preoccupied with the simplifying assumptions that must be invoked to solve the electronic structure of even a single atom. In contrast, the author has adopted the experimental alternative as a device for introducing the student to solid-state physics. Accordingly, this book is intended to enable the student to learn how qualitative and quantitative information concerning the electronic structure of matter is obtained from important classes of experiments; how various types of measurements are combined to permit the synthesis of electronic models; and of equal significance, what limitations are encountered in this procedure.

The net result of this approach is that the student becomes immersed in solid-state physics in a realistic and stimulating way. It seems especially appropriate for non-physics majors who have a firmer footing in measureable phenomena than in pure theory.

Massachusetts Institute of Technology

MORRIS COHEN
Ford Professor of Materials Science and Engineering

PREFACE

After 25 years the original concept for *Solid State Physics for Metallurgists* still seemed sufficiently valid to issue this edition under the current title. University metallurgy departments have broadened their outlook and changed their names to materials engineering, but this reflects the recognition that solid materials of all descriptions drive our technology.

Composites, lasers, optical fibers, and optoelectronics have changed our society in this quarter century and metals, per se, are no longer sacrosanct. One traditionally defined metals as electrical conductors but ceramic superconductors have become a household word. No seer could have anticipated these developments at the time this book was written; one can only teach the past and rely on accidental discovery and the marvels of human ingenuity to shape the future.

This brief introduction is only meant to give the flavor of the last 25 years and I would like to begin with a little story. About five years after the first edition appeared, I was taken by surprise when I received a Russian translation in the mail. Examining the package I discovered that the Moscow publisher decided to omit both royalties and page xiii (p. xxi, this edition). On a subsequent conference trip to the USSR an Academician friend dug into his "personal" account to give me some rubles to offset the failure of the publisher to pay royalties. I used this money (it had to be spent in Russia) to organize a roundtable discussion among the 20 westerners attending the conference. We gathered in a private lounge at the hotel and each ordered vodka. The waiters appeared with twenty uncorked bottles and my rubles disappeared! As to page xiii, I hope that *glasnost* will restore it in the next Russian reprint.

Composites. These materials are based on the combination of strong fibers held together by a glue (matrix). In normal service, the fibers sustain the load and the weaker matrix is only necessary to keep the composite together. Wood, banana skins, hair, and celery stalks are a few examples of natural composites while wattle and daub or straw and dried cow dung are ancient human-produced examples. The principal inservice stresses are along the fiber axis, although there are composites employing short (chopped) fibers in random orientations where the load is uniformly distributed in three directions.

Common examples of human-produced composites are fiberglass in epoxy and graphite in epoxy. Both are distinguished by high strength to weight ratios with the former reasonably inexpensive. Other than keeping the costs of the fibers down, the principal technological problem is to ensure that the fiber and matrix do not delaminate.

Graphite fibers have a high modulus of elasticity along the fiber axis. These are fabricated from hydrocarbon (polymeric) precursors by driving the hydrogen off at elevated temperatures and leaving the pure carbon crystallites aligned along the strong *a* axis. When imbedded in a metal matrix like aluminium, these composites provide a high strength-to-density ratio desirable in an airborne scenario. The military have been the principal consumers of such composites, since the high cost of hundreds of dollars per kilogram limits its commercial application.

Boron, sapphire, and SiC are further examples of expensive fibers whose development is supported by the military. Nonetheless, the philosophy of combining materials with differing properties to produce a synergistic end product is aggressively pursued today and should continue to provide the modus operandi in future efforts.

Lasers. When Einstein produced his theory of stimulated emission in the early part of this century, no one (including Einstein) could have anticipated the technological revolution it would achieve. The laser was invented in the early 1960s, a few years after Einstein died, and promises to outshine even the transistor in its sociological impact.

If one excites an atom to a higher energy state and a photon of that energy passes close to that atom, it will hasten its return to the ground state. The photon emitted in this return will be in phase with the stimulating photon and travelling in the same direction, so that a box full of excited atoms will undergo a chain reaction as photon after photon induce their clones. Suddenly all the atoms have returned to their ground state and a pulse of light will emerge that is well collimated, monochromatic, and in phase. The photon that performs the stimulation merely acts as a catalyst and is not altered in any discernible way.

Lasers of all shapes and sizes are commercially available and have been employed in such diverse technologies as fingerprint detection, healing of skin burns, surgery, weapons, engraving, printing, non destructive testing, gyroscopes, and machining. Soon to be realized is optical computing, destined to enhance memory storage and speed by many orders of magnitude.

Optical fibers. An optical fiber consists of a core, and an outer cladding whose index of refraction is about one percent less. When laser light is launched down the fiber the cladding acts as an optical reflector and prevents the photons from escaping. High purity silica fibers can transmit light over kilometer distances without appreciable attenuation and a systematic replacement of copper telephone lines is underway.

If the core is of small diameter relative to the cladding, the fiber is monomode and will transmit a single frequency, while a large diameter core is described as multimode and can transmit a range of frequencies.

The advantages of optical fibers over copper wires for transmitting information are numerous. Photons can be launched in both directions at the same time, can maintain separate frequencies, can be polarized, are insensitive to electromagnetic interference, and can maintain their phase. This variety of parameters provides a versatility that is only beginning to be realized.

The field of optical fiber sensors is growing rapidly. If optical fibers are subjected to changes in temperature, stress, pressure, chemical environment, etc., this alters the local index of refraction and affects the optical transmission. By appropriate design of the fibers this technology can be employed to measure anything imaginable.

Silica (glass) is not the only fiber material. ZrF_2, sapphire, quartz, alkali halides etc. are all considered possibilities as one tailors the fiber to the need. It is interesting that in most schools it has been the electrical engineering department that has added optical fibers to its curricula.

Superconductivity. This phenomenon has been with us for over 75 years but has found limited applicability due to the low temperatures required. The possibilitry of finding a room temperature superconductor is perceived to have multibillion dollar capabilities and is all the fashion in research.

Superconductors have zero resistance because this is the ground state of the system. The current carrying electrons are in constant motion, the applied electric potential merely straightens out some of these paths. It is a group cooperative motion, like magnetic order in a ferromagnet, and must be part of the so-called zero-point motion of the atoms—the quantum mechanical ground state of constant motion even at absolute zero (p. 76). The superconducting behavior is locked into this zero point motion in such a way that it is isolated from the phonons that are excited as one raises the temperature. This contrasts with the normal metallic state in which the electrons must be accelerated up to speed with the applied potential.

The BCS theory (p. 264) has not succeeded in providing guidelines as to what materials will be superconductors, and with the discovery of the high temperature oxide superconductors, theoreticians believe that a separate theory is required. It can only mean that the electrons are too confused to know what is happening to them.

Nonetheless, the prospects for a room temperature superconductor in the next 25 years must be high. If so, we might find optical fibers replaced by superconductors! (More likely a clever combination of the two will emerge). After the extensive effort to find this blessed material subsides, physicists may reexamine the precise electronic details of the superconducting state and produce a clear enough picture to present in a book like this.

Optoelectronics. Semiconductors like GaAs can be doped to create the conditions for stimulated emission of excited electronic states. This produces a material in which the photons and electrons are intimately coupled. One interesting possibility is that the excitation of electrons to the excited state alters the index of refraction. This topic comes under the umbrella of nonlinear optics—the light impinging on the GaAs excites the electrons, changes the index of refraction, and alters the optical properties for subsequent transmitted photons, i.e., the optical properties depend on the light intensity. A vast variety of materials with interesting optoelectronic interactions are being developed for optical computing applications.

Richard J. Weiss

ACKNOWLEDGEMENT

I wish to thank John Antal, David Chipman, John DeMarco, Arthur Freeman, Ralph Harrison, Anton Hofmann, Lawrence Jennings, Al Marotta, Mary Norton and Kenneth Tauer for their help and contributions to this volume. I should also like to thank Mr. J. Gilkes for the art work.

On Mount Gerizim in Samaria dwelt the "miracle maker" called Ramuth Gilead. There came to him the first born of old and ailing King Cyrus, young Bezer, who begged for infinite wisdom to lead his people.

"Infinite wisdom I cannot grant, but I bestow upon you all the wisdom of mankind. Go forth, and use your knowledge wisely."

And so Bezer returned to his people and he was crowned King upon the death of his father.

Many years passed and his people prospered and they blessed Ramuth Gilead for the wisdom of their King. But in the fourteenth year of his reign King Bezer again made the journey to Mount Gerizim, and he came before the "miracle maker" saying,

"In my sweet youth I knew not of the Heavens and the Earth and of the science and so I did sleep well. But I have drunk deeply of the wisdom you have bestowed upon me and have looked into the unknown. The many questions I now ask myself keep me from my sleep."

And answered Ramuth Gilead,

"That is the price you must pay for wisdom."

And King Bezer, who was now full grown of beard, thought deeply and asked,

"How long must this endure?"

And Ramuth Gilead answered,

"Until you know nothing."

CHAPTER I

THE ATOM

A. Quantum Mechanics and the Hydrogen Atom

Introduction to Quantum Mechanics

A logical place for a metallurgist to begin his study of solid state physics is with the rudiments of quantum mechanics. Quantum mechanics was developed because classical mechanics (i.e. Newton's laws of motion) failed to explain the behaviour of atoms. As an example consider the hydrogen atom consisting of one electron and one proton with an attractive Coulomb potential between them. Newton's laws of motion would allow the electron any value of energy in its orbit about the proton as long as the Coulomb attractive force which varies inversely as the square of the distance between the proton and electron is balanced by the outward centrifugal force of the electron. However, experiments on the hydrogen atom inform us that the electron appears to have only very specific or discrete values of energy. A further problem arises in regard to some other classical equations of physics (Maxwell's equations) which have to do with moving charges and electromagnetic fields. These equations predict that the negatively charged electron whirling around the proton would continuously radiate energy (in the form of light) and gradually be drawn into the proton as it slowed down. This is contrary to the fact that hydrogen atoms can exist in nature without continuously emitting light. After much trial and error quantum mechanics was discovered in 1926 and explained these discrepancies.

The Schrödinger Equation

In quantum mechanics we replace the classical equations of physics (like $\mathbf{F} = m\mathbf{a}$) with the Schrödinger equation. The discovery of this equation followed the revolutionary suggestion of de Broglie that not only light but all matter, electrons, neutrons and even metallurgists, possessed a wavelength.*

* The de Broglie wavelength is given by $\lambda = h/mv = h/p$ where p is the momentum, v the velocity, m the mass and h Planck's constant. A proton travelling with the velocity of a rifle bullet would have a wavelength of approximately $1{\cdot}8 \times 10^{-8}$ cm or some 10,000 times greater than its size whereas a metallurgist shot out of a cannon would have a wavelength $\sim 10^{-35}$ cm considerably shorter than most metallurgists. The wave nature of particles has been demonstrated for cases where the wavelength is greater than the size of the particle as in neutron and electron diffraction.

The Schrödinger equation is in fact a wave equation for particles. In this equation the particle possesses a wave function, Ψ, which is a function of position x, y, z and which has the following properties.

1. The momentum of the particle in the three directions x, y, z is proportional to the slope of the wave function divided by the wave function. More specifically

$$p_x = -i\hbar \frac{\partial \Psi}{\partial x}\Big/\Psi; \quad p_y = -i\hbar \frac{\partial \Psi}{\partial y}\Big/\Psi; \quad p_z = -i\hbar \frac{\partial \Psi}{\partial z}\Big/\Psi$$

where p_x, p_y, p_z are the momenta, $i = \sqrt{-1}$ and $\hbar$ is Planck's constant divided by 2π.

2. The kinetic energy of the particle, $(p_x{}^2 + p_y{}^2 + p_z{}^2)/2m$ is proportional to the second derivate or curvature of the wave function divided by the wave function

$$\text{K.E.} = -\frac{\hbar^2}{2m}\left(\frac{\partial^2 \Psi}{\partial x^2} + \frac{\partial^2 \Psi}{\partial y^2} + \frac{\partial^2 \Psi}{\partial z^2}\right)\Big/\Psi \tag{1}$$

3. The absolute magnitude squared of the wave function $|\Psi|^2$ (i.e. square the real and the imaginary parts of Ψ separately and add them) is equal to the probability per unit volume that the particle is at the position x, y, z.

The third property is in contrast to classical mechanics which fixes the position of a particle at a given time. Unfortunately, one of the limitations of quantum mechanics is that it only tells us the fraction of time (or saying the same thing the probability) that a particle will be at a given position x, y, z. For example referring back to the hydrogen atom it is possible to scatter X-rays from atomic hydrogen gas. Upon analyzing the data we would find the electrons in a whole range of positions relative to their nuclei, some positions more often than others. We express this mathematically by assigning a probability to each position.

The Hydrogen Atom

Let us now apply the Schrödinger equation to the hydrogen atom. Since the proton is 1847 times heavier than the electron it remains relatively fixed and we place it at the centre of our co-ordinate system and focus our attention on the electron.* The Schrödinger equation is simply written as the kinetic energy of the electron as expressed in eqn. (1) plus the potential energy of the electron and set equal to the total energy. For this problem it is con-

* Strictly speaking the centre of the coordinate system is at the *centre of mass* and both proton and electron are considered relative to it. The electron mass is corrected by a factor $(1 - m_0/M)$ where M is the mass of the nucleus and m_0 is the rest mass of the electron. The correction is always small.

siderably simpler to use spherical co-ordinates, r, θ, ϕ instead of cubic, x, y, z and the Schrödinger equation becomes

$$-\frac{\hbar^2}{2m\Psi}\left(\frac{\partial^2\Psi}{\partial r^2}+\frac{2}{r}\frac{\partial\Psi}{\partial r}+\frac{1}{r^2\sin^2\theta}\frac{\partial^2\Psi}{\partial\phi^2}+\frac{1}{r^2}\frac{\partial^2\Psi}{\partial\theta^2}+\frac{1}{r^2\tan\theta}\frac{\partial\Psi}{\partial\theta}\right)-\frac{e^2}{r}=E \quad (2)$$

where $-e^2/r$ is the attractive Coulomb potential energy due to the proton's positive charge, E is the total energy of the electron, and $m = m_0(1 - m_0/M)$.

Having written down the Schrödinger equation we must now solve it in order to determine Ψ as a function of r, θ, ϕ. The solution of eqn. (2) appears quite formidable but if we make the mathematical substitution $\Psi = R\cdot\Theta\cdot\Phi$ where R only depends on r, Θ only on θ and Φ only on ϕ the mathematics reduces to relatively simple equations in R, Θ and Φ.

Now there are many solutions to this equation but we discard most of the solutions because the wave functions, Ψ, are not what are termed "well behaved". Since the wave function enables us to determine the probability of finding the electron at some position r, θ, ϕ we impose the sensible restrictions that it should only have a single value for each position, it should never become infinite anywhere and it should approach zero as r, the distance from the electron to the nucleus, approaches infinity.* Only a few of the many wave functions which are solutions to the Schrödinger equation obey these restrictions and only for specific values of the energy E. The wave functions which are "well behaved" are called *eigenfunctions* and each of the specific energy values which yield "well behaved" wave functions are called *eigenvalues*. (While the corresponding values of momenta are also called eigenvalues we shall limit our discussion to energy eigenvalues.)

Since it will be useful to us, we shall write down the complete solution for a somewhat more general case than the hydrogen atom, that of the hydrogenic atom in which the charge on the nucleus is allowed to vary but in each case retaining only one electron. This corresponds to solutions for the series H, He^+, Li^{+2}, Be^{+3} etc.

$$R=\left(\frac{4[(n-l-1)!]Z^3}{[(n+l)!]^3n^4a_1^3}\right)^{1/2}\left(\frac{2Zr}{na_1}\right)^l e^{-Zr/na_1}\left[L_{n+1}^{2l+1}\left(\frac{2Zr}{na_1}\right)\right] \quad (3)$$

$$\Theta=\left(\frac{(2l+1)[(l-|m_l|)!]}{2[(l+|m_l|)!]}\right)^{1/2}\sin^{|m_l|}\theta[P_l^{|m_l|}(\cos\theta)] \quad (4)$$

$$\Phi=\frac{e^{im_l\phi}}{\sqrt{2\pi}} \quad (5)$$

* This arises from the restriction that the total probability of finding our electron somewhere must be 1 or stated mathematically

$$\int_0^{\pi}\int_0^{2\pi}\int_0^{\infty}|\Psi|^2r^2\sin\theta\,d\theta\,d\phi\,dr=1.$$

If $|\Psi|^2$ does not approach zero faster than $1/r^2$ for large r, the integral cannot be satisfied.

where Z is the charge on the nucleus ($Z = 1$ for hydrogen), a_1 is a constant commonly called the *Bohr radius* ($a_1 = \hbar^2/me^2 = 0{\cdot}529 \times 10^{-8}$ cm) and the remaining symbols are given below. The conditions placed on the "well-behaved" wave functions ($\Psi = R \cdot \Theta \cdot \Phi$) can be satisfied by restricting the values of l, n, and m_l in eqns. (3), (4) and (5) (called *quantum numbers*) to integers with the further restrictions that:

1. l can take any positive integer including zero. This restriction arises from the Θ-part of the wave function. A code derived by physicists designates each integral value of l by a letter $0 = s$, $1 = p$, $2 = d$, $3 = f$, $4 = g$, $5 = h$, etc. It seems silly but one gets used to it.

2. n can take any positive integer one greater than l. This restriction on n keeps the R part of the wave function "well behaved". For any given pair of values of n and l the code used by physicists is to write down the number n followed by the code letter for l. For example if $n = 4, l = 3$, we have $4f$.

3. m_l can take any integer from l to $-l$, i.e. $l, l-1, l-2, l-3 \ldots -l$. These values of m_l keep the Φ-part of the wave function well behaved. This m_l is a number and should not be confused with the electron mass, m.

4. The eigenvalues E are

$$E = -\frac{me^4Z^2}{2n^2\hbar^2} \tag{6}$$

n being the integer defined in 2. The value of $me^4/2\hbar^2$ is 13·598 eV and is commonly called a *Rydberg*.

5. The values of $\left[L_{n+1}^{2l+1}\left(\frac{2Zr}{na_1}\right)\right]$ are

n	l	Designation	$\left[L_{n+1}^{2l+1}\left(\frac{2Zr}{na_1}\right)\right]$, Laguerre Polynomials
1	0	$1s$	-1
2	0	$2s$	$\frac{2Zr}{a_1} - 4$
3	0	$3s$	$-\frac{4Z^2r^2}{3a_1^2} + \frac{12Zr}{a_1} - 18$
4	0	$4s$	$\frac{Z^3r^3}{2a_1^3} - \frac{12Z^2r^2}{a_1^2} + \frac{72Zr}{a_1} - 96$
2	1	$2p$	-6
3	1	$3p$	$\frac{16Zr}{a_1} - 96$

n	l	Designation	$\left[L_{n+1}^{2l+1}\left(\frac{2Zr}{na_1}\right)\right]$, Laguerre Polynomials
4	1	$4p$	$-\frac{15Z^2r^2}{a_1^2}+\frac{300Zr}{a_1}-1200$
3	2	$3d$	-120
4	2	$4d$	$\frac{360Zr}{a_1}-5760$
5	2	$5d$	$-\frac{2016}{5}\frac{Z^2r^2}{a_1^2}+35280\frac{Zr}{a_1}-105{,}840$
4	3	$4f$	-5040
5	3	$5f$	$16{,}128\frac{Zr}{a_1}-322{,}560$

6. The values of $[P^{|m_l|}(\cos\theta)]$, are

l	m_l	$[P^{|m_l|}(\cos\theta)]$, Associated Legendre Polynomials
0	0	1
1	1, −1	1
1	0	$\cos\theta$
2	2, −2	3
2	1, −1	$3\cos\theta$
2	0	$\frac{1}{2}(3\cos^2\theta-1)$
3	3, −3	15
3	2, −2	$15\cos\theta$
3	1, −1	$\frac{3}{2}(5\cos^2\theta-1)$
3	0	$\frac{1}{2}(5\cos^3\theta-3\cos\theta)$

The constants in the square root brackets in eqns. (3), (4) and (5) are normalization factors and ensure that the total probability of finding the electron is unity, i.e.

$$\int_0^{\pi}\int_0^{2\pi}\int_0^{\infty}|\Psi|^2r^2\sin\theta\,\mathrm{d}r\,\mathrm{d}\theta\,\mathrm{d}\phi=1.$$

Let us now look at some of these solutions by investigating the probability distribution, $|\Psi|^2$. The absolute magnitude of the wave function squared can be obtained by multiplying the wave function by its *complex conjugate*. The complex conjugate of a function is obtained by replacing i everywhere by $-i$. This is generally simpler than separately squaring the real and imaginary parts. The complex conjugate is normally designated with an asterisk so that

$|\Psi|^2 = \Psi\Psi^*$. Fig. 1 is a visual representation of the probability distribution for various electron designations (often called *configurations*) commonly met in physics. All the distributions are symmetric about the ϕ axis which is vertical and in the plane of the paper, thus permitting a two dimensional picture.

While it is not revealed in these pictures of the electron distributions of the hydrogen atom, the electron possesses an *orbital angular momentum*, $\hbar\sqrt{[l(l+1)]}$. This is the average value of the angular momentum as the electron whirls about the proton. While this average value can be experimentally measured we haven't the faintest notion of the precise path the electron follows nor its point to point angular velocity. It is even more perplexing when we state that an s electron ($l = 0$) has zero angular momentum. How then does it get about to yield the probability distributions shown in Fig. 1? Unfortunately, we again encounter a limitation of quantum mechanics in that it tells us the answer we shall get if we measure something but is unable to reveal any of the details.*

Now how do we measure this angular momentum? We do this by utilizing the charge on the electron. If a charged particle traces out a circular path it gives rise to a magnetic dipole moment (similar to a current in a wire loop). For electrons the value of this magnetic dipole moment, μ_l, is equal to the orbital angular momentum times $e/2m_0c$.

$$\mu_l = \frac{e\hbar}{2m_0c}\sqrt{[l(l+1)]} \tag{7}$$

Magnetic dipole moments are measured by application of a suitable magnetic field and such measurements confirm eqn. (7).

Another property not revealed in the probability distributions is that the electron itself has a magnetic dipole moment, μ_s, whose average value is

$$\mu_s = \frac{e\hbar}{m_0c}\sqrt{[s(s+1)]} \tag{8}$$

where s is a quantum number that gives the right value of the electron's intrinsic angular momentum, $\hbar\sqrt{[s(s+1)]}$.† This s, which is equal only to $\frac{1}{2}$, should not be confused with the code letter for $l = 0$. We shall use this notation only to keep it consistent with common usage and assure the reader that most people eventually get the "hang" of it. To explain this intrinsic

* One might envisage an s electron as moving in one direction, slowing down and then reversing its direction so that the net effect of the angular momentum is cancelled, particularly if there are many reversals in the time it takes to make a measurement. We shall probably never know the answer nor does it appear important if we do not.

† If a magnetic field is applied to a free electron the component of magnetic moment in the direction of the field is $(e\hbar/m_0c)m_s$ for the intrinsic magnetic moment. The value of this component is called the "*Bohr magneton*" ($= e\hbar/2m_0c$), *denoted* μ_B and m_s can equal either $+\frac{1}{2}$ or $-\frac{1}{2}$.

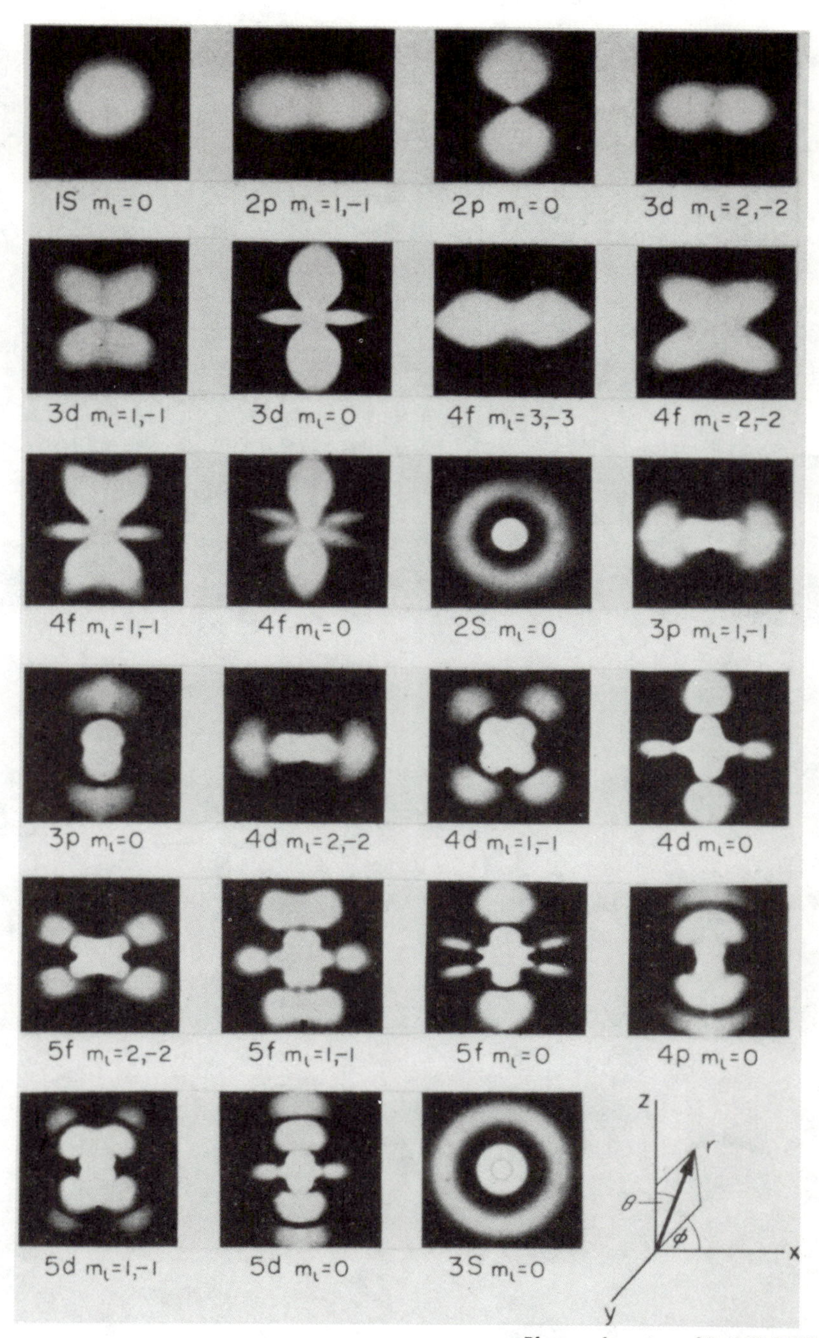

Photograph courtesy of Prof. H. E. White

FIG. 1. THE probability distribution, $|\Psi|^2$, of the hydrogen atom in several of its electronic states. These distributions are symmetric about ϕ, i.e. about the z-axis in the plane of the paper. The radial scales of these distributions are 1 Bohr radius, a_1, ($a_1 = 0{\cdot}529 \times 10^{-8}$ cm) equals approximately the following distances for each figure: 1*s*, 5·7 mm; 2*s*, 2 mm.; 3*s*, 1 mm; 2*p*, 1·8 mm; 3*p*, 1 mm; 4*p*, 0·6 mm; 3*d*, 0·9 mm; 4*d*, 0·75 mm; 5*d*, 0·4 mm; 4*f*, 0·46 mm; 5*f*, 0·35 mm.

Facing page 6

angular momentum, the electron can be envisaged as a negatively charged cloud (instead of a mathematical point) spinning on its own axis. The negative charge and mass of this cloud is spread out over a radius of about $\frac{1}{4} \times 10^{-12}$ cm. but with the negative charge spread out somewhat further than the mass such that the intrinsic magnetic dipole moment of the electron is given by an expression just twice as large as that for its orbital magnetic moment (cf. eqns. 7 and 8). This intrinsic angular momentum, commonly called the *spin* plays essentially no role in the energy or wave function of the hydrogen atom but is extremely important for atoms with two or more electrons. In these cases the spin quantum number is designated as m_s and takes the values of $+\frac{1}{2}$ or $-\frac{1}{2}$. When combined with the quantum numbers n, l and m_l these four values serve to identify each electron.

The principal experimental verification of quantum mechanics applied to the hydrogen atom arises from the ability to predict all the discrete energies the electron may possess. This is observed as the electron makes a transition from one energy state to a lower energy state. It does this by emitting a photon whose energy just equals the difference in energy of the two states. The energy of all such photons emitted from hydrogen atoms has been measured and excellent agreement is obtained with eqn. (6). While the probability distribution, $|\Psi|^2$, of the electrons is also predicted by eqn. 3, 4 and 5 and in principle could be measured with X-rays the experiment is rather difficult and has not been satisfactorily performed as yet.

Heisenberg Uncertainty Principle

We shall complete this section on the fundamentals of quantum mechanics with the *Heisenberg Uncertainty Principle.* It states that any simultaneous measurement of the position, x, and momentum, p (or energy, E, and time, t) of a particle must have an error in the product of at least $\hbar$ (Planck's constant divided by 2π) or stated mathematically,

$$\begin{aligned} \Delta p \cdot \Delta x &\gtrsim \hbar \\ \Delta E \cdot \Delta t &\gtrsim \hbar \end{aligned} \qquad (9)$$

This uncertainty arises since any measurement made on a particle to determine these quantities must disturb the particle. For example, if you wish to measure the position and momentum of an electron coming towards you, you might scatter photons from it (in order to see it). If the photon has a wavelength λ then we cannot locate the electron any more accurately than say $\Delta x \sim \lambda/2\pi$. On the other hand the photon has a momentum $p = h/\lambda$ and in the scattering process will transfer a good fraction of it to the electron so that the uncertainty in the electron's momentum is $\Delta p \sim h/\lambda$. This is just the extent to which the electron was disturbed in making the measurement. The product $\Delta x \cdot \Delta p \sim (\lambda/2\pi)(h/\lambda) \approx \hbar$.

If we make the wavelength shorter in order to measure the position more accurately then the momentum of the photon is proportionately greater and it will disturb the electron even more. Practically every physicist has tried to concoct an experiment that defies the uncertainty principle but no one has succeeded yet.

Problems

1. Devise an experiment to measure simultaneously the energy ($E = \frac{1}{2}Mv^2$) and time for a proton without the use of photons, and verify the uncertainty principle.

2. Using eqn. 6 determine the wavelength of the photons that are emitted when a hydrogen atom in its excited states ($n = 2, 3, 4$) falls to the ground state ($n = 1$). Verify this in the *Handbook of Chemistry and Physics.*

3. (For the mathematically minded). A property of eigenfunctions often used is that they are *orthogonal* which means:

$$\int_0^{2\pi}\int_0^{\pi}\int_0^{\infty} \Psi^*_{n',l',m'_l,m'_s}\Psi_{n,l,m_l,m_s} r^2\, dr \sin\theta\, d\theta\, d\phi \begin{cases} = 1 & \text{If } n = n',\ l = l',\ m_l = m'_l,\ m_s = m'_s. \\ = 0 & \text{If any of the quantum numbers differ.} \end{cases}$$

Show that if Ψ^*_{n',l',m'_l,m'_s} is the complex conjugate of the $1s$ hydrogen wave function and Ψ_{n,l,m_l,m_s} is the $3s$ hydrogen wave function that the integral vanishes, whereas the integral equals unity if $n = n' = 3$, $l = l' = 0$, $m_l = m'_l = 0$, $m_s = m'_s$.

Show that the integral vanishes for $n = n' = 2$, $l = l' = 1$, $m_l = 0$, $m'_l = +1$ and $m_s = m'_s$.

(Remember that m_s does not enter the hydrogen wave function. It is written down since it applies to all atoms or molecules involving more than one electron).

4. (For the mathematically minded). Quite often the physicist wishes to know the average value of some quantity like the average value of the kinetic energy of a $3s$ hydrogenic electron, etc. The average value of some quantity Q is given as

$$\langle Q \rangle = \int_0^{2\pi}\int_0^{\pi}\int_0^{\infty} \Psi^*_{n,l,m_l,m_s} Q_{op} \Psi_{n,l,m_l,m_s} r^2 \sin\theta\, dr\, d\theta\, d\phi$$

where Q_{op} under the integral is called an *operator* and for the following quantities whose averages are sought becomes:

Quantity sought	*Operator under integral*
1. ⟨Kinetic energy⟩	$\frac{1}{r^2}\frac{\partial}{\partial r}\left(r^2\frac{\partial}{\partial r}\right) + \frac{1}{r^2 \sin\theta}\frac{\partial}{\partial\theta}\left(\sin\theta\frac{\partial}{\partial\theta}\right) + \frac{1}{r^2\sin^2\theta}\frac{\partial^2}{\partial\phi^2}$ (spherical co-ordinates)
2. ⟨Potential energy⟩	Coulomb potential, i.e. for hydrogen $-e^2/r$
3. ⟨Energy⟩, total; ⟨H⟩	Total of potential and kinetic energy operators, commonly called the *Hamiltonian*
4. $\langle r \rangle$	r
5. $\langle r^2 \rangle$	r^2

Quantity sought	*Operator under integral*
6. $\langle r^n \rangle$	r^n
7. Angular momentum about z axis, $\langle M_z \rangle$	$-i\hbar \dfrac{\partial}{\partial \phi}$
8. Angular momentum about x axis, $\langle M_x \rangle$	$i\hbar\left(\sin \phi \dfrac{\partial}{\partial \theta} + \cot \theta \cos \phi \dfrac{\partial}{\partial \phi}\right)$
9. Angular momentum about y axis, $\langle M_y \rangle$	$i\hbar\left(-\cos \phi \dfrac{\partial}{\partial \theta} + \cot \theta \sin \phi \dfrac{\partial}{\partial \theta}\right)$
10. Linear momentum in x, y, or z direction, $\langle p_x \rangle$, $\langle p_y \rangle$, $\langle p_z \rangle$	$-i\hbar \dfrac{\partial}{\partial x}$; $-i\hbar \dfrac{\partial}{\partial y}$; $-i\hbar \dfrac{\partial}{\partial z}$

The relationship between the r, θ, ϕ axes and the x, y, z axes are given in Fig. 1. The brackets $\langle \ \rangle$ about the quantity designates the average value and is generally referred to as the *expectation value* since it is the value that is measured. The partial differentials that appear in the right-hand column differentiate the wave function Ψ_{n,l,m_l,m_s}. As an example suppose we wish to find the expectation value of the angular momentum about the z axis for the $2p$ wave function with $m_l = +1$. We have

$$\Psi = R \cdot \Theta \cdot \Phi = -\frac{r}{2\pi a_1{}^{5/2}} e^{-r/2a_1} \cdot \left(\frac{3}{4}\right)^{1/2} \sin \theta \cdot \frac{e^{i\phi}}{\sqrt{2\pi}}$$

$$\langle M_z \rangle = -i\hbar \int_0^{2\pi}\int_0^{\pi}\int_0^{\infty}\left[-\frac{r}{2\pi a_1{}^{5/2}} e^{-5/2a_1} \cdot \left(\frac{3}{4}\right)^{1/2} \sin \theta \cdot \frac{e^{i\phi}}{\sqrt{2\pi}}\right] \frac{\partial}{\partial \phi}$$

$$\left(-\frac{r}{2\pi a_1{}^{5/2}} e^{-5/2a_1} \cdot \left(\frac{3}{4}\right)^{1/2} \sin \theta \cdot \frac{e^{i\phi}}{\sqrt{2\pi}}\right) r^2 \sin \theta \, dr \, d\theta \, d\phi$$

$$\langle M_z \rangle = \hbar$$

Show that the average value of r for the $1s$, $2s$, $3s$ etc., hydrogenic wave functions is

$$\langle r \rangle = (3/2)\, a_1\, n/Z$$

B. The Helium Atom

Introduction to the Helium Problem

The next element, helium, contains two electrons, each in the Coulomb attraction of the nucleus (now double that of hydrogen). If this were the only force on the electrons then the problem would be as simple to solve as the hydrogen atom but the negative charges of the electrons produce an additional repulsive force between them. The Coulomb potential in eqn. (2) becomes $-(2e^2/r_1) - (2e^2/r_2) + (e^2/|\mathbf{r}_1 - \mathbf{r}_2|)$ where r_1 and r_2 are the distances of electron one and two from the nucleus and $\mathbf{r}_1 - \mathbf{r}_2$ the distance between the electrons. *This Coulomb repulsion between the electrons prevents us from solving the Schrödinger equation exactly (for mathematical reasons) and is the basic reason for inaccuracy in the electron theory of atoms, molecules and solids.*

As such we shall dwell somewhat on the approximate techniques the physicist uses to solve the Schrödinger equation for helium.

The proper wave function, for helium would contain six variables r_1, θ_1, ϕ_1, r_2, θ_2, ϕ_2, three for each electron. The term in brackets in eqn. (2) would appear twice, once for each electron, the potential term would be altered as stated in the above paragraph and the right-hand side of eqn. (2) would remain unaltered. The first approximation the physicist makes is to replace the proper wave function, Ψ, with the product of two wave functions $\Psi = \psi_1 \cdot \psi_2$, one for each electron. Since the absolute magnitude squared of the wave function tells us the proability distribution of the electrons, this replacement is approximately equivalent to stating that the probability of finding electron one at some position is independent of the position of electron two (the probability of two independent events is the product of their individual probabilities). This assumption is untrue since the Coulomb repulsion of the electrons tends to keep them out of each others way. However, unless this approximation is made, the mathematics is hopelessly difficult. The justification for such an approximation is that it gives good (but not exact) answers to many problems. It also provides an excuse (and a justifiable one) for the theoretical physicist whose calculation does not agree with experiment. The proper wave function for any system is called the *many-electron wave function* while the individual wave functions are called *one-electron wave functions.*

Hartree Self Consistent Field

If we now substitute this product of one-electron wave functions into the Schrödinger equation for helium there still exists great mathematical difficulty due to the repulsive potential between the electrons and we must add a second approximation developed by D. R. Hartree. In the Hartree approximation the true potential $-(2e^2/r_1) - (2e^2/r_2) + (e^2/|\mathbf{r}_1 - \mathbf{r}_2|)$ is replaced by some average potential which we call $\bar{V}$. Hartree reasoned as follows: Imagine yourself sitting on electron number one held at rest at some fixed position, R, on the He atom. You would find yourself in an attractive potential $-(2e^2/R)$ due to the nucleus and if you waited about 10^{-16} sec at this position electron number two would have whirled about the nucleus several hundred times. If we assume that electron two has a spherically symmetric probability distribution then whenever it traces out a sphere between you on electron one and the nucleus its acts as though it were a negative charge concentrated at the nucleus. During these times the attractive potential experienced by electron one would be reduced to $-(e^2/R) = [-(2e^2/R) + (e^2/R)]$. At other times when electron two traces out a sphere further away from the nucleus than electron one, the effective Coulomb potential due to electron two is zero and the total potential of electron one is still $-(2e^2/R)$. If we knew the fraction of the time electron two spends inside R and the fraction outside R for all values of R we would then known the average potential, $\bar{V}$, to substitute into the Schrödinger equation and could solve it and obtain the wave function of electron one. Of course, the fraction of the time electron two is inside and outside of R can be obtained from the probability distribution of electron two, but if we knew the probability distribution of electron two wouldn't the problem have been solved? This may seem like trying to lift one's self by the bootstrap but Hartree continued as follows. He assumed some spherically symmetric probability distribution for electron two which then gave him the potential, $\bar{V}$, for all values of r (or R) and solved the Schrödinger equation for electron one. (This has to be done numerically.) He then has an eigenfunction or probability distribution for electron one in the form of a table of values of ψ or $|\psi|^2$ versus r. Attention is then shifted to electron two whose Schrödinger equation is solved using the average potential, $\bar{V}$, obtained from the probability distribution just calculated for electron one. If one is very clever or lucky this solution for electron two will yield a probability distribtuion that is identical with the one assumed at the beginning. If not, a better guess is made by using an assumed probability distribution midway between the one originally assumed and the one calculated. This procedure continues until the assumed potential and the calculated one agree (or come as close as one desires). This is the principle of the *Hartree self consistent field* method.

In solving the Schrödinger equation with the spherically symmetric Hartree potential one makes the same mathematical substitution as in the case of the hydrogen atom, i.e. $\Psi_1 = R_1 \cdot \Theta_1 \cdot \Phi_1$ and $\Psi_2 = R_2 \cdot \Theta_2 \cdot \Phi_2$ where R is only a function of r, Θ of θ, and Φ of ϕ. Since the Hartree potential only depends on r we obtain the same solutions for Θ and Φ as for the hydrogen atom (eqn. (4) and (5)) and the principal quantum numbers l and m_l and the values of $[P_l^{|m_l|}(\cos\theta)]$ are the same as given in items 1, 3 and 6 on pages 2, 3, and 4. Only the R part differs and in fact must be calculated numerically in each case. The principal quantum number n which originates from the R part of the hydrogenic solution no longer appears nor does the factor $[L_{n+1}^{2l+1}(2Zr/na_1)]$ appear. Yet when we plot the Hartree probability distributions we find that they appear similar in *shape* to the hydrogenic solutions. We therefore "label" these solutions with the same quantum numbers n as their hydrogenic counterparts, i.e. a $3s$ electron would have 3 radial peaks in the probability distribution (Figs. 1, 3) and would be spherically symmetric since the angular solutions are identical with the hydrogenic solutions. The eigenvalues for the Hartree solutions would not be given by eqn. (6) but would appear as tabulated energies for each electron.

Before we plot the results of the Hartree calculation for helium we should mention that some attempts have been made to solve the Schrödinger equation for helium properly in order to obtain a many-electron wave function (i.e. one which specifically includes the interelectronic co-ordinates) as well as the correct eigenvalue. This can be done by a very powerful method called *the variational principle* which states that the lowest expectation value of the energy that can be calculated for any wave function is the observed (hence correct) energy and that wave function which yields the correct energy is the correct eigenfunction. All other wave functions yield higher expectation values of energy. This calculation is performed by assuming some wave function (for example a power series in r_1, r_2 and $|\mathbf{r}_1 - \mathbf{r}_2|$) and then calculating the total energy with the potential $-(2e^2/r_1)-(2e^2/r_2)+(e^2/|\mathbf{r}_1 - \mathbf{r}_2|)$ as in problem 4 on page 8. The coefficients of the terms in r_1, r_2 and $|\mathbf{r}_1 - \mathbf{r}_2|$ are varied until the minimum energy is reached. In a recent calculation Pekeris,[1] using several thousand terms, obtained an extremely accurate eigenvalue and a very good eigenfunction but to obtain similar accuracy for an atom like copper (29 electrons) would be impossible.*

In Fig. 2a we have plotted the Hartree probability distribution for helium and for hydrogen in comparison. Both distributions are spherically symmetric. In Fig. 2b is plotted a probability distribution which specifically attempts to solve for the many-electron wave function of helium. In this latter case the probability distribution is compressed slightly in comparison with the Hartree probability distribution but the major effect is the tendency of the two electrons to repel each other. While the Hartree probability distribution yields identical

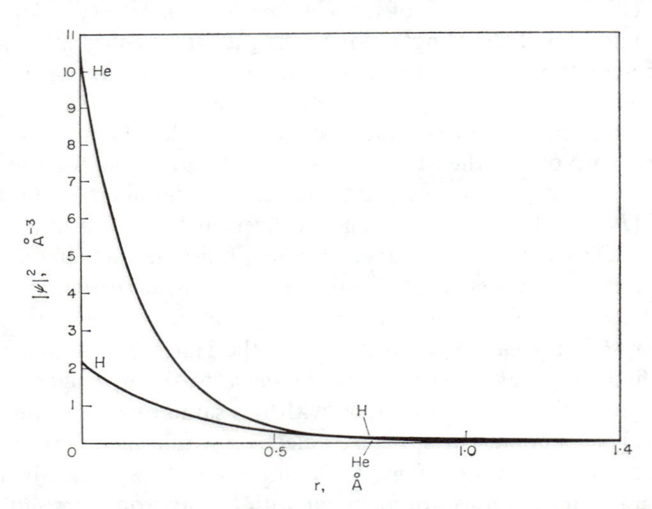

FIG. 2a. The probability per cubic Angstrom as a function of distance from the nucleus for finding a 1*s* electron on the hydrogen atom and on the helium atom. Both distributions are spherically symmetric and the helium atom contains two such electrons. The hydrogen solution is exact whereas the helium solution is based on the Hartree approximation which averages out the interelectronic co-ordination. The error in the Hartree eigenvalue is $\frac{1}{2}$ eV per electron.

* Hartree has pointed out that to itemize an accurate calculation of a many electron wave function for an element like copper would require a set of tables so long that it would require all the atoms in the universe to fabricate this table.

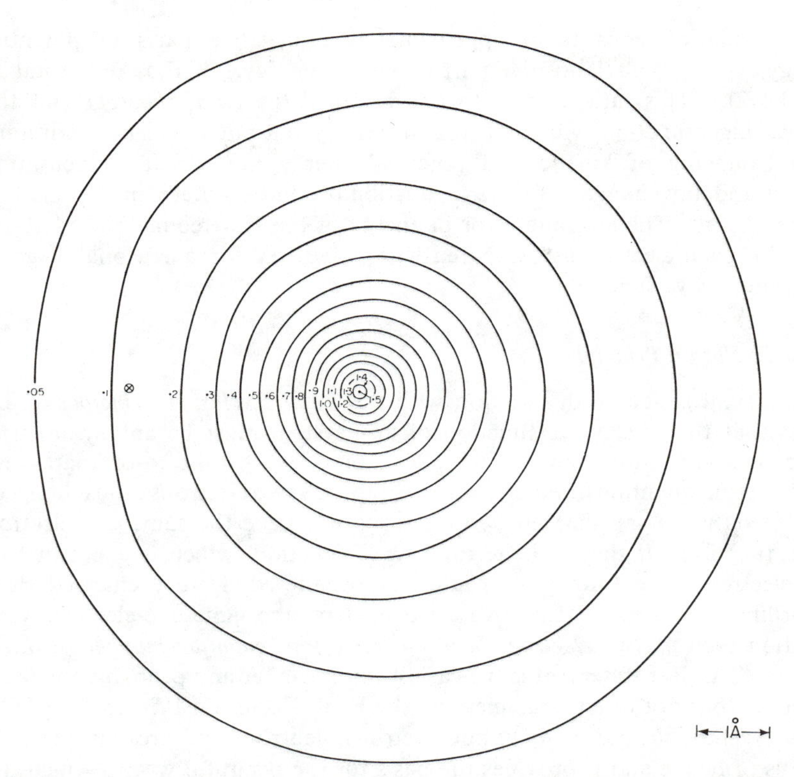

FIG. 2b Equal probability per cubic Angstrom contours for finding electron two in He when electron one is at its most probable distance from the nucleus marked by ⊗. The solution is based on an approximate many electron wave function whose energy is in error by 0·017 eV.

While the probability per unit volume is a maximum at the nucleus the most probable distance is at the point marked ⊗ due to the r^2 weighting factor in the volume integration. The probability scale is relative and is marked in the figure next to each contour line.

curves for each electron independent of the position of the other (since the Coulomb repulsion of the other electron is averaged out) the many-electron wave function must be plotted for each electron by first assigning a specific position for the other electron. We have chosen the most probable radial distance for this electron which is marked with an ⊗ in Fig. 2b. The correlation or tendency of the electrons to avoid each other is quite apparent and indicates in a general way the errors that may be made in comparing the Hartree theory with any experimental measurement that depends on the probability distribution. On the other hand the total energies (eigenvalues) of the electrons calculated by the two methods are within $1\frac{1}{2}\%$ of each other. However, the variational principle is known to give good energies even with poor wave functions. This is due to the fact that the average total energy in any calculation consists of approximately 2 negative parts of potential energy, $-\langle V\rangle$, and 1 positive part of kinetic energy, $\langle K.E.\rangle$, or a total of $-(\langle V\rangle/2)$. (This ratio of two to one is called the *virial theorem* and the correct eigenfunction will fulfill this necessary but not sufficient condition). This balancing of kinetic and potential energy has a self-compensating feature and any change in the wave function produces a much smaller change in the energy. The absolute error in energy of the Hartree method is about $\frac{1}{2}$ eV for each electron as compared with $\sim 0{\cdot}01$ eV for a reasonably good correlated wave function.

Pauli Exclusion Principle

We shall conclude this section with the *Pauli Exclusion Principle* which states that the total wave function of any system must be anti-symmetric, meaning that it must change sign $+ \rightarrow -$ or $- \rightarrow +$ if the co-ordinates and intrinsic spin quantum numbers, $m_s = \pm \frac{1}{2}$, of any two electrons are exchanged. This leads us to say that no two electrons will have the same one-electron wave function. If they did the total wave function, which is a product of one-electron wave functions, would not be altered if we exchanged their co-ordinates and spin. If no two electrons have the same one-electron wave function then *no two electrons have the same four quantum numbers* n, l, m_l and m_s. This last statement is the easiest to remember and probably the most practical (but not exact) statement of the Pauli Exclusion Principle. *We do not understand the reason for it* but electrons, neutrons and protons all abide by this principle and it provides the basis for the beautiful way in which the electronic structure of the elements is built.

Problems

1. On the same graph plot the kinetic energy ((eqn. 1)), the potential energy, the total energy, the wave function and probability distribution as a function of r (the distance from the nucleus) for the $2s$ hydrogen electron. While the potential energy is infinite at $r = 0$ the total energy remains finite when the kinetic energy is included.

2. (For the mathematically minded). An excellent method for ascertaining how close a trial wave function is to the true eigenfunction is to check the constancy of the total energy as a function of r, θ, and ϕ. As a trial wave function for the $1s$ electron in hydrogen try $\psi = A/(B + r^2)$ where A and B are constants. One of the constants can be determined by setting the total probability equal to 1, i.e.

$$\int_0^{2\pi}\int_0^{\pi}\int_0^{\infty} |\psi|^2 r^2 \, dr \sin\theta \, d\theta \, d\phi = 1$$

This is called *normalization*. Using the variational principle determine the other constant by minimizing the expectation value of the total energy (see problem 4 on p. 8). Plot the total energy as a function of r and compare it with the exact solution. Plot the wave function and compare it with the exact wave function. How much is the expectation value of the energy and r^2 in error? Is the virial theorem satisfied for the trial wave function?

3. Show that the average value of $1/r^3$ for the $3d$ hydrogenic wave function is $27Z^3/15a_1^3$.

C. The Periodic Table

Building Up the Periodic Table

We have already indicated that the quantum numbers n, l and m_l are only exactly defined in the case of hydrogen (for it is the only case for which we have an exact solution of the Schrödinger equation) but that the one-electron wave function approximation and the Hartree self-consistent method yield wave functions which have the same qualitative shapes as the hydrogenic wave functions. We now wish to show that the use of the quantum numbers n, l, m_l and m_s coupled with the Pauli exclusion principle and the rules concerning the allowed values of n, l, m_l and m_s on pages 3, 4 and 7 *provide the key to the electronic configuration of the elements.* We will confine our attention to the lowest energy or ground state.

For hydrogen, $n = 1, l = 0, m_l = 0$ and $m_s = \pm\frac{1}{2}$, the latter choice meaning that the electron will be found with equal probability with its spin $+\frac{1}{2}$ and $-\frac{1}{2}$. Any measurement of its magnetic dipole moment such as by application of a weak magnetic field will reveal half the hydrogen atoms to have their magnetic moments pointing in one direction and half in the opposite direction. In the case of helium, if one electron is designated $n = 1, l = 0, m_l = 0, m_s = +\frac{1}{2}$ the other electron must be $n = 1$, $l = 0$, $m_l = 0$, $m_s = -\frac{1}{2}$ in order to satisfy the Pauli exclusion principle. Lithium is next and we can assign to two of the three electrons the same quantum numbers as He, i.e. $n = 1$, $l = 0$, $m_l = 0$, $m_s = +\frac{1}{2}$ and $n = 1$, $l = 0$, $m_l = 0$, $m_s = -\frac{1}{2}$ but the Pauli exclusion principle forces the third electron into a higher quantum state $n = 2$, $l = 0$, $m_l = 0$, $m_s = \pm\frac{1}{2}$. This third electron has a considerably higher energy and is much further away from the nucleus than the two $1s$ electrons. The value of the principal quantum numbers from hydrogen to neon as well as aluminium, iron, gadolinium and thorium are given in Table I.

After beryllium we begin to use the p orbitals ($l = 1$). The values of m_l and m_s are chosen so that the algebraic sum of all the m_s is a maximum independent of sign (*commonly referred to as the maximum multiplicity*) and the Pauli principle is obeyed. This filling of the orbitals to the maximum multiplicity is called *Hund's rule* and is obeyed in free atoms. One of the reasons for this rule is that it keeps the electrons in different orbitals (different m_l). This keeps them out of each others way and reduces the repulsive energy between them. Referring to Fig. 1 the electron in $m_l = 0$ and $m_l = \pm 1$ avoid each other quite nicely as far as p orbitals are concerned. Thus in nitrogen by assigning the last three electrons all $m_s = +\frac{1}{2}$, the m_l must be -1, 0 and $+1$ respectively not to violate the Pauli principle.

Even though the m_s are chosen to yield the maximum multiplicity this does not uniquely determine the m_l values. The only restriction so far stated is that no two electrons can have the same four quantum numbers. Consider the case of carbon in which the last two electrons are assigned $m_s = +\frac{1}{2}$.

TABLE I

THE PRINCIPAL QUANTUM NUMBERS, n. l. m_l AND m_s FOR THE ELECTRONS IN THE GROUND STATE OF THE FREE ATOMS OF H THROUGH Ne AND Al, Fe, Gd AND Th. THE CONFIGURATION AND THE SPECTROSCOPIC STATE DETERMINED FROM THE SUM OF m_l AND m_s ARE ALSO GIVEN

Element	n	l	m_l	m_s	State
H (1)	1	0	0	$\pm\frac{1}{2}$	$1s$; $^2S_{1/2}$
He (2)	1	0	0	$+\frac{1}{2}$	$1s^2$; $^1S_0$
	1	0	0	$-\frac{1}{2}$	
Li (3)	1	0	0	$+\frac{1}{2}$	
	1	0	0	$-\frac{1}{2}$	$1s^22s$; $^2S_{1/2}$
	2	0	0	$\pm\frac{1}{2}$	
Be (4)	1	0	0	$+\frac{1}{2}$	
	1	0	0	$-\frac{1}{2}$	$1s^22s^2$; $^1S_0$
	2	0	0	$+\frac{1}{2}$	
	2	0	0	$-\frac{1}{2}$	
B (5)	1	0	0	$+\frac{1}{2}$	
	1	0	0	$-\frac{1}{2}$	
	2	0	0	$+\frac{1}{2}$	$1s^22s^22p$; $^2P_{1/2}$
	2	0	0	$-\frac{1}{2}$	
	2	1	-1	$+\frac{1}{2}$	
C (6)	1	0	0	$+\frac{1}{2}$	
	1	0	0	$-\frac{1}{2}$	
	2	0	0	$+\frac{1}{2}$	
	2	0	0	$-\frac{1}{2}$	$1s^22s^22p^2$; $^3P_0$
	2	1	-1	$+\frac{1}{2}$	
	2	1	0	$+\frac{1}{2}$	
N (7)	1	0	0	$+\frac{1}{2}$	
	1	0	0	$-\frac{1}{2}$	
	2	0	0	$+\frac{1}{2}$	
	2	0	0	$-\frac{1}{2}$	$1s^22s^22p^3$; $S_{3/2}$
	2	1	-1	$+\frac{1}{2}$	
	2	1	0	$+\frac{1}{2}$	
	2	1	$+1$	$+\frac{1}{2}$	
O (8)	1	0	0	$+\frac{1}{2}$	
	1	0	0	$-\frac{1}{2}$	
	2	0	0	$+\frac{1}{2}$	
	2	0	0	$-\frac{1}{2}$	
	2	1	-1	$+\frac{1}{2}$	$1s^22s^22p^4$; $^3P_2$
	2	1	0	$+\frac{1}{2}$	
	2	1	$+1$	$+\frac{1}{2}$	
	2	1	$+1$	$-\frac{1}{2}$	

TABLE I—*continued*

Element	n	l	m_l	m_s	State
F (9)	1	0	0	$+\frac{1}{2}$	
	1	0	0	$-\frac{1}{2}$	
	2	0	0	$+\frac{1}{2}$	
	2	0	0	$-\frac{1}{2}$	
	2	1	-1	$+\frac{1}{2}$	$1s^22s^22p^5$; $^2P_{3/2}$
	2	1	0	$+\frac{1}{2}$	
	2	1	$+1$	$+\frac{1}{2}$	
	2	1	$+1$	$-\frac{1}{2}$	
	2	1	0	$-\frac{1}{2}$	
Ne (10)	1	0	0	$+\frac{1}{2}$	
	1	0	0	$-\frac{1}{2}$	
	2	0	0	$+\frac{1}{2}$	
	2	0	0	$-\frac{1}{2}$	
	2	1	-1	$+\frac{1}{2}$	
	2	1	0	$+\frac{1}{2}$	$1s^22s^22p^6$; $^1S_0$
	2	1	$+1$	$+\frac{1}{2}$	
	2	1	$+1$	$-\frac{1}{2}$	
	2	1	0	$-\frac{1}{2}$	
	2	1	-1	$-\frac{1}{2}$	
	10-neon core electrons				
Al (13)	3	0	0	$+\frac{1}{2}$	
	3	0	0	$-\frac{1}{2}$	$1s^22s^22p^63s^23p$; $^2P_{1/2}$
	3	1	-1	$+\frac{1}{2}$	
	18-argon core electrons				
Fe (26)	3	2	-2	$+\frac{1}{2}$	
	3	2	-1	$+\frac{1}{2}$	
	3	2	0	$+\frac{1}{2}$	
	3	2	$+1$	$+\frac{1}{2}$	$1s^22s^22p^63s^23p^63d^64s^2$; $^5D_4$
	3	2	$+2$	$+\frac{1}{2}$	
	3	2	$+2$	$-\frac{1}{2}$	
	4	0	0	$+\frac{1}{2}$	
	4	0	0	$-\frac{1}{2}$	
	54-xenon core electrons				
Gd (64)	4	3	-3	$+\frac{1}{2}$	
	4	3	-2	$+\frac{1}{2}$	
	4	3	-1	$+\frac{1}{2}$	
	4	3	0	$+\frac{1}{2}$	$1s^22s^22p^63s^23p^63d^{10}4s^24p^64d^{10}5s^2$
	4	3	$+1$	$-\frac{1}{2}$	$5p^65d4f^76s^2$; $^9D_2$
	4	3	$+2$	$-\frac{1}{2}$	
	4	3	$+3$	$-\frac{1}{2}$	
	5	2	-2	$+\frac{1}{2}$	
	6	0	0	$+\frac{1}{2}$	
	6	0	0	$-\frac{1}{2}$	
	86-radon core electrons				
Th (90)	7	0	0	$+\frac{1}{2}$	
	7	0	0	$-\frac{1}{2}$	$1s^22s^22p^63s^23p^63d^{10}4s^24p^64d^{10}5s^2$
	6	2	-2	$+\frac{1}{2}$	$5p^65d^{10}4f^{14}6s^26p^67s^26d^2$; $^3F_2$
	6	2	-1	$+\frac{1}{2}$	

We actually have several possible assignments of m_l for these two electrons, i.e. $m_l = -1$, $m_l = 0$; $m_l = -1$, $m_l = +1$; $m_l = +1$, $m_l = 0$. However, the choice $m_l = -1$, $m_l = +1$ is not as favourable as the other two in the matter of the electrons keeping out of each others way for they occupy the same space on the atom (Fig. 1). Of the remaining two choices the rule is that the opposite sign is chosen between the sum of the m_s and the m_l when we deal with less than a half filled group of orbitals (in this case there are six orbitals for $n = 2, l = 1$), and the same sign for more than a half filled group or orbitals. This arises from a magnetic interaction between the magnetic moment of the electron due to its intrinsic spin and the magnetic field created by the apparent motion of the nucleus relative to the electron. For example, imagine yourself sitting on the last electron in boron $n = 2, l = 1, m_l = -1$, $m_s = +\frac{1}{2}$. Your intrinsic spin and magnetic moment are pointing upward but your orbital motion about the nucleus is creating a magnetic field downward at the nucleus due to your negative charge. However, as far as you are concerned you might just as well be considered standing still with the nucleus rotating about you and creating an upward magnetic field at your position due to its positive charge. This upward magnetic field is energetically favoured relative to your upward magnetic moment. Thus one always tries to keep the individual m_l and m_s of opposite sign for each electron but only after first satisfying Hund's rule. This coupling of the spin and orbital motion is called *spin-orbit coupling*.

While the pictures in Fig. 1 do not make it apparent the total probability distribution which contains one electron in each m_l orbital is spherically symmetric and such a distribution is called a *half closed shell*. For *p* electrons this consists of one electron in $m_l = 0$, $m_l = +1$ and $m_l = -1$; for *d* electrons this consists of one electron in $m_l = +2$, $m_l = +1$, $m_l = 0$, $m_l = -1$ and $m_l = -2$ or a total of five electrons; and for *f* electrons a total of seven electrons in $m_l = +3 \ldots m_l = -3$. A *closed shell* is just double these numbers one for $m_s + \frac{1}{2}$ and $m_s = -\frac{1}{2}$ for each m_l.

Physicists employ a code to designate the state (Table I) or configuration of the atom. All the m_l are added algebraicaly and the magnitude of the sum is designated with a capital letter (just as the individual l are designated by small letters) $S = 0$, $P = 1$, $D = 2$, $F = 3$, etc. For example, the total m_l for fluorine is $+1$ which is designated by the letter P. The superscript above and to the left is obtained by adding the m_s, multiplying its absolute magnitude by 2 and adding 1, i.e. the total m_s for fluorine is $+\frac{1}{2}$ yielding a superscript 2. The subscript to the right is obtained by adding the m_l and m_s algebraically and then dropping the sign. The sum of the m_l represents the quantum mechanical addition of the orbital angular momentum of the individual electrons. The expectation value of this total orbital angular momentum is $\hbar\sqrt{[L(L + 1)]}$ where L is the absolute magnitude of the sum of the m_l. Likewise the absolute magnitude of the sum of the m_s represents the quantum mechanical addition of the intrinsic angular momenta of the individual electrons and is designated by S. Its expectation value is $\hbar\sqrt{[S(S + 1)]}$. The absolute magnitude of the sum of the m_s and m_l represents the quantum mechanical sum of all the angular momenta and is called J. The expectation value of this total angular momentum is $\hbar\sqrt{[J(J + 1)]}$. Thus the configuration code is ${}^{2s+1}L_J$.

The entire periodic table is given in Table II, but only with the designations n and l since the m_l and m_s invariably follow Hund's rule and the rule for spin-orbit coupling. The actual designations given in the table are the ground states as determined from an analysis of the spectrographic lines emitted by the free atoms. We have also listed the room temperature form of the elements, i.e. gas, liquid or crystal and, if crystalline, the crystal structure and lattice parameters. The crystals are designated m for metal, s for semiconductor and sm for semi-metal.

The electronic configuration of the unfilled shells of outer electrons (and in some cases the last completed shell also) provide the key to the properties of the elements. Only up to argon do the electrons follow a strict sequence of hydrogenic levels. Beyond that we find deviations such as the $5s$ and $5p$ levels filling before the $4f$ levels. The reason for this is discussed on p. 54.

The Properties of the Elements in the Periodic Table

Each column in Table II contains elements with similar outer electron configurations in their free atom state and they invariably exhibit similar behaviour. The following general comments elaborate on this point for each column.

Column 1. The completion of a group of p orbitals (six in all) represents an extremely stable configuration and these elements are indicated in Table II as *inert gases*. (Helium is a special case in that it has a completed $1s$ shell but is still an inert gas. This arises from the lack of $1p$ electrons forbidden as an acceptable solution to the Schrödinger equation.) The inert gases do not form compounds and solidify only at very low temperatures.

Column 2. All elements with one electron outside the stable inert gas shell are designated as *alkali metals* and, as metals, exhibit weak binding, are quite corrosive and are very soft. Hydrogen is a special case and is discussed elsewhere in the text.

Column 3. All elements with two electrons outside the stable inert gas shell are designated as *alkaline earth metals* and exhibit stronger binding than the alkali metals, less tendency to corrosion, and are somewhat harder.

Column 4. All elements with one d and two s electrons outside an inert gas shell do not have a common designation. They appear to have chemical properties similar to rare earth metals and we designate them as "rare earth like". They oxidize quite readily but tend to be harder than either the alkali or alkaline earth metals.

Columns 5 *to* 18. All elements with an unfilled (or just filled) f shell are called *rare earths*. They oxidize quite readily, exhibit medium strength and hardness, and have pronounced magnetic properties due to the f electrons. They are all similar in chemical behaviour causing them to be difficult and costly to separate. To date they are principally magnetic "curiosities".

Columns 19 *to* 25. All elements with two or more electrons in an unfilled d shell are called *transition metals*. The unfilled d shells give rise to very strong binding, high melting points, hardness, etc. In some cases like iron, cobalt and nickel the d electrons exhibit magnetic behaviour. After the wheel and the thumb they have been the greatest contribution to man's progress.

Column 26. The elements copper, silver and gold contain a just completed d shell plus one electron outside this d shell. They are relatively non-corrosive and soft but gain some strength from the just filled d shells in contrast with the next group. They are called *noble metals*.

Column 27. Zinc, cadmium and mercury are very soft and weak since the combination of a closed d shell and s shell produces a relatively stable configuration. They are called group II B elements.

Column 28. All elements with outer electron configuration s^2p are designated group III elements. Except for the hard metal, boron, they are soft and metallic.

Columns 29 *to* 32. Except for lead and one of the allotropic modifications of tin all elements with two or more electrons in the unfilled p shell are non-metallic. The elements carbon, silicon, germanium and tin all crystallize in the diamond structure and are well known *semi-conductors*. In some other cases these elements behave somewhere between semi-conductors and metals and are designated semi-metals.

With all these clues staring us in the face one might expect that the physicist would have a rather good understanding of the electronic structure of metals but until the physicist can calculate the observed properties of a metal he is unable to say that he understands. As we shall see these calculations are exceedingly difficult and to date have had little success.

In the next sections we shall discuss the difficulties in calculating the electronic structures of molecules and crystals but before we do that we want to say something further about the Hartree solutions for atoms other than hydrogen, and to show that they give wave functions that appear hydrogen-like. As we have already said this permits us to assign the principal quantum numbers n, l, m_l and m_s to them quite readily even though an exact solution of the Schrödinger equation does not yield such quantum numbers.

The early Hartree solutions did not properly account for the Pauli exclusion principle in that the total wave functions were not antisymmetric (page 14). This later taken into account by Hartree and Fock in a mathematical way suggested by Slater by writing the one-electron wave functions in the form of a determinant.* If one calculates the total energy by the Hartree–Fock (Slater) method and the original Hartree method, extra terms between pairs of electrons appear in the Hartree–Fock method which are called *exchange terms* and are due to the presence of products of wave functions of each electron at the position of some other electron times the wave function of the other electron at the position of the first electron. It appears as though the electrons have switched or exchanged positions but this is only a mathematical switch. Physicists frequently refer to the Hartree–Fock calculation as a self-consistent field calculation with exchange. It does give better energy values than the Hartree method. (A typical exchange term can be obtained by squaring any of the equations 10–12, taking the cross term and multiplying it by the Coulomb potential between the electrons.)

In Fig. 3 we have plotted the radial parts of the wave functions for a Hartree–Fock calculation of the free atom of iron. Since the Hartree and Hartree–Fock methods assume the potentials to be spherically symmetric the angular parts of the wave functions Θ and Φ are identical to the hydrogenic as given in eqns. (4) and (5), and the principal quantum numbers l and m_l still apply. From the shape of the radial parts of the wave function we see the

* It always seems to take this author a considerable amount of time to explain the use of the *Slater determinant* so I shall simply illustrate the results for the simplest case, helium. In the Hartree method the wave function is simply written as $\Psi = \psi_1(r_1)\psi_2(r_2)$ where r_1 and r_2 are the positions of electron 1 and 2 and ψ_1 and ψ_2 their wave functions. In the Hartree-Fock method there are several ways of writing the antisymmetric wave function. First, for the case where the exchange of the spin quantum numbers $m_s = \pm\frac{1}{2}$ of the electrons results in the sign change of the wave function.

$$\Psi = [\psi_1(r_1)\psi_2(r_2) + \psi_2(r_1)\psi_1(r_2)](\alpha\beta - \beta\alpha) \tag{10}$$

where α denotes $m_s = +\frac{1}{2}$ and β, $m_s = -\frac{1}{2}$. If α and β as well as r_1 and r_2 are exchanged we have

$$\Psi = [\psi_1(r_2)\psi_2(r_1) + \psi_2(r_2)\psi_1 r_1)](\beta\alpha - \alpha\beta) \tag{11}$$

which is the negative of the above wave function. This is called the singlet state.

For the case where the exchange of positions r_1 and r_2 yields a sign change there are three possible wave functions:

$$\begin{aligned} \Psi &= [\psi_1(r_1)\psi_2)r_2) - \psi_2(r_1)\psi_1(r_2)](\beta\beta + \beta\beta) \\ \Psi &= [\psi_1(r_1)\psi_2(r_2) - \psi_2(r_1)\psi_1(r_2)](\alpha\alpha + \alpha\alpha) \\ \Psi &= [\psi_1(r_1)\psi_2(r_2) - \psi_2(r_1)\psi_1(r_2)](\alpha\beta + \beta\alpha) \end{aligned} \tag{12}$$

The proper wave function is the sum of all three and is called the triplet state. For more complicated atoms see Seitz *Modern Theory of Solids*. Chapter VI, McGraw-Hill Co., 1940.

If we are dealing with the ground state of He then both electrons are 1*s* electrons and their approximate one-electron wave functions are $\psi_1 = A\,e^{-ar_1}$; $\psi_2 = A\,e^{-ar_2}$ where $a \cong 27/16$ and $A \cong 1{\cdot}237$. Thus only the singlet state remains giving $\Psi = A^2\,e^{-ar_1}\,e^{-ar_2}$, since the triplet state vanishes $[A\,e^{-ar_1}\,A\,e^{-ar_2} - A\,e^{-ar_2}\,A\,e^{-ar_1} = 0]$. On the other hand the excited state of He in which one electron is 2*s* and the other 1*s* has both a singlet and triplet state with different energies and total probability distributions, $|\Psi|^2$. Once the proper antisymmetric state is formed the spin quantum numbers α and β can then be dropped if we wish to make any calculations.

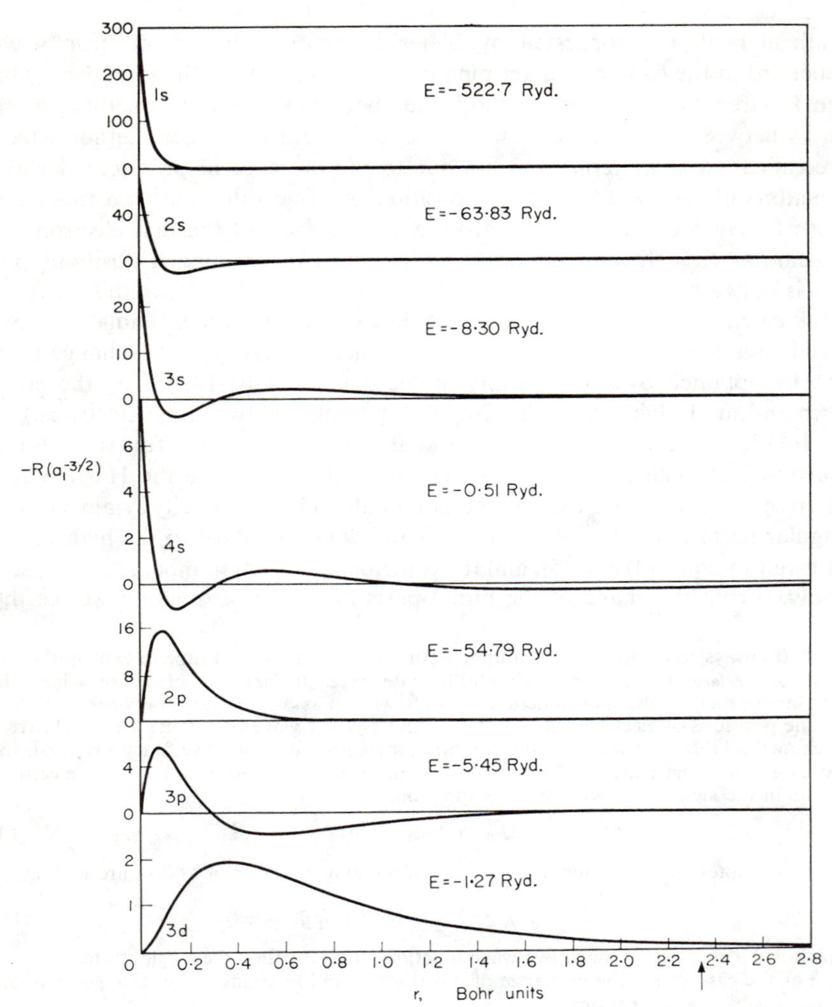

FIG. 3. The radial part, R, of the $1s$, $2s$, $3s$, $4s$, $2p$, $3p$ and $3d$ one electron wave functions of the free atom of iron ($3d^6 4s^2$) plotted as a function of the distance from the nucleus. The unit of distance is the Bohr radius, a_1, (1 Bohr radius = 0·529 Å) and the unit of wave function is $a_1^{-3/2}$. Thus the square of the wave function is the probability per cubic Bohr radius. The wave functions are normalized so that $\int_0^\infty 4\pi R^2 r^2\, dr = 1$ and the angular wave functions Θ and Φ are the hydrogenic wave functions eqns. (4) and (5). These wave functions for iron were determined by the Hartree-Fock method in which the total wave function of the atom is chosen to be the proper anti-symmetrized one. Thus the square of the individual wave functions does not directly yield the probability of finding the individual electrons but the square of the properly anti-symmetrized products of all the wave functions does give the probability of finding the entire 26 electrons of iron at specific distances from the nucleus. The wave functions depicted in Fig. 3 are still useful as a visual guide to the radial extent of the electron distribution of the free atom of iron. The eigenvalues, E, are given in the figure.

$1s$, $2s$, $3s$ and $4s$ wave functions to be similar to the hydrogenic in that the $1s$ has no nodes, the $2s$ one node, the $3s$ two nodes, etc. A node occurs when the wave function changes sign and, of course, the wave function and probability distribution are zero at this point. The number of nodes enables us to assign the principal quantum number n ($n = l + 1 +$ number of nodes). While m_s does not appear in the hydrogenic solution its assignment is made consistent with the Pauli principle by employing the determinantal form of the wave function.

Problems

1. Plot the density, melting point and cost per pound as a function of atomic number for the elements in the fifth row of Table II ($Z = 58$–85). Are there any regularities that can be related to the electron configuration?

2. Do the rare earths form solid solutions with each other? Do the transition metals form solid solutions with each other? Are there many cases of two elements in different columns of Table II which form a single phase solid solution over the entire range? Are there many cases in which this occurs for two elements in the same column? Would you say that the occurrence of solid solutions is more prevalent amongst elements with similar outer electron configuration?

3. (For the mathematically minded.) Employing the hydrogenic wave functions show that the total probability distribution containing one electron in each of the five d orbitals, $m_l = 2, 1, 0, -1, -2$ is spherically symmetric.

4. The Hartree–Fock solutions employ the hydrogenic angular wave functions Θ and Φ, eqns. (4) and (5), and yield a numerical solution for the radial function, R. Some attempts have been made to use the hydrogenic radial function for R (eqn. (3)) by replacing Z in this equation by an "*effective Z*", Z_{eff}, which is no longer restricted to integral values. Z_{eff} can be determined from the variational principle by minimizing the total energy or by adjusting it in eqn. (6) to yield an energy equal to the measured energies required to remove the electron from the atom.

Adjust Z in eqn. (6) to yield the eigenvalues given in Fig. 3 for each of the electrons on the free atom of Fe. These values are close to the measured values. Using these values of Z_{eff} in eqn. (3) plot the radial functions on semi-transparent co-ordinate paper (10×10 to the half inch) with the scales adjusted to coincide with the scale in the book so that the curves can be superimposed. Why do they yield a better fit for the $1s$ electron than for the $3d$ electron? Adjust Z to give what you consider to be the best fit for the $3d$ radial wave function and calculate the error in the energy.

The "effective Z" method utilizing the observed energy values from X-ray absorption edges can be employed for one electron wave functions of heavy elements where Hartree solutions are unavailable. They are, of course, rather approximate.

Chapter Summary

1. The fundamental equation of quantum mechanics is the Schrödinger equation. The well-behaved solutions are the eigenfunctions each for a specific energy value called an eigenvalue. Energy, momenta, probability distributions, etc., are calculated from the eigenfunctions.

2. The eigenfunctions for the individual electrons in an atom can be categorized by the four principal quantum numbers n, l, m_l and m_s. The Pauli principle forbids any two electrons from having the same four quantum

numbers and the periodic table is determined from these quantum numbers. Elements with the same quantum numbers for the outer electrons have similar properties in the solid state.

3. The eigenfunctions of an atom are calculated by the Hartree–Fock method which replaces the many-electron wave function by a product of one-electon wave functions. This fails to account for the correlation between electrons which leads to errors in the energies of at least $\frac{1}{2}$ eV per electron.

CHAPTER II

THE MOLECULE AND THE SOLID

A. Molecules

The Hydrogen Molecule

In Chapter I we discussed the electronic structure of free atoms as a prelude to a discussion of the electronic structure of solids. Since all elements form solids at sufficiently low temperatures there must be some force causing these atoms to stick together. Two isolated free atoms of hydrogen for example are both neutral if separated by about 1 cm, yet if allowed to approach to within several Angstroms of each other a weak attractive force appears, and if allowed to approach to within 1 Å of each other they can form the strongly bound H_2 molecule. In fact, the energy favouring the formation of the H_2 molecule from two free atoms of hydrogen is 4·72 eV. It is instructive to consider the mechanism which bonds the H_2 molecule for it provides some of the framework for considering the bonding of atoms in a solid.

If two hydrogen atoms are 1 cm apart they are essentially unaware of each other's presence and the ground state configurations of each of the electrons on their respective atoms is $n = 1$, $l = 0$, $m_l = 0$ and $m_s = \pm\frac{1}{2}$. The electron distributions are spherically symmetric about their respective nuclei with a radial distribution as shown in Fig. 2a. If we move the atoms to within about six Angstroms of each other and assume that each electron is at its most probable distance from its nucleus, $a_1 = 0{\cdot}529$ Å, we can postulate many instantaneous arrangements of the electrons, some of which lead to attractive potentials between the atoms (left side of Fig. 4) and some of which lead to repulsive potentials (right side of Fig. 4). For example, configuration number 1 in Fig. 4 has a total potential energy of

$$V = e^2/R + e^2/2[(R/2)^2 + a_1^2]^{1/2} - e^2/(R^2 + a_1^2)^{1/2} - e^2/(R^2 + a_1^2)^{1/2}$$

where R is the distance between the nuclei (6 Å) and a_1 is the Bohr radius ($a_1 = 0{\cdot}529$ Å). The first two terms in the potential represent the nuclear and electronic repulsive terms respectively while the third and fourth represent the attractive potentials between the electrons and the opposing nuclei. Of course the attractive potential of the electrons for their own nuclei still exists but we are only looking for additional potential terms that will cause the hydrogen atoms to stick together. Since $R^2 \gg a_1^2$ the potential is approximately given as $V \cong -e^2 a_1^2/R^3$ or an attractive potential of $\sim 10^{-3}$ eV.

Now if the electrons were to move about their respective protons independently of each other the average potential would be zero. On the other hand if the electrons tend to follow each other such that they favour instantaneous positions such as the left side of Fig. 4 rather than the right side we should have a net attractive potential. Indeed the electrons do this and in the limit of large distances between the hydrogen atoms we can calculate this effect quite accurately and find the total attractive energy to be

$$E = -6{\cdot}5 e^2 a_1^5/R^6 \tag{13}$$

or only about 3×10^{-5} eV for the two hydrogen atoms 6 Å apart. This weak attractive potential arising from the instantaneous preferential arrangement of electrons at relatively large interatomic distances is called the *van der Waals attraction* and is exhibited by all atoms.

FIG. 4. Some possible instantaneous configurations of the electrons (circles) at various separations of two hydrogen atoms. The protons are the points. Positions 1 to 6 are about 6 Å apart, the first three positions favouring the attractive forces while positions 4 to 6 favour the repulsive. Position 7 places the nuclei 0·74 Å apart, the equilibrium distance for the H_2 molecule, and illustrates a configuration favouring the attractive forces. In position 8 the two nuclei are 0·2 Å apart and their repulsive force outweighs the electron-nuclear attraction.

Let us now bring the two hydrogen atoms to within about 1 Å of each other, close enough so that the two electron orbits intermesh appreciably. The problem is basically the same as before in that the electrons will again try to take up instantaneous arrangements favouring the attractive terms over the repulsive terms but now the Pauli principle (antisymmetrization of the wave function) becomes extremely important. If the two electrons have opposite m_s values ($+\frac{1}{2}$ and $-\frac{1}{2}$) then without violating the Pauli exclusion principle the electrons can intermesh and seek the most favourable position, but with identical m_s values the Pauli principle keeps them so far apart that it becomes virtually impossible for the electrons to assume positions that will lower the potential energy.* One such favourable position for opposite m_s values is shown in configuration number 7 in Fig. 4.

Finally, if the atoms are brought so close together that the nuclei are only ~0·2 Å apart the repulsion of the nuclei becomes greater than any attraction of the electrons for the opposite nuclei and the atoms will now repel each other, as shown in configuration number 8 in Fig. 4. The theoretical curve in Fig. 5 shows how the total energy varies smoothly with interatomic

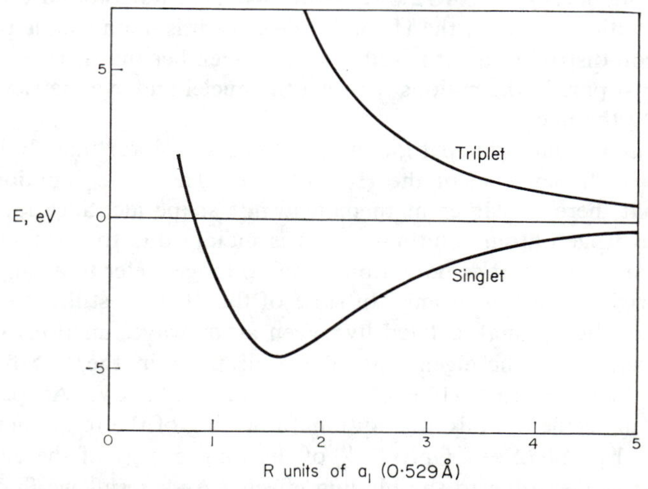

FIG. 5. The total energy in eV versus interatomic separation in Bohr units for the singlet state (antiparallel spins) and triplet state (spins parallel) of the H_2 molecule.

separation, the distances of 5–8 Å being dominated by the weak van der Waals attraction (not shown), the distances 0·5 to 5 Å being dominated by the electron-nuclear attraction, and the distances below 0·5 Å being dominated by the repulsion of the nuclei. Just as in the helium atom the configuration in which the electrons have opposite spin is called the *singlet state* and parallel spin the *triplet state*. We see from Fig. 5 that the singlet state has an energy minimum − 4·72 eV at a distance of 0·74 Å which corresponds to the ground state energy (binding energy) and observed interatomic separation of the hydrogen molecule. On the other hand, the Pauli principle prevents the triplet state from having a stable distance. The type of curves in Fig. 5 can be calculated either by the methods of the one electron approximation of

* There are special cases involving other than *s* electrons in which the state with identical m_s values may be slightly preferred. These cases do not occur too often but will be discussed under the theory of magnetism.

Hartree–Fock in which case one is in error by about $\frac{1}{2}$ eV in the ground state energy, or else one can use the many-electron wave function for which one can obtain as accurate an answer as possible, depending on the amount of effort one is willing to expend. In both cases the energy is calculated by the variational principle.

To summarize then, the binding of the H_2 molecule arises from electron probability distributions consistent with the Pauli exclusion principle such that the attraction of the electrons and the opposite nuclei outweigh the repulsion of the two nuclei and two electrons. In fact, the virial theorem tells us that this probability distribution corresponds to a net attractive potential of $-9{\cdot}44$ eV and a net kinetic energy of electron motion of $+4{\cdot}72$ eV or a total binding energy of $-4{\cdot}72$ eV. While we have not plotted the electron probability distribution of the H_2 molecule since it is not a simple product of one electron distributions, it is sufficient to remember that it is concentrated for the most part in the regions between the nuclei and symmetric about the line joining the nuclei.

What about our principal quantum numbers? Unfortunately they have been lost in the solution of the H_2 molecule. The wave functions are so altered that there is neither mathematical nor shape identification with the original hydrogen atom solutions. This is mainly due to the fact that the wave functions are part of both atoms and no longer refer to a single mathematical origin as in free atoms. In spite of this there is still a considerable influence of the original isolated hydrogen atom wave functions in the H_2 molecule solution. The eigenvalue of the electrons in the two free atoms is one Rydberg apiece ($-13{\cdot}6$ eV) a total of $-27{\cdot}2$ eV. As part of the hydrogen molecule there is an additional lowering of the total energy of the two atoms by $-4{\cdot}72$ eV. Only 15% of the total energy of the electrons in the H_2 molecule is due to the binding effects. As a result we find that the eigenvalues of the free atoms still serve as a guide in molecular structure.

Other Molecules

For example, suppose we bring a third hydrogen atom up close to the H_2 molecule then its electron's m_s value would have to be the same as one of the electrons on the molecule and the Pauli principle would prevent them from coming close enough to gain any attractive energy from the nuclei. Thus we do not have a stable hydrogen molecule, H_3. In fact, if we continue to use the original free-atom quantum numbers, we can, in many cases, decide the composition of molecules. *We must be careful to remember, though, that there exists no mathematical justification for doing so but only a qualitative argument that the energy gained in bonding is generally small compared with the original eigenvalues of the electrons.* Consider for example the carbon atom, $1s^22s^22p^2$. If we bring a hydrogen atom near to it the hydrogen electron overlaps the outer electrons of the carbon atom and during this overlap time it must act like a carbon $2p$ electron in order not to violate the Pauli principle. In the remaining time it can be *envisaged* as a $1s$ electron as it whirls close to its own nucleus. Likewise, any $2p$ electrons of the carbon which overlap the hydrogen electron will act like a hydrogenic $1s$ electron during the time when it is close to the hydrogen nucleus. Since it requires four additional electrons to fill the $2p$ shell of carbon we can bring four hydrogen atoms to within overlap distance of the carbon and arrange them on a tetrahedron so that they do not overlap each other. This forms the stable CH_4 or methane molecule. No further hydrogen atoms are accepted by the carbon atom since the overlapping electron would have to appear $3s$-like not to violate the Pauli principle and this requires too much energy. The filling of the orbitals to form closed shells and stable molecules is called *saturation* and is commonly observed in molecules.

Theoretical Approach to More Complicated Molecules

Even though this liberal use of free-atom quantum numbers enables us to predict saturation in many, but not all molecules, this still falls far short of the sort of answer we should like a theory to yield, i.e. eigenfunctions and eigenvalues. In order to secure these we must develop a somewhat different mathematical approach than the Hartree–Fock scheme which assumed a spherical potential around the origin of the co-ordinate system, because we now have several origins for a co-ordinate system since we have several nuclei. This is commonly referred to as the mathematical problem of the "*many centres*" and is attacked by two principal methods, the *Heitler–London* method and the *Hund–Mulliken (molecular orbital)* method. In the former, one deals with the interaction of atomic-like wave functions, one chosen for each atom, while in the latter, one makes up combinations of the sum and differences of atomic wave functions and treats the resulting wave function as a whole. This is illustrated in Fig. 6 for the hydrogen molecule. While the H_2 molecule is simple enough to be treated more elegantly, the one-electron Heitler–London and Hund–Mulliken schemes have been adopted for more complicated molecules. To date the Hund–Mulliken scheme is preferred for the more complicated molecules but the errors between calculated and observed energies are quite large. Self-consistency is rarely possible and a molecule containing about 100 electrons is considered a prodigious effort.

What then have we learned about the electronic structure of molecules that is important in understanding the electronic structure of solids?

Summary

1. Atoms stick together by virtue of the attractive energy terms between electrons and opposite nuclei outweighing the repulsive energy terms between

electrons and between nuclei. An antisymmetric wave function with the overlapping electrons having opposite m_s values is generally more effective in accomplishing this.

2. The total energy versus interatomic distance as given in Fig. 5 for the singlet state is the type of curve we obtain predominantly between atoms that stick together to form molecules or solids.

3. The theoretical calculation of eigenfunctions and eigenvalues for molecules is rather difficult with errors in energy frequently a large percentage of the experimental binding energy.

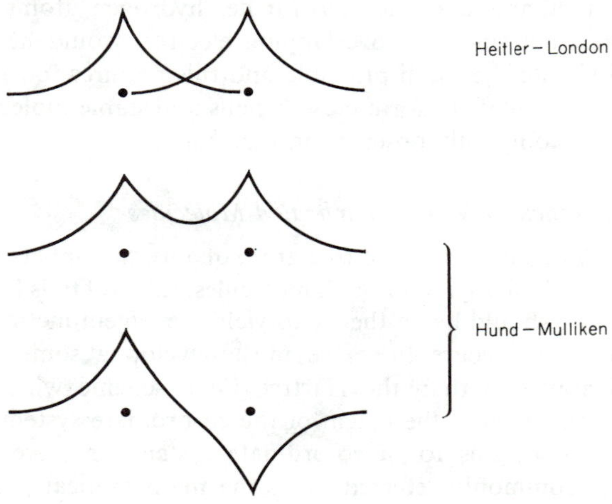

FIG. 6. The type of wave functions employed in the two principal one electron treatments of the H_2 molecule. In the Heitler–London method there is a separate wave function centred on each atom. In the Hund–Mulliken (molecular orbital) method each wave function is associated with the molecule as a whole. In this case they were selected by taking the sums and differences of hydrogenic free atom 1*s* wave functions, although the specific choice and combinations may vary (a good answer justifies the choice).

Problems

1. Try writing down the formula for molecules composed of hydrogen and each of the elements from He to Cl. In each case use as many hydrogen atoms as are necessary to complete the outer shell of the second element. Are all of these molecules known to be stable? Now try the same with oxygen instead of hydrogen using the outer electrons of the second element to complete the outer shell of oxygen. Does the idea of saturation work often enough to be useful? Can you relate the filling of closed shells to the chemist's concept of valency? Are there many stable molecules which do not obey the idea of saturation?

2. Look up the binding energy (sometimes called the energy of dissociation) of the following molecules and determine the percentage by which the free atom energies of the overlapping electrons are lowered in forming the molecule. Determine the approximate free atom energies of the outer electrons from their ionization potential, i.e. the energy required to remove one electron from the atom. (Note: divide the binding energy equally amongst the overlapping electrons.) NO, NaCl, HF, N_2. Are the percentages small? Do these molecules follow the rule for saturation? Can you find a molecule for which the centage is large ($>25\%$)?

3. The inert gases do not form stable molecules since their electronic shells are already filled although they do form very weak attractive bonds as a result of the van der Waals interactions. An accurate calculation of the energy of interaction of two neon atoms gives an expression which can be approximated by $E_1 = 1{\cdot}18 \times 10^4 e^{-r/0{\cdot}395}$ for $r > 3$ Å where E_1 is in eV and r in units of a_1, the Bohr radius ($a_1 = 0{\cdot}529$ Å). In addition the van der Waals attractive energy is given by $E_2 = -174{\cdot}5/r^6$ where E_2 is in eV and r in units of a_1. Determine the total energy $E = E_1 + E_2$ as a function of r in the range $4a_1$ to $7a_1$; determine the value of r giving the minimum value of E. Compare this equilibrium value of r with the nearest neighbour distance in solid neon. Plot the Hartree–Fock radial wave function of the $2p$ electrons in neon given approximately by $R = [10{\cdot}06r/\sqrt{(\pi)}](e^{-4{\cdot}48r} + 0{\cdot}198e^{-1{\cdot}91r})$, where r is in units of a_1 and R in units of $a_1{}^{-3/2}$, and ascertain from inspection whether there is much overlap at this equilibrium value of r.

The heat of vapourization is the energy required to convert a liquid to a gas at the boiling point and is approximately equal to the binding energy or *cohesive energy* (as it is often called). This value is $\sim$21 cal/g for neon. Does this agree with the calculated van der Waals binding energy per pair of atoms? (1 eV = 23,060 cal/g atom weight). Does this agreement indicate that the theory of van der Waals' forces is reasonably satisfactory?

B. Solids

Electron theory of Solids, Lithium Metal

So far we have discussed the electronic structure of atoms and simple molecules. Attempts to apply quantum mechanics to these systems by utilizing the Hartree–Fock approximation for atoms and the Heitler–London or Hund–Mulliken extensions of the Hartree scheme for molecules have been limited to systems containing about one hundred electrons or less (*circa* 1960). This limitation arises from the capabilities of our calculating machines plus the discouraging feature that these approximate methods tend to give poorer results as the number of electrons increases.

Let us now turn our attention to solids and in particular to one of the simplest solids we know, lithium metal, containing two $1s$ electrons and one $2s$ electron in the free atom. If we consider the smallest piece of lithium metal upon which we can conveniently make most physical measurements ~ 1 mm^3 we find that we are dealing with $\sim 10^{20}$ electrons, a rather prodigious jump from the one hundred electrons treated in the theory of molecules. In addition, the "saturation" properties of molecules are not obeyed since we can have a piece of lithium metal any size we desire. Thus the free atom principal quantum numbers are not as clear cut a guide to the electronic structure of solids as they were in the case of some molecules.

To date, the electron theory of solids has concerned itself with trying to make sense out of the experimental observations of the solid state physicist or metallurgist. Let us approach the electron structure of lithium in this fashion. X-ray diffraction tells us that lithium forms a body-centred cubic metal with lattice parameter $a_0 = 3{\cdot}509$ Å, Fig. 7. The atom at the centre of the cube in Fig. 7 has eight nearest neighbours along the body diagonals at a distance of 3·04 Å and six second nearest neighbours along cube directions 3·509 Å away. Suppose we plot the free atom radial probability distributions of the spherically symmetric 1*s* and 2*s* electrons, Fig. 8. It is quite clear that the 1*s* electrons barely overlap, and cannot contribute to the binding (you recall the cohesive energy of the H_2 molecule comes from the overlap of the elec-

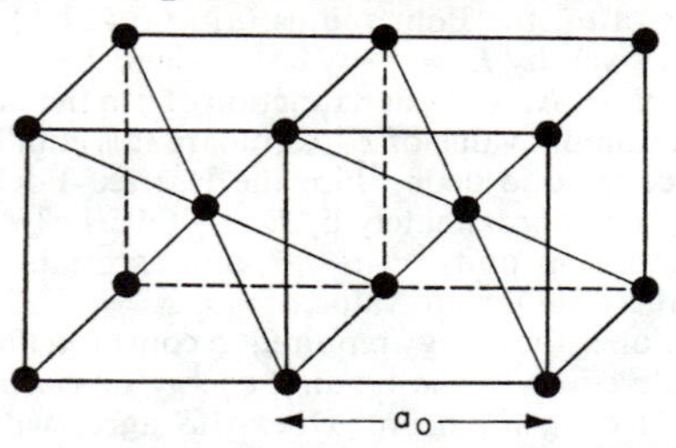

FIG. 7. The body-centred cubic structure. There are eight nearest neighbours at $\sqrt{(3)}a_0/2$ and six second nearest neighbours at a distance a_0.

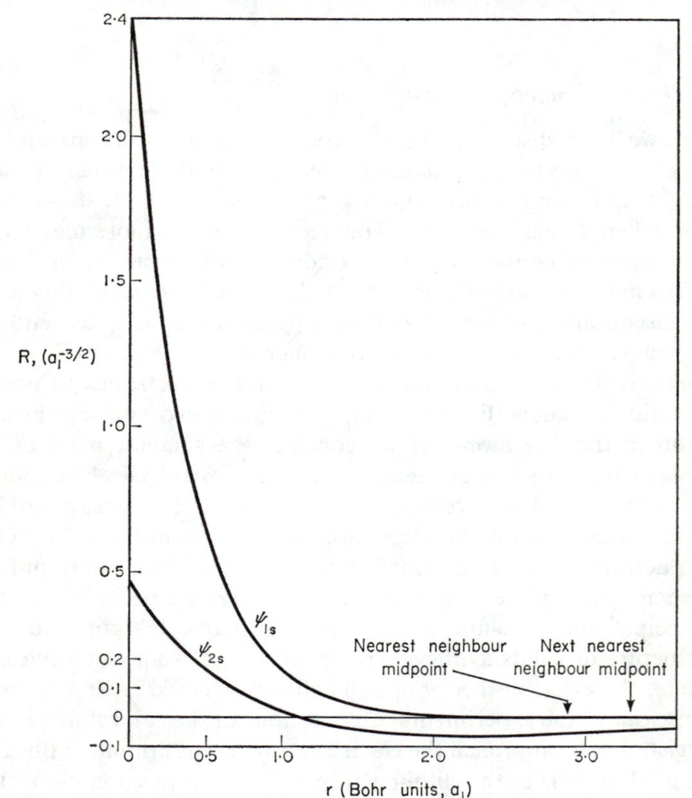

FIG. 8. The 1*s* and 2*s* free atom wave functions of lithium in Bohr units (a_1 = 0·529 Å). The nearest neighbour and next nearest neighbour midpoints of b.c.c. lithium are indicated, showing the negligible overlap of the 1*s* and the large overlap of the 2*s* wave functions. The 1*s* electrons provide the repulsion and the 2*s* electron the attraction between lithium atoms.

trons enabling them to gain attractive energy from the opposite nuclei). In fact, the $1s$ electrons do not overlap for the same reasons as in helium, i.e. the $1s$ shell is filled. On the other hand, the one $2s$ electron shows a large overlap with the nearest neighbours and a smaller overlap with the second nearest neighbours. The cohesion of the metal arises from the overlap of the $2s$ electron, but the presence of eight rather than one overlapping electron is obviously inconsistent with the saturation approach in molecules and prevents the liberal use of free atom quantum numbers. On the other hand, the fact that the $1s$ electrons do not overlap is consistent with the Pauli principle and the use of free atom quantum numbers. Thus we require a new approach for the overlapping electrons. The basis of this new approach is called *Bloch's theorem.* The important clue to this approach is that every lithium atom is identical since they all have an identical number of electrons and identical surroundings (except for those at the surface, and these are generally a negligible fraction). This means that the electron probability distribution (and wave function) must repeat itself just as the atoms repeat in the crystal, and that the potential to be substituted into the Schrödinger equation for a lithium crystal must repeat regularly (periodically). The Bloch theorem states that the solution of the Schrödinger equation with a periodic potential is a wave function of the form

$$\Psi = e^{i\mathbf{k}\cdot\mathbf{r}}\varphi_k(\mathbf{r}) \tag{14}$$

where $\hbar\mathbf{k}$ turns out to be a momentum, $\mathbf{r}$ the position in the crystal and $\varphi_k(\mathbf{r})$ is a function of $\mathbf{r}$ (slightly different for each value of $\mathbf{k}$) and is periodic with the same periodicity as the crystal. Instead of dealing with the probability distribution of 10^{20} electrons we need only deal with the probability distribution of the electrons on each atom, the solution repeating itself to make up the whole crystal. While this makes the problem more tractable mathematically we are still plagued with the same problem we first encountered in the helium atom. We are unable to solve the Schrödinger equation exactly because of the Coulomb repulsion of the electrons (now considerably increased over the free atom of lithium because of the overlap of all the surrounding lithium electrons). We must again resort to the Hartree scheme and replace the many-electron wave function in eqn. (14), by a product of one-electron wave functions, one for each electron in the crystal. The next step is to assume a trial probability distribution for each atom and from it obtain the average potential, $\bar{V}(r)$, which is to be substituted into the Schrödinger equation.

Let us try to guess this average potential, $\bar{V}(r)$. We can gain some clue by inspecting Fig. 9. (Since we are dealing with a cubic crystal we shall express everything in terms of the cubic co-ordinates x, y, z.) Along the line AB, cutting through the cube face closest to the reader, the potential is approximately zero along the x direction since we are at the exact midpoint between the atoms (the positive nuclear charge Z is approximately balanced by the Z electrons). Along the line CD the potential is a minimum closest to the atoms while along the line EF passing through the nuclei of the atoms the potential becomes infinitely negative at the nuclei. These potentials are shown in the bottom of Fig. 9. Similar curves, of course, exist in all three cubic

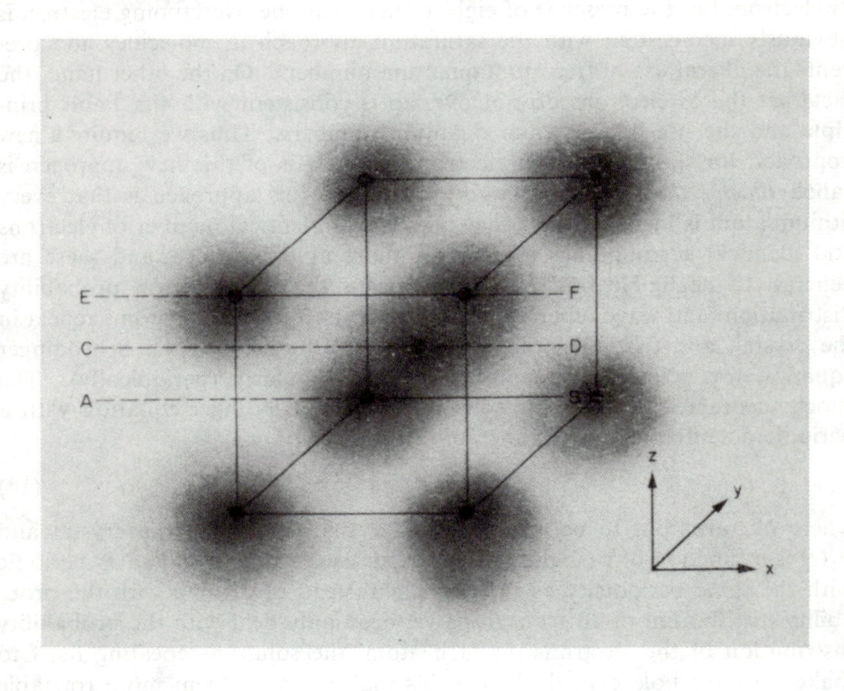

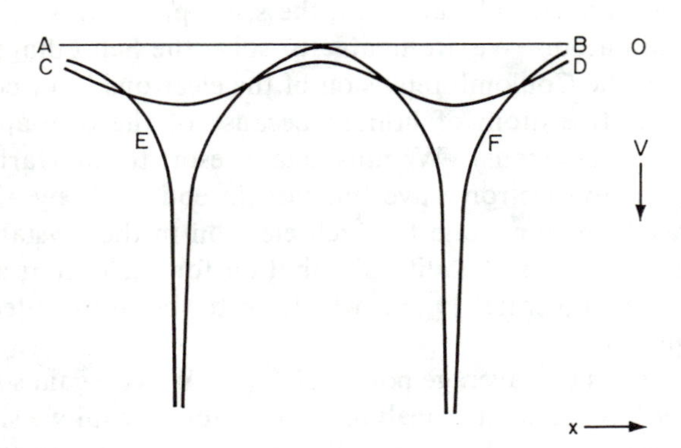

FIG. 9. Qualitative potential curves along the paths AB, CD, and EF in b.c.c. lithium All three paths are in the cube face nearest the reader. The potential minima along the path CD are about $-3{\cdot}5$ eV ($a_0 = 3{\cdot}509$ Å for lithium).

directions. Let us now restrict our attention to these regions and substitute in turn each of the three potentials of Fig. 9 in the Schrödinger equation. Since the solution is identical in all three directions we need only consider one direction for this will suffice for us to understand the types of solutions we shall get. Firstly, for the case AB ($V \cong 0$) we can solve the Schrödinger equation exactly (as such we shall write it in all 3 dimensions)

$$\Psi = b\, e^{ik_x x}\, e^{ik_y y}\, e^{ik_z z} \tag{15}$$

for the eigenfunction, with the eigenvalues

$$E = \frac{\hbar^2}{2m}(k_x^2 + k_y^2 + k_z^2) \tag{16}$$

and the momentum of the electron in each direction

$$p_x = \hbar k_x; \quad p_y = \hbar k_y; \quad p_z = \hbar k_z \tag{17}$$

This shows that $\mathbf{k}$ can be associated with the momentum $\hbar\mathbf{k}$. $\mathbf{k}$ is actually called a *wave number* since it has the dimensions of reciprocal length and from the expression relating momentum and wavelength we have $|k| = 2\pi/\lambda$. The electron probability distribution $\Psi\Psi^*$ is a constant, $\mathbf{b}^2$, and is independent of $\mathbf{k}$.

Secondly, for the case CD the solutions are somewhat more complicated and are shown graphically in Fig. 10. The solutions are in fact different for different values of $\mathbf{k}$ and the probability distribution and eigenvalues for

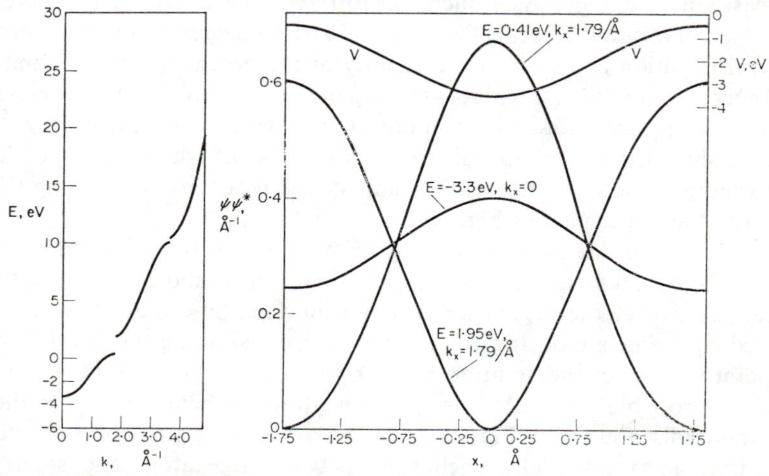

FIG. 10. Energy in eV versus wave number k_x in reciprocal angstroms for the one dimensional periodic potential $V(\text{eV}) = [-3{\cdot}05 \sin^2(\pi x/3{\cdot}509\ \text{Å}) - 0{\cdot}5]$ labelled V in this figure and CD in Fig. 9. The electron probability distribution per Å, $\Psi\Psi^*$, is shown for three values of the energy. The first for the bottom of the band at $k_x = 0$, the second and third at the bottom and top of the first energy gap respectively. The probability distribution repeats every 3·509 Å. While the energy gaps in one dimension occur at one half the values shown, we have used the values corresponding to the three-dimensional lithium crystal. (See also Fig. 11).

various values of k_x are given. The principal feature of these solutions is the presence of a broad *energy band* separated by *energy gaps*. These gaps are values of the energy which are forbidden, just as in the case of free atoms, because they do not yield well-behaved solutions. These energy gaps occur at values of the wave number ($|k| = 2\pi/\lambda$) determined from the conditions for the Bragg diffraction of electrons and X-rays. The particular condition is that $\mathbf{k} \cdot \boldsymbol{\tau}/2 = \boldsymbol{\tau}^2/4$ where $\boldsymbol{\tau}$ is a vector normal to the Bragg plane and its magnitude is $1/d$, d being the spacing of the planes. Thus $|\mathbf{k}| = |\tau|/2$ when $\mathbf{k}$ and $\boldsymbol{\tau}$ are in the same direction. As an example, an electron travelling in the direction CD would diffract from the (200), (400), (600), etc., planes of the lithium crystal or according to the Bragg condition for a b.c.c. crystal $|\tau| = 2\pi/d = 2\pi (h^2 + k^2 + l^2)^{\frac{1}{2}}/a_0$. h, k, l are the so-called Miller indices (in this case $h = 2, 4, 6; k = 0; l = 0$) and a_0 is the lattice spacing 3·509 Å. Substituting these values we get $k_x = 1{\cdot}79$/Å, 3·58/Å, 5·37/Å for the (200), (400) and (600) planes respectively. The probability distribution varies smoothly from its almost constant value along CD for $k_x = 0$ to one which has a maximum at the point along CD closest to the atoms ($x = 0$). This energy gap occurs at $k_x = 1{\cdot}79$/Å, $E = 0{\cdot}41$ eV. Across the gap at $E = 1{\cdot}95$ eV the probability distribution undergoes a marked change such that it is now concentrated at points along the line CD midway between the atoms. We can also see that the wave functions at $k_x = 0$ must have the least curvature, hence on the average the least kinetic energy. As k_x increases to 1·79/Å the kinetic energy increases since the curvature increases; the potential energy decreases since the probability distribution increases in the vicinity of the potential minima; and the total energy increases. As we cross the gap to $E = 1{\cdot}95$ eV the curvature hence kinetic energy does not change appreciably but the potential energy must increase since the probability distribution now peaks where the potential is a maximum along CD. The actual value of the potential in eV along CD is shown in the upper part of Fig. 10.

If we now consider solutions along EF we find that the solutions near the nucleus (in three dimensions) look like the free atom $1s$ and $2s$ wave functions. Indeed we expect the region close to the nucleus ($r < 0{\cdot}5$ Å) to be little affected by the neighbouring atoms. As we proceed along the line EF to its midpoint we expect the solution to look like the solution along AB, i.e. a constant probability distribution. This is shown schematically for the $2s$ probability distribution in Fig. 11. The $1s$ probability distribution is omitted since it remains essentially unchanged between free atom and solid. At $k_x = 0$ the $2s$ probability distribution looks similar to the free atom except that it is now periodic. At $k_x = 1{\cdot}79$ Å the $2s$ solution peaks up slightly near the origin for values of the energy corresponding to the lower part of the energy gap and peaks up at the midpoint for the upper part of the energy gap. The low energy region below -9 eV is only slightly broadened into a band of energies.

From the solutions along AB, CD, and EF we can now see the type of solution we should obtain if we were to solve this problem for all regions and all directions in the crystal. It would require considerable space and effort to do this and it suffices to say that the solutions are all combinations of the types of solutions we have shown. We can in fact now summarize the principal features of the one electron solutions (eigenfunctions, eigenvalues and probability distributions) for lithium metal. The general features of these solutions apply to all metals.

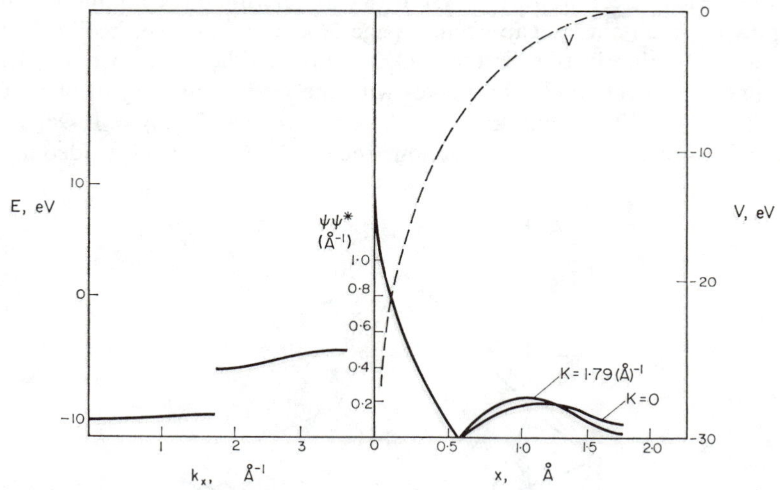

FIG. 11. Energy in eV versus wave number k_x in reciprocal Angstroms for the one dimensional periodic potential labelled EF in Fig. 9 and V in this figure. The electron probability distribution is assumed "2*s* like" as in lithium, and is shown for $k = 0$ at the bottom of the band, and $k_x = 1{\cdot}79$/Å at the bottom of the first energy gap.

1. *In the vicinity of the nuclei* the probability distribution is not very different from the free atom. Specifically for lithium the two 1*s* electrons are identical in energy and wave functions and the 2*s* electron is practically identical in its wave function. The total energy of the 2*s* electron is altered by the way it behaves in the overlap regions. As a general rule then, the closed shell configurations can be assumed to be almost identical in the solid as in the free atom and these closed shell configurations are those specifically omitted for each element in the periodic chart in Table II. For example, the $1s^2 2s^2 2p^6 3s^2 3p^6$ configuration in the first group of transition metals, etc.

2. *In the overlap region* the 2*s* wave functions are significantly altered compared to the free atom. For values of **k** close to zero the probability distribution is somewhat constant while for values of **k** at the energy gaps (those values determined from the Bragg conditions) the probability distribution tends to concentrate around the atoms for the low energy side of the

energy gap, and the probability distribution tends to concentrate in the regions between atoms on the high energy side of the energy gap.

3. The eigenvalues of the overlapping electrons are spread out into a range of values depending on **k** (the wave number), and at values of **k** determined from the Bragg conditions an energy gap or forbidden range of energies exists. The forbidden energies correspond to solutions of the Schrödinger equation whose periodicity does not coincide with the periodicity of the lattice and must be discarded.

4. For each direction in the crystal we can determine the minimum value of $|\mathbf{k}|$ at which an energy gap occurs (page 36). For example, in Fig. 12 the minimum $|\mathbf{k}|$ value in the direction OX (centre of the cube to point X) is determined from the angle OX makes with the normal to the 110 plane OZ ($\theta \cong 32°$ in Fig. 12) so that $|\mathbf{k}| = \pi\sqrt{(1 + 1 + 0)}/\cos 32°(a_0) = 1{\cdot}494/\text{Å}$. If these values are plotted for all directions they will give the twelve-sided figure

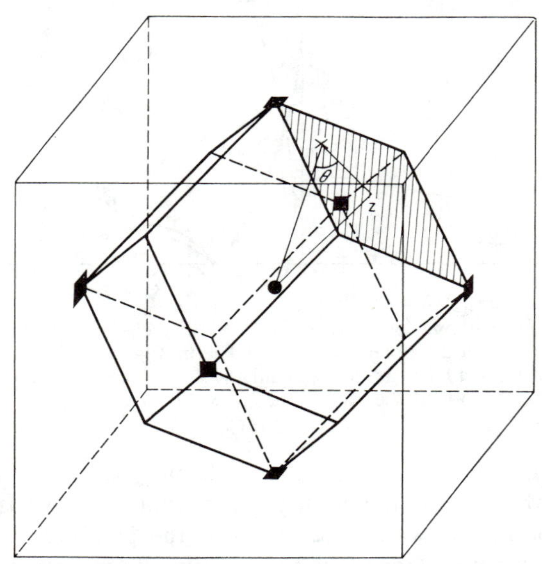

FIG. 12. The first Brillouin zone in the b.c.c. lattice showing the twelve 110 planes each at a distance $k = \pi\sqrt{(2)}/a_0$ from the centre. The twelve planes are inside the six 200 planes each at a distance $2\pi/a_0$ from the centre.

shown in Fig. 12 and this is called the *first Brillouin zone*. It is a figure in **k** space showing the values of **k** at which energy gaps occur. In the b.c.c. structure it is a twelve-sided figure representing diffraction from the 110 planes.

So far we have not mentioned the limitations imposed by the Pauli exclusion principle. You will recall that in the case of the free atoms the Pauli exclusion principle provided a guide in filling the energy levels (eigenvalues). In the case of the crystal the eigenvalues are quite closely spaced. For example, from the lowest energy in lithium at $\mathbf{k} = 0$ to the first energy gap for values of $\mathbf{k}$ at the first Brillouin zone there are N energy levels where N is the number of atoms in the crystal $\sim 10^{20}$ in our case. The Pauli principle permits each level to hold two electrons (one with $m_s = +\frac{1}{2}$ one with $m_s = -\frac{1}{2}$). Except for the quantum numbers m_s the crystal solutions do not give rise to simple quantum numbers as in the case of free atoms. From $\mathbf{k} = 0$ to the edge of the first Brillouin zone the overall energy is 5 to 6 eV in lithium and with 10^{20} energy levels in the crystal we would have an unwieldy number of quantum numbers if we numbered each level. (In addition the spacing between energy levels would be very small $\sim 10^{-19}$ eV.) Instead we use the *density of states* which is the number of energy levels per atom per energy interval, a number that generally runs from about 0·1 to 10.

It is quite difficult to calculate an accurate density of states curve but an approximate curve can be obtained from eqn. (25) (on page 44) for lithium. The density of states curve represents the manner in which the energy levels are filled consistent with the Pauli exclusion principle. The highest energy to which the electrons are filled at absolute zero is called the Fermi energy or *Fermi level.* The Fermi level corresponds to different values of $|k|$ depending on the direction in the crystal and if we plot the values of $\mathbf{k}$ at the Fermi level (called $\mathbf{k}_F$) we trace out a surface in $\mathbf{k}$ space called the *Fermi surface.* If we plotted the Fermi surface for lithium in Fig. 12 it would probably appear almost spherical with a radius slightly smaller than the distance OZ, thus being entirely inscribed within the first Brillouin zone. Each of the energy levels from $\mathbf{k} = 0$ to $\mathbf{k}_F$ would be represented by a sphere within the Fermi surface and the number of spheres per energy interval per atom would be obtained from the density of states curve. Because of the very close spacing of energy levels in this theoretical approach both the Fermi energy and Fermi surface are quite sharp. The energy level spacing is specified to within 10^{-19} eV and the momentum at the Fermi surface to within 10^{-29} g cm/sec.

This outline of the approach to the one-electron theory of solids and the qualitative results for the case of lithium still omits several crucial points necessary for an accurate theory of solids. Firstly, having assumed a probability distribution in order to obtain our starting potential we should compare it with the probability distribution derived from the calculated eigenfunctions, just as the Hartree method does for free atoms, and continuously adjust these until self consistency is obtained. However, this procedure involves such mathematical difficulties that it has not yet been done for any crystal. Secondly, we have omitted the specific interelectronic repulsion in the potential (by replacing it with an average potential). Theoreticians are currently giving much thought to this formidable problem of correlation since they believe it might seriously affect some of the theoretical results. If we imagine a many-electron wave function solved with the specific interelectronic co-ordinates and interelectronic repulsion we might expect the $2s$

electrons in lithium metal to keep out of each others way as compared to the approximate Bloch one electron solutions in which the interelectronic Coulomb repulsion is average out. The simplest way for the electrons to do this is for each one to keep to the region around one atom. In the many-electron wave function the probability distribution would be a maximum for each lithium atom to have one electron, unlike the Bloch solutions which do not confine electrons to specific atoms but only confine the electrons to within the volume of the crystal. According to the Heisenberg Uncertainty Principle however, confining an electron to the region around an individual atom would give rise to an uncertainty in momentum

$$\Delta p \cong \frac{\hbar}{\Delta x} \cong \frac{\hbar}{3{\cdot}5\text{Å}} = 0{\cdot}3 \times 10^{-19} \text{ g cm/sec} \tag{18}$$

This is considerably greater than the 10^{-29} g cm/sec obtained from the Bloch functions for which Δx is the length of the crystal. This is quite important, as we shall later see, to experimental observations like cyclotron resonance, de Haas van Alphen effect, etc., which appear to measure a "sharp" momentum of the electrons at the Fermi surface. While the uncertainty principle in eqn. (18) suggests that the Fermi surface is "fuzzy" or broadened when each electron is considered separately, the collective behaviour in a many electron wave function may restore the sharpness. (We really do not know.)

Various Methods for Solving the Schrödinger Equation for Crystals

We now turn to some of the approximate methods for simplifying the Schrödinger equation for crystals. The metallurgist's interest in these arises partly because these methods are frequently employed and he should know something of the techniques and limitations, and partly because these simplified solutions are employed in the theoretical approach to physical properties of interest to the metallurgist such as resistivity, specific heat, phase diagrams, etc. While everything said so far indicates a good qualitative understanding of crystal wave functions and eigenvalues, quantitative calculations are in error by an appreciable fraction of the cohesive energy of a solid so that the methods described below require considerable improvement. None of the methods are self-consistent nor include correlation.

1. *The Cellular Method.* In this method, each atom is enclosed in a multisided figure (called a *polyhedra*) formed by the intersection of planes halfway between the neighbours and normal to the line joining the neighbours. Such a figure is shown in Fig. 13 drawn for a b.c.c. lattice. It is a fourteen-sided figure, the planes bisecting the lines joining the nearest and next nearest neighbours. By stacking these polyhedra, each with its one atom at the centre, one can build up the body-centred cubic lattice. We solve the Schrödinger

equation inside this polyhedra by the Hartree method assuming, as in the free atom, a spherically symmetric potential. We do not discard those solutions which do not approach zero as **r** approaches infinity as we did in the free atom (see page 3) since we are only interested in values of **r** up to the surface of the polyhedra. We now require that the one electron wave functions join smoothly at the surface of the polyhedra to the wave functions in the adjoining cell.* In addition the wave function must repeat with the lattice periodicity.

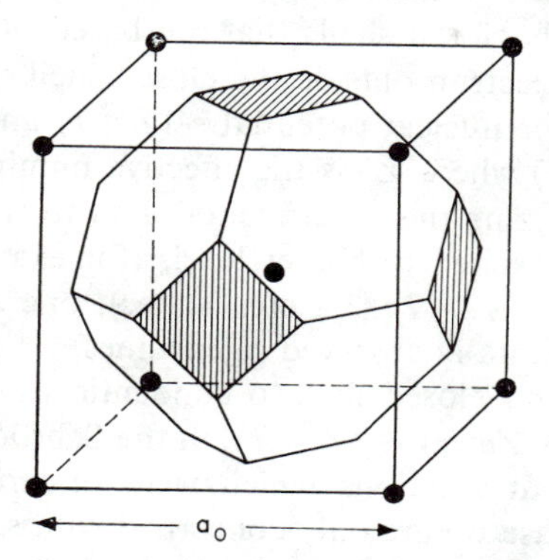

FIG. 13. The body-centred cubic polyhedra formed by the eight planes which are normal to and bisect the line joining the eight nearest neighbours (111 direction), and the six planes (shaded) which are normal to and bisect the line joining next nearest neighbours. Each polyhedra contains one atom at the centre and they will stack so as to fill completely the space of a b.c.c. lattice.

One need only solve this for one polyhedra since the crystal is merely a repetition of identical polyhedra. The current limitations of this method are:

(*a*) In general, the smooth joining of wave functions at the boundaries only occurs at a few points. Furthermore we assume the potential to be spherically symmetric rather than polyhedra shaped so as to permit the separation of ψ into $\theta \cdot \phi \cdot R$. This prevents the solution from being self-consistent.

(*b*) It is too difficult to use antisymmetrized wave functions (Hartree–Fock) so that exchange is not included.

2. *Augmented Plane Wave Method* (and related methods). In this method, each atom is surrounded by a sphere generally just big enough to touch the sphere of the nearest neighbour. Inside the sphere the potential is chosen to be spherically symmetric, and outside the sphere it is chosen to be a constant. This leads to atomic like solutions inside the sphere and plane wave solutions (sines and cosines) outside the spheres. One then selects those plane wave solutions which join smoothly at the surface of the sphere to the solution

* We require the smooth joining of the wave function ψ to ensure that the probability distribution, $|\psi|^2$, and momentum $(-i\hbar\partial\psi/\partial r)/\psi$ *is* continuous everywhere.

inside the sphere. Variations of this method are concerned mostly with the mathematical form of the solutions and the size of the sphere employed. This method is possibly the most widely employed. Its limitations are:

(*a*) The wave functions cannot be joined smoothly over the entire sphere.

(*b*) The solutions are not antisymmetrized (exchange omitted), nor self-consistent.

3. *Quantum Defect Method.* This is a semi-empirical method which employs solutions outside closed shells that are based on Coulomb potentials, i.e. one assumes the electron outside the closed shell experiences a potential which is the sum of the nuclear potential $-(Ze^2/r)$ and the potential of the closed shell $+(Z'e^2/r)$ where Z' is the effective number of electrons in the closed shell. By analyzing the experimental data for the free atom in which the outer electron is excited to higher levels (for example in lithium the $2s$ electron is excited to $3s$, $4s$, $2p$, $3p$, etc., states), one adjusts Z' to give the best fit to the experimentally observed eigenvalues.

We then assume this closed shell to be identical in the solid and employ the same potential $[-(Ze^2/r) + (Z'e^2/r)]$ in the Schrödinger equation for the solid. We now permit solutions which were discarded in the free atom, since in the present case we are only concerned with solutions outside closed shells and up to the surface of the atomic polyhedra surrounding each atom in the crystal. We then discard those solutions which do not join smoothly with their mirror images at the surface of the polyhedra. Its limitations and advantages are:

(*a*) The method is good only when the experimental eigenvalues of the free atom can be empirically fitted to a Coulomb potential. This is only a good approximation for the alkali metals.

(*b*) It is not useful for telling us the nature of the solutions inside the closed shell.

(*c*) It gives remarkably good results for the cohesive energy and lattice spacing of the alkali metals; being in fact the most successful of the techniques to date.

4. *Tight Binding Method.* A less accurate method for metals than the above three is the tight binding method applicable to cases where the electrons do not overlap very much (quite rare for a metal). With small overlap the wave functions are assumed to be those of the free atom multiplied by $e^{i\mathbf{k}\mathbf{r}}$ to give them the Bloch form. With these wave functions and the free atom potential we calculate the energy for each value of $\mathbf{k}$ to obtain E versus $\mathbf{k}$. While this method is simple enough to include anti-symmetrization (exchange) the method has limited use for metals.

All the above methods have been used frequently, and no doubt many more methods and variations of these methods will be invented. The approach in all cases is to construct solutions to the Schrödinger equations which are everywhere smooth and periodic with the lattice and are reasonably close to the truth. The methods all entail approximations in order to ease the mathematical and computational difficulties, *but until a method is developed which gives accurate results we shall not have an electron theory of solids.* Nevertheless, it should not take many generations before calculating machines overcome most of the limitations.

Free Electron Theory

We now wish to introduce a gross oversimplification of the outer electron wave functions of a metal which is used *ad nauseam* to make calculations on more complicated problems such as the effect of electric fields giving rise to electric currents, the effects of magnetic fields giving rise to magnetic susceptibility, the effects of temperature giving rise to specific heat, etc. This simplification is called the *free electron theory* and is applied to metals by assuming the outer electrons move in a constant potential. If the value of this constant potential is $-V_0$ and the metal is considered to be in the form of a cube of side L the eigenfunctions are

$$\Psi = \left(\frac{8}{L^3}\right)^{1/2}\left(\sin\frac{n_x \pi x}{L}\sin\frac{n_y \pi y}{L}\sin\frac{n_z \pi z}{z}\right) \tag{19}$$

where n_x, n_y and n_z are the principal quantum numbers which can assume any positive integral value other than zero. The functions are "well behaved" in that they vanish at the surface of the metal ($x, y, z = 0$ or L) thus forcing the electrons to remain inside the metal.* The eigenvalues are

$$E = \frac{\hbar^2}{2m}\left(\frac{n_x^2\pi^2}{L^2} + \frac{n_y^2\pi^2}{L^2} + \frac{n_z^2\pi^2}{L^2}\right) - V_0 \tag{20}$$

The components of momentum in the three directions are

$$p_x = \frac{\hbar n_x \pi}{L}, \qquad p_y = \frac{\hbar n_y \pi}{L}, \qquad p_z = \frac{\hbar n_z \pi}{L} \tag{21}$$

and wave numbers

$$k_x = \frac{n_x \pi}{L}, \qquad k_y = \frac{n_y \pi}{L}, \qquad k_z = \frac{n_z \pi}{L} \tag{22}$$

The Pauli principle is satisfied by adding the quantum numbers $m_s = \pm\frac{1}{2}$ to n_x, n_y and n_z. The energy levels are filled accordingly, with the first few levels shown in Table III.

* Strictly speaking, this is true only for a constant potential $-\infty$ but as we are not presently concerned with details near the surface this approximation will suffice.

TABLE III

n_x	n_y	n_z	m_s	E
1	1	1	$\frac{1}{2}$	$\dfrac{3\hbar^2\pi^2}{2mL^2} - V_0$
1	1	1	$-\frac{1}{2}$	
2	1	1	$\frac{1}{2}$	$\dfrac{6\hbar^2\pi^2}{2mL^2} - V_0$
2	1	1	$-\frac{1}{2}$	
1	2	1	$\frac{1}{2}$	
1	2	1	$-\frac{1}{2}$	
1	1	2	$\frac{1}{2}$	
1	1	2	$-\frac{1}{2}$	
2	2	1	$\frac{1}{2}$	$\dfrac{9\hbar^2\pi^2}{2mL^2} - V_0$
2	2	1	$-\frac{1}{2}$	
2	1	2	$\frac{1}{2}$	
2	1	2	$-\frac{1}{2}$	
1	1	2	$\frac{1}{2}$	
1	1	2	$-\frac{1}{2}$	
2	2	2	$\frac{1}{2}$	$\dfrac{12\hbar^2\pi^2}{2mL^2} - V_0$
2	2	2	$-\frac{1}{2}$	

The first four eigenvalues for an electron in a constant potential $-V_0$ enclosed within a cube of side L.

When two or more different sets of quantum numbers give identical eigenvalues for an electron they are called *degenerate levels*. The lowest level has a degeneracy of two, the next level has a degeneracy of six, etc. If we fill all the energy levels until we run out of electrons we reach the topmost or Fermi level and this is

$$E_F = \frac{\hbar^2\pi^2}{2m}\left(\frac{3N}{\pi L^3}\right)^{2/3} - V_0 \tag{23}$$

where N is the number of free electrons in the sample (N/L^3 is the number of free electrons/cm^3) . The average energy of all the free electrons is obtained by averaging over all the levels from the first to the Fermi level weighting each according to its degeneracy. It is

$$\bar{E} = \frac{3\hbar\pi^2}{10m}\left(\frac{3N}{\pi L^3}\right)^{2/3} - V_0 \tag{24}$$

The number of energy levels per atom per energy interval (density of states) is

$$\frac{dn}{dE} = \frac{4\pi mL^2}{Nh^2}(n_x^2 + n_y^2 + n_z^2)^{1/2} = \frac{4\pi L^3(2m)^{3/2}}{Nh^3}(E + V_0)^{1/2} \tag{25}$$

The density of states at the Fermi level is

$$\left(\frac{dn}{dE}\right)_{E=E_F} = \frac{4\pi m}{(N/L^3)^{2/3}}\left(\frac{3}{\pi}\right)^{1/3}\frac{1}{h^2} \tag{26}$$

The momentum of the electrons at the Fermi level is

$$p_F = \frac{h}{2}\left(\frac{3N}{\pi L^3}\right)^{1/3} \tag{27}$$

Since the potential $-V_0$ is constant, we have removed the crystal periodicity and the momentum and wave number of the electrons are independent of direction. The Fermi surface is thus a sphere in **k** space with radius

$$k_F = \pi\left(\frac{3N}{\pi L^3}\right)^{1/3} \tag{28}$$

If we apply these results to a lithium metal cube of side 1 mm with constant potential, $-V_0 = -9$ eV we obtain:

Lowest energy electron (bottom of band)

$$E = -9 \text{ eV}$$

Highest energy electron (top of band, Fermi energy)

$$E_F = -4{\cdot}3 \text{ eV}$$

Band width

$$E_F - E = 4{\cdot}7 \text{ eV}$$

Density of states at Fermi level

$$\left(\frac{\mathrm{d}n}{\mathrm{d}E}\right)_{E=E_F} = 0{\cdot}319/\text{eV per atom}$$

Density of states at bottom of band

$$\left(\frac{\mathrm{d}n}{\mathrm{d}E}\right) = 1{\cdot}57 \times 10^{-7}/\text{eV per atom}$$

Principal quantum numbers at Fermi level

$$n_x = n_y = n_z \cong 20,500,000$$

Wave number at Fermi surface

$$k_F = 1{\cdot}11 \times 10^8/\text{cm}$$

Wave number of closest Brillouin zone surface (110)

$$= 1{\cdot}262 \times 10^8/\text{cm}$$

Wavelength of electrons at Fermi level

$$\lambda_F = \frac{2\pi}{k_F} = 5{\cdot}65 \text{ Å}$$

Wavelength of electrons at bottom of band

$$\lambda = 2L = 0{\cdot}2 \text{ cm}$$

In this free electron approximation the spherical Fermi surface is small enough not to touch the first Brillouin zone (Fig. 12).

Now we know that the correct solution for the metal eigenfunctions must have the lattice periodicity but in this free electron approximation the lowest energy electron has a wavelength some ten million times larger than the lattice periodicity whereas the electrons at the Fermi surface have a wavelength within a factor of two of the lattice periodicity. It is this latter accidental fact that enables the free electron theory to give reasonable results (in some cases) *but only near the Fermi surface*. If one applies the free electron theory it should be limited to the electrons near the Fermi surface.

In order to show graphically the probability distribution of the electrons in the free electron approximation and in addition to show how electrons with the same quantum numbers stay out of each others way (exchange) and how electrons tend to avoid each other due to their repulsion (correlation), we shall simplify the problem somewhat further by artificially restricting the electron to one dimension, from $x = 0$ to L. The problem can be made a bit more realistic by making L, the length of the one dimensional box, equal to 7·0 Å or twice the lattice spacing of lithium and placing two electrons in this box. In order to determine the probability distribution we must take the proper antisymmetrized product of one electron wave functions. The lowest energy state is the singlet composed of the antisymmetric product of the one electron wave functions with principal quantum numbers $n_x = 1$, $m_s = +\frac{1}{2}$; $n_x = 1$, $m_s = -\frac{1}{2}$ (in one dimension n_y and n_z are absent). The next highest state is the triplet composed of the antisymmetric product of one electron wave functions with principal quantum numbers $n_x = 1$, $m_s = +\frac{1}{2}$; $n_x = 2$, $m_s = +\frac{1}{2}$. In order to show the probability distribution in a one dimensional plot we must specify a position for one of the electrons and we assume that electron to be at $x = L/4 = 1{\cdot}75$ Å. In Fig. 14 we have plotted the probability distribution for the singlet (P_0) and triplet (P_1) states without correlation and with an approximate correlation correction ($P_0{}^*, P_1{}^*$). As one can see, the Pauli principle causes the electrons to avoid each other if they have the same spin (triplet). The effect of correlation is equally marked but tends to be over-exaggerated in one dimension for the electrons cannot go around each other as in three dimensions. The first excited singlet state ($n_x = 1$, $m_s = +\frac{1}{2}$; $n_x = 2$; $m_s = -\frac{1}{2}$) is labelled P_2'. Even though the same one electron wave functions are used as in the triplet state, anti-symmetrization yields different distributions. A more realistic picture of correlation is given in Fig. 2b for helium. However, the distributions in Fig. 14 do give a qualitative picture of the effect of correlation and anti-symmetrization on the probability distribution obtained in the free electron approximation. By choosing the length of the box to be just twice the lattice spacing of lithium the effect is qualitatively appropriate to the free electrons at the Fermi surface in lithium whose wave length, 5·65 Å, and energy, $-4{\cdot}3$ eV, is comparable to the above one-dimensional case, 7 Å and $-8{\cdot}2$ eV respectively, for the wavelength and energy.

Now that we have seen the nature of the solutions of the Schrödinger equation for crystals and the problems associated with obtaining these solutions, let us see if we can say any more about the elements by examining the periodic table.

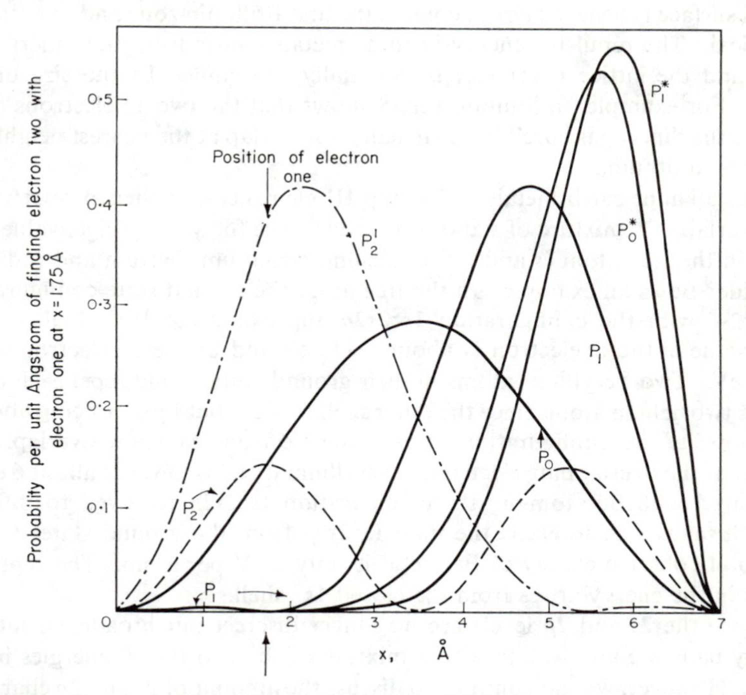

FIG. 14. The approximate electron probability distribution for two electrons in a one dimensional box 7 Å long. One of the electrons is permanently positioned at 1·75 Å indicated by the arrow. P_0 is the probability per unit Å of finding electron two if its spin is antiparallel to electron one and the Coulomb repulsion between the electrons is neglected. P_1 is the same for parallel spins. P_0* and P_1* are respectively the same as P_0 and P_1 but include the Coulomb repulsion (correlation). P_2 (dashed curve) is the probability distribution for only *one* electron in the box in the first level above the ground level. P'_2 (dash-dot curve) is the probability distribution of finding electron two if its spin is antiparallel to electron one and the system is in the first excited state (Coulomb repulsion neglected).

The Bonding Electrons in the Periodic Table

We have already said that the inert gases do not form real chemical bonds due to their stable filled *p* shells. The use of free atom quantum numbers and the Pauli principle appears applicable in these cases. The electrons can only overlap by being excited to the next level $2p \rightarrow 3s$, $3p \rightarrow 4s$, etc., but this requires too much energy, ~16 eV for argon and 21 eV for neon, considerably greater than can be gained by cohesion.

The alkali metals are bonded by the overlap of one *s* type electron which is bonded to the free atom by about 5 eV. In the solid the overlapping of these outer electrons spreads the eigenvalues over an energy range of about 5 eV (band width) and lowers the average energy by about 1 – 2 eV. The Fermi surface is believed not to contact the first Brillouin zone and to be fairly spherical. The repulsive energy in these metals comes from the "inert gas" cores and the lattice parameter is essentially determined by the size of the cores. For example, in lithium, Fig. 8 shows that the two 1*s* electrons comprising the "inert gas core" have virtually no overlap at the nearest neighbour distance in lithium.

The alkaline earth metals and group III elements owe their properties to the overlap of a mixture of *s* and *p* electrons since the *s* and *p* eigenvalues are close in the free atom (*s* and *d* for calcium, strontium, barium and radium). Consider Be as an example. In the free atom the ground state configuration is $1s^2 2s^2$ with the configuration $1s^2 2s2p$ approximately 3 eV higher. The eigenvalue of the 1*s* electron is about -112 eV and of the 2*s* electron about $-9{\cdot}2$ eV. Two beryllium atoms in their ground state would repel each other just as two helium atoms since the outer shell is filled, but by sacrificing about 3 eV to excite the configuration $1s^2 2s2p$ the electrons can now overlap. The energy of the overlapping electrons in beryllium metal is lowered about 6 eV as compared with Be atoms in the configuration $1s^2 2s2p$ removed to infinity. Since it takes 3 eV to excite the state $1s^2 2s2p$ from the ground state, $1s^2 2s^2$, the total cohesive energy of Be metal is only 3 eV per atom. The repulsive terms in the energy arises from the closed $1s^2$ shell.

Since the 2*s* and 2*p* levels are no longer discreet but broadened into an energy band we are dealing with a mixture of the two for all energies in the band. However, we can continue to discuss the amount of 2*s* and 2*p* character in the band by examining the electron wave functions inside the atomic volume i.e. the volume of the atom in the neighbourhood of the nucleus where the influence of the neighbouring atoms is small. This would range from $r = 0$ to $r \cong 0{\cdot}5$ Å in beryllium. Within this volume the 2*s* wave function is spherically symmetric, has a finite value at the nucleus and has one node at about 0·3 Å. The 2*p* wave function is zero at the nucleus, has no nodes, and has an angular distribution given by the Legendre polynomials on p. 5. The qualitative band and zone structure of Be is shown in Fig. 15. The first Brillouin zone is a twenty-sided figure with the six 100 faces at a distance 1·59/Å from the centre at $\mathbf{k} = 0$, the two 002 faces at a distance 1·76/Å and the twelve 101 faces at a distance 1·815/Å. The radius of the Fermi sphere in the free electron theory is $|k| = 1{\cdot}94$/Å and is shown protruding from the first Brillouin zone in Fig. 15. If one were dealing only with discrete energy values as for the 1*s* electrons then we would have a separate Brillouin zone for each shell and each zone would be filled when the shell is filled. But a band structure is a mixture of many states and it is difficult to say how many electrons are required to fill a Brillouin zone. In the free electron theory the volume of the sphere containing two electrons per atom just equals the volume of the first Brillouin zone in the h.c.p., f.c.c. and b.c.c. metals, but this can only serve as a guide. If we limit our consideration to the 2*s* and 2*p* states in beryllium metal, Fig. 15 gives qualitative *E* versus **k** curves in the three principal crystallographic directions, normals to the 100, 002, 101 planes. (The 1*s* electrons have a discrete energy, −112 eV, independent of **k** and their Brillouin zone is filled.) The 2*s* states in the band are filled till a value of **k** in the 100 direction is reached which corresponds to the 100 zone faces. The

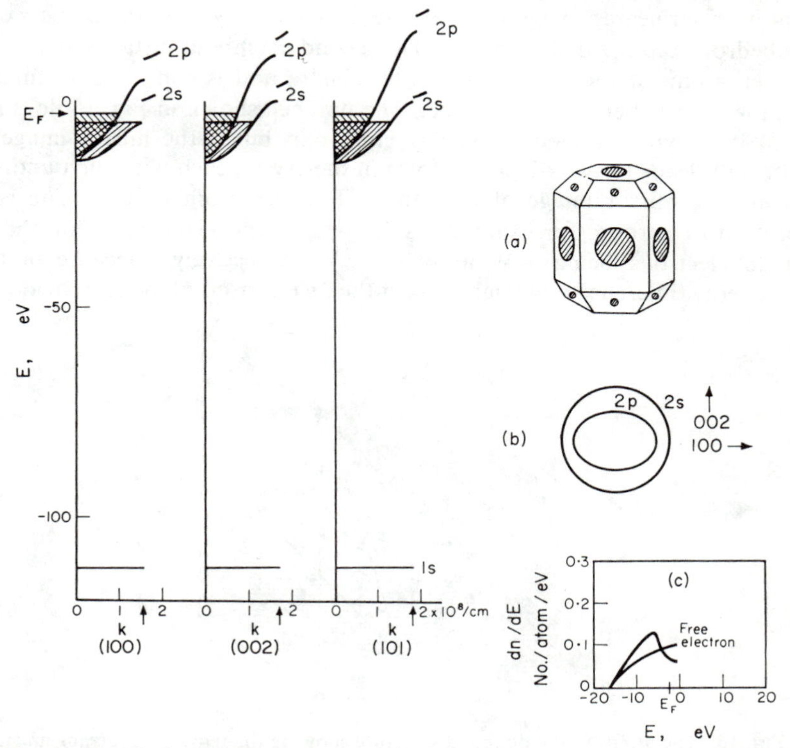

FIG. 15. Approximate *E* versus *k* curves for the 1*s*, "2*s*" and "2*p*" states in the three principal directions which are normal to 100, 002 and 101 planes in h.c.p. beryllium metal. The Fermi level is identified as E_F. The first Brillouin zone is shown, ***a***, with the free electron Fermi surface protruding through the six 100 faces, the two 002 faces and just barely through the twelve 101 faces. The maximum values of *k* attained for the "2*s*" and "2*p*" states in the directions normal to the 100 and 002 planes are shown in ***b***. The density of states (number of levels per atom per eV) versus energy in eV is shown in ***c*** as well as that derived from the free electron theory.

energy gap prevents further filling of the 2*s* states and the 2*p* states are then filled until we exhaust our supply of electrons at the Fermi surface. The width of the 2*p* band of states is greater than the 2*s* since they overlap the neighbours more. Herring and Hill[2] have calculated a density of states curve in beryllium by the orthogonalized plane wave method and it is given in Fig. 15. It appears to be peaked in the vicinity of the filling of the $2s$ states. Their cohesive energy is about 40% lower than observed.

The columns of the periodic table containing oxygen and fluorine are mostly non-metals. In these cases the outer p electron eigenvalues are increasingly higher than the outer s electron eigenvalues for the free atoms, and there is less tendency to use s and p states combined. Carbon, silicon, germanium and tin all form the *cubic diamond type* crystal structure shown in Fig. 16. Each atom has four nearest neighbours at $\sqrt{(3)}a_0/4$ and they form the corners of a tetrahedron, a four-sided pyramid. The second neighbour distance is $a_0/\sqrt{(2)}$ between atoms at the corners of the tetrahedra and is considerably further than the nearest neighbour distance. The unit cell shown in Fig. 16 does not repeat itself when rotated about any cubic axis but is the mirror image of itself. This leads to two different atoms in the crystal each with surroundings that are the mirror image of the other. The free atom ground state configuration of carbon, for example, is $1s^2 2s^2 2p^2$, the eigenvalues of the $2s$ and $2p$ electrons being -19 and -11 eV, respectively. Because of this large energy difference one might expect the $2p$ electrons alone to provide the

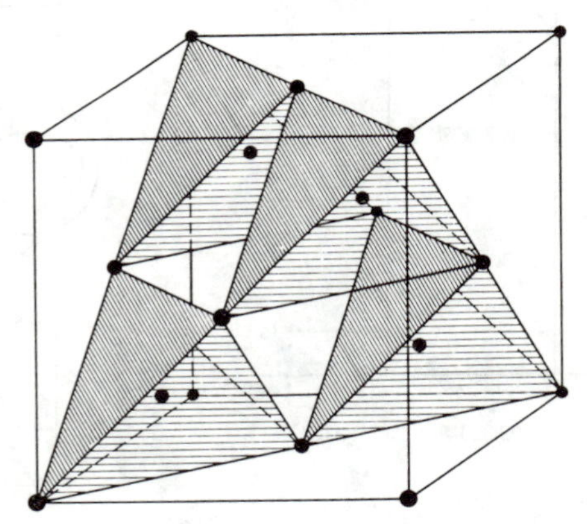

FIG. 16. The atoms in the diamond structure showing the tetrahedral arrangement of neighbours around each atom. The four atoms not connected by lines are in the centre of the tetrahedra.

binding, the filled $2s^2$ shell providing the repulsive energy. However this limits us to two bonding electrons per atom. We can gain sufficient bonding energy to offset this 8 eV difference by promoting a $2s$ electron to a $2p$ and thus have four bonding electrons per atom. If we add the four wave functions $2s$, $2p\ (m_l = 0)$, $2p(m_l = 1)$, $2p(m_l = -1)$ and square the sum to obtain the probability distribution we find 4 lobes pointing in the directions of the atoms on the tetrahedra. This process of adding wave functions to obtain a new wave function is called *hybridization* and they can be added in many combinations. In this particular case (sp^3) each of the lobes in the electron probability distribution overlaps the lobe from the nearest neighbour with opposite m_s value to form a strong "singlet type" of bond. In the overlap region all eight of the orbitals are present and this fills the entire 2*s* and 2*p* group of orbitals. This is the basic reason for the non-metallic character of these diamond structures for there are no energy states available very close to the highest filled energy state.* However, in the diamond structures of carbon, silicon, germanium and tin there are higher energy electronic states from less than one volt higher in tin to several volts in carbon. These higher states are not normally occupied in the pure crystals but by heating, electrons can be excited to these states and the crystals will conduct. This is the nature of a *semiconductor*.

The low temperature form of tin has the diamond structure but it transforms to a metallic tetragonal form just below room temperature. The energy difference of the two modifications is small ($\sim$0·01 eV) in spite of a 25% decrease in atomic volume for the tetragonal phase. This involves a pronounced electron structure change from sp^3 hybridization and four nearest neighbours in the diamond structure to a metallic mixture of *s* and *p* orbitals and twelve nearest neighbours (neglecting the tetragonal distortion) in the metallic phase. Lead only forms a f.c.c. metallic phase.

The *semi-metals* in the columns under nitrogen and oxygen are principally bound by *p* electrons and all have complex crystal structures. The free atom ground state configuration for nitrogen is $1s^2 2s^2 2p^3$ with the 2*s* eigenvalue $\sim$13 eV lower than the 2*p*. Unlike carbon it is too costly to promote the 2*s* electron to a 2*p* orbital and, since the 3*s* eigenvalues are about 12 eV higher than 2*p*, the cohesion is limited to the 2*p* orbitals. This still forms a very stable N_2 molecule. The $2p^3$ electrons on one nitrogen atom overlap the $2p^3$ electrons on the second nitrogen atom with opposite m_s values to form a bound singlet state just as the H_2 molecule for 1*s* electrons. In all cases of the semi-metals little other than *p* electrons are available for bonding and this leads to complicated crystal structures. The atoms have difficulty in attaching to each other since the *p* orbitals are not spherically symmetric.

The *halides* are almost "inert" since they have five *p* electrons. They can only bond with a relatively small overlap of the *p* electrons, since six *p* electrons form a closed shell. The stablest structures formed with the halides are compounds with the alkali metals and are called *alkali halides* (like NaCl). The outer electron of the alkali metal moves towards the halide neighbour and assumes *p* character in the overlap region tending to fill the *p* shell. The

* A metal is distinguished by the relative ease by which it conducts electrons. When an electric potential is applied to the ends of a metal wire the electrons are accelerated along the wire and so increase their energy. If there are no empty energy states available very close to the Fermi level then the electrons cannot absorb energy from the electric potential and the metal will not conduct.

transfer of an electron from the alkali atom to the halide leaves the former positively charged (since the nucleus has one more proton than the atom now has electrons) and the latter negatively charged (since it has one more electron). This Coulomb attraction holds the atoms together and is called *ionic bonding*. Actually this is a somewhat simplified picture since the electron from the alkali atom does not leave its parent atom completely although a simple model of positively and negatively charged atoms (for example Na^+, Cl^-) in the crystal gives fairly good values for the cohesive energy, elastic constants, etc.

The transition metals have *d*, *s* and *p* electronic states energetically close while the rare earth metals have *f*, *d* and *s* states close. In the transition metals the *d*, *s* and *p* states overlap and provide the bonding. In the rare earths the radially compact 4*f* electrons do not bond but do produce the magnetic properties of these metals. The bonding is provided by the outer 5*d* and 6*s* states. We shall discuss these two groups more fully in the next section.

Crystal Field Theory; Liquids

We shall conclude this section with a brief mention of two subjects, *crystal field theory* and *liquids*.

Crystal field theory tells us that the electron probability distribution around each atom must possess what is called the crystal symmetry. We saw this in the case of the tetrahedra which form the diamond structure in that the electron probability distribution was also tetrahedral (see Problem 5, p. 63). If we apply this to the case of a body-centered cubic structure we have eight nearest neighbours in the 111 directions. If the electron probability distribution points toward any one of these eight neighbours it must similarly point toward the other seven. The symmetry of a crystal refers to the number of ways you can rotate, translate, reflect, or produce a mirror image of the crystal so that the crystal appears identical again. For example, a body-centred cubic crystal is identical under any number of 90° rotations about any cubic axis, or if reflected in a mirror with its 100, 110 or 111 planes parallel to the mirror. Crystal field theory tells us that any such operation which makes the crystal appear identical must also make the electron distribution appear identical. This is quite sensible since the crystal is built up from the electron distribution. The main reason for considering crystal symmetries is that it helps decide which combination of orbital wave functions one may use in order to construct the electron probability distribution of a crystal.

The cases most frequently discussed in crystal field theory are *d* electrons in cubic crystals, hexagonal crystals and crystals with nearest neighbour tetrahedral symmetry. In the photograph in Fig. 17 are models of the angular part of the electron probability distribution formed from *d* orbitals in a cubic crystal. The t_{2g} group consists of three 3*d* electrons, one in $(m_l = -1)$,

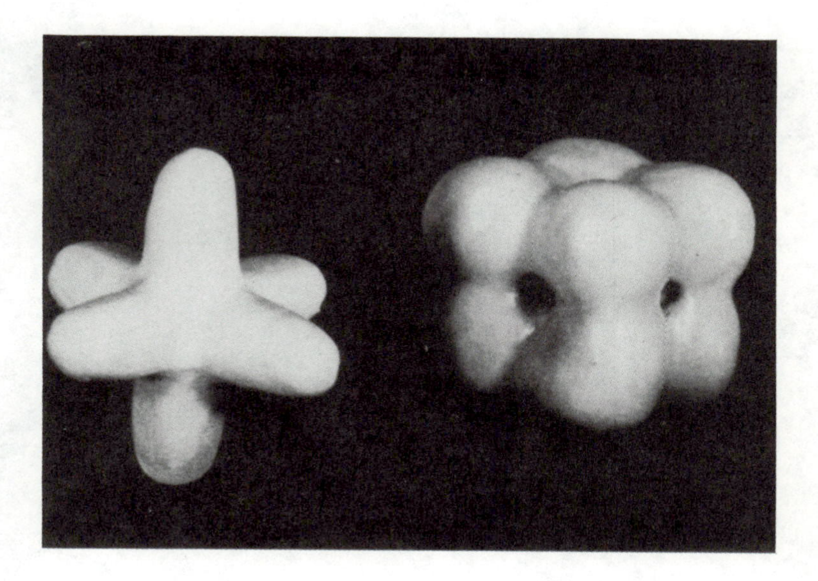

FIG. 17. Clay models of the e_g and t_{2g} electron probability distribution around an origin which is at the centre of each figure. The relationship to the cubic directions is apparent. The value of the angular part of the electron probability distribution is obtained from the intersection of a line originating from the centre of the figure with the surface of the figure. The total electron probability distribution is obtained by multiplying this with the radial probability distribution.

Models by Mrs. A. Harvey, U.S. Army photograph.

Facing page 52

one in $(m_l = +1)$ and one in a linear combination of the difference $(m_l = +2) - (m_l = -2)$. It has eight lobes in the 111 directions. The e_g group contains two $3d$ electrons, one in $(m_l = 0)$ and one in the linear sum $(m_l = +2) + (m_l = -2)$. It has six lobes in the 100 directions. If all five states are filled the electron distribution is spherically symmetric and this, of course, always has the symmetry of any crystal, since a sphere is the same under any rotation or reflection.

The subject of the electron structure of a liquid has received little attention. While the Bloch theorem tells us the type of solutions of the Schrödinger equation for a perfectly repeating lattice, it does not tell us the type of solutions for a liquid which is not periodic. A visual image of the atoms in a liquid is a box full of sticky steel balls in constant motion. The centres of any two neighbouring atoms do not approach any closer than twice their radii and the arrangement in space is random. Some attempts have been made to solve the Schrödinger equations for a very crude model of a liquid. It appears that energy bands and energy gaps do exist but the gaps are smaller than in a crystal. The first Brillouin zone still exists but is somewhat fuzzy (see chap. XII). This is not a surprising result since many properties of metals such as resistivity, atomic volume, colour, etc., do not change radically on melting. However, the mathematical problems are formidable and the question of the electronic structure of liquids is generally swept under the rug. (See chap. XII).

C. Transition Metals and Rare Earth Metals

The Bonding Electrons in the Transition Metals and the Rare Earth Metals

Metallurgists have expended considerable effort on the transition metals, particularly iron, cobalt and nickel, yet the theoretical solid state physicist has more difficulty with these elements than any others. Part of the problem arises from the overlapping d electrons and their angular factors which are more complicated than for s or p electrons. In addition, some of the overlapping electrons are always s and p like since the next s shell has always about the same energy as the previous d shell, and the next p shell is only a few volts higher ($4s$, $4p$ for the $3d$; $5s$, $5p$ for the $4d$; $6s$, $6p$ for the $5d$). We are thus dealing with the overlap of between five to nine orbitals (five d three p, one s) if we limit ourselves to "singlet type" bonding in which the overlapping orbitals are paired (opposite m_s values). Even the rare earth metals do not present as much of a problem as the transition metals though the f electrons have seven orbitals. The reason is that the f electrons *do not overlap* because their radial factors are quite compressed compared with the s and d electrons of the same energy ($6s$, $5d$) for the $4f$; $7s$, $6d$ for the $5f$). These outer s and d electrons provide all the overlap, and in the rare earth metals there are only three or in some cases only two of them. The next figure (Fig. 18) gives the $4f$, $5p$, and $6s$ Hartree–Fock free atom radial wave functions for gadolinium. The nearest neighbour midpoint is at 3·4 atomic units and is indicated by an arrow. The $4f$ wave functions are quite compact and do not overlap the nearest neighbours. In fact, they are well inside the closed $5p$ shell which provides the repulsion in the solid.* The $5d$ and $6s$ wave

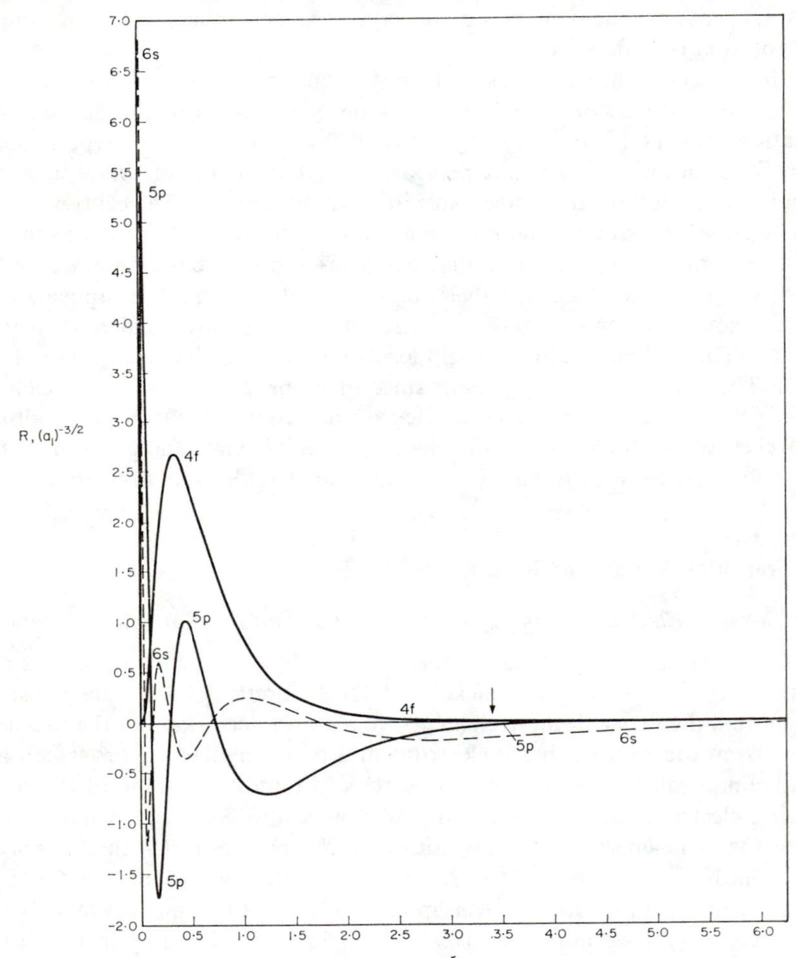

FIG. 18. The $4f$, $5p$, and $6s$ radial wave functions for the free atom of gadolinium. The nearest neighbour interatomic midpoint in h.c.p. gadolinium is indicated by the arrow demonstrating the "isolation" of the $4f$ electrons (units are the Bohr radius, a_1).

* The reason the $4f$ electrons are closer to the nucleus than the $5p$, and yet have a higher energy, arises from the higher Coulomb interelectronic repulsion between each $4f$ electron and all other electrons. Its complicated angular factors with their many angular nodes make it difficult for them to fit into the rest of the gadolinium electron distribı tion without coming close to other electrons.

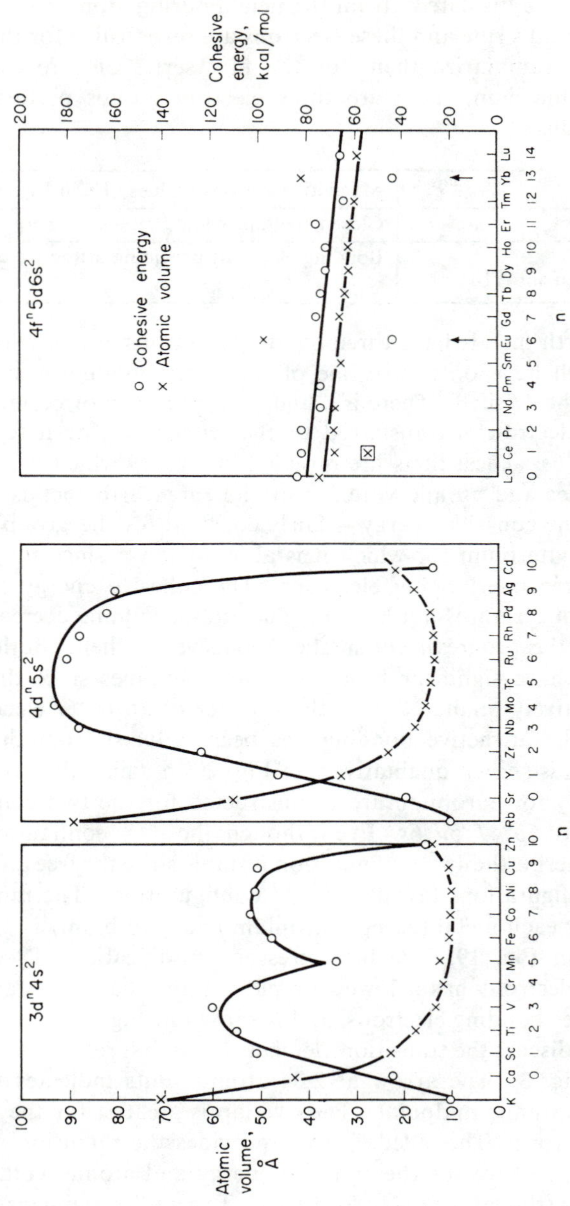

FIG. 19. The atomic volume in Å³, crosses, and the cohesive energy in kcal/mole, circles, for the metals of the 3*d*, 4*d* and 4*f* electron groups. *n* is the number of 3*d*, 4*d* or 4*f* electrons for the configurations above each figure. In case of Eu the stable free atom configuration is $4f^7 6s^2$ and for Yb $4f^{14} 6s^2$.

functions overlap the neighbours considerably and provide the bonding. Since the $4f$ electrons are "isolated" from the neighbouring atoms the $4f$ shell fills according to Hund's rule and these electrons are responsible for the magnetic properties. To summarize then, for the first series of rare earth metals lanthanum to lutecium, there are three electron groups performing three separate functions:

(1) $4f^n$	Magnetic electrons (values of n in Fig. 19)
(2) $5p^6$	Closed shell, providing repulsive energy
(3) $\sim 5d^2 6s$ ($5d$, $6s$ for Eu and Yb)	Bonding electrons providing attractive energy

All the rare earth metals have three bonding electrons except europium and ytterbium which have only two, one of the outer bonding electrons being transferred to the $4f$ shell. There is a high pressure form of cerium in which the single $4f$ electron is transferred to the $5d$ shell giving it four bonding electrons $\sim 5d^3 6s$. These facts are revealed in Fig. 19 which is a plot of the cohesive energies and atomic volumes of the rare earth metals lanthanum to lutecium. The cohesive energy is fairly constant for the group except for europium and ytterbium for which it is about $\frac{1}{3}$ lower since they have two rather than three overlapping electrons. The cohesive energy of the high pressure form of cerium is not known. The atomic volume decreases slightly with Z due to the progressively smaller repulsive $5p$ shell. Both europium and ytterbium have significantly larger atomic volumes since the repulsive $5p$ shell is relatively unaltered with the transfer of an outer electron to the $4f$ shell, but the attractive bonding has been reduced from three to two electrons. This is shown qualitatively in Fig. 20 in which the attractive and repulsive energy for europium are reconstructed for the two configurations $5p^6 4f^7 5d 6s$ and $5p^6 4f^6 5d^2 6s$. Even though the $4f^6$ configuration has a lower crystal energy, the $4f^7$ configuration is stable since the free atom energies of the two configurations favour the $4f^7$ configuration. The number, n, of $4f$ electrons for each metal (except europium and ytterbium) is given under each element in Fig. 19. The high pressure modification of cerium with four bonding electrons has a lower atomic volume than the stable modification with three bonding electrons, and is shown in Fig. 19.

In order to discuss the transition elements let us first refer to the iron wave functions in Fig. 3. The arrow at 2·34 atomic units indicates the nearest neighbour interatomic midpoint. The overlap is greatest for the $4s$ and less for the $3d$ electrons. The filled $3p$ shell provides the repulsion and has no overlap. In Fig. 19 we see the cohesive energy and atomic volume for the $3d$ and $4d$ series (the $5d$ is similar to the $4d$). Except for manganese in the $3d$ group the cohesive energy has a broad maximum near the middle of each of the d groups. In addition the atomic volume is a minimum in the same vicinity. These facts clearly indicate that the *d* electrons are strongly involved in the bonding, since the transition metals have the largest cohesive energies in the periodic table. Just as the rare earth metals europium and ytterbium give evidence for a different electronic structure in the solid than the other rare earth metals, by virtue of their cohesive energy and atomic volume not fitting a smooth curve for their group (Fig. 19), so manganese gives evidence for a different electronic structure. However, the electron structure change involved in the case of manganese is not as simple as in the rare earth metals europium and ytterbium. We can only say it appears to have fewer bonding electrons.

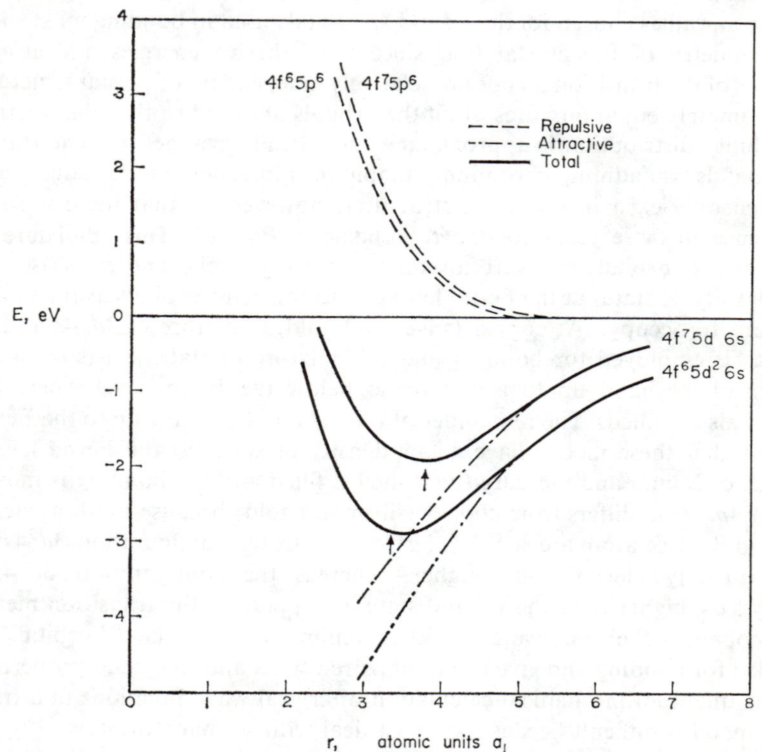

FIG. 20. The approximate total energy versus interatomic separation of metallic europium decomposed into its repulsive and attractive parts. The cohesive energy of the $4f^6$ configuration is greater than the $4f^7$ since it has more bonding electrons. The $4f^7$ state is preferred in the metal, though, because the free atom energy is lower by more than the gain in cohesive energy.

There are two points to be noted in regard to Fig. 19 for the transition metals. Firstly, it is known that the atomic volumes are practically independent of crystal structure and, secondly, the cohesive energies are adjusted to the same free atom configurations $3d^n4s^2$ and $4d^n5s^2$. For example, in regard to this latter point the observed cohesive energy of chromium is 80 kcal/mole, but the free atom ground state is $3d^5 4s$. It takes about 20 kcal/mole to excite it to $3d^4 4s^2$ and so the cohesive energy in Fig. 19 is given as 100 kcal/mole.

An examination of the free atom configurations in the $3d$ (and $4d$) transition metal groups indicates that the $3d$ and $4s$ eigenvalues are very close, and the $4p$ not much higher. In a typical case such as iron the free atom configurations $3d^6 4s^2$, $3d^7 4s$, $3d^6 4s 4p$, and $3d^7 4p$ are all within a few volts of each other so that we can expect the metallic configuration to consist of approximately six or seven $3d$ orbitals with the remaining $4s$ and $4p$. The specific m_l values chosen for the $3d$ and $4p$ orbitals used in bonding must show the symmetry of the crystal but, since the cohesive energies and atomic volumes of the transition metals are relatively independent of crystal structure, approximately equal mixtures of all the orbitals are used so that the electron probability distribution is approximately spherically symmetric. The transition metals vanadium, chromium, niobium, molybdenum, tantalum and tungsten only exist in the b.c.c. structures, however, so that there is some preference in these cases for the t_{2g} orbitals in Fig. 17. This admixture of $3d$, $4s$ and $4p$ orbitals exists from scandium through nickel and gives rise to a high density of states at the Fermi level due to the many orbitals available for electrons to occupy. At copper (silver and gold) a mixture of $3d$, $4s$ and $4p$ orbitals is employed for bonding and this mixture of states exists from the bottom of the band up to a volt or so below the Fermi level where the $3d$ orbitals are filled. The remainder of the band is $4s$, $4p$ like up to the Fermi level so that these metals have a low density of states at the Fermi level.* In zinc, cadmium and mercury the d shell is filled and the bonding is mostly $4s$ and $4p$. Zinc differs from copper (silver and gold) because the low energy states in the free atom are $3d^{10} 4s^2$, $3d^{10} 4s 4p$ with the configuration $3d^9 4s^2 4p$ approximately eleven volts higher, whereas the configuration $3d^9 4s 4p$ is only 5 eV higher than the ground state in copper. In the transition metals iron, cobalt, nickel, manganese, and chromium some of the $3d$ orbitals are not used for bonding and give rise to unpaired spins and magnetic properties.

The quantum mechanical calculation of crystal wave functions in a transition metal is difficult because we must deal with so many orbitals. Fig. 21 shows some results of John Wood for b.c.c. iron using the augmented plane wave method. The starting potential employed in the Schrödinger equation and the potential derived from the resulting eigenfunction is shown for the (100) direction and indicates that noticeable differences still exist in the overlap region. The $3d$ and $4s$ wave functions near the bottom and top of the band

* If all the $3d$ orbitals are utilized in copper, one may wonder how they contribute to the bonding, since filled shells normally repel each other. In copper there are eleven outer electrons, and there are at least 12 orbitals available for them (ten $3d$, two $4s$, and some $4p$). The copper wave functions are some mixture of all three, near the bottom of the band mostly $3d$ and mostly $4s$, $4p$ near the Fermi level.

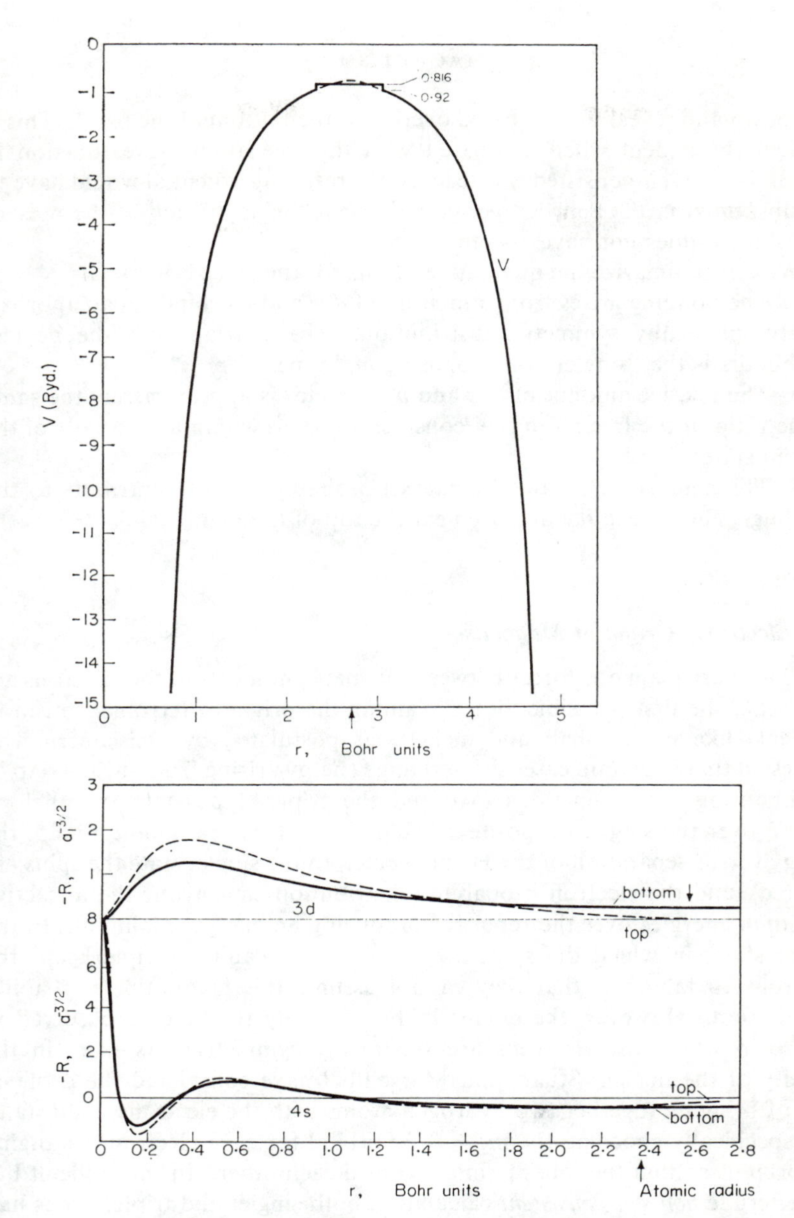

FIG. 21. The 3*d* and 4*s* wave functions near the top and bottom of the band in b.c.c. iron as calculated by the APW method from the periodic potential labelled *V* in the upper figure. The wave functions at the bottom of the band overlap considerably more than at the top. The band width is about 9·5 eV. The potential, *V*, is spherically symmetric inside spheres which just touch their eight nearest neighbours and is a constant between the spheres. The potential in the upper figure is along a line joining second nearest neighbours, the discontinuity occurring at the surface of the sphere. The dotted curve in the upper figure is the approximate potential derived from the calculated wave functions and demonstrates the lack of self consistency. (1 Rydberg = 13·6 eV).

Courtesy of Dr. J. H. Wood.

are shown and reveal the increased overlap at the bottom of the band. This is particularly evident when compared with the free atom wave function in Fig. 3. Before self-consistency is reached the resulting potential would have to be substituted in the Schrödinger equation but this is difficult at the present time since it does not have a simple form.

We can summarize the qualitative results for the transition metals.

1. The bonding arises from a mixture of *d*, *s* and *p* orbitals in an approximately spherically symmetric distribution. The overlap with the nearest neighbours is the "singlet" or "opposing m_s" type.

2. The relative amount of *d*, *s* and *p* character is approximately the same as the ratio in the free atom (i.e. consider all levels within about 5 eV of the ground state).

3. The magnetic electrons (unpaired) probably do not contribute to the bonding. They are *d* like and are near the top of the band.

D. Magnetism

The Electronic Origin of Magnetism

Since pure magnetic forces between magnetic moments of the electrons are too weak, the first plausible idea explaining the origin of ferromagnetism in elements like iron, cobalt and nickel was postulated by Heisenberg who suggested that in certain cases the exchange energy arising from the overlap of neighbouring wave functions favoured the triplet type state (parallel m_s values) over the singlet (opposite m_s values). Let us re-examine Fig. 5, the energy versus separation of the H_2 molecule. In the singlet state the spins are "paired" and the electron probability distribution can favour the attractive Coulomb energies over the repulsive producing an energy minimum. In the triplet state, in which the spins are parallel, the Pauli principle keeps the electrons so far apart that they cannot assume this favourable probabilty distribution. However, the curves in Fig. 5 apply to the case observed in nature in which the electrons are spherically symmetric "1*s* like" in the vicinity of the nuclei. Stuart and Marshall[3] have considered the artificial case of bringing together two hydrogen atoms with the electrons in 3*d* states and specifically choosing the $m_l = 0$ 3*d* orbital for each electron, arranging the orbitals so that the lobes pointed toward each other. In this difficult but rather crude *non self-consistent* calculation both singlet and triplet states had energy minima with the triplet energy about 1% lower than the singlet. *It appears that by confining the electrons to paths strongly concentrated along the lines joining the atoms the energy difference between the singlet and triplet states can be small, and the triplet may even lie lower, but more realistic calculations always reveal the singlet to be lower.**

* In a crude way, the triplet state may be lower when the electrons are confined to paths concentrated between the atoms, since the Pauli principle helps to keep the electrons

Furthermore, *the exchange energy* (i.e. the difference between the singlet and triplet energy) was too small in the Stuart and Marshall calculation to explain the ferromagnetism of the metals. Finally, we know that all molecules composed of identical atoms form a singlet ground state (except O_2) so that paired spins are overwhelmingly preferred in bonding. If we consider the metals, we know that some of the 3*d* wave functions overlap in iron metal (Fig. 21) but evidence to be discussed in the section on neutron diffraction indicates that the unpaired or "magnetic" 3*d* wave functions have little overlap and are presumably near the top of the band. In addition the 4*f* wave functions in ferromagnetic gadolinium do not overlap as seen in Fig. 18. Thus the original Heisenberg suggestion of a triplet state lower than the singlet due to the overlap of unpaired electrons on neighbouring atoms does not appear correct in detail. Instead it is believed that the mechanism aligning the spins in a ferro or antiferro-magnetic metals is via an "intermediary" such as the 4*s* electrons, or the 3*d* electrons at the bottom of the band in iron. These electrons overlap neighbouring atoms as well as the unpaired 3*d* electrons on each atom (which are near the top of the band). In b.c.c. iron, for example, these overlapping electrons interact differently with the unpaired "magnetic" 3*d* electrons depending on whether their spin is parallel or antiparallel to them, a consequence of anti-symmetrization or the Pauli principle. The energy of the crystal is lowered by arranging the iron atoms ferro-magnetically so that these overlapping electrons can assume a probability distribution that is more favourable in regard to their *Coulomb repulsion with the magnetic* 3*d electrons*. Thus magnetism is electrostatic in origin.

However, independent of the electronic origin of the exchange energy, which we shall denote by J, the Heisenberg theory postulated the well-known *vector model* which has more general applicability. The vector model stated that the magnetic exchange energy of a crystal is the sum of the magnetic exchange energies between all neighbours and is given as

$$W_{\mathrm{EX}} = -2 \sum_{\text{pairs}} J_{ij} \mathbf{S}_i \cdot \mathbf{S}_j{}^* \tag{29}$$

---

apart and reduces the Coulomb repulsion between them. This is much more important here than in the spherical case since the electrons here must come quite close as they switch atoms. More recent calculations of Freeman and Watson (*Phys. Rev.* **124**, 1439 (1961)) show that in the realistic case of two cobalt atoms the singlet lies lower.

*The classical dot product of two vectors is $\mathbf{A} \cdot \mathbf{B} = AB \cos \theta$ where θ is the angle between the vectors. Quantum mechanically, only certain discrete values are permitted for this dot product. These values are obtained by algebraically adding the total m_s value for the two atoms and then continually subtracting integral values until minus the sum of m_s is reached. For two gadolinium atoms the total $\Sigma m_s = \frac{7}{2}$ for each atom so that we have $\frac{7}{2} + \frac{7}{2} = 7$ allowing the fifteen values 7, 6, 5, 4, 3, 2, 1, 0, −1, −2, −3, −4, −5, −6, −7. In the case of two atoms each with total $\Sigma m_s = \frac{1}{2}$ we have 1, 0, −1. The direction of the magnetic moment determines the direction of the vector.

where $|S_i|$ and $|S_j|$ are the total m_s values for the neighbouring atoms i and j, and J_{ij} is the exchange energy between atom i and j independent of its precise electronic origin. As a simple example, let us apply this equation to a single H_2 molecule. There is one electron on each atom so that $m_s = \pm\frac{1}{2}$ for each atom, and if J is negative for the particular electron configuration being considered, the energy of the triplet state ($S_i = S_j = \frac{1}{2}$) is

$$W_{\text{EX}} = -2(-|J|)(\tfrac{1}{2})(\tfrac{1}{2}) = |J|/2$$

and of the singlet state

$$(S_i = \tfrac{1}{2}, S_j = -\tfrac{1}{2}) \quad \text{is} \quad W_{\text{EX}} = -2(-|J|)(\tfrac{1}{2})(-\tfrac{1}{2}) = -|J|/2.$$

Thus the singlet has a lower energy if J is negative and the triplet is lower if J is positive. The energy difference between the triplet and singlet is $-|J|/2 - |J|/2 = -|J|$, the exchange energy.

Applying this to the case of N atoms of h.c.p. gadolinium, considering only the twelve first and six second nearest neighbours ($S = \frac{7}{2}$), we have

$$W_{\text{EX}} = -2\frac{N}{2}J_1(12)(\tfrac{49}{4}) - 2\frac{N}{2}J_2(6)(\tfrac{49}{4}) = (-147NJ_1 - 73{\cdot}5NJ_2.)$$

Since we cannot count each pair twice the factor $N/2$ is used rather than N. In general the magnitude of J falls off rapidly as the neighbour distance increases and second nearest neighbours are sufficient for most cases. In early work only nearest neighbours were considered but many magnetic metals show evidence for second or further nearest neighbour interactions. For metals like nickel in which the value of S does not correspond to an integral number of electrons the vector model becomes tenuous. It is a reasonable approximation in such a case to assume nickel to be composed of 60% of the atoms with $m_s = \frac{1}{2} = S$ and 40% with $S = 0$ so that

$$W_{\text{EX}} = -2(0{\cdot}6)\frac{N}{2}J_1(12)(\tfrac{1}{4}) - 2(0{\cdot}6)\frac{N}{2}J_2(6)(\tfrac{1}{4}) = -1{\cdot}8NJ_1 - 0{\cdot}9NJ_2.$$

The average value of J is about 10^{-4} eV in gadolinium and about 10^{-2} eV in nickel, although we do not know the values of J_1, and J_2 separately. The important thing to remember, though, is that the vector model is a phenomenological theory and J can be obtained experimentally even if we cannot calculate it. A fundamental theory of magnetism, however, is one capable of either verifying the Heisenberg vector model and calculating the values of J or else replacing it with a new theory.

The following (Table IV) is a list of *elemental* magnetic substances. The table includes the number of unpaired electrons per atom, the type of unpaired wave functions and the magnetic structure.

TABLE IV

Element	Structure	Number of unpaired electrons per atom	Type of unpaired electrons	Magnetic structure
O	molecule	1	$2p$?	ferro
Cr	b.c.c.	~0·4	$3d$	anti-ferro
Mn	complex α	various values in different positions	$3d$	anti-ferro
Mn	f.c.c.	2·6	$3d$	anti-ferro
Fe	b.c.c.	2·22	$3d$	ferro
Fe	f.c.c.	~0·5	$3d$	anti-ferro
Co	h.c.p.	1·73	$3d$	ferro
Co	f.c.c.	1·8	$3d$	ferro
Ni	f.c.c.	0·6	$3d$	ferro
Ce	f.c.c.	1	$4f$	anti-ferro
Pr	hex	2	$4f$	anti-ferro
Nd	hex	3	$4f$	anti-ferro
Sm	hex	5	$4f$	anti-ferro
Eu	b.c.c.	7	$4f$	anti-ferro
Gd	h.c.p.	7	$4f$	ferro
Tb	h.c.p.	6	$4f$	ferro; anti-ferro
Dy	h.c.p.	5	$4f$	ferro; anti-ferro
Ho	h.c.p.	4	$4f$	ferro; anti-ferro
Er	h.c.p.	3	$4f$	ferro; anti-ferro
Tm	h.c.p.	2	$4f$	ferro; anti-ferro

Problems

1. Plot the melting and boiling points of the elements from $Z = 1$ to $Z = 72$. Is there any correlation with the various electron groups? Is there any correlation with the number of orbitals available for bonding in each group?

2. The h.c.p. metals beryllium, magnesium, strontium, zinc and cadmium all have a free atom outer electron s^2 configuration outside a closed shell. The closed shell provides the repulsion in the metal and the two outer electrons the cohesion. Compare the c/a ratios with the ideal (1·63).

Is there a correlation with the types of closed shell? Do h.c.p. metals in the same column in the periodic table have the same c/a ratio? Do the rare earth metals?

3. Try to predict the crystal structure, lattice parameter, density and melting point of element $Z = 106$.

4. Look up the crystal structure of several dozen compounds between transition elements and oxygen and determine the distance between transition element-oxygen nearest neighbours. Assume each oxygen atom utilizes two of the transition element electrons for bonding. Are these nearest neighbour distances related to the number of bonding electrons? (The unpaired $3d$ electrons are not utilized in bonding).

5. Make up a one electron sp^3 hybridized orbital by taking a linear sum of one s and three p orbitals ($m_l = 0, 1, -1$). Show the angular part of the electron probability distribution to be pointed toward the corners of a tetrahedron.

6. (For the mathematically minded.) Assume the $2s$ electron in lithium to be hydrogenic. Look up the ionization potential of lithium and assume this to be the eigenvalue of

the $2s$ electron. From eqn. (3) determine an effective nuclear charge, Z_{eff}. Construct a hydrogenic $2s$ wave function with this Z_{eff}. To simulate crudely conditions in the metal cut off this wave function at a value of r which is the radius of a sphere equal in volume to the atomic volume in the metal. Renormalize the $2s$ wave function. Calculate the expectation value of the effective Coulomb potential energy ($V = -Z_{eff}e^2/r$) and dividing by two to get the approximate total energy (virial theorem) compare this with its energy in the free atom. This lowering of the energy is a crude means of estimating the cohesive energy. How does this compare with the observed cohesive energy? Ans.: E (calc.) = $-4{\cdot}64$ eV/atom, E (observed) = $-3{\cdot}78$ eV.

Summary

1. Atoms in molecules and solids stick together by virtue of the overlapping electron probability distribution. This distribution adjusts itself to maximize the Coulomb attraction between electrons and opposite nuclei and minimize the electron–electron and nuclear–nuclear repulsion. Only distributions which obey the Pauli principle are permitted.

2. In a metallic crystal the energy eigenvalues of the overlapping electrons have a continuous band of values with gaps in energy at values of momentum determined from the crystal structure. (Brillouin zones). The electrons fill these levels up to the Fermi level and a three-dimensional plot of the electron momentum at the Fermi level is termed a Fermi surface.

3. There are many approximate techniques for calculating the eigenfunctions and eigenvalues of electrons in a metal but the calculations are so formidable that the errors are quite large and prevent an accurate calculation of the energy differences between crystal structures.

4. The periodic table and properties of the elements in their solid state show excellent correlation with the free-atom electron quantum numbers. In general the more bonding electrons the higher the cohesive energy and the smaller the interatomic distance.

5. While the precise electronic origin of the mechanism causing magnetic moments to align in magnetic solids is unknown, the phenomenological Heisenberg vector model (eqn. 29) is made the basis for many theoretical calculations. The exchange energy, J, becomes a parameter that is determined experimentally.

CHAPTER III

TEMPERATURE AND PRESSURE

A. Temperature

The Meaning of Temperature

The most common outside influence to which elements are subjected is the introduction of thermal energy. As far as the metallurgist is concerned, most measurements are influenced by the temperature of the sample. Furthermore, the practical problems of metallurgical processing and the technological use of any piece of metal are influenced by temperature. Let us see if we can understand the atomic behaviour of a metal as a function of temperature, after first recalling what we mean by temperature.

Suppose we have a very large tank of pure ice and water constantly being stirred to give the contents ample opportunity to mix thoroughly. Neglecting the effects of the wall of the tank we define the temperature of this mixture as exactly 0°C, the melting point of pure ice. If we place a small piece of aluminium into the tank and wait a minute or so it will assume the temperature 0°C. Repeating this procedure with a tank of water and steam we define the temperature 100°C. If we employ something we believe changes its length *linearly* in this range of temperatures, like a column of mercury in a capillary tube, and call this a thermometer, we can divide its length change between 0°C and 100°C into 100 equal parts and call each part one degree centigrade. We are now prepared to measure temperatures between 0°C and 100°C. For other temperatures we use other standards and other thermometers and try to join them smoothly to the melting and boiling points of ice and water. *The temperature of a substance, then, is merely the reading of an acceptable thermometer when the two are brought into good thermal contact and equilibrium is established.* The question of acceptable thermometers is periodically reviewed since thermometers depend on the change of some measureable physical property with temperature such as the pressure of a gas, the change in the length of a solid, etc. *In the final analysis the temperature dependence of this particular property is assumed by theoretical intuition.*

The Harmonic Oscillator

The first clue to the behaviour of atoms in thermal agitation in a crystal comes from the observation that the force, **F**, necessary to stretch or compress a piece of metal elastically (i.e. it returns to its original length after the force is removed) is just proportional to the change in length. This is identical to the behaviour of a mass, m, attached to a spring and is usually written

$$F = -\kappa\mathbf{x} \tag{30}$$

κ being the spring constant and $\mathbf{x}$ the displacement. The minus sign refers to the fact that the restoring force is opposite to the direction the spring is stretched or compressed. The potential energy of the mass is $V = \frac{1}{2}\kappa x^2$. Such a system is called a *harmonic oscillator* and if allowed to oscillate freely will do so with a natural or resonant frequency,

$$\nu = \frac{1}{2\pi}\sqrt{\frac{\kappa}{m}} \tag{31}$$

One is naturally led to assume that the bonding electrons between atoms simulate a spring even though the quantum mechanical calculations of the variation of the force between atoms under tension or compression is exceedingly difficult. The elastic behaviour of all metals, though, is only demonstrated for changes of about 1/10,000 of their interatomic distance since larger changes deform the metal. As the thermal excitation of atoms involve changes of as much as 10% of the interatomic distance, an idea of the variation of the interatomic force (or potential) is required over larger distances. Such a variation can be seen by inspecting the energy versus separation curve for the H_2 molecule, singlet state Fig. 5. The energy minimum, $-4{\cdot}72$ eV, is at an equlibrium distance of 0·74 Å.

If we stretch or compress the H_2 molecule a distance x from its equilibrium distance, the potential energy change is just the difference between the lowest energy and the energy at that distance. This appears to be more for compression than expansion whereas it appears to be the same for a piece of metal displaced elastically. The point is that the change in interatomic distance in the metal is only $\sim 10^{-4}$ Å, a very small distance on the scale in Fig. 5. In fact, over such small distances the H_2 molecule can be approximated as two atoms connected by a spring with potential energy, $V = \frac{1}{2}\kappa x^2 = 17{\cdot}5x^2$ eV/Å $= 2{\cdot}8 \times 10^5 x^2$ erg/cm^2.

Anharmonic Oscillator

The energy versus interatomic distance given in Fig. 5 for the H_2 molecule is obtained from an approximate solution of the Schrödinger equation *for the electrons*, keeping the nuclei fixed at each specific distance. At each internuclear distance the two electrons have some specific probability distribution and some specific eigenvalue. We can now forget about the electrons since they will always adjust themselves quite rapidly to any change in the nuclear separation (they are not as sluggish as the heavier nuclei) and apply the Schrödinger equation to the motion of the two hydrogen nuclei relative to each other. We use for the potential energy in the Schrödinger equation the

Morse potential which is an approximate mathematical fit around the minimum of the singlet curve in Fig. 5,

$$V(r) = D\{e^{-2a(r-r_e)} - 2\,e^{-a(r-r_e)}\} \tag{32}$$

where r is the internuclear separation, r_e the equilibrium distance 0·74 Å, a is a parameter (= 1·93/Å) adjusted to fit the curve, and D is the minimum energy value at the equilibrium distance (−4·72 eV). The eigenvalues obtained from this potential are approximately given by

$$E_n = h\nu(n + \tfrac{1}{2})\left[1 - \frac{h\nu(n + \frac{1}{2})}{4D}\right] \tag{33}$$

$$\nu = \frac{1}{2\pi}\sqrt{\frac{2\kappa}{M}}$$

$$E_n \text{ (for } H_2) \cong (0{\cdot}535 \text{ eV})(n + \tfrac{1}{2})[1 - 0{\cdot}0283(n + \tfrac{1}{2})] \tag{34}$$

where M is the mass of the hydrogen atom, n any integer including 0, and $\kappa = 2Da^2$ is the spring constant for small displacements although the solution applies to large displacements. *Unlike classical mechanics which permits all values of energy, the oscillator in quantum mechanics has quantized eigenvalues.* Only the first term in eqn. (33) would appear if the potential were identical to the simple spring or harmonic oscillator. The second and generally smaller term is due to the faster rise of the potential energy in compression. An oscillator with this type of potential curve is called the *anharmonic oscillator* and all atoms in metals behave in this way. In spite of the anharmonicity the natural frequency of the hydrogen molecule is identical with the classical harmonic oscillator $\nu = 1/2\pi\sqrt{(2\kappa/M)} = 12{\cdot}9 \times 10^{13}$ c/s the factor of two under the square root being due to the two masses at the ends of the spring.

The energy levels of the hydrogen molecule oscillator are shown in Fig. 22, as well as those for the harmonic oscillator term only, and we see appreciable deviations at about the fifth energy level ($n = 5$). Because of the $\frac{1}{2}$ term present in eqn. (33) even the lowest energy state of an oscillator, $n = 0$, is in constant vibration. If we assume a solid like iron metal has the same shaped potential curve as the H_2 molecule, and obtain the minimum energy D from the cohesive energy, and the value of κ from the Young's modulus for iron ($\kappa \cong 6{\cdot}8 \times 10^4$ ergs/cm^2) we find the deviations from the harmonic oscillator to be $\sim[1 - 0{\cdot}0018(n + \frac{1}{2})]$ (eqn. (34)). Even for the hundredth energy level the deviation is only 20%, so for many purposes (but not all) we can treat the forces between atoms in metals as harmonic.

Coupled Harmonic Oscillators

A crude picture of a metal, then, is a regular array of atoms connected by springs, probably stronger for the nearest neighbours with their large electron overlap and weaker for the next nearest neighbours, etc. The springs attached to one iron atom and extending as far as second nearest neighbours are schematically shown in Fig. 23. With such a complicated array of springs

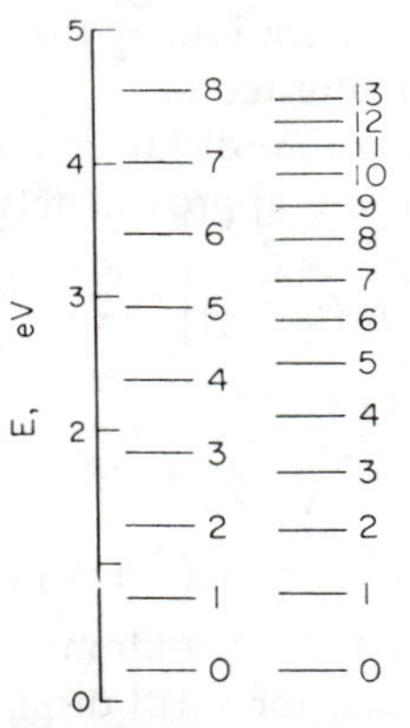

FIG. 22. The low energy vibrational eigenvalues for the H_2 molecule obtained from the Morse potential (right) and the levels obtained from the harmonic term alone (left). The value of the quantum number, n, is to the right of the level.

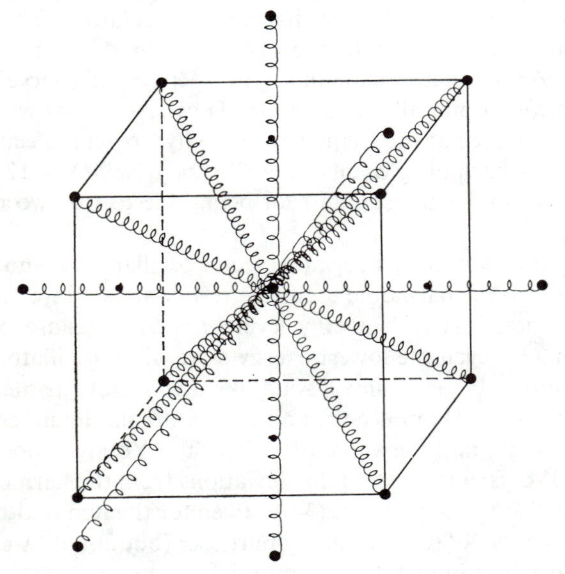

FIG. 23. A schematic model of the "springs" connecting an atom in the bcc lattice with its eight nearest neighbours and six next-nearest neighbours.

one might expect that the Schrödinger equation for the motion of atoms in a crystal would be insoluble. It is, even if we know the values of the spring constants. However, we can see our way clear by first solving the problem of the coupled harmonic oscillators shown schematically in Fig. 24. There are three springs each with spring constant κ and two indentical masses M_1 and M_2, separated by the lattice spacing a, which oscillate in the x direction only. The walls at the end are fixed. The potential energy depends not only on the displacement of each mass but their relative displacements and is given as

$$V = \tfrac{1}{2}\kappa x_1^2 + \tfrac{1}{2}\kappa x_2^2 + \tfrac{1}{2}\kappa(x_1 - x_2)^2 \qquad (35)$$

where x_1 and x_2 are the displacements of M_1 and M_2. The Schrödinger equation is soluble exactly and yields the eigenvalues

$$E = h\nu(m + \tfrac{1}{2}) + h\sqrt{(3)}\nu(n + \tfrac{1}{2}) \qquad (36)$$

where m and n are quantum numbers and $\nu = 1/2\pi\sqrt{(\kappa/M)}$. The two frequencies ν and $\sqrt{(3)}\nu$ do not correspond to any frequencies one might measure if the masses were photographed with a high-speed movie camera. However, if you imagined yourself on a small platform always positioned at the average of the algebraic sum of the two displacements you would vibrate with frequency ν. For example, if M_1 moved 0·1 Å to the right and M_2 0·2 Å to the right you would have moved 0·15 Å to the right of your equilibrium position at the midpoint. On the other hand, if you arranged to be always at the

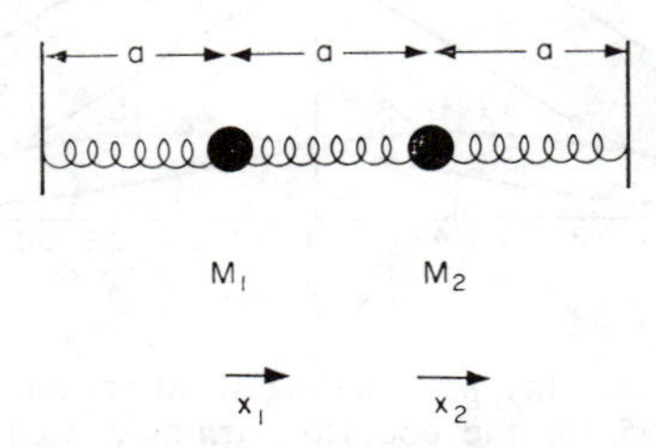

Fig. 24. A model of the coupled harmonic oscillators. The two atoms M_1 and M_2 and the rigid walls are connected by springs with identical spring constants, κ. The two normal modes of the system have frequencies ν and $\sqrt{(3)}\nu$ [$\nu = (1/2\pi)\sqrt{(\kappa/M)}$].

algebraic difference of the two displacements you would vibrate with frequency $\sqrt{(3)}\nu$. This system of co-ordinates (the sum and difference in this case) is called *the normal co-ordinates* or *normal modes* and greatly facilitates the mathematical solution of Schrödinger's equation in this case.

Let us now look at the probability distribution for the atoms obtained from the eigenfunctions of the coupled harmonic oscillator in its ground state $E = h\nu[\tfrac{1}{2} + \sqrt{(3)/2}]$, $m = n = 0$. Figure 25 represents the relative probability of finding M_1 at various values of x_1 for specific values of the displacement, x_2, of M_2. Specific values of x_2 must be specified since the solution depends on both co-ordinates and is not easily demonstrated in a two dimensional plot. However, the solution for the system of normal co-ordinates $\xi = (x_1 + x_2/2)$ and $\eta = (x_1 - x_2/2)$ is readily shown since the probability distribution is merely the product of wave functions (one for ξ, one for η). *Just as in the case of electrons, Fig. 25 shows that quantum mechanics does not specify the position of an atom exactly.**

If we solve the coupled harmonic oscillator in Fig. 24 by classical methods we find the same natural frequencies $\nu = 1/2\pi[\sqrt{(\kappa/M)}]$ and $\sqrt{(3)}\nu$ in the system of normal co-ordinates. The energy, however, is not quantized but can assume any value. The position of each coupled oscillator can be specified *exactly* as a function of time. It spends most of its time at the end of its displacement when it slows down to reverse itself, and the probability of finding it at a specific position is inversely proportional to the velocity at that position. The classical and quantum mechanical probability distributions for the normal

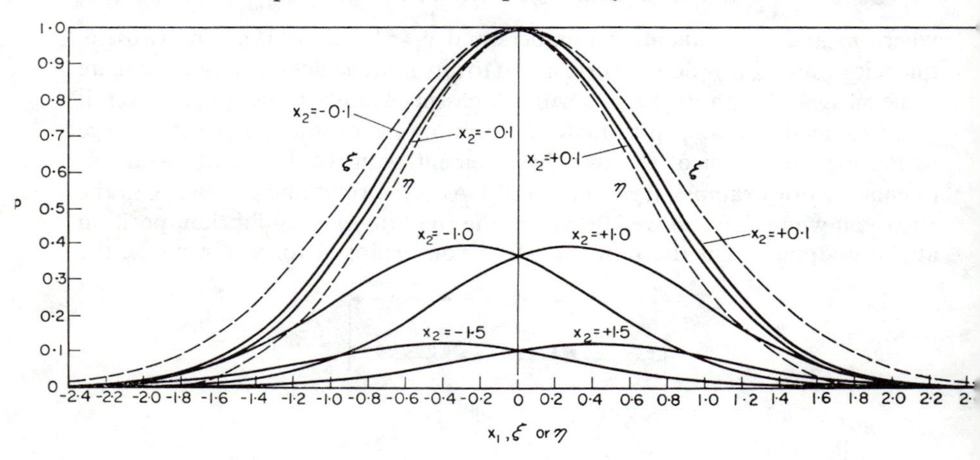

FIG. 25. The relative probability, p, of finding M_1 at various values of x_1 and M_2 at $x_2 = \pm 0{\cdot}1$; $\pm 1{\cdot}0$; $\pm 1{\cdot}5$ for the coupled harmonic oscillator of Fig. 24 (solid curves). The dashed curves are the probability of various displacements of the normal co-ordinates $\xi = (x_1 + x_2)/2$ and $\eta = (x_1 - x_2)/2$. The displacements are all less than the equilibrium spacing between M_1 and M_2.

co-ordinate, ξ, of the coupled harmonic oscillators with identical energy are shown in Fig. 26 for the ground state ($m = 0$) and the tenth level ($m = 10$). The classical probability distribution approaches the quantum mechanical as the energy or quantum number increases.† A comparison of the absolute magnitude of the average displacement $\langle|\xi|\rangle$ in the quantum mechanical

* This also follows from the uncertainty principle. For example, the ground state wave function of the coupled harmonic oscillator yields an expectation value of the displacement in normal co-ordinates $\langle|\xi|\rangle = \sqrt{(2E/\kappa\pi)}$ and of the momentum $\langle|p|\rangle = \hbar/\sqrt{(2\pi E/\kappa)}$. Since the uncertainties are of this order of magnitude we have $\Delta\xi\Delta p \cong \langle\xi\rangle\langle|p|\rangle = \hbar/\pi$.

† This is an application of a general rule called the *Correspondence Principle* stating that the classical and quantum mechanical solutions become identical as the quantum numbers become infinitely large.

and classical solutions shows the classical to be 13% greater for the ground state, $m = 0$, 2% greater for $m = 1$ and quite negligible thereafter. In addition the average values $\langle \xi^2 \rangle$ and the square of the momentum $\langle p^2 \rangle$ are identical for all values of m.

How to Determine the Eigenvalues of a Crystal

By combining the results for the anharmonic oscillator with those for the coupled harmonic oscillators, we can now set down a practical approach to the problem of determining the eigenvalues, momenta and probability distribution of the coupled oscillators in a real crystal.

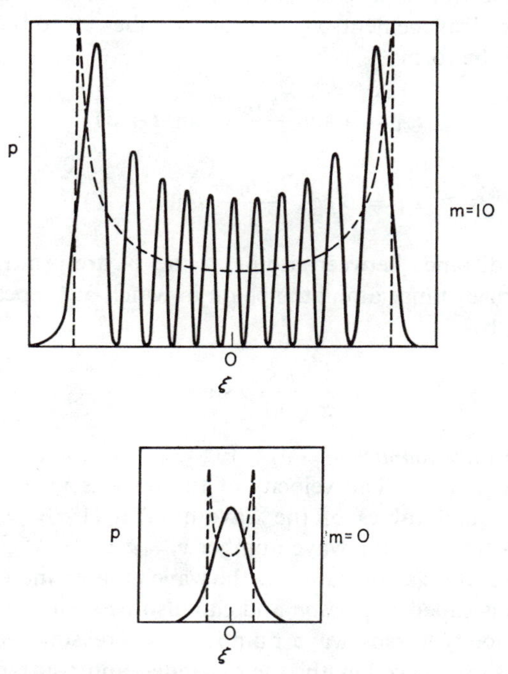

FIG. 26. The relative probability of finding the coupled harmonic oscillator of Fig. 24 at various values of the normal co-ordinate ξ (ξ is the average displacement of M_1 and M_2). The solid lines are the quantum mechanical solutions for the quantum number $m = 10$ (upper curve) and $m = 0$ (lower curve). The dashed lines are the probability distributions for the classical solutions of the same energy.

1. The energy levels are quantized and are given by eqn. (33) for each frequency, ν, of the normal modes.* These specific frequencies can be determined classically.

* Actually an anharmonic potential will not yield independent normal modes for the coupled oscillators, but the excitation levels of one will depend on the excitation of the other. However, for most (but not all) work this can be neglected.

2. Except for the zero point oscillations (ground state) the displacements and momenta can be determined classically.

One Dimensional Chain of Atoms

Let us see how we determine the classical displacements and frequencies in a more realistic system than we have been treating but still one that is relatively simple. We consider a long one dimensional chain of atoms each of mass, M, connected by identical springs with spring constant κ. We have merely added many more atoms to the coupled oscillators in Fig. 24. By employing Newton's laws of motion ($\mathbf{F} = m\ddot{x}$) we find the classical solutions for the displacement, $x_{\mathcal{N}}$ of atom number $\mathcal{N}$ to be wavelike and a sum of waves of the form

$$x_{\mathcal{N}} = A \sin \frac{2\pi a_0 \mathcal{N}}{\lambda} \sin(\nu t + \delta) \tag{37}$$

$$x_{\mathcal{N}} = B \cos \frac{2\pi a_0 \mathcal{N}}{\lambda} \sin(\nu t + \delta')$$

where a_0 is the distance between atoms, δ and δ' are arbitrary phases, λ is the wavelength or distance along the chain in which $x_{\mathcal{N}}$ repeats and ν is the frequency given by

$$\nu = \frac{1}{\pi}\sqrt{\frac{\kappa}{M}} \cdot \sin \frac{\pi a_0}{\lambda} \tag{38}$$

If we define the *wave number* $\sigma \equiv 1/\lambda$, Fig. 27 is a plot of frequency versus wave number, eqn. (38). The velocity of the wave is given by the slope of eqn. (38). For small values of the wave number ($1/\sigma = \lambda \gg \pi a_0$) the frequency is proportional to the wave number $\nu = a_0\sigma\sqrt{(\hbar/M)}$, and the velocity $v = \nu/\sigma = a_0\sqrt{(\kappa/M)}$ is a constant. As the wave number increases the velocity decreases. This is called *dispersion* and can also be seen in Fig. 27 where we have plotted velocity versus wave number. The reason for this dispersion effect is that at short wave lengths the expanded and contracted parts of the wave are so close that they interfere with each other. When the wavelength is so small that it just equals twice the lattice spacing the expanded and contracted parts cancel each other and the wave goes nowhere (velocity equals zero). This provides the upper limit to the wave number and frequency of the vibrational waves.

$$\sigma_{\text{max}} = \frac{1}{\lambda_{\text{min}}} = \frac{1}{2a_0} \tag{39}$$

$$\nu_{\text{max}} = \frac{1}{\pi}\sqrt{\frac{\kappa}{M}}$$

Which of all the frequencies from $\nu = 0$ to $\nu = 1/\pi[\sqrt{(\kappa/M)}]$ correspond to the normal co-ordinates? We must know this so that we can determine the eigenvalues from eqn. (33). However, there are as many frequencies as there are atoms in our chain and it becomes impractical to identify each one. Instead we employ the same device we did for the electronic states in a crystal, i.e. the density of states, denoted now by $q(\nu)$. This is the number of states or allowed frequencies as a function of the frequency ν. (Since the energy is simply related to the frequency by eqn. (33) we use frequency for convenience.) It turns out that the number of normal modes is a constant for each interval of wave number and that the density of states, $q(\nu)$, is for N atoms in the chain

$$q(\nu) = Na \frac{d\sigma}{d\nu} = N \sqrt{\frac{M}{\kappa}} \Big/ \sqrt{[1 - (\pi^2 \nu^2 M/\kappa)]} \qquad (40)$$

and is shown in Fig. 27.

The lowest frequency normal mode corresponds to a standing wave whose wavelength is twice the length of the linear chain. The next corresponds to one whose wavelength equals the length of the chain, the next to $\frac{2}{3}$ the length, then $\frac{1}{2}$ the length, etc. The frequency of each normal mode is given by eqn. (38).

Normal Modes in Three Dimensions

In three dimensions the classical solutions reveal the same characteristics as in one dimension but the mathematics is exceedingly difficult for we are dealing with the many springs that are assumed to exist between the nearest, next nearest, etc., neighbours. The relative value of the spring constants are generally selected intuitively so that we are dealing with a phenomenological theory. (A fundamental theory of the vibrations of a solid is one that calculates the behaviour of the springs from the electron wave functions.) There are $3N$ normal modes, one longtidudinal and two transverse for each atom. *The transverse modes* correspond to displacements at right angles to the direction of propagation while the *longitudinal modes* are parallel. The maximum wave number in any direction is determined from the First Brillouin zone constructed exactly as for electrons (see p. 36). By assuming values of the spring constants, Leibried[(4)] has calculated the density of states for b.c.c. lithium, solid line Fig. 28. We cannot give any simple physical picture for the peaks that appear in Leibfried's calculation but it is typical of the type of curves that are calculated and measured.

Thermodynamics

Once we have the density of states or frequency distribution of the normal modes of a crystal we can resort to thermodynamics to calculate the temperature dependence of many of the observed physical characteristics. The eigenvalues for each allowed frequency are given by eqn. (33) and thermodynamics tells us that the probability of finding a normal mode in any of its quantized energy levels denoted by E_n is

$$P = e^{-E_n/kT} / \sum_n e^{-E_n/kT} \tag{41}$$

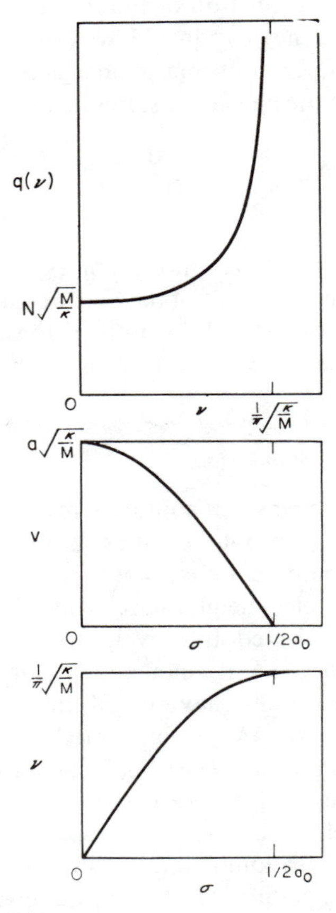

FIG. 27. The frequency distribution of normal modes for the linear chain, i.e. a one dimensional array of identical atoms connected by identical springs (top curve); the velocity versus wave number (central curve); and the frequency versus wave number or dispersion curve (lower curve).

where k is Boltzman's constant and T is the absolute temperature (0°C is 273·2°K). The denominator of eqn. (41) is called the *partition function*. By weighting each of the eigenvalues with its probability we can calculate the average energy of each normal mode as a function of temperature, and by integrating all the normal modes we can calculate the average total energy of the crystal. An approximate solution* for the *total energy*, *U*, of a crystal is

$$U \cong \left(1 + \frac{kT}{2D}\right) \int_0^\infty \frac{h\nu}{e^{h\nu/kT} - 1}\, q(\nu)\, d\nu + \frac{1}{2}\int_0^\infty h\nu q(\nu)\, d\nu \tag{42}$$

The factor $kT/2D$ arises from the anharmonic correction to the eigenvalues in eqn. (33) and is small at low temperatures. The second term in eqn. (42) is a

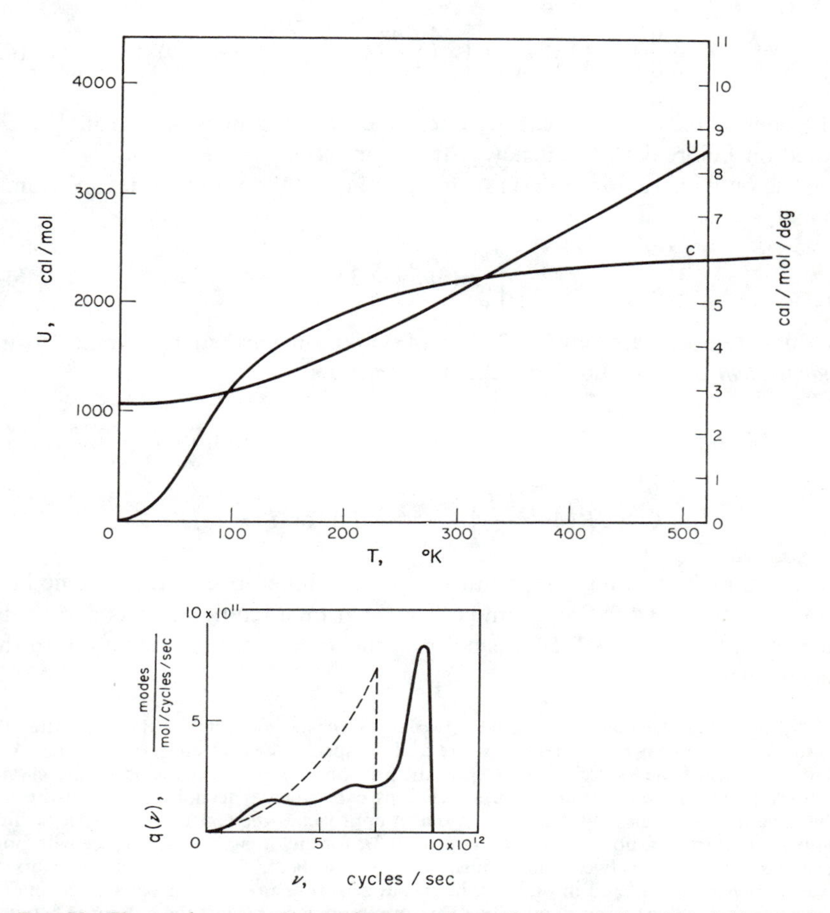

FIG. 28. The total energy, *U*, in cal/mole and specific heat, *C*, in cal/mole/deg as a function of *T* obtained from the calculated frequency spectrum, $q(\nu)$, for b.c.c. lithium (solid line). The dashed curve is the Debye frequency spectrum for lithium ($\theta = 360°K$).

* Many of the expressions to be given are approximate because of the anharmonic terms. We have used the Morse potential to calculate them and we really do not know how good this is. In addition, the anharmonic terms cause $q(\nu)$ to change with temperature. As the temperature is increased there is a gradual increase in the lower frequencies.

constant independent of temperature and is called the *zero point energy.* It arises from the $n = 0$ term in eqn. (33) and *is present in all crystals even at absolute zero.* It represents a large fraction of the energy of a crystal even at the melting point but is not directly measureable since it is essentially independent of temperature. *The specific heat*, C,* of a crystal is the energy required to raise the temperature a unit temperature interval and is given by

$$C = \frac{dU}{dT} \cong \left(1 + \frac{kT}{D}\right) \int_0^\infty k\left(\frac{h\nu}{kT}\right)^2 \frac{e^{h\nu/kT}}{(e^{h\nu/kT} - 1)^2} q(\nu)\, d\nu \tag{43}$$

The energy and specific heat for one mole† of lithium is shown in Fig. 28 based on Leibreid's[4] frequency distribution, $q(\nu)$.

The only restriction on $q(\nu)$ is that there are $3N$ normal modes (N atoms) in a crystal so that

$$\int_0^\infty q(\nu)\, d\nu = 3N \tag{44}$$

At high temperature eqns. (42) and (43) take on particularly simple forms *independent of* $q(\nu)$, and applicable to all metals

$$U \cong 3RT\left(1 + \frac{kT}{2D}\right) + \text{zero point energy} \qquad \sim \text{valid for } T > T_m/2 \tag{45}$$

$$C \cong 3R\left(1 + \frac{kT}{D}\right) \qquad \sim \text{valid for } T > T_m/3 \tag{46}$$

where T_m is the melting temperature. The continued rise of the specific heat above the value $3R$ (5·958 cal/mole/deg K) at high temperature is due to the correction term $(1 + kT/D)$ arising from the anharmonic contribution to the eigenvalues.‡

* Invariably the reader encounters two types of specific heat in the literature, C_p meaning specific heat at constant pressure and C_v specific heat at constant volume. Virtually all measurements are made at constant pressure since the expansion of the crystal with temperature would require some ingenious experimental technique to measure C_v. The advantage of even considering C_v is that it approaches the constant value $3R$ at high temperature. In fact, both C_p and C_v approach $3R$ for any imaginary substance with pure harmonic oscillators between the atoms. The specific heat, C, that we utilize refers to measurements encountered in practice, i.e. C_p at zero pressure. It approaches $3R$ times a correction factor due to anharmonicity. This correction factor is just the difference between C_p and C_v, i.e. $C_p = C_v(1 + kT/D)$.

† One mole of a substance is a gram molecular weight and contains $6{\cdot}028 \times 10^{23}$ atoms. This times Boltzman's constant is the commonly used *gas constant*, R, equal to 1·9865 cal/mole/deg K. (i.e. $N_0k = R$)

‡ Classical thermodynamics tells us that $k/D = 3\alpha\gamma$ where α is the coefficient of linear expansion and γ is the Gruneisen constant. $\gamma = 3\alpha V/CK$ where V is the volume of one mole, C the lattice specific heat and K the compressibility. Some values of γ are given on p. 159. In general $\gamma \cong 2{\cdot}0$.

At very low temperatures only the low frequency normal modes are excited since the probability factor (eqn. (41)) prevents the excitation of the higher frequency modes which are spaced further apart in energy. (The spacing between energy levels is $\sim h\nu$.) The lowest frequency mode which can be excited in a crystal above the zero point energy is one whose wavelength is just equal to twice the length of the crystal in the three independent directions, the next is equal to the length in one of the directions and twice in the other two. If we continue in this fashion and count up the first million or so we will find that the number per frequency interval is proportional to ν^2. (The curve is not so smooth for the first hundred or so.) This low frequency distribution for ($q \propto \nu^2$) is true for all metals up to about $\nu = 5 \times 10^{11}$ c/s and gives

$$U \cong \frac{6{\cdot}5Ak^4T^4}{h^3} + \text{zero point energy} \qquad \text{valid for } T < T_m/100 \quad (47)$$

$$C \cong 26Ak\left(\frac{kT}{h}\right)^3 \qquad \text{valid for } T < T_m/100 \quad (48)$$

where A is the constant of proportionality between $q(\nu)$ and ν^2 and would have to be determined separately for each metal. The approximation $q(\nu) = A\nu^2$, however, cannot be used for the zero point normal modes since they are all excited even at absolute zero.

Debye Theory

An extremely useful and widely used approximation was first suggested by Debye. He extended the $q(\nu) = A\nu^2$ low frequency approximation to cover the entire frequency spectrum. This is shown in Fig. 28 (dashed curve). It is cut off at some maximum frequency, ν_{max}, which becomes an empirical parameter for each metal. By convention, though, this maximum frequency in the Debye theory is multiplied by Planck's constant and divided by Boltzman's constant so that it has the dimensions of temperature and is called the *Debye temperature or characteristic temperature*, denoted by Θ

$$\Theta = h\nu_{max}/k$$

$$q(\nu) = 9N\nu^2/\nu_{max}^3 \quad (49)$$

Th e integrals in eqns. 42 and 43 as well as the entropy

$$S = \int_0^T (C/T)\,\mathrm{d}T^*$$

are tabulated in appendix I for this Debye approximation. The zero point

* The entropy is an important thermodynamic property. It is proportional to the logarithm of the average number of levels excited for each normal mode.

energy becomes $\frac{9}{8}R\Theta$. The high temperature limits are still given by eqns. (45) and (46) and the low temperature limits, eqns. (47) and (48), become

$$U = 58{\cdot}5RT\left(\frac{T}{\Theta}\right)^3 + \tfrac{9}{8}R\Theta \tag{50}$$

$$C = 234R\left(\frac{T}{\Theta}\right)^3 \tag{51}$$

Even though the Debye theory is based on a crude approximation for the frequency spectrum one can always adjust the parameter Θ so that the observed specific heat of a metal can be adequately fitted for most metallurgical purposes. The reason for this is that the specific heat is rather insensitive to the precise frequency spectrum. At high temperatures, for example, the specific heat is given by eqn. (46) which is independent of the frequency spectrum. Historically the bulk of metallurgical interest has been at relatively high temperatures where thermodynamic equilibrium is easily and quickly obtained.

Frequency Spectrum for Vanadium

We can clarify the picture even further by examining the relative contribution of the various normal modes to the energy, specific heat, and entropy. These curves are shown in Fig. 29 as well as the measured frequency spectrum of vanadium. At 40°K the thermal contribution to the specific heat entropy and energy is limited to the very low frequency modes. The total energy is principally zero point energy but we have not included this in Fig. 29 since it is an additive constant. The Debye Θ for vanadium is 338°K and at $T = 338$°K the specific heat has about equal contribution from all normal modes while the entropy still favours the low frequency modes. At $T = 3380$°K the specific heat has not changed, the entropy is now more uniformly sensitive to the frequency spectrum and the energy is independent of the frequency. This last condition is called *equipartition of energy*, the energy of each mode being identical and just equal to kT. While 3380°K is well above the melting point for vanadium it does serve as a guide to the very high temperature behaviour. In order to obtain the actual contribution to the specific heat, energy and entropy the functions in Fig. 29 must be multiplied by $q(\nu)$. This has been done in Fig. 29 for the free energy, $F = U - TS$ at $T = 338$°K. The free energy is an important thermodynamic function since it governs the equilibrium of different phases. As the metallurgist principally deals with metals at or above their Debye temperature Fig. 29 reveals to us that each frequency contributes to the free energy an amount approximately proportional to the number of normal modes present at that frequency.

Phonons

The word *phonon* has come into common usage to designate each normal mode. It is derived by combining the prefix phono, referring to sound, with the word photon. Sound waves in a metal are very long wavelength phonons. If one strikes a piece of metal the audible sound is transmitted as a normal mode with frequency equal to the sound frequency. An audible

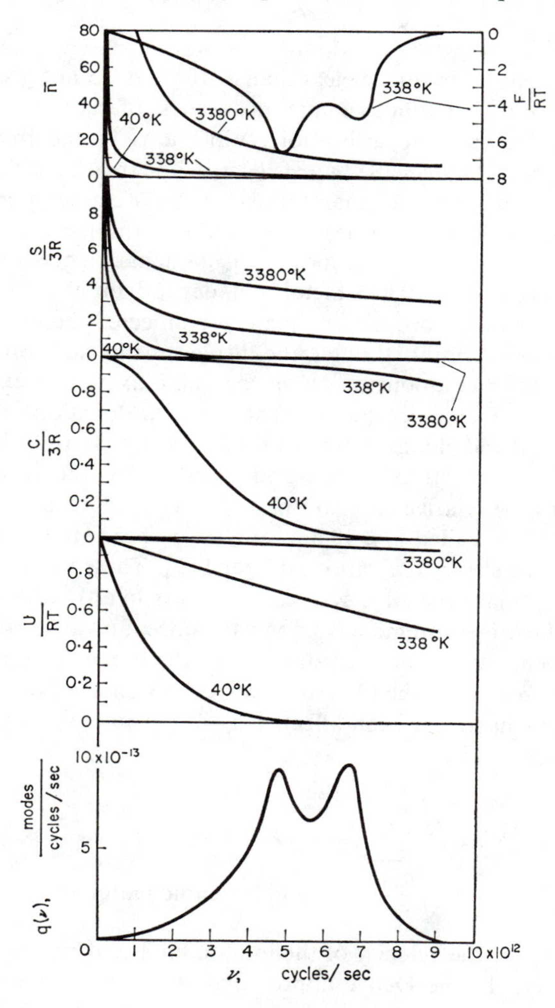

FIG. 29. The contribution of the various frequencies to the energy, specific heat and entropy at 40°K, 338°K, and 3380°K and the average quantum number $\bar{n}$; the contribution of the various frequencies to the free energy of vanadium at 338°K (uppermost curve) based on the measured frequency spectrum (lowest curve). The Debye Θ of vanadium is 338°K.

frequency is at least 10^9 times smaller than most phonon frequencies, thus having little practical consideration in matters relating to temperature, but it does explain the origin of the word. The photon aspect arises from the similarity in their energy relationships, $h\nu$ for the photon and $(n + \frac{1}{2})h\nu$ for the phonon, and the similarity in momentum, $h\nu/c$ for the photon and $(n + \frac{1}{2})h\nu/v$ for the phonon.

Elastic Constants

While phonons of frequency less than $\sim 10^{11}$ c/s do not give appreciable contribution to the thermodynamic properties of a metal, considerable effort is expended in their utilization. Vibrations in the frequency range 10^5 to 10^9 c/s are normally called *ultrasonics*, while the range down to essentially zero c/s comes under the general classification of elasticity measurements. One might ask if one cannot at least obtain the harmonic part of the spring constants from these low frequency measurements. Not so directly, since the force required to stretch a metal will depend on the direction in the crystal, as the springs are approximately connected between the atomic centres. In a cubic crystal, though, the displacements and forces can always be reduced to their components along the cubic axes. For example, if one applies a stress, **X** (force per unit area) along the x direction of a cubic single crystal the crystal will elongate in the x direction and contract in the y and z direction. (The contraction in the y and z direction is identical.)* The ratio of this stress to the relative elongation, $\Delta x/x$, is called c_{11} and to the relative contraction $\Delta y/y$ is called c_{12}. If one applies a shear stress in the yz plane and measures the shear, the ratio is denoted c_{44}. These three coefficients are sufficient to calculate the effects of elastic stress in any direction in a cubic crystal. The shear measurement is important since it provides the mechanism for the transverse waves in a crystal. From these elastic constants we can determine the velocity of the ultrasonic waves as well as of the low frequency phonons which have not been affected by dispersion. For example, in the (200) direction.

$$v_l = \sqrt{\frac{c_{11}}{\rho}} \qquad (52)$$

$$v_t = \sqrt{\frac{c_{44}}{\rho}} \qquad \text{valid for cubic materials}$$

where v_l and v_t are the velocity of the longitudinal and transverse waves and ρ is the density. In the Debye approximation for the frequency spectrum,

* The reason a linear stress applied in the x direction of a cubic single crystal like lithium causes a contraction in the y and z direction is that the springs between the atoms are not confined to the cube axes but are also between nearest neighbours in the 111 direction. *Poisson's ratio* is the ratio of the contraction in the y (or z) direction to the elongation in the x direction. It is generally less than $\frac{1}{2}$.

the above relations for the velocity allow us to calculate an "average" Debye Θ for a cubic crystal.

$$\Theta = \frac{h}{k}\left(\frac{3N}{4\pi V}\right)^{1/3}\left(\frac{1}{3}\sqrt{\frac{c_{11}}{\rho}}+\frac{2}{3}\sqrt{\frac{c_{44}}{\rho}}\right) \tag{53}$$

where the weighting factors $\frac{1}{3}$ and $\frac{2}{3}$ refer to the one longitudinal and the two transverse waves. N/V is the number of atoms per cm^3 and ρ is the density. Table V gives the elastic constants for lithium at 78, 155 and 195°K. The decrease with temperature is presumably due to the anharmonic potential shifting the frequency spectrum to lower frequencies.

TABLE V
ELASTIC CONSTANTS OF LITHIUM METAL IN 10^{11} dyn/cm^2

T°K	c_{11}	c_{12}	c_{44}	10^{11} (dyn/cm^2) $(c_{11}-c_{12})/2c_{44}$
78	1·485	1·253	1·08	0·107
155	1·398	1·176	1·00	0·111
195	1·345	1·127	0·96	0·113

A rather important point that is revealed by the elastic constant values in Table V is that the forces between the lithium atoms are *non-central*, i.e. the forces tending to return the atoms to their equilibrium positions after they have been displaced do not act along a line joining the atomic centres. If they were central, the ratio c_{12}/c_{44} would be unity for a cubic material. This ratio is called the *Cauchy relation* and is virtually never obeyed. *This means that the forces between atoms can not be simulated by connecting the atoms with springs.* It is an approximation made for mathematical practicality. It reflects the more complicated nature of the forces between atoms arising from the overlapping electrons. If the forces between atoms are independent of direction (isotropic) then the last column of Table V would be unity. This rarely occurs.

Electronic Specific Heat

So far we have considered the thermal excitation of atoms in terms of the nuclear vibrations. In a metal it is also possible to excite electrons thermally to higher energy states, but the details are considerably different than for the normal modes of a crystal. Consider, for example, helium gas just above its boiling point at 4·2°K. All the atoms are in their ground state $1s^2$, the first excited state $1s2s$ is about 20 eV higher in energy. If we heat the gas to 3000°K ($kT \cong 0{\cdot}25$ eV) the helium atoms will move about quite rapidly making many collisions but the fraction of atoms excited to the $1s2s$ configuration is one in 10^{35} (eqn. (41)). We can thus ignore thermal excitation of the electrons in this case. Consider now the free atom of iron in its ground state $1s^2 2s^2 2p^6 3s^2 3p^6 3d^6 4s^2$. The eigenvalues of each of the electrons is given approximately in Fig. 3. The $1s^2 2s^2 2p^6 3s^2$ electron groups are all filled and the Pauli principle prevents any of these electrons from being excited because the next higher level is already completely filled. We did not concern ourselves with such considerations for phonons since they are not restricted by the Pauli principle. The only unfilled shells in iron are $3d$, $4p$ and higher. The next higher state in iron is $3d^7 4s$ requiring about 1·25 eV to excite it. At 4000°K the number of iron atoms excited to this higher state is 1 in 45. Here again we can ignore the thermal excitation of the electrons at most temperatures but now we must remember that the Boltzman equation (eqn. (41)) can not be applied to the excitation of the $1s$, $2s$, $2p$ and $3s$ electrons to their next higher levels due to the Pauli principle. In the metal the same considerations apply to these innermost electrons but now the band structure produces closely spaced energy levels for the outer electrons. The Pauli principle still prevents the excitation of most of these outer electrons, since they occupy all the levels up to the Fermi level, but it is possible to excite electrons near the Fermi level to the empty levels above it.

Fermi–Dirac Function

To consider properly the thermal excitation of electrons, we must replace the Boltzman probability function with a new one that includes the Pauli principle. *We do not know what this function should be for a metal* but presumably it should have the following properties.

1. For all eigenvalues well below the Fermi level it should give a probability of unity ($\frac{1}{2}$ for electrons with $m_s = +\frac{1}{2}$ and $\frac{1}{2}$ for those with $m_s = -\frac{1}{2}$).
2. It must be discontinuous at the Fermi level *if the Fermi level is sharp*.
3. It should satisfy the equipartition of energy at very high temperatures and approach the classical distribution for high energy (large quantum numbers).

Such a function is qualitatively shown in Fig. 30 for several temperatures. At $T = 0$ all eigenvalues well below the Fermi level are filled ($P = 1$) and all eigenvalues well above the Fermi level are empty ($P = 0$). Close to the Fermi level there is a statistical fluctuation (not thermal) due to the uncertainty principle that yields a distribution of energies (eqn. (18)). As the temperature is increased, electrons from below the Fermi surface are thermally excited to above the Fermi surface, and at very high temperatures ($kT \sim E_F$) the classical Boltzman limit is approached ($e^{-E/kT}$).

It may be that the precise shape of the distribution function may vary for different metals but at least in one approximation a specific function has been derived. This is called the *Fermi–Dirac* function and is applicable to the free electron theory although it has had rather widespread use in other cases. It is

$$P = \frac{1}{e^{(E_n - E_F)/kT} + 1} \qquad \text{valid for } kT \ll E_F \tag{54}$$

where P is the probability that an eigenvalue, E_n, is occupied by two electrons ($m_s = \pm\frac{1}{2}$) and E_F is the Fermi energy. This function is plotted in Fig. 30 at $kT = 0$, $kT = 0{\cdot}01\, E_F$ and (with a slight modification for normalization) at $kT = E_F$. In this latter case the approach to equipartition of energy (average energy per electron equals $\frac{3}{2}\, kT$) can be seen as well as the approach to the classical Boltzman distribution $e^{-E/kT}$ for large energy.

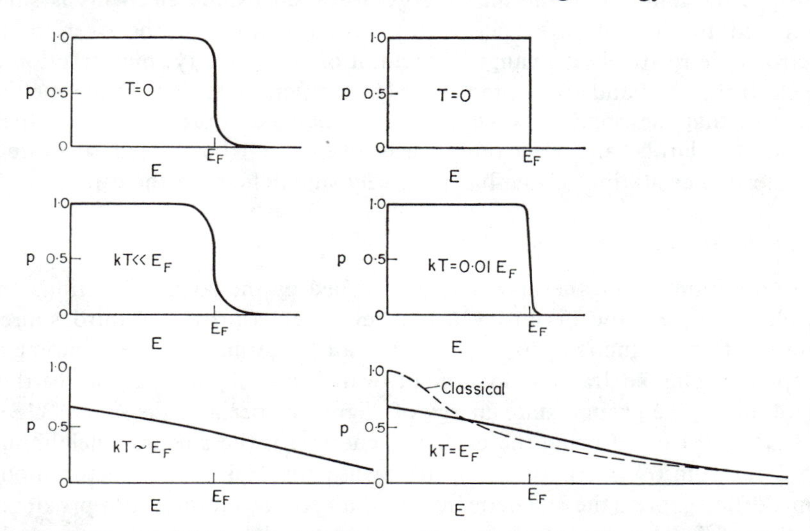

FIG. 30. The probability, P, of finding an electron in an energy state E (relative to the Fermi energy, E_F) at $kT/E_F = 0, 0{\cdot}01$, and 1. The curves on the left are estimates for real metals while those on the right are based on the Fermi–Dirac function applicable to free electrons. The classical curve (dashed) is based on the Boltzman function and shows the approach to classical behaviour at high energies.

Since we cannot calculate the distribution function for a real metal the Fermi–Dirac function is commonly used. We do not know the extent of the errors introduced. It leads to the following approximate expressions for the energy and specific heat of the electrons in a metal.

$$U = 2N\int_0^{E_F} E\left(\frac{\mathrm{d}n}{\mathrm{d}E}\right)\mathrm{d}E + \frac{kR}{3}(\pi T)^2\left(\frac{\partial n}{\partial E}\right)_{E=E_F} \tag{55}$$

$$C = \tfrac{2}{3}\pi^2 RkT\left(\frac{\partial n}{\partial E}\right)_{E=E_F} = \gamma T \quad \text{valid for } kT \ll E_F \tag{56}$$

$(\partial n/\partial E)_{E=E_F}$ is the density of states *at the Fermi level.* The first term in the energy expression is the total cohesive energy of the electrons while the second term is due to thermal excitation. The first term is independent of T and can be treated as an additive constant. The electronic specific heat is linear in temperature, a rather well confirmed experimental observation. As an example, if we apply the free electron theory to lithium we have $(\partial n/\partial E)_{E=E_F} = 0{\cdot}319$/eV/atom (p. 43), so that the *electronic specific heat coefficient* (eqn. (56)) $\gamma = 3{\cdot}65 \times 10^{-4}$ cal/mole/deg^2 which compares favourably with the observed coefficient $\gamma = 4{\cdot}18 \times 10^{-4}$ cal/mole/deg^2. The electronic contribution to the total specific heat of lithium is about 75% at 4°K but only 0·2% at room temperature.

Since the amount of thermal energy absorbed by the electrons is small compared to the cohesive energy the wave functions of the overlapping electrons are relatively unchanged. In addition, only a very small fraction of the electrons in a band are excited ($\sim$1% at temperatures close to the melting point) so that the bonding forces have not changed appreciably. For these reasons the lattice and electronic specific heats can be treated separately. The thermal excitation of one has relatively small effect on the other.

Magnetic Specific Heat, Spin Waves

In addition to the thermal energy absorbed by the lattice vibrations and the electrons near the Fermi level, magnetic metals provide a third source. Consider ferromagnetic iron. At $T = 0$°K all the atomic magnetic moments except a negligible fraction in domain walls* are aligned parallel to their neighbours. The ground state energy of alignment per atom is $\sim 2(z/2)J\mathbf{S}\cdot\mathbf{S} = 8JS^2 \cong 0{\cdot}1$ eV, J being the exchange energy for the z nearest neighbours and S the spin ($S \cong 1{\cdot}1$ for iron). If we use the Heisenberg vector model (eqn. (29)) to express the magnetic interaction between atoms, the approximate solution of the Schrödinger equation for nearest neighbours only leads to the quantized eigenvalues (above the ground state)

$$E = nh\nu = 2nSJ\left(\frac{4\pi a_0}{\lambda}\right)^2 = 2nSJ(4\pi a_0\sigma)^2$$
$$\nu = 32\pi^2 SJa_0^2\sigma^2/h \quad \text{valid for } T < T_c/5 \tag{57}$$

where n is the quantum number which takes on integral values including zero, a_0 the lattice parameter in body-centered or face-centered cubic metals, σ the wave number and T_c the Curie temperature. There is a negligible zero point energy. There are two distinct differences between the allowed modes in the ferromagnetic case and the normal modes of the lattice vibrations in a

* Pure iron at $T = 0$°K consists of magnetic domains about one micron in size and oriented somewhat at random such that the net magnetization is zero. Within each domain the atomic magnetic moments are all aligned. A relatively weak magnetic field of a few hundred gauss will remove the domain structure leaving a completely magnetized sample.

crystal. Firstly, the frequency is now proportional to the square of the wave number rather than the first power as for low frequency phonons. Secondly, the modes are not vibrational waves but are *spin waves* schematically illustrated in Fig. 31. As a model for spin waves one can envisage each spin coupled to its neighbour by a torsional spring. The angular displacement of a spin out of line with its neighbours gives rise to a torque proportional to this angular displacement and tends to return it. This torque is only proportional to the angular displacement provided the angular displacement is small, but as long as this is the case the spin waves execute harmonic motion. As the angular displacement gets large (n large) the problem becomes rather difficult to solve and eqn. (57) is no longer valid. This limits us to low temperatures, i.e. $T < T_c/5$. The low temperature frequency distribution of spin waves for N atoms is considerably different than for phonons,

$$q(\nu) = \frac{N}{(2\pi)^2} \frac{\nu^{1/2}}{b} \left(\frac{h}{2JS}\right)^{3/2} \tag{58}$$

which leads to the energy per mole

$$U = \frac{0{\cdot}045RT}{b} \left(\frac{kT}{2JS}\right)^{3/2} \tag{59}$$

and specific heat per mole

$$C = \frac{0{\cdot}113R}{b} \left(\frac{kT}{2JS}\right)^{3/2} \tag{60}$$

FIG. 31. A schematic representation of a spin wave.

where $b = 2$ for b.c.c. and $b = 4$ for f.c.c. and h.c.p. metals. At low temperature the average value of the quantum number, n for all oscillators is

$$\bar{n} = \frac{0{\cdot}0587}{b} \left(\frac{kT}{2JS}\right)^{3/2} \tag{61}$$

or only about $\sim 0{\cdot}06$ at room temperature in iron. Each spin wave reduces the total magnetization, $\mathscr{M}$, of the sample. If $\mathscr{M}_0$ is the magnetization at $T = 0°\mathrm{K}$ (all spins aligned), then the magnetization at other temperatures is

$$\mathscr{M} = \mathscr{M}_0 \left[1 - \frac{0{\cdot}0587}{bS} \left(\frac{kT}{2JS}\right)^{3/2}\right] \tag{62}$$

This is called the Bloch $T^{3/2}$ law.

Above $T_c/4$ the spin wave modes become anharmonic since the torque tending to return them is no longer proportional to the angular displacement for the large displacements encountered, but is progressively less.* As the temperature is increased further the angular displacements become very large and eventually at the Curie temperature there is no net magnetization. We can see this from a simple illustrative diagram, Fig. 32. If we take an instantaneous picture of a ferromagnetic material and select an atom at random, we can plot the probability that the magnetic moments of the surrounding

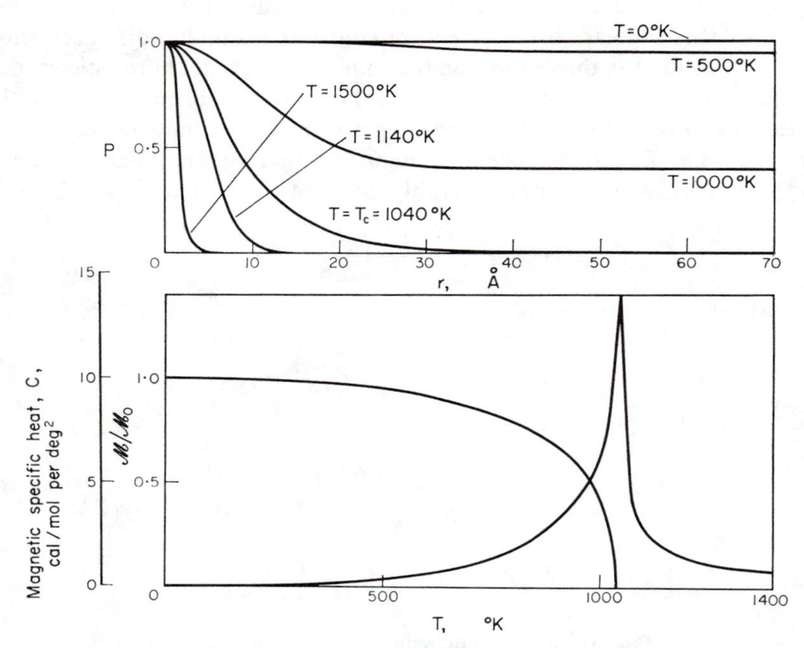

FIG. 32. The probability at various temperatures of finding the magnetic moment of an atom parallel to the magnetic moment of an arbitrarily selected atom in iron, as a function of distance from the selected atom (upper curve). The limiting value of P (at r large) times the magnetization at absolute zero is the magnetization at each temperature. In the lower curves are plotted the observed magnetization and magnetic specific heat of iron as a function of temperature. Spin wave theory is applicable below ~ 200°K.

* This can be seen classically by replacing $\mathbf{S}\cdot\mathbf{S}$ by $S^2 \cos\theta$ where θ is the angle between the spins. The energy difference per atom between the ground state in iron and the excited state is $8JS^2 - 8JS^2\cos\theta \simeq 8JS^2[(\theta^2/2) - (\theta^4/24) + \ldots]$. The θ^2 term is analogous to the harmonic oscillator but the $-\theta^4$ term reduces this when θ becomes large. As a crude model, if the potential $-2JS^2\cos\theta$ is substituted into the Schrödinger equation the solutions are similar to the one dimensional periodic potential considered in Fig. 10. The potential $-2JS^2\cos\theta$ is periodic in that it repeats for $\theta = 2\pi, 4\pi, 6\pi\ldots$ etc. This leads to energy gaps as in Fig. 10 except that we need only consider states up to the first gap. The eigenvalues are spaced progressively closer as the energy increases up to its maximum value. If these eigenvalues are substituted into eqn. (41), it leads to an energy curve which rises rapidly with T, has an inflection point at the Curie temperature and then slowly approaches the constant value $2JS^2$ at $T \gg T_c$. The specific heat has a peak at the Curie temperature.

atoms are parallel to this atom as we move away from it. While the surrounding atoms are at discrete distances, we have drawn smooth curves through these points. The curves are approximately correct for b.c.c. iron, $T_c = 1040°K$. The value of P is of course unity for the atom selected. At large distances P approaches a constant value, which if multiplied by the magnetization at $T = 0°K$ gives the average magnetization of the sample at that temperature. The Curie temperature is that temperature at which P just approaches zero for large distances. The magnetization versus T is plotted in Fig. 32. Above the Curie temperature there still exists some short range order, i.e. the spins are still aligned over short distances but average to zero at larger distances. The thermal energy absorbed by the excitation of the magnetic states is shown by the magnetic specific heat experimentally observed for iron, Fig. 32. It reaches a peak at the Curie temperature, but due to the short range order above the Curie temperature, energy is still absorbed up to $\sim 1800°K$. Only when the magnetic specific heat falls to zero has all short range order disappeared. The transition from an ordered state in iron at 0°K to a disordered state at $\sim 1800°K$ is called a *second order transition.* It is generally identified by the absorption of energy over a wide temperature range with a peak in the specific heat at the Curie temperature.

We do not have a theory for the eigenvalues of a ferromagnetic metal above $\sim T_c/5$ nor do we have a theory for the energy levels of an antiferromagnetic metal at *any temperature.* In the chapter on thermodynamic measurements, however, some empirical results will be presented to aid the metallurgist in analyzing observations on magnetic metals.

We have treated the thermal excitation of the magnetic moments as independent of the lattice vibrations and electronic excitations. Experience indicates that this is reasonable, but our failure to have a good theory for the electronic origin of the exchange energy leaves the question open.

Thermodynamics of Phase Changes

The last mechanism for the absorption of energy that the metallurgist must consider are *phase changes or first-order transitions,* like ice to water or b.c.c. iron to f.c.c. iron. Thermodynamics tells us that a first order transition occurs at temperatures at which the Gibbs free energies, G, of the two phases are equal.

$$G = U + PV - TS \tag{63}$$

where P is the external pressure on the sample, V the volume, U the energy and S the entropy. At zero pressure $U = \int C\,dT$; $S = \int C/T\,dT$ so that in its most convenient form

$$G = \int_0^T C\,dT - T\int_0^T \left(\frac{C}{T}\right) dT \tag{64}$$

The sum $U + PV$ is called the *enthalpy*, denoted H. As we transform from one phase to another the substance will absorb or release energy. This energy is called the *latent heat*, denoted ΔH, and the entropy change during this transformation is $\Delta S = \Delta H/T'$ where T' is the transformation temperature. The stable phase at any other temperature is the one with the minimum free energy.

B. Pressure

Effects of Pressure

At absolute zero all the normal modes of a crystal are in their zero point states ($h\nu/2$) and each atom is essentially vibrating about the minimum in its potential. If a uniform compression is placed on the sample the volume will decrease. If this is a cubic material we can determine the *compressibility*,* K, from the elastic constants,

$$K = -\frac{1}{V}\frac{\mathrm{d}V}{\mathrm{d}P} = 3/(c_{11} + 2c_{12}) \tag{65}$$

In this compressed state the atoms are vibrating about a new position that is shifted toward a higher potential. In a typical case such as iron a pressure of 100,000 atm (1 atm = $1{\cdot}01 \times 10^6$ dyn/cm^2) decreases its volume by $\sim 6\%$ or about 2% in length ($\Delta l/l \cong (\frac{1}{3})\Delta V/V$). On the other hand, iron has increased its length about 3% from $T = 0°$K to the melting point. Thus the effect of anharmonicity is as readily observed in thermal measurements. But thermal oscillations average over both the repulsive and attractive parts of the potential whereas pressure measurements at low temperatures directly yield the repulsive part of the potential curve. This provides a clear advantage to pressure measurements.

What changes in the properties of solids can we expect if we increase their pressure?

1. The band structure will change since we have more overlap of the outer wave functions. In general this is small, since the energy introduced into the sample by compressing it to pressures ordinarily attainable in the laboratory is less than 0·1 eV or considerably smaller than the cohesive energy.

2. The magnetic properties may change more drastically since these have comparable energies $\sim$0·1 eV.

3. The elastic constants increase, for we are in the region of stronger repulsive forces.

4. The effects of anharmonicity should be more pronounced, like the linear term in the high temperature specific heat, etc.

Actually it is difficult to imagine any property that will not change with pressure, but except for the alkali metals we are doomed to small changes

* The reciprocal of the compressibility is called the *bulk modulus*.

because of the difficulty of attaining high laboratory pressures. If we could attain pressures as high as the centre of the earth where iron is ~*2·5 times as dense* then we should have some interesting results.

A more dramatic effect than those described in 1 to 4 are pressure induced phase changes. Let us consider several possible cases where these might occur. It is almost obvious at the outset that a pressure induced phase change, from phase *A* to phase *B*, must be accompanied by a volume decrease for phase *B* since this is the only way that phase *B* can lower its total energy relative to *A*. In case I, Fig. 33, phase *A* is more stable than phase *B* although

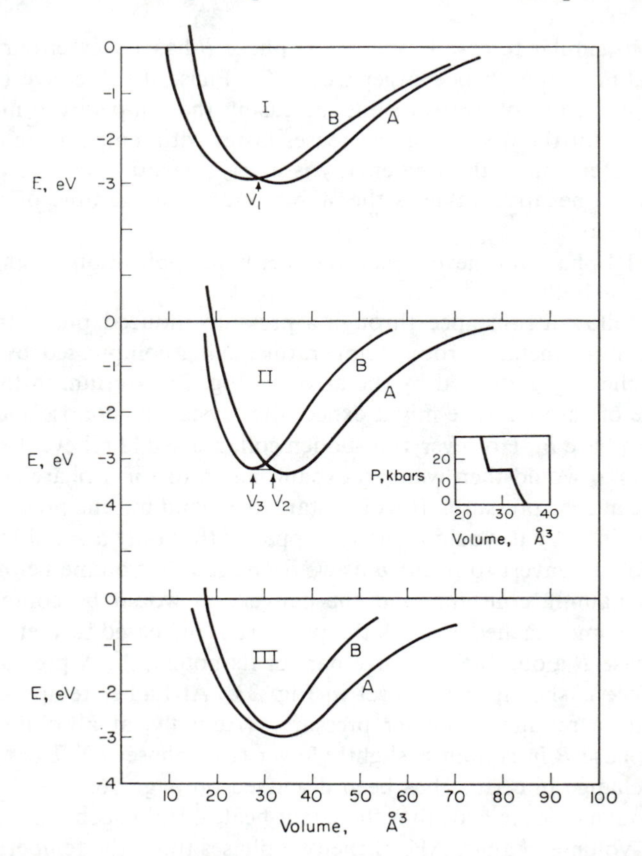

FIG. 33. Approximate energy in eV versus atomic volume of several types of allotropic phases. In case I the higher energy phase is stabilized by both temperature and pressure (like tin). In case II the higher energy phase is stabilized by pressure but the lower energy phase is stabilized with increasing temperature (like cerium near room temperature). In case III the higher energy phase is not stabilized by either pressure or temperature.

The inset in case II is the approximate pressure in kilobars versus volume in Å³ of cerium at room temperature, showing the isobaric phase transition.

phase B has an equilibrium volume less than A. In addition, phase B has a smaller curvature at the energy minimum hence smaller elastic constants and lower Debye Θ (eqn. (53)). If the volume at $T = 0°K$ of phase A is decreased by pressure to below V_1 then phase A will convert to phase B since the total energy of the substance will be lowered. If the pressure is applied at some higher temperature phase A will convert to phase B at lower pressure, since the lower Debye Θ of phase B lowers its free energy, G, more than phase A. In fact, at zero pressure, temperature alone is sufficient to convert phase A to phase B. An example of case I is the diamond and tetragonal phases of tin.

Case II is similar to case I except that phase B has a greater curvature at its potential minimum, hence larger Debye Θ. Phase A will convert to phase B under application of pressure but increasing the temperature makes this more difficult for the free energy decreases faster with temperature for phase A than B. (Remember the free energy is always negative so that the phase with the more negative value is the stable one.) An example of case II is europium metal.

In case III phase B is never reached either with application of temperature or pressure or both.

Let us follow a substance through a pressure induced phase transition. Consider cerium metal at room temperature being compressed by a piston until it reaches V_2, indicated by the arrow in Fig. 33. Cerium in this case is an example of case II. One might expect the crystal to collapse suddenly to V_3 forming phase B. However, this sudden collapse would relieve the pressure and the crystal would then want to expand again to form phase A which is more stable at zero pressure. It would start to expand but the pressure would build up again! What would actually happen is that only a small fraction of phase A would convert to phase B as we decreased the volume below V_2, the pressure remaining constant, and the conversion would be complete only when the volume reached V_3. As the pressure is increased further we would squeeze phase B along the repulsive part of its potential. A pressure versus volume curve is shown in the inset in Fig. 33. At higher temperatures the transformation requires a higher pressure. (Actually, at absolute zero the energy of phase B in cerium is slightly lower than phase A.) The reason for the phase change in cerium has been discussed on page 56.

Thermodynamics tells us that the latent heat, ΔH, in such a phase change equals the volume change, ΔV, of the two phases times the temperature and the change of pressure with temperature.

$$\Delta H = T \frac{\mathrm{d}P}{\mathrm{d}T} \Delta V \tag{65}$$

This is called the *Clapeyron equation.*

C. Liquids

Liquids

Just as the electronic structure of liquids is in an unsatisfactory state so is the question of the thermodynamic properties. We have already said that a conceptual picture of a liquid is a box full of steel balls in constant motion. If one imagines oneself sitting on an atom, say of copper, then in the solid there are twelve nearest neighbours, six second nearest neighbours, etc., and one is oscillating in an approximately harmonic potential created by these neighbours. While there are approximately twelve nearest neighbours in the liquid the further neighbours are not at such fixed points as in a crystal but are in a somewhat random array. In addition during the time it takes to make one oscillation ($\sim 10^{-13}$ sec) one of the twelve nearest neighbours in the liquid may have moved away or interchanged with another atom so that the spring constant in the approximately harmonic potential will be varying. The precise way this effects the quantized energy levels (eqn. (33)) and the frequency distribution of normal modes is difficult to say. However, from the latent heat of fusion we do know that the energy of the liquid is higher than the energy of the solid and this is due mostly to the fact that the nearest neighbours are somewhat further apart in the liquid. In addition from free energy considerations (eqn. (63)) we know that the entropy of the liquid is higher than the solid and this is presumably due to the varying spring constant increasing the number of quantized energy levels.

In general the entropy change upon melting averages several entropy units (cal/mol/deg) while the latent heat of melting is a few percent of the cohesive energy. (See problem 6 at end of chapter.) Since liquids will not resist a shear stress they do not transmit transverse waves but do transmit longitudinal waves. (It is this property which has enabled geologists to ascertain the structure of the earths' centre through analysis of seismic waves.)

Problems

1. Build a model of the two coupled harmonic oscillators. Employ masses and springs that give a frequency of about 3 c/s. Paint an observable mark on the centre of the spring connecting the two atoms and look for the $\xi = (x_1 + x_2)/2$ mode. The centre of the spring moves approximately as the average of the displacements. Can you devise a *simple* method for observing the $\eta = (x_1 - x_2)/2$ co-ordinate?

2. Look up the specific heats of a dozen various liquids including metals and compare them with $3R$. Can you put an upper limit on the Debye Θ for water? Estimate the specific heat per gram of a dry martini including the olive (or onion).

3. (For the mathematically minded.) The average value of a function can be obtained from eqn. (41) by multiplying it by the function and integrating over all energy levels. Assume the energy levels to be that of the harmonic oscillator ($E = (n + \frac{1}{2})h\nu$) and $kT \gg h\nu$,

calculate the semi-classical average value of x^2, where x is the displacement. Hint: for n large replace the sums in eqn. (41) by integrals, and $E = \frac{1}{2}\kappa' x^2$ (classically), so that

$$\overline{x^2} = \frac{\int_{-\infty}^{\infty} x^2 e^{-\frac{1}{2}\kappa' x^2/kT}\,dx}{\int_{-\infty}^{\infty} e^{-\frac{1}{2}\kappa' x^2/kT}\,dx} \qquad \text{valid for } n \text{ large.}$$

Using the Morse potential, calculate the average value of the displacement $\bar{x}$ for small displacements (expand the exponential to terms in x^3). Using the value of $\kappa = 6{\cdot}8 \times 10^4$ ergs/cm^2 for iron as the spring constant and $D \cong 3$ eV show that the expansion coefficient is within a factor of two of that observed for iron. Will the metal expand if the forces are harmonic?

4. Frequently a metallurgist measures the temperature of a metal by its colour. As the temperature of a metal is increased the electrons vibrating with the atoms radiate photons since they are being accelerated and decelerated in the electric fields caused by the neighbouring atoms. Thus the metal becomes "full of photons", but they are quickly reabsorbed by the electrons near the Fermi level so that an equilibrium is set up between the photons and the electrons. The eigenvalues of the photons are those of the harmonic oscillator since this is the electromagnetic nature of photons. Their frequency distribution of normal modes are similar to the long wavelength normal modes of the atomic oscillators and is $q(\nu)\,d\nu = 8\pi\nu^2\,d\nu/c^3$ per unit volume. Using eqn. (41) calculate the average energy of the photons of frequency ν and multiply by their frequency distribution to determine the energy distribution of those that escape from the surface before being reabsorbed. Plot the energy versus frequency at $T = 100°$K, 500°K, 1000°K and 3000°K and indicate on the frequency scale the colour of visible light appropriate to that frequency. Does this appear to explain the colour of light radiated by a metal?

How hot do you estimate is the W filament of an incandescent bulb? From your frequency distributions how efficient do you estimate the bulb to be for converting electrical power into visible light? Look up a response curve of the eye for light of various wavelengths and estimate the most efficient temperature to operate an incandescent bulb.

5. (For the mathematically minded.) Often a physicist encounters the problem of a system containing just two eigenvalues with an energy separation, ΔE. Assume the ground state has a degeneracy g_0 and the upper level a degeneracy g_1, and use eqn. (41) to calculate the energy, specific heat, free energy and entropy as a function of T. Show that the total energy absorbed per mole from $T = 0°$K to $T = \infty°$K is $N_0\Delta E/(1 + g_0/g_1)$ and the total entropy is $R \ln(1 + g_1/g_0)$. Hint: (for a two-level system with degeneracy the partition function, f, is $(g_0 + g_1 e^{-\Delta E/kT})$ where the subscript zero refers to the ground state and the subscript 1 to the upper state). Plot the specific heats for $g_0 = g_1$; $g_0 = (g_1)/25$. This type of two level excitation function is called a *Schottky function.*

Answers:

$$U = N_0\Delta E\Big/\left(\frac{g_0}{g_1}e^{\Delta E/kT} + 1\right); \quad C = R\left(\frac{\Delta E}{kT}\right)^2 \frac{g_0}{g_1}e^{\Delta E/kT}\Big/\left(\frac{g_0}{g_1}e^{\Delta E/kT} + 1\right)^2;$$

$$S = R\left\{\frac{\Delta E/kT}{\frac{g_0}{g_1}e^{\Delta E/kT} + 1} + \ln(g_0 + g_1 e^{-\Delta E/kT})\right\}; \; F = -RT\ln(g_0 + g_1 e^{-\Delta E/kT})$$

Calculate the energy and free energy directly from the partition function, f.

$$U = \frac{N_0 d\ln(1/f)}{d\left(\frac{1}{kT}\right)} \qquad ; F = -RT\ln f.$$

6. (For the mathematically minded.) As a crude model for a liquid assume each atom spends half its time surrounded by twelve nearest neighbours and half its time surrounded by eleven nearest neighbours, the twelfth site being vacant. This represents approximately

a 4% decrease in density over a solid (one vacancy in 24) and increases the zero point energy of the liquid by ~8% of the cohesive energy. (There are $Nz/2$ bonds broken in determining the cohesive energy while the removal of 4% of the atoms breaks $0{\cdot}04Nz$ bonds; z = number of nearest neighbours and N = number atoms). Each atom is thus subjected to two spring constants κ_1 and κ_2 where $\kappa_2 = 11/12\kappa_1$ so that we have two natural frequencies $\nu_1 = 1/2\pi\surd(\kappa_1/m)$ and $\nu_2 = 1/2\pi\surd(\kappa_2/m)$. In the high temperature limit the partition function of the liquid becomes

$$f = kT\frac{h\nu_2 + h\nu_1}{h\nu_1 h\nu_2} \cong 2{\cdot}045\ kT/h\nu.$$

Determine the free energy of the oscillator in the liquid and solid states directly from the partition function and use a Debye spectrum for both to average over the frequency distribution (the ratio of the Debye Θ values of the solid to the liquid are $\sim 2\nu_1/(\nu_1 + \nu_2)$). Setting the free energies equal at the melting point, T_m, show that

$$T_m \cong \frac{0{\cdot}08H}{3R \ln 2{\cdot}095}$$

where H is the cohesive energy per mole and R is the gas constant. Select twenty or so metals at random and ascertain whether the above expression relating melting temperature and cohesive energy is valid to ~20%.

Summary

1. The basic assumption in the solid is that the bonding electrons between atoms act as nearly harmonic oscillator springs being somewhat stiffer in compression than expansion. With this assumption, quantum mechanics predicts quantized energy levels (eqn. (33)).

2. The quantum-mechanical displacements and frequencies of the atoms, however, differ very little from the classical ones so that they can be calculated classicially. If it were not for this, the problem would be virtually insoluble.

3. In a solid of N atoms there are $3N$ natural frequencies of vibration, called normal modes or phonons, of which N are longitudinal and $2N$ transverse. In a longitudinal mode the atoms are displaced in the same direction as the mode is propagating while in a transverse wave the displacement is perpendicular to the direction of propagation.

4. The true frequency distribution of normal modes in a solid is rather complicated so that the Debye simplification (ν^2 distribution) is often used especially since most thermodynamic properties are relatively insensitive to fine details of the frequency distribution.

5. The electrons in a metal can absorb energy from the thermal motion of the atoms, but only at the Fermi level above which there are unoccupied levels. It is difficult to calculate the amount of energy to be absorbed but a simplified procedure using Fermi–Dirac statistics enables one to relate the observed result to the spacing of the energy levels at the Fermi level (density of states eqns. (55), (56)).

6. Magnetic solids also absorb energy through the disorientation of the magnetic moments. This occurs over a wide temperature range with a peak

in the specific heat at the Curie temperature. Only for a ferromagnetic metal at low temperature can one make a quantum mechanical calculation of the energy absorbed (eqns. (59), (60)).

7. Many substances will undergo a change in crystal structure under pressure. With pressures currently attainable in the laboratory (~ 1000 kilobars) the stable phase at high pressure must have a smaller volume, and a free energy less than a few kilocalories per mole higher than the phase stable at zero pressure.

8. There is no good approximation to the quantized energy levels of a liquid. We are thus unable to calculate its thermodynamic properties.

CHAPTER IV

THE NUCLEUS

Static Properties of the Nucleus

Since the various properties of the nucleus have been used to gain insight into the behaviour of solids we shall enumerate some of these properties. We do not have an adequate theory of the nucleus since we lack knowledge of the potentials that cause nuclear particles (protons, neutrons) to attract each other. In addition, we are not certain that the Schrödinger equation (or an improved version that includes relativity) is exact when applied to the nucleus although some success has been achieved. But in spite of this we can still measure many properties of the nucleus and utilize them. The important properties are:

1. *Charge*, Ze. The number of protons in the nucleus determines the charge Ze. Each proton has a positive charge *exactly* equal in magnitude to the negative charge of the electron.

2. *Atomic Mass, A*. The atomic mass is the sum of the number of protons, Z, and neutrons, N, contained within the nucleus (they are collectively called *nucleons*). An element is distinguished only by the number of protons since this in turn determines the number of electrons and hence the chemical properties. *Isotopes* are nuclei with the same Z and different N. They have the same chemical properties but different masses. In general, the number of neutrons equals the number of protons for light elements but the ratio N/Z gradually increases to $\sim 1{\cdot}6$ for the heavy nuclei. A nucleus is identified as follows, for example

$$_{53}\mathrm{I}^{127} \rightarrow \text{Iodine}, Z = 53, A = 127, N = A - Z = 74$$

3. *Mass, M*. Even though the mass of the neutron and proton are known, 1·008982 amu and 1·008142 amu respectively, (1 amu = 1 atomic mass unit = $1{\cdot}66 \times 10^{-24}$ g) the mass of a nucleus is not merely the sum of the proton masses and neutron masses. The sum is always slightly more and the difference is the amount of mass that has been converted into the binding energy of the nucleus through the well known *Einstein relation.*

$$E = Mc^2 \tag{66}$$

where 1 amu = 931 MeV (1 MeV = 10^6eV). For example, the measured mass of $_{53}\mathrm{I}^{127}$ is 126·94528 amu whereas 74 neutrons plus 53 protons have a total mass of 128·096194 amu. The difference, 1·151 amu = 1·072 BeV (1 BeV = 10^9 eV) is the total binding energy of the Iodine nucleus. Since metals are only bound with a few eV the mass loss is only a few parts in 10^{-9} and is neglected. This loss of mass is converted into the potential energy of the attractive fields holding the neutrons and protons together. In the case of metals the mass loss goes into the electrostatic Coulomb field.

4. *Spin, I.* The neutron and proton each have an intrinsic spin of $s = \frac{1}{2}$, $m_s = \pm\frac{1}{2}$ as does the electron. This intrinsic spin is associated with an intrinsic angular momentum $\hbar\sqrt{[s(s+1)]}$. The total angular momentum of the nuclei is designated by the quantum number I, $(\hbar\sqrt{[I(I+1)]})$, and this depends on the manner in which the neutrons and protons build up their shell structure. The neutrons and protons in a nucleus have an intrinsic spin angular momentum and an orbital angular momentum just as electrons in atoms. However, the neutrons and protons appear to pair off so that every *pair of protons* and every *pair of neutrons* give a resultant $m_s = 0$ and $m_l = 0$. This leads to the following rules:

(*a*) All nuclei with an even number of neutrons *and* an even number of protons (called even-even nuclei) have an $I = 0$.

(*b*) All nuclei with an even number of neutrons and an odd number of protons or vice versa have an $I \neq 0$. The value of I must be determined experimentally but is always half integral, $\frac{1}{2}$, $\frac{3}{2}$, $\frac{5}{2}$, $\frac{7}{2}$, $\frac{9}{2}$, etc.

(*c*) All nuclei with an odd number of neutrons *and* an odd number of protons have $I \neq 0$ (called odd-odd). The last neutron and the last proton have their m_s parallel so that I is integral, 1, 3 etc. There are very few cases of odd-odd nuclei since this is generally an unstable condition in the nucleus.

5. *Size, R.* The size of atoms can be determined from the scattering of X-rays of comparable wavelength (~1Å). The size of a nucleus is determined from the scattering of high energy neutrons and electrons and its radius, R, is

$$R \cong 1{\cdot}37 \times 10^{-13}(A)^{1/3}\ \text{cm} \tag{67}$$

where A is the atomic number. The volume occupied by each nucleon is $\frac{4}{3}\pi R^3/A \cong 11{\cdot}5 \times 10^{-39}$ cm^3 independent of A. Each nucleon has a constant volume with a nearest neighbour distance of $\sim 3 \times 10^{-13}$ cm. At this distance the Coulomb repulsion of two protons is ~0·3 MeV whereas the binding energy of a pair of protons in a nucleus is ~17 MeV. Thus the Coulomb repulsion has a small effect on nuclear structure. It accounts in part for the excess of neutrons over protons at high Z since the binding energy of a neutron is about the same as a proton but one does not have to overcome the Coulomb repulsion.

6. *Magnetic Moment.* The magnetic moment of all even-even nuclei ($I = 0$) is zero. In all other cases the magnetic moment is not simply a

constant factor times I but must be measured in each case. In a manner similar to electrons (see p. 6) where

$$\mu_B = \frac{e\hbar}{2m_0c} = 9{\cdot}273 \times 10^{-21} \text{ erg/gauss} = \text{“Bohr magneton”}$$

we write

$$\mu_N = \frac{eh}{2M_pc} = 5{\cdot}05 \times 10^{-24} \text{ erg/gauss} = \text{ nuclear magneton} \qquad (68)$$

where M_p is the proton mass. However, the proton does not have a magnetic moment of one nuclear magneton but 2·7934 nuclear magnetons. To explain this value requires a specific theory for the nuclear structure of a proton which we do not have. However, for solid state physics applications it is only necessary to know its value, not its origin. The magnetic moment of the neutron is $-1{\cdot}9135$ nuclear magnetons, opposite to the proton. The ratio of the observed magnetic moment to $e\hbar I/2M_pc$ is called the *nuclear g factor*, (I is the spin quantum number for the particular nucleus).

7. *Quadrupole moment*, eQ. All nuclei with $I \leq \frac{1}{2}$ have a spherically symmetric distribution of protons and all those with $I > \frac{1}{2}$ are not spherical. If we define the direction of the magnetic moment as an axis the electric quadrupole moment tells us the extent to which the probability distribution of the protons is elongated along the axis as compared with a direction normal to the axis.* If the probability distribution is elongated along the axis, the quadrupole moment is positive and is called “cigar shaped” , and if contracted along the axis, it is negative called “watch-shaped” . In general these departures from spherical symmetry are small.

8. *Nuclear shell structure*. A qualitive theory of the nucleon configurations in nuclei leads to a model which predicts closed shells at the “magic numbers” 2, 8, 20, 28, 40, 50 and 82. Any nucleus which contains a magic number of neutrons or protons appears to be particularly stable (high binding energy).

9. *Positrons*. There are a whole swarm of particles of varying mass, spin, etc. known and observed in nuclear physics (mesons and that sort of thing). Most of them provide considerable thought for the theoretical nuclear physicist who attempts to explain their properties and existence. In the main, they are of “no practical value” to the metallurgist except possibly the positron.

The positron is not a nuclear particle but is sometimes created by nuclei. It is identical to the electron except that it has a positive charge and magnetic moment.

In paragraphs 1–9 we have listed the stable or static properties of nuclei. While neutrons and positrons live only about ten minutes and less than a fraction of a second respectively before they are converted into protons and γ-rays respectively, they do live long enough to be useful.

*By definition the electric quadrupole moment is $eQ = e\iiint \rho(3\cos^2\theta - 1)r^4 \, sin\, \theta d\, \theta d\phi dr$ where $\rho(r, \theta, \phi)$ is the probability distribution of the protons and θ is the angle ρ makes with the axis of the magnetic moment. If ρ is spherically symmetric, the integral is zero.

Dynamical Properties of the Nucleus

Dynamical observations of nuclei are frequently important tools for metallurgists. This includes radio-active decay in which a nucleus undergoes some observable change. Such dynamical changes are listed below.

1. *Beta decay.* In this case one of the neutrons in the nucleus converts into a proton. The extra energy gained by the reduction in mass of the nucleus (and most of this reduction is due to the neutron being heavier than the proton), and the extra negative charge is emitted from the nucleus as a high energy electron (called a β-ray) plus a neutrino n_0. The energies are in the range $\frac{1}{2}$ to 4 MeV and this energy is shared between the electron and neutrino. The neutrino is a very light neutral particle that is virtually impossible to detect because it exerts an extremely weak potential for nuclei or electrons and passes through substances undisturbed.

A typical energy distribution for β-ray and neutrino is shown in the next figure. *These particles do not reside in the nucleus but are created and emitted*

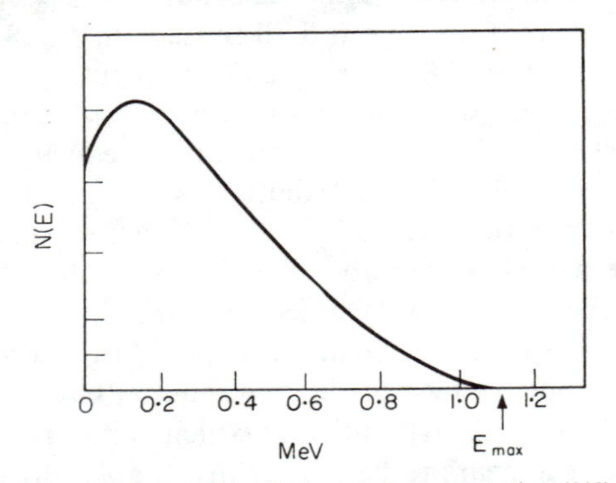

FIG. 34. The relative number $N(E)$ of β-rays from RaE($_{83}$Bi^{210}) as a function of energy in MeV. A neutrino is emitted with each β-ray and its energy is the difference between the upper energy limit, 1·11 MeV, and the β-ray energy.

to conserve energy and charge. For example, consider the reaction,

$$_{13}\text{Al}^{29} \rightarrow {}_{14}\text{Si}^{29} + \beta + n_0 \qquad E = 2{\cdot}5 \text{ MeV}$$

By converting one of the neutrons in the $_{13}$Al^{29} nucleus to a proton we have a $_{14}$Si^{29} nucleus plus excess energy which is emitted as a beta ray, β, and neutrino, n_0. One reason we are certain that a β-ray does not reside in the nucleus arises from the uncertainty principle. An electron confined to a region $\Delta x \sim 3 \times 10^{-13}$cm would have a momentum $\Delta p \sim \hbar/\Delta x$ or an energy $\Delta E \sim (\Delta p)^2/2m = \hbar^2/2m(\Delta x)^2$. This amounts to $\sim$100 BeV (10^{11} eV) or about a hundred times greater than the binding energy of a typical nucleus.

If one had a sample of aluminium with a certain fraction of $_{13}$Al^{29} atoms one would find that the β-rays would be emitted with some intensity (number of β-rays/sec) and that this intensity would be reduced to half its value in 6·7 min and half again every 6·7 min. The time, τ, required for half the atoms to decay is called the *half life* and the intensity follows the formula

$$I = I_0 e^{-\lambda t} \tag{69}$$

where t is the time, I the intensity at any time t, and I_0 the intensity at $t = 0$ i.e. the time at which the measurements began. λ is called the decay *constant* and is related to the half life by

$$\lambda = 0{\cdot}693/\tau \tag{70}$$

Half lives for β-decay vary over very wide ranges from a fraction of a second to many centuries. One might expect that such decay should happen instantly since the energy of the nucleus is lowered by virtue of the β-emission. While it would be inappropriate to attempt to answer the question here it will suffice to say that conditions in the nucleus have to be just right for β-emission. The nucleus can be envisaged as a dynamic system with the nucleons in constant motion and it must wait for a rather definite arrangement of nucleons to emit the β-ray and neutrino. The frequency of internal nuclear motion is something like 10^{19}/sec so that it takes more than $\sim 10^{20}$ rearrangements of the nucleons in order to emit the β ray in ${}_{13}Al^{29}$. (You could open any combination lock with this number of tries.)

2. *γ-rays, isomers, internal conversion.* A second method of radioactive decay is by emission of a γ-ray i.e. a high energy photon. Since the γ-ray carries no charge the nucleus is unchanged in Z, N or A although its mass is decreased by the energy of the γ-ray. An excited state of a nucleus which decays by emitting a γ-ray is called an *isomer*. In most cases one finds the sequence; (1) nucleus A emits a β-ray and becomes nucleus B in an excited state, (2) nucleus B then emits a γ-ray ending up in its ground state. For example, this is a typical sequence of steps:

$${}_{51}Sb^{124} \rightarrow {}_{52}Te^{124*} + \beta + n_0 \qquad (\tau = 60 \text{ days})$$

$${}_{52}Te^{124*} \rightarrow {}_{52}Te^{124} + \gamma \qquad (\tau = \text{instantaneous})$$

The asterisk over the atomic number refers to an excited state. In some cases the γ-ray is not emitted instantaneously, for example

$${}_{26}Fe^{57*} \rightarrow {}_{26}Fe^{57} + \gamma \qquad \tau = 10^{-7} \text{ sec} \qquad E = 14{\cdot}4 \text{ keV}$$

$${}_{48}In^{115*} \rightarrow {}_{49}In^{115} + \gamma \qquad \tau = 4{\cdot}5 \text{ hr} \qquad E = 0{\cdot}3 \text{ MeV}$$

and these cases are more frequently alluded to as isomers rather than the Te^{124*} above. The distinction is only relative.

Quite frequently the nucleus will decay by giving this energy to one of the atomic electrons (generally the ls since it has the largest probability of being at the nucleus) rather than emit a γ-ray. This is called *internal conversion.*

The observer will detect a high energy electron but since this is identical to a β-ray he must distinguish the two processes by whether the nuclear charge Ze is unchanged as in internal conversion or whether it is changed by one charge as in β-emission. The γ-ray emission and the internal conversion are competing processes and the fraction of time a $1s$ electron is ejected as compared to a γ-ray emitted is called the *K internal conversion coefficient* (if a $2s$ electron is involved it is called the L internal conversion coefficient). In a typical case the K internal conversion coefficient ranges from 1 for a γ-ray energy of $\sim 0{\cdot}2$ MeV and decreases smoothly to $\sim 10^{-3}$ for a γ-ray energy of 2 MeV. The energy of the internal conversion electron is equal to the energy of the γ-ray minus the energy required to remove the electron from the $1s$ shell. Internal conversion is followed by the emission of an X-ray as the $2p$ electron falls into the hole left by the ejected $1s$ electron, etc.

3. *α-decay.* Some nuclei decay by emitting a helium nucleus, i.e., a group of two protons and two neutrons called an α-particle. This occurs in heavy nuclei ($Z > 60$) and like β-decay, generally is of long half life. Unlike β-decay the α-particle does reside in the nucleus before it is ejected. The energies of α particles are $\sim 1 - 5$ MeV.

4. *K capture.* In some cases the $1s$ electron is consumed by the nucleus and one of the protons is neutralized and becomes a neutron. This is called K capture, as for example,

$$_{26}\mathrm{Fe}^{55} + K \rightarrow {}_{25}\mathrm{Mn}^{55} + n_0 \qquad \tau = 4 \text{ years}$$

In this case a $_{26}\mathrm{Fe}^{55}$ nucleus will sit around on the average for four years before it annihilates one of its $1s$ electrons. The excess energy in this process is given to the neutrino since the resultant nucleus ($_{25}\mathrm{Mn}^{55}$ in this case) will never have an energy exactly equal to the initial nucleus and the $1s$ electron. K capture is followed by the emission of an X-ray as the $2p$ electron falls into the hole left by the $1s$ electron. This X-ray is characteristic of the final nucleus ($_{25}\mathrm{Mn}^{55}$ in this case).

5. *Fission.* In this process a nucleus will split into two large parts plus some neutrons. The most famous case is $_{92}\mathrm{U}^{236}$. There are several possible ways of splitting and a typical case is

$$_{92}\mathrm{U}^{236} \rightarrow {}_{38}\mathrm{Sr}^{97} + {}_{54}\mathrm{Xe}^{137} + {}_0n^1 + {}_0n^1 + \gamma \qquad \begin{array}{l} \tau = \text{instant.} \\ E \cong 200 \text{ MeV} \end{array}$$

where $_0n^1$ is a neutron. The energy is split four ways most of it going to the heavy nuclei. As $_{92}\mathrm{U}^{236}$ does not exist in nature, since it would decay instantaneously via fission, it must be produced. This is accomplished by adding a neutron to $_{92}\mathrm{U}^{235}$ which does occur in nature. If one has a mass of $_{92}\mathrm{U}^{235}$ and introduces one neutron it will convert one of the $_{92}\mathrm{U}^{235}$ nuclei to $_{92}\mathrm{U}^{236}$ which then fissions, releasing 200 MeV energy as well as two neutrons. These two neutrons may in turn cause two more $_{92}\mathrm{U}^{235}$ nuclei to be converted into $_{92}\mathrm{U}^{236}$ which would then emit a total of four neutrons, etc. If the mass of

$_{92}U^{235}$ is large enough so that the neutrons do not escape from the sample but are always captured by a $_{92}U^{235}$ nucleus we have a chain reaction. There has been considerable interest in this process.

6. *Positron emission; creation and annihilation.* A mirror process to β-emission is positron emission denoted, β^+. Except for the opposite sign of the positron which decreases the nuclear charge by one unit the process is identical in other details to β-emission. A typical case is

$$_{10}Ne^{19} \rightarrow {}_{9}F^{19} + \beta^+ + n_0 \qquad \begin{matrix} \tau = 20{\cdot}3 \text{ sec} \\ E = 2{\cdot}2 \text{ MeV} \end{matrix}$$

A positron, though, is not stable. Ultimately it finds an electron and the pair annihilate themselves converting their total mass into two γ-rays each of 0·5 MeV (on some occasions 3 photons). Since the electron and positron are oppositely charged they attract each other and can form a "quasi hydrogen atom" called *positronium.* In the case of the hydrogen atom the nucleus is practically stationary since it is so much heavier than the electron but in the case of positronium the two particles whirl about the centre of gravity between them. The Schrödinger equation yields the same solutions as for the hydrogen atom except that we must replace the mass, m, by $m_0/2$ and remember that the origin refers to the centre of gravity. The solution with both spins antiparallel is called the singlet and has a half life of $\sim 10^{-10}$ sec, decaying into two γ-rays. The solution with spins parallel is called the triplet (the Pauli principle does not apply to non-identical particles); it has a half life of $1{\cdot}5 \times 10^{-7}$ sec and decays into three γ-rays each about $\frac{1}{3}$ MeV.

The reverse of positron annihilation is *pair production* in which a γ ray converts itself into an electron and positron. The γ-ray must have enough energy to create the total mass of the pair which is $\sim$1 MeV. In order to conserve momentum the γ-ray needs a second body (generally an atom). Pair production occurs only when a γ-ray of energy greater than 1 MeV is near an atom.

Neutrons, protons, deuterons and α-particles are all used to cause nuclei to undergo some change by directing a beam of these particles against a sample containing some particular nuclei of interest. We shall now discuss some aspects of these processes.

1. Neutron Interaction

A neutron can interact with other nuclei by being scattered or absorbed and both these processes depend on the neutron energy and the particular nucleus. Let us consider a particular example. The next figure is a plot of the neutron cross section of vanadium metal from 0 to 7,000 eV. The cross section determines the rate at which a beam of neutrons of some particular energy is absorbed in traversing a piece of vanadium metal according to

$$I = I_0 e^{-N_0 \sigma x} \tag{71}$$

where I is the intensity of the neutron beam after traversing a distance x into the metal, N_0 is the number of vanadium atoms per cm^3 and σ is the cross section. The units of cross-section are barns (10^{-24} cm^2) actually derived from an old American expression "you can't hit the side of a barn", presumably with a baseball.

In the absorption process the neutron is absorbed and a γ-ray is emitted, a new isotope being created. Thus

$$_{23}Va^{51} + {_0n^1} \rightarrow {_{23}Va^{52*}} \rightarrow {_{23}Va^{52}} + \gamma \qquad \begin{array}{l}\tau = \text{instantaneous} \\ E = 1 \text{ MeV}\end{array}$$

This is called an n, γ-process. The $_{23}Va^{52}$ nucleus is unstable and so the above reaction is followed by

$$_{23}Va^{52} \rightarrow {_{24}Cr^{52*}} + \beta + n_0 \qquad \begin{array}{l}\tau = 3{\cdot}8 \text{ min} \\ E \cong 2{\cdot}5 \text{ MeV}\end{array}$$

which is followed by

$$_{24}Cr^{52*} \rightarrow {_{24}Cr^{52}} + \gamma \qquad \begin{array}{l}\tau = \text{instantaneous} \\ E = 1{\cdot}4 \text{ MeV}\end{array}$$

the $_{24}Cr^{52}$ nucleus being stable. This is typical of one of the methods for creating radioactive isotopes.

In the neutron scattering process called an *n, n-process* the neutron leaves the nucleus in the same state but its own direction has been altered. The process can be written

$$_{23}Va^{51} + {_0n^1} \rightarrow {_{23}Va^{52*}} \rightarrow {_{23}Va^{51}} + {_0n^1}$$

Both n, γ- and n, n-processes occur over the entire range of neutron energies in Fig. 35 and the relative amounts of each are shown. There are pronounced peaks at 4200 eV and 6500 eV called *resonances* and neutrons of these energies have a relatively high probability of suffering an n, γ- or n, n-collision. The cross section, as a function of energy for the two processes, is expressed by the Breit-Wigner formula for each resonance

$$\sigma(n, \gamma) = \frac{\pi g}{k^2} \frac{\Gamma_n \Gamma_\gamma}{(E - E_0)^2 + [(\Gamma_n + \Gamma_\gamma)/2]^2} \tag{72}$$

$$\sigma(n, n) = \frac{4\pi g}{k^2} \left| \frac{\Gamma_n/2}{(E - E_0) + i(\Gamma_n + \Gamma_\gamma)/2} + e^{ikR} \sin kR \right|^2 \tag{73}$$

where g is a quantum mechanical statistical weight factor giving the relative probability of a neutron being scattered or absorbed with its spin parallel or antiparallel to the $_{23}Va^{51}$ nuclear spin, $I = \frac{7}{2}$.

$$g = \tfrac{1}{2}[1 \pm 1/(2I + 1)] \tag{74}$$

plus sign for parallel, minus sign for antiparallel ($g = 1$ for $I = 0$).

$k = 2\pi/\lambda$ where λ is the wavelength, E_0 is the energy of the resonance and Γ_n and Γ_γ are the so-called *resonance widths*. They are expressed in eV and give a measure of the energy range over which the n, n- and n, γ-process are important in the vicinity of the resonance. They are perhaps better thought to be parameters which give the relative probability as a function of energy for the n, n- and n, γ-process. R is the effective nuclear radius as seen by the neutron, $R = 1{\cdot}37 \times 10^{-13}\, A^{1/3}$cm and E is the neutron energy. At energies less than $\sim$10,000 eV $\sigma(n, n)$ can be written (since kR is small).

$$\sigma(n,n) = 4\pi g \left| \frac{\Gamma_n/2}{k(E - E_0) + ik(\Gamma_n + \Gamma_\gamma)/2} + R \right|^2 \quad \text{valid for } E < 10^4 \text{ eV} \qquad (75)$$

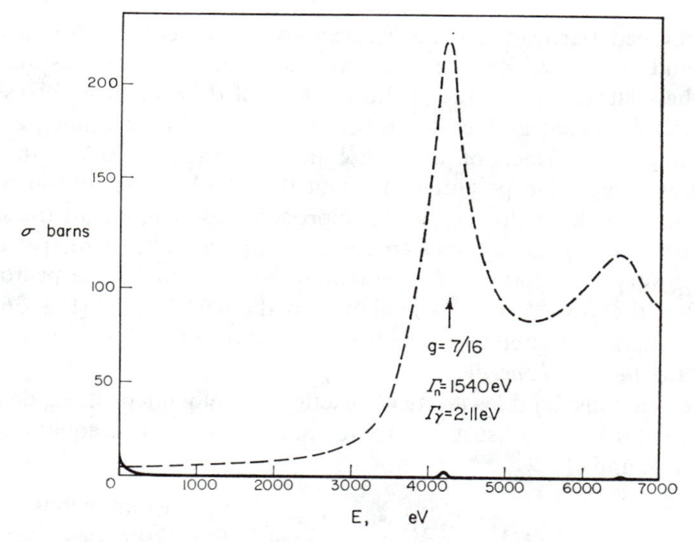

FIG. 35. The total neutron scattering cross-section (dashed-line) and total neutron absorption cross-section (solid line) of vanadium as a function of energy in eV. The resonance at 4200 eV has the Breit-Wigner parameters given in the curve.

At energies other than resonance Γ_n is proportional to k and Γ_γ is independent of k. An analysis of the vanadium resonance at 4200 eV when coupled with scattering and absorption cross-section measurements at 0·025 eV informs us that the resonance at 4200 eV is due solely to the times when the neutron is scattered or absorbed with its spin antiparallel to the vanadium nucleus. While the resonances at 4200 eV and 6500 eV are principally scattering rather than absorption, resonances at lower energies show an increased proportion of absorption. In general, resonances below about 10 eV are practically all absorption.

Other neutron processes are n, p and n, α examples of which are (p = proton)

$$_7N^{14} + {_0n^1} \rightarrow {_6C^{14}} + {_1p^1} \qquad \tau = \text{instantaneous} \quad E = 0{\cdot}5 \text{ MeV}$$

$$_5B^{10} + {_0n^1} \rightarrow {_3Li^7} + {_2He^4} \qquad \tau = \text{instantaneous} \quad E = 2{\cdot}5 \text{ MeV}$$

This latter process provides us with a method for detecting neutrons. One uses a neutron counter filled with boron ($_5B^{10}$) trifluoride gas, the α-particle ejected in the process producing a detectable electrical signal.

2. Proton, Deuteron and α-Particle Interactions

All charged particles can be accelerated in a cyclotron (or any other similar and expensive device) and then directed into various materials. Unlike the neutral neutron the positive charge of these particles gives rise to a Coulomb repulsion as they approach a nucleus. For example, a proton approaching the surface of a $_{23}Va^{51}$ nucleus (radius $\sim 0{\cdot}5 \times 10^{-12}$ cm) experiences a repulsive potential of about 0·15 MeV. The proton must be accelerated to at least this energy to approach close enough to the nucleus for the attractive nuclear forces to come into play. The attractive nuclear forces are very short range. The potential experienced by the proton as it approaches the $_{23}Va^{51}$ nucleus is shown in the next figure. (Fig. 36). The minimum energy required by the proton to overcome this Coulomb repulsion is called the *threshold energy*.

There is a considerable variety of reactions employing protons, deuterons and α-particles and we list a few representative ones. The equations must balance in Z and A.

1. $$_3Li^6 + {_1H^2} \rightarrow {_2He^4} + {_2He^4} \qquad \tau = \text{instantaneous} \quad E = 22{\cdot}17 \text{ MeV}$$

2. $$_{11}Na^{23} + {_1H^2} \rightarrow {_{11}Na^{24*}} + {_1p^1} \qquad \tau = \text{instantaneous}$$

$$_{11}Na^{24*} \rightarrow {_{12}Mg^{24*}} + \beta + n_0 \qquad \tau = 14{\cdot}8 \text{ hr} \quad E = 1{\cdot}39 \text{ MeV}$$

$$_{12}Mg^{24*} \rightarrow {_{12}Mg^{24}} + \gamma_1 + \gamma_2 \qquad \tau = \text{instantaneous} \quad E = 1{\cdot}4;\ 2{\cdot}8 \text{ MeV for } \gamma\text{-rays}$$

3. $$_7N^{14} + {_2He^4} \rightarrow {_8O^{17}} + {_1p^1} \qquad \tau = \text{instantaneous} \quad E = -1{\cdot}26 \text{ MeV}$$

This is the first experiment in artificial transmutation performed by Lord Rutherford in 1919. The energy is negative and requires an α-particle with at least 1·26 MeV to make this reaction "go" since the mass of the particles on the right are heavier by 1·26 MeV. Rutherford used α-particles from a natural α-emitter.

4. $${}_4Be^9 + {}_2He^4 \rightarrow {}_6C^{12} + {}_0n^1 \qquad \tau = \text{instantaneous}, \quad E = 5{\cdot}7 \text{ MeV}$$

This reaction led Chadwick to discover the neutron in 1932.

5. $${}_3Li^7 + {}_1p^1 \rightarrow {}_2He^4 + {}_2He^4 \qquad \tau = \text{instantaneous}, \quad E = 17{\cdot}1 \text{ MeV}$$

This reaction was used in the first Cockroft–Walton machine in 1932.

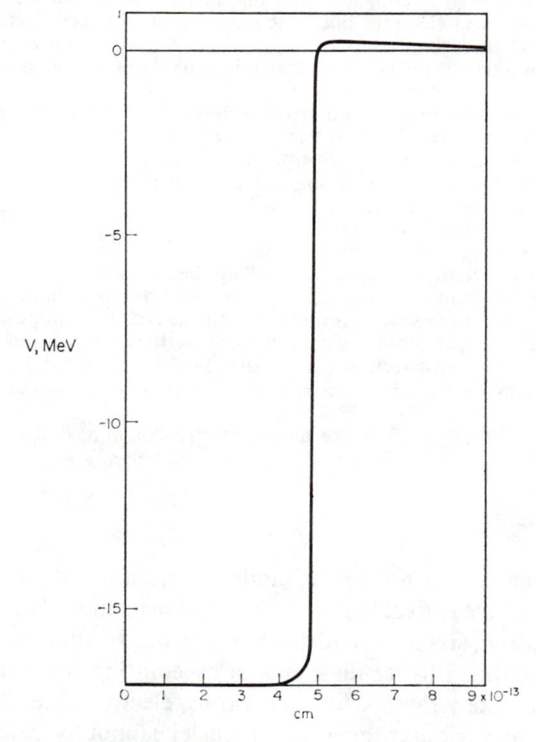

FIG. 36. The potential in MeV experienced by a proton as a function of distance from the centre of a vanadium nucleus. The distances greater than $\sim 5 \times 10^{-13}$ cm are dominated by the Coulomb repulsion, while at distances less than the nuclear radius the attractive nuclear forces dominate.

Problems

1. Assume that you have a sample of gallium containing some ${}_{31}Ga^{65}$ and could detect **all** the particles emitted by this sample. Make a list of at least 20 of them, identify each (X-ray, γ, β^+, β, light, etc.) and give its wavelength or energy.

2. How much uranium is consumed by a 35,000 ton nuclear powered ship in crossing the Atlantic? If one could make a power source from a lithium deuteride compound ($_3Li^6$), the reaction on page 104 would provide considerable energy. However, the Coulomb repulsion of the Li-nucleus makes it difficult for the deuteron to penetrate into the nucleus. To what temperature must one heat the lithium deuteride so that the thermal energy ($\sim kT$) of the deuterons could overcome this repulsion? How would you build such an oven?

3. Show that the wavelength of a neutron is related to its energy by the expression

$$\lambda = 0{\cdot}283/\sqrt{(E)}$$

where λ is in Å and E in eV. What is the wavelength of a neutron whose energy is equal to kT (T = room temperature)? What is the energy of a neutron whose wavelength is the radius of a Li-nucleus?

4. (For the mathematically minded.) Newton's third law of motion states that every action has an equal and opposite reaction. A rocket ship in outer space can only increase its forward momentum by ejecting an equal amount of momenta in the opposite direction. For example, it can eject electrons but these electrons are lost and decrease the weight of the rocket ship. Considering the following possible sources of energy, which appears the most practical for a trip to the moon and back in a 1000 ton rocket ship? Remember that the weight of fuel is important.

(*a*) Nuclear reactor plus anything you wish for momentum conservation.
(*b*) Any β-ray source (specify which).
(*c*) Any γ-ray source (specify which).
(*d*) Any conventional chemical source (specify which).
(*e*) Lithium deuteride source.
(*f*) Any conventional electron gun.
Estimate the weight of the power plant.

5. (For the mathematically minded). Look up the total cross-section of cadmium from $0 - 1$ eV and see if you can fit it to eqn. (72), since $\sigma(n, \gamma) \gg \sigma(n, n)$ in this range ($\Gamma_\gamma \gg \Gamma_n$). The large cross-section in this range is due solely to the $_{48}Cd^{113}$ isotope ($I = \frac{1}{2}$) so that the observed cross-section per atom must be divided by the isotopic abundance of $_{48}Cd^{113}$ (0·1226) to obtain the cross-section per $_{48}Cd^{113}$ nucleus. The total cross-section of an element is the sum of the cross-sections of each isotope weighted with the isotopic abundances.

Assume $\Gamma\gamma = 0{\cdot}1$ eV at resonance and determine $\sigma(n, n)$ at 0·025 eV for both values of g (eqn. 74). Ans., $\sigma = 85{\cdot}6$ barns, $g = \frac{1}{4}$; $\sigma = 12{\cdot}2$ barns, $g = \frac{3}{4}$.

Summary

Even though we do not understand the structure of the nucleus, it has many useful and measureable properties. These include charge, weight, magnetic moment, spin, size and quadrupole moment. In addition there are many unstable nuclei which lower their energy by emitting some identifiable high energy particles like γ-rays, X-rays, positrons, electrons (beta rays) and alpha particles. These various properties of nuclei cannot be calculated but the measured values are tabulated (in some cases to very high precision). Most of these properties are utilized in learning something about the electronic structure of solids (chap. XI).

CHAPTER V

EXPERIMENTAL TECHNIQUES

Introduction

There is a wide variety of experimental techniques which yield information about the behaviour of metals, but a great chasm exists between the experimental results and the ability of the theoretician to predict these results. Of course the experimentalist has the upper hand in this game (assuming his experiment is properly done) since he reports the behaviour of nature, and nature is always right.

The fundamental theoretical interest in solids has been the straightforward (although approximate) solutions of the Schrödinger equation and these yield eigenfunctions and eigenvalues. One can also solve for the cohesive energy and with successively less accuracy calculate the details of the Fermi surface, the lattice parameter and the elastic constants. However, to obtain the more practical answers of interest to the metallurgist such as mechanical properties, diffusion coefficient, thermal and electrical conductivity, etc., presents another order of magnitude of difficulty, particularly since the metallurgist rarely deals with pure metals and the solution of the Schrödinger equation is much more difficult for alloys. Thus we have the wide gap between theory and experiment.

The emphasis in the next few chapters will be on those experimental techniques that either afford direct experimental verification of the solutions of Schrödinger's equation for solids or indirectly serve as a guide to the nature of the solutions. The chapters each cover rather broad branches of experimental techniques, since most experimentalists associate themselves with definite techniques. The following table lists the various techniques to be discussed and the information they yield.

Sample Preparation

In any measurement an important consideration is the sample and it would do well to remark about this even though this may be a review for metallurgists. So often relatively time-consuming and careful measurements have been made on samples literally taken off a stockroom shelf and marked for example "pure iron". It may be pure iron, but all too often this assumption is unwarranted. Furthermore, it may be pure enough for some measurements but not others. Ample consideration should be

TABLE Va
A GUIDE TO THE VARIOUS EXPERIMENTAL TECHNIQUES USED IN SOLID STATE PHYSICS AND AN INDICATION OF THE INFORMATION THEY YIELD

Measurement	Information	Comments
X-ray diffraction	Total electron probability distribution	Only feasible method for solids
Neutron diffraction	Unpaired electron distribution	Only method
X-ray and neutron diffraction	Crystal structure determinations	Order– disorder, short range order, phase diagrams
Neutron diffraction	Atomic magnetic moment arrangement in crystals	Order– disorder, short range order
X-ray and neutron diffraction	Frequency spectra and dispersion curves of crystals	Only method
Neutron diffraction	Spin wave spectra and dispersion curves	Only method
X-ray diffraction	Momenta of electrons in crystals	Compton scattering—very difficult measurement
Soft X-ray spectroscopy	Eigenvalues of electrons in solids	Very difficult measurement
Cyclotron resonance	Shape of Fermi surface	Requires very pure metals
Electrical resistivity	Sensitive to everything, impurities, cold work, order–disorder, etc.	Very easy measurement, difficult to make quantitative
Hall effect	Sign of current carriers (electron or holes)	Difficult to interpret
De Haas van Alphen effect	Details of the Fermi surface	Requires pure single crystals
Thermal conductivity	Anharmonic forces in crystals	Difficult to interpret
Electron diffraction	Crystal structure of surfaces	Only method for study of atomic arrangement on surface
Thermo-electric power	Details of Fermi surface	Interpretation difficult, very sensitive
Ferro-magnetic resonance	g Factor	Very sensitive

TABLE Va—*continued*

Measurement	Information	Comments
Nuclear resonance	Effective magnetic fields at nucleus—electric field gradient at nucleus—relaxation times	Very useful for self diffusion coefficients also
Mössbauer effect	Effective magnetic fields at nucleus—electric field gradient at nucleus	Very simple measurement
Nuclear specific heat $< 1°$	Effective magnetic fields at nucleus—electric field gradient at nucleus	Costly—cumbersome equipment
Anomalous skin effect	Details of the Fermi surface	Requires very pure single crystals
Ultra-soņic velocity	Elastic constants	Difficult at high temperature
Ultra-sonic attenuation in magnetic fields	Shape of Fermi surface	Very sensitive
Optical properties	Energy levels in bands	Difficult
Magneto-resistance	Shape of Fermi surface	Requires pure single crystals
Susceptibility dia-	$\langle r^2 \rangle$ of electron distribution	Easy to interpret
Susceptibility para- (non-magnetic metals)	Density of states at Fermi surface	Difficult to interpret
Susceptibility para- (magnetic metals)	Effective magnetic moment	Moderately sensitive
Magnetization (low temp.)	Number of unpaired spins —exchange energy	Easy to interpret
Low temp. specific heat 1 – 4°K	Electronic specific heat coefficient	Gives density of states
Positron annihilation	Momentum of electron-positron pair	Not sensitive to band structure
Intermediate temp. specific heat 4°K to $T \cong 300°$K	Debye Θ—exchange energy for rare earths	Easy to interpret
Radioactive tracers	Atomic indentification	Good for self-diffusion
Heat of solution—entropy of mixing, etc.	Cohesive energy of alloys	No generally applicable technique available for these measurements
High Temp. specific Heat $T > 300°$K	Electronic specific heat coefficient–exchange energy for transition metals	Difficult measurement

given to the problem of sample analysis consistent with the requirements of the measurement.

Generally the problems of sample analysis divide into two categories of samples: (*a*) *Pure elements* in order to make measurements on them or use them to make alloys. (*b*) *Alloys or compounds* of very definite composition. The following rules can serve as a guide for these two categories:

1. No analyzing technique is capable of yielding a quantitative analysis for *all* elements.

2. All pure metals contain at least one foreign atom per 10,000 atoms (*circa* 1960).

3. It is a good policy to invest at least 5% of the time intended for the study of a material to investigate the constitution of the material.

4. Metallic alloys (not compounds) probably vary in solute concentration by at least 0·1% over their volume.

5. Whenever possible, two techniques of analysis are desirable to check each other.

The following table lists the various techniques of analysis and indicates their limitations. However, to clarify the matter let us select two practical situations, one concerned with pure metals, the other with alloys, and discuss the problems of analysis for each.

A. Pure Metals

Analysis of Pure Metals. Suppose we wish to obtain some pure vanadium in order to determine the solubility of magnesium in b.c.c. vanadium. Since the solid solubility is probably not high and since magnesium is known to be available in high purity, we need not concern ourselves with the magnesium purity but turn to the major problem of the vanadium purity. If we investigate the sources of pure vanadium we find that the supplier reports vanadium to be ~99·6% pure with oxygen, nitrogen, carbon and hydrogen the principal impurities. The supplier may be willing to provide evidence for this analysis. It will probably be a typical vacuum fusion analysis.*

If the vanadium is important to you, you might request the supplier to provide a special analysis for the sample supplied to you. In any event, it is a good idea to make friends with a metallographer, for he can polish and etch the vanadium and detect the presence of phases other than b.c.c. vanadium which etch differently. He may also have enough experience to specifically

* In vacuum fusion, the sample is placed in a graphite crucible and heated to the melting point so that the oxygen in the metal will combine with the carbon to form CO and the hydrogen and nitrogen are liberated in their pure state. These gases are collated and separated, and if all the oxygen in the sample has formed CO then a measure of the volume and pressure of the three gases yields a measure of the nitrogen, hydrogen and oxygen in the sample. Generally the CO is converted to CO_2 and frozen out, while the H_2 is converted to water, the nitrogen then determined by difference.

identify the foreign phases (some definite oxide for example) and even estimate their amounts. If he advises you that the vanadium contains $\sim\frac{1}{2}\%$ of a second phase (probably oxide) then it is possible that the b.c.c. vanadium is saturated with oxygen. If the metallographic examination reveals no second phase, then assume you have a good start in the direction of pure vanadium. Your next step is to determine the other metallic impurities. The optical spectroscopic method vapourizes some sample and records the wavelengths of the various photons emitted by the free atoms of the materials as the excited electrons drop back into their lower energy levels. This technique is not quantitative unless some standards are provided for comparison (a standard is a sample containing a known amount of impurity). The reason for the standard is that all elements are not excited as readily and the intensity of photons is not an exact measure of the amount present. However even in the absence of standards a competent spectroscopist can probably estimate the various quantities to within a factor of ten. As another check one can use X-ray fluorescence in which the K (1s) or L (2s, 2p) electrons are ejected from their shells. This is followed by the emission of X-rays of wavelength appropriate to the element when higher energy electrons drop into these empty states. This technique is quite good for elements of $Z > 22$ (Ti). The X-rays emitted by lower Z elements are strongly absorbed in the sample and in the air and the sensitivity of the technique is reduced. By surrounding the X-ray path with a chamber containing He, the air absorption can be removed and the technique can be used down to $Z \cong 13$. Nevertheless, this method also requires standards for quantitative analysis although a competent X-ray physicist can probably estimate the quantities to within a factor of ten. (If he is that competent, though, he is probably not interested in making such an estimate unless it is for his own work.) In any event, the above techniques will probably suffice for samples containing at least 0·1 % impurities.

Suppose we obtained a sample with a suggested impurity content 100 times less ($\sim$99·996 % purity) as revealed from optical microscopy, vacuum fusion, optical spectroscopy and X-ray fluorescence. You may still be desirous of extending the estimate of impurity concentration and more sophisticated methods could be tried. Firstly, the resistance ratio of the sample at 295°K and 4°K (or to just above T_c if it is a superconductor) is sensitive on a relative basis and increases in approximately inverse proportion to the impurity content. As a crude guide a ratio of room temperature resistance to liquid helium temperature resistance of about 1000 indicates an impurity content of $\sim$0·01 %. Secondly, the mass spectrograph could be used particularly for oxygen, nitrogen and hydrogen. The sample is vapourized and each atom ionized (i.e. an electron is removed) and accelerated through an electric potential. The atoms are then deviated in a magnetic field and can be separated and detected since atoms of different masses have different curvatures. Only a small number of atoms are required to give a detectable signal

TABLE VI

TECHNIQUES OF ANALYSIS

A Guide to various techniques of analysis for sample purity (1) and alloy content (2). The symbols M, C, P refer to metallurgist, chemist and physicist respectively indicating the principal contributors to its development.

1. PURE ELEMENTS (1)(2)—*Detection of Impurities*

Technique	Sensitivity	Comments
M Optical microscopy	Good for presence of second phases	Highly recommended as a general technique for determining presence of other phases and general condition of sample.
C Optical spectroscopy	Good for metallic impurities, poor for O, N, H, etc.	Semi-quantitative generally. A good technique when combined with others.
M X-ray spectroscopy (fluorescence)	Good for elements $Z > 22$ (Ti) Electron probe good for small areas	Very sensitive for elements $Z \cong 22$ to $\cong 50$. Incapable of detecting element of Z lower than ~ 13. Quantitative only with standards.
C Chemistry—wet analysis	Poor for low concentrations	Part of sample must be destroyed. Generally good for high concentrations such as alloy analysis.
P Resistivity ratio, room temp. to liquid He.	Very high for metals	An excellent technique for comparative purity. The higher resistivity ratio, the purer the metal. Does not indicate the specific impurity.
M X-ray diffraction	Fair for presence of second phase	Only detects the presence of a second phase, and then only to $\sim 1\%$.
P Neutron or reactor activation	High for some impurities	Not good for O, N, H and many other elements. Only good when impurity becomes much more radioactive than the pure material.
C Mass spectrograph	Very high for some impurities	Difficult to make quantitative except when used with enriched isotopes. Requires a good physical chemist to analyze the results.
P Nuclear reactions, proton, deuteron and α-bombardment	Good for some impurities	Requires expensive apparatus like a Van de Graaf accelerator. Still being developed. May prove to be good for O, N, H, etc.
C Vacuum fusion	Good in some cases, high in others	Technique currently in common use to detect O, N, H, etc., in metals. Quantitative in some cases.

TABLE VI—*continued*

2. ALLOYS—*Determination of Alloy Content and Homogeneity*

Technique	Sensitivity	Comments
M Preparational	Can be high	Always advisable to keep track of relative amount of elements used in preparation.
C Chemistry—wet analysis	Good for high concentrations	A good general-purpose technique
M X-ray diffraction	Good for concentrations $> 1\%$	Good technique for homogeneity (line shape). Good technique for solute concentration (lattice parameter).
M X-ray spectroscopy (fluorescence)	Good for elements $Z > 22$	Requires standards to be quantitative.
C Optical spectroscopy	Good for most elements	Quantitative only when standards are available. Not good for O, N, H, etc.
M Optical microscopy	Relatively insensitive to solute concentration	Good for detection of foreign phases.
C Vacuum fusion	Good in some cases for detection of O, N, H	Technique in general use for O, N and H as solute.
C Mass spectroscopy	Can be high	Requires standards and a good physical chemist.

(1) A rather complete list of American Producers of High Purity Metals is available for $1·50 by writing to the Office of Technical Services, U.S. Dept. of Commerce, Wash. 25, D.C. and requesting TID–4500.

(2) Recommended Reading is Chap. 2 (Preparation and Purification of Materials) *Methods of Exp. Physics*, Vol. 6—Academic Press, 1959.

($\sim 10^{-10}$ g) so the sensitivity can be high. Thirdly, if a reactor were available one might activate a piece of the sample. The radioactivity induced in the vanadium has a half life of 3·74 min so that after 5 hr only one atom in 10^{24} is still radioactive. This is sufficiently small so that impurities like manganese and cobalt with longer half lives could be detected by the energy and type of their radioactivity. In special cases this method can detect impurity concentrations down to $\sim 0{\cdot}0001\,\%$. These three techniques require a high initial purity for their success since their high sensitivity would give large and confusing backgrounds for impure samples.

B. Alloys

Analysis of Alloys. Suppose we are cognizant of the purity of our elements and we wish to make a specific alloy from them. We should like to know the alloy concentration of the samples after they have been prepared as well as their homogeneity (i.e. variation in concentration over the volume of the sample). Continuing with the vanadium-magnesium problem, we might, for example, prepare the alloy by first arc melting some pure vanadium in vacuum and weigh the sample before and after melting. Any loss in weight can be assumed to be due to vapourization and subsequent arc melting of the vanadium will result in the same percentage loss. If we now add a known quantity of magnesium (say $\sim 1\,\%$) and arc melt the mixture, the loss in weight can be proportioned between the magnesium and vanadium and the alloy concentration determined from the initial weights of magnesium and vanadium. This method of preparational weighing is not absolutely accurate because gases may be absorbed by the sample if the vacuum is not perfect, or the sample may react with the crucible or electrode, but it is a good starting point (the use of a vanadium electrode is a good idea). We can now do several things. Firstly, a chemist can dissolve some of the sample in acid and separate (and weigh) the vanadium and magnesium through various chemical reactions. This will generally give fair results and can be compared to the results of preparational weighing. If the chemist is supplied with specimens from different parts of the sample he can estimate the homogeneity, Secondly, we might use an X-ray lattice parameter measurement to determine the alloy concentration provided someone has made an independent determination of lattice parameter versus concentration (generally this is not the case), or we might use X-ray fluorescence if we had some standards. In the vanadium-magnesium system, however, neither of these techniques are suitable since the lattice parameters are not recorded in the literature* nor is magnesium suitable for X-ray fluorescence. Thirdly, we can provide an optical spectroscopist with a sample, but he too requires a standard and the problem of making a

* One can use Vegard's law as an approximation, i.e. the lattice spacing or cube root of the atomic volume varies linearly from pure vanadium to pure magnesium.

standard is identical to the problem we are now discussing (pulling oneself up by the bootstrap!). We shall probably have to make do with the chemist's wet analysis and the preparational estimates. As such you might send the chemist a second sample to determine the reproducibility of his determination!

In order to ascertain the presence of a second phase in our vanadium-magnesium sample, optical microscopy is probably the most convenient test. In addition it will reveal the general condition of the sample as a result of the melting process (dirt, slag, voids, etc.). X-ray diffraction can be used to detect the presence of a second phase, but generally at least 1% must be present. Since we are only adding 1% magnesium this is unlikely.

The next problem is one of homogeneity assuming that the wet chemistry and preparational weighing give good agreement and the metallographic examination does not reveal any second phase. The best technique to complement the chemist's determination of homogeneity is lattice parameter variation as determined from X-ray diffraction at high angles. If one chooses an X-radiation which diffracts at high angles any variation in lattice parameter (hence composition) over the surface of the sample will cause the X-ray line to broaden or shift by an angle.

$$\Delta\theta_B = (\Delta a_0/a_0)\tan\theta_B \tag{76}$$

where θ_B is the Bragg angle and a_0 the lattice parameter. In the vanadium-magnesium (1%) b.c.c. alloy one could use Cu $K\alpha$ and the (321) peak and obtain an estimate of uniformity to $\sim$0·1%. The X-ray line should be sharp and this can be improved by annealing, although the broadening due to variations in composition may not be eliminated by annealing.

The last problem concerns the preparation of the sample for the particular measurement to be performed. If the sample is ductile, and our vanadium-magnesium (1%) sample would probably be so, then all the normal operations of machining, rolling, swaging, etc., at room temperature will not alter the purity or analysis and may even improve the homogeneity. Such ductility should enable us to prepare specimens for most measurements. In the absence of room temperature ductility the preparation of specimens with specific dimensions presents a problem. It is always possible to obtain ductility in any metal by heating to a sufficiently high temperature and many fabricating operations on the sample can be performed at elevated temperatures with special equipment. If such is the case then an analysis of the sample should be made after the high temperature operation. (A technique for making resistivity measurements on irregularly shaped samples is discussed in the chapter on transport properties.)

As one can see, the problems of analysis are manifold and require considerable objectivity. We can do something, though, about improving the purity of a material supplied by a commercial company. But this is generally neither simple nor apparent for if it were, the supplier might have undertaken the task himself. In the case of metals, and vanadium is possibly typical, the only general method worth trying is the floating zone purification technique.[5] An induction coil or electron beam is slowly moved along a rod-shaped sample of metal which is in vacuum and a small fraction of the metal is melted, this molten zone passing from one end of the rod to the other. Impurities are generally more soluble in the liquid and the melted zone "floats" the impurities from one end to the other. When this method works, the end of the rod initially melted is purer and can be cut off and used. (N.B. This does not always work.) Other techniques like distillation and thermal decomposition fall into the category of pure empiricism.

In many experiments single crystals are required and their production frequently requires some witchcraft. The simplest solution is to buy them* but if they are unavailable one can attempt to grow them and so some reading would be desirable but not necessary.

In the main, metal crystals can be grown by slow cooling from the melt and this is generally not too difficult if one has pure single-phase materials. Crystals have also been grown by the strain anneal technique by which the metal is cold worked and then annealed, but the precise conditions for working and annealing always seem to be rather critical. The former method is generally more desirable, while the latter method can be used for substances that undergo phase changes between room temperature and the melting point, thus destroying the single crystal.

Problems

1. Make a list of the types of measurements you have made as a metallurgist, and indicate the information gained from each measurement. Taking each measurement in turn, would a knowledge of the electronic structure of the samples have been useful in the planning or the interpretation of the experiment?

2. An irregularly shaped homogeneous piece of metal with a thin coat of black paint and weighing about 25 g requires an analysis for metallic elements. If you must return the sample exactly as given to you (except for some slight wear of the paint on the surface) how would you make this analysis? Remember this may be an alloy! Can you think of an alloy that would defy such detection?

3. A rolling mill producing various types of metal foils (brass, lead, aluminium, iron) by continuous rolling, requires a continuously operating device for checking the uniformity in thickness. Devise methods for such inspection which are accurate to

(*a*) 10% in thickness;
(*b*) 1% in thickness;
(*c*) 0·1% in thickness.

Estimate the cost of each.

4. Determine the cost of stocking a research laboratory with one pound of each of the elemental metals in the purest form available commercially.

* 1. Metals Res. Ltd., 91 King Street, Cambridge, England.
2. Virginia Inst. for Scientific Research, 326 North Blvd., Richmond 20, Virginia, U.S.A.

Summary

Table Va lists the various experimental techniques to be discussed in this book and indicates the information to be gained from each technique.

The problems of knowing the atomic constitution of a sample are generally difficult to solve. There are no clear-cut solutions. They require considerable objectivity in their solution, specifically an assessment of the sensitivity of the measurement to the state of the sample. Samples are invariably less pure and less homogeneous than we measure.

CHAPTER VI

DIFFRACTION

What can be Learned from Diffraction

The three diffraction techniques, X-ray, neutron and electron, have specific advantages over each other and these advantages have been quite adequately explored. X-rays are scattered by the electrons surrounding each nucleus and provide the best of the three techniques for determining the electron probability distribution. The neutrons are primarily scattered by the nuclei and are only weakly scattered by the electrons except when the electrons are unpaired and produce magnetic fields on the atoms. Since the neutron is itself a magnet (magnetic dipole moment = $-1{\cdot}9135$ n.m.) it is scattered by the magnetic fields of the unpaired electrons. This enables us to determine the probability distribution of the unpaired electrons as well as the direction of the magnetic dipole moments relative to the crystal axes. In electron diffraction the electrons are scattered by the electrostatic fields of the atoms since the electron has a charge $-e$. While the electron also has a magnetic moment and is scattered by the magnetic fields of the unpaired electrons this magnetic scattering is so weak relative to the electrostatic scattering as to be unobservable.

All three techniques are capable of determining crystal structures and except for special cases, it is most sensible to employ X-ray diffraction for this purpose. The reason is that neutron diffraction is expensive and not readily available and electron diffraction is rather difficult in the matter of sample preparation and accurate intensity measurements. The special cases are compounds and alloys whose constituents differ greatly in Z as the X-rays then fail to reveal the lighter element relative to the heavy element. For such cases neutron diffraction is probably useful since its scattering properties are almost independent of Z, depending more on special properties of the individual nuclear species. (As electron scattering depends on Z in the same general way as X-ray scattering it runs into the same difficulty.) Another special case is the crystal structure of surface layers such as oxide coatings or any specimen too thin (< 100 Å) to give sufficient X-ray scattering. Electron diffraction is appropriate for such a problem since electrons are scattered much more strongly than X-rays.

Both neutrons and X-rays have been used for ascertaining the frequency spectra of the normal modes of crystals by measuring their dispersion curves (ν versus σ) since they are both capable of exciting one of these modes by

FIG. 37. The monochromating attachment to the Norelco unit designed and built by D. Chipman. The X-ray tube and LiF monochromator are on the left and the divergence slit is at the centre of the photograph. The powder specimen is mounted on a specimen spinner and the counter is just visible in the upper right corner. The distance from the left to the right end of the photograph is about two feet.

The inset shows the unit in operation (U.S. Army photograph).

transferring energy to it (or de-exciting one by absorbing energy). Both techniques have yielded good experimental results although the neutron technique has certain advantages.

We can separate the problems attacked by diffraction techniques into three broad categories.

(A) Crystal structures, position of atoms, position and magnitude of atomic magnetic moments, sizes of atoms.

(B) Unpaired and paired electron probability distributions.

(C) Frequency spectra and dispersion of the normal modes of lattice vibrations and magnetic excitations.

We shall discuss the application of the three diffraction techniques to these three categories of problems.

X-RAYS

Equipment

X-ray diffraction equipment is available commercially and is fairly reliable. For routine work it can be used as received from the supplier, but any reasonably accurate intensity work requires monochromatic radiation and such adaptation is not available commercially. The next photograph (Fig. 37) reveals the requirements for making this adaptation on the Norelco equipment. The X-ray beam produced by the electron bombardment of a cathode (like copper) in an X-ray diffraction tube contains a background of wavelengths superimposed on the $K\alpha$ and $K\beta$ lines.* The intense $K\alpha$ line can be selected by diffraction from a crystal like LiF bent plastically to focus the beam.

The arrangement in Fig. 37 is shown schematically in Fig. 38. Such an arrangement yields fairly high intensity monochromatic X-rays capable of resolving peaks fifteen minutes apart. The intensity of monochromatic X-rays impinging on the sample is $\sim 2 \times 10^7$/sec. For high resolution work the diffractometer can be used without this attachment thus sacrificing beam purity, but in most cases these two methods complement each other rather well.

A. Crystal Structures, Atomic Positions and Atomic Sizes

Crystal Structure Determinations

The problem of determining the crystal structure of a substance utilizing X-ray diffraction has been investigated quite extensively. If one has need to

* The $K\alpha$ line is emitted when a $2p$ electron drops into the hole left by an ejected $1s$ electron, the $K\beta$ when a $3p$ electron falls into the hole. The $2s$, $3s$ or $3d$ electrons cannot fall into this hole because the X-ray carries off one unit of angular momentum $\hbar\sqrt{l(l+1)}$ where $l = 1$. To conserve this angular momentum the electron which fills the $1s$ hole changes its angular momentum by one unit from 1 to 0, or p to s.

attack such a problem a good reference is the book by H. P. Klug and L. E. Alexander, *X-ray Diffraction Procedures*, John Wiley and Sons, N. Y., (1954). We shall cover only those aspects pertinent to the more general metallurgical interest.

The intensity of Bragg reflections is proportional to the square of the absolute magnitude of the structure factor, F,

$$I \propto |F|^2 \equiv FF^*$$

$$F = \sum_j f_j \, e^{2\pi i(hx/a + ky/b + lz/c)} \tag{77}$$

where a, b, c are the lengths of the sides of the X-ray unit cell, x, y, z the positions of the atoms in the X-ray unit cell, and h, k, l the so called *Miller indices* which collectively designate a particular diffracting plane. f_j is the atomic scattering factor of atom j in the unit cell and depends principally on the electron probability distribution of the atom. The structure factor represents the interference between the various atoms in the crystal since the wave character of the X-rays determines the diffraction pattern.

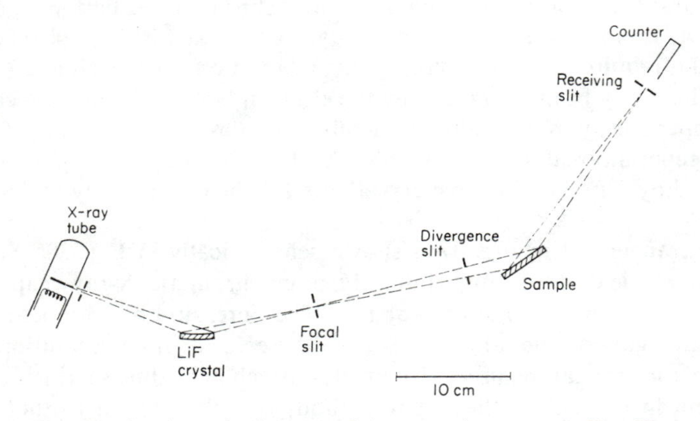

FIG. 38. A schematic diagram of the monochromator in Fig. 37.

Once the structure factor is known the positions of the atoms in the unit cell are known and the crystal is a repetition of the unit cells. For each of the diffracting planes the Bragg conditions require the angle of incidence of the X-ray to be equal to the angle of reflection. This angle, θ_B, is called the *Bragg angle* and is given by the following expressions for several crystal structures of interest to the metallurgists

$$\sin \theta_B = \frac{\lambda}{2a_0}(h^2 + k^2 + l^2)^{1/2} \tag{78a}$$

cubic $(a_0 = a = b = c \text{ in } (77))$

$$\sin\theta_B = \lambda\left(\frac{h^2 + hk + k^2}{3a_0^2} + \frac{l^2}{4c_0^2}\right)^{1/2} \qquad (78b)$$

hexagonal $(a_0 = a = b;\ c_0 = c$ in (77))

$$\sin\theta_B = \frac{\lambda}{2}\left(\frac{h^2}{a_0^2} + \frac{k^2}{b_0^2} + \frac{l^2}{c_0^2}\right)^{1/2} \qquad (78c)$$

orthorhombic $(a_0 = a;\ b_0 = b;\ c_0 = c$ in (77))

$$\sin\theta_B = \frac{\lambda}{2}\left(\frac{h^2 + k^2}{a_0^2} + \frac{l^2}{c_0^2}\right)^{1/2} \qquad (78d)$$

tetragonal $(a_0 = a = b;\ c_0 = c$ in (77))

In crystal structure determinations one first "indexes a pattern", i.e. tries to account for the angles of all the reflections by trying expressions like those in eqn. (78) until one obtains a fit for some specific values of the lattice parameters a_0, b_0 and c_0. Bragg reflections may not be present for all combinations of integral numbers for h, k, l, but that does not affect the indexing so long as there are no unaccounted for reflections. Of course, if two or more phases are present in your sample the problem of indexing tends to be rather messy. Once the pattern is indexed you then determine which combinations of h, k, l are missing (they would be missing because the structure factor, F, was zero for this combination),* and look into the *International Tables for X-ray Crystallography.*[6] This tells one the position x, y, z of the atoms in the unit cell. In some cases, the positions are given exactly as some specific fractions of a_0, b_0 and c_0 while in others there is an additional parameter. Once we know a_0, b_0 and c_0 and the positions x, y, z for each atom our last step is to substitute these values into the structure factor, F, eqn. (77) for each allowed value of h, k, l, determine the intensity of each reflection, and compare it to the experimental observations. If there is an adjustable parameter in the positions of the atoms then one adjusts it until the intensities agree. While the atomic scattering factor, f, must be known in order to determine the structure factor and this requires a knowledge of the electron probability distribution, only approximate electron probability distributions are generally necessary. Hartree or Hartree–Fock free atom electron probability distributions or some other approximation is generally adequate and the atomic scattering factors obtained from them are available in the literature.[7][8] If the crystal

* For example, in the simple structures the following rules tell us the "allowed reflections":

Face-centered cubic h, k, l	All even or all odd (0 is even)
Body-centered cubic k, k, l	Sum even ($h + k + l$ even)
Hexagonal close packed h, k, l	All h, k, l except if $h - k = 0, 3, 6, 9,$ etc., then l must be even

The positions of the atoms in the unit cell of these structures are:

f.c.c. $0, 0, 0;\ \frac{1}{2}, 0, \frac{1}{2};\ 0, \frac{1}{2}, \frac{1}{2};\ \frac{1}{2}, \frac{1}{2}, 0$ — 4 atoms $a = b = c = a_0$

b.c.c. $0, 0, 0;\ \frac{1}{2}, \frac{1}{2}, \frac{1}{2}$ — 2 atoms $a = b = c = a_0$

where the three numbers for each atom are values of x/a_0, y/a_0, z/a_0 respectively

h.c.p. $0, 0, 0;\ \frac{1}{3}, \frac{2}{3}, \frac{1}{2}$ — 2 atoms $a = b = a_0;\ c = c_0$

where the three numbers for each atom are values of x/a_0, y/a_0, z/c_0 respectively.

structure is very complicated so that it becomes impossible to resolve all the diffraction lines (the more complicated the structure the more diffraction lines) or the intensities of some of the lines are too weak to observe, it may be necessary to use single crystals. The vast bulk of crystal structure determinations of interest to metallurgists has been compiled by W. B. Pearson.[9]

In spite of the vast literature on crystal structure determinations, the theoretician has not been helped much in solving the Schrödinger equation for a crystal. The reason is that the difference in eigenfunctions and eigenvalues for various metallic crystal structures of a particular element is at least a factor of ten smaller than the accuracy of his calculations. Even the vast quantity of information concerning the arrangement of atoms surrounding a given atom in a metal crystal has not provided the clues to which orbitals are used even though the combinations of orbitals is limited to those which have the local symmetry of the crystal. Some success has been achieved for non-metals, but the mixtures of orbitals always present in metals has left the problem unsolved.

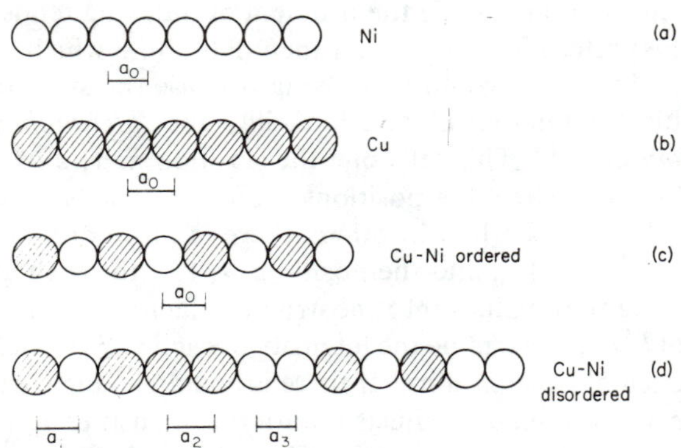

FIG. 39. A schematic illustration of the packing of atoms along a cube edge. *a* and *b* are the regular packing in the pure metals copper and nickel and *c* is the regular packing in the 50–50 alloy if it were ordered. In all three cases there is a unique distance between atomic centres but in the disordered alloy *d* at least three distances are involved—Cu-Ni, a_1; Cu-Cu, a_2; Ni-Ni, a_3.

Atomic Sizes in Alloys

Once the crystal structure is solved and the arrangement of atoms is known we can turn our interest to the size of atoms in crystals or more precisely how closely atoms approach each other. Presumably, this is related to the overlap of the outer electrons. In a pure element of known crystal structure the distance between atoms can be determined from the lattice parameters. For example, in b.c.c. structures the nearest neighbour distance (distance between atomic centres) is $\sqrt{(3)}a_0/2$. Even in an ordered alloy the distance can be determined from the lattice parameter. However, in a disordered alloy the situation is considerably more complicated as is revealed in Fig. 39

for copper and nickel in their pure states and in an alloy. In the pure metals and ordered alloy (if it existed) the unit of distance is a_0 but in the disordered alloy there are at least three; a_1 the copper-nickel distance; a_2 the copper-copper distance; and a_3 the nickel-nickel distance, and even these distances may fluctuate depending on the local surroundings of each atom. With such a mixture of lattice spacings one may wonder what the X-ray diffraction pattern looks like. One still obtains Bragg reflections and these occur at angles corresponding to the lattice spacing averaged over approximately several thousand atomic distances. For example, in the cupronickel alloy one starts at any atom, proceeds several thousands atom distances along the cube-axis, measures the total distance and divides by the number of atoms, less one.

If the electron overlap was unaffected in the alloying process and was merely the sum of the overlaps of the pure elements we would expect the average lattice parameter of the alloy to be merely the sum of the lattice parameters

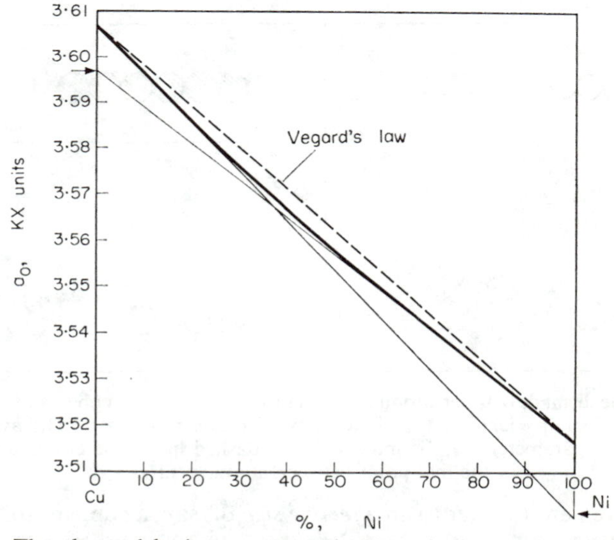

FIG. 40. The observed lattice parameters in KX units for the f.c.c. copper-nickel system (solid line) and Vegard's law (dashed line). The arrows at 3·597 and 3·500 indicate the intercept of the tangent lines for small dilution and determine the effective sizes of the solute atoms in the dilute alloys. (1 Å = 1·002 KX units).

of the pure elements weighted by their abundances in the alloy. In effect this means the copper-copper and the nickel-nickel distances are the same in the alloy as in the pure metals and the copper-nickel distance is the average of the nickel-nickel and copper-copper distance. If this occurs it is called *Vegard's law*. Figure 40 gives the observed lattice parameters of the f.c.c. copper-nickel system, the straight line connecting the pure copper lattice parameter 3·603 Å to that of pure nickel 3·517 Å being that expected from Vegard's law. Since the atoms appear closer together on the average than expected by Vegard's law the electron overlap has increased slightly in the alloy. In detail, how does this happen? If we draw tangents to either end of the observed lattice parameter curves these reveal the effect of small percentages of copper in nickel and nickel in copper. At either end of the curve, for example, 1% copper in nickel or 1% nickel in copper, a typical array of atoms along a cube-axis might appear as in Fig. 41. Small percentages have been chosen so that the solute atoms are not too close. The models in Fig. 41 are a great oversimplification since the problem must be considered in three dimensions. Furthermore, we may find marked differences along different crystalline directions. The difficulty is that neither the experimental technique nor the theory are sufficiently advanced to enable us to draw accurate curves like those in Fig. 41 for *all directions*. We only know that the average lattice parameter in *3 dimensions* must correspond to that observed by the X-rays.

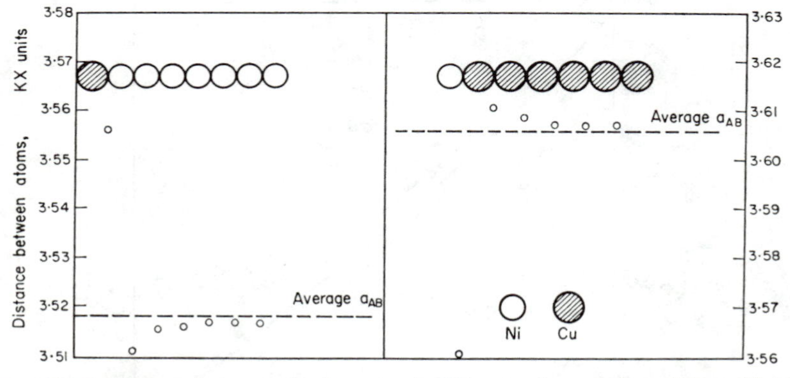

FIG. 41. The distance between atomic centres in KX units along a cube-axis for a dilute alloy of copper in nickel and a dilute alloy of nickel in copper. The average or X-ray lattice parameter, a_{AB}, is indicated as a dashed line. The circles are placed at the mid-point between atomic centres.

It is convenient to refer to an *effective size* of, say, a copper atom in nickel not as the distance from the copper to the nickel but as the averaged effect on the lattice parameter of the addition of small amounts of copper to nickel. For example, if the addition of 1% copper in nickel changed the lattice parameter from 3·5170 to 3·5178 Å then 0·99(3·5170 Å) + 0·01(a_{Cu}) = = 3·5178 Å, yielding an effective size corresponding to a_{Cu} = 3·597 Å. The effective size of copper in nickel can also be obtained from the value of the intercept of the tangent line with the lattice parameter axis at 0% nickel marked in Fig. 40. Of course the same is true for nickel in copper. As we increase the concentration of the solute so that solute atoms become second nearest or even nearest neighbours the size effect is considerably more complicated since one cannot state to what extent the copper atoms have increased the lattice parameter and the nickel atoms have decreased the parameter. In principle, though, X-rays can tell us this.

However, since this has not been done for copper-nickel we shall consider the case of copper-gold. The lattice parameters for this f.c.c. system are shown in Fig. 42 and are seen to be the reverse of copper-nickel in that they show a positive deviation from Vegard's law. From the intercepts of the tangents drawn at each end of the curve we see that copper dissolves in gold with an effective size corresponding to a lattice parameter $a_{Cu} = 3{\cdot}67$ Å compared to $a_0 = 3{\cdot}603$ Å for pure copper, while gold dissolves in copper with an effective size corresponding to $a_{Au} = 4{\cdot}17$ Å compared to $a_0 = 4{\cdot}07$ Å for pure gold. In using X-rays to obtain the values of the effective sizes one measures the

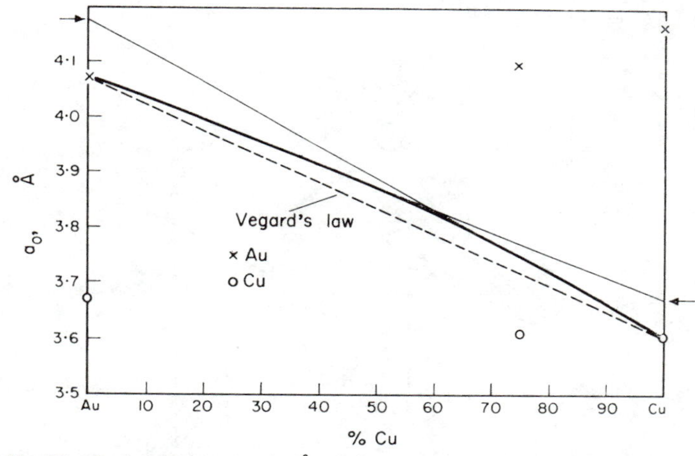

FIG. 42. The lattice parameter in Å of f.c.c. copper-gold alloys, solid line, and Vegard's law, dashed line. The arrows indicate the effective sizes of the solute atoms in the dilute alloys as determined from the intercept of the tangent lines at small concentrations. The crosses, x, and the circles, O, indicate the variations of the effective size of gold and copper respectively at three concentrations. The values at Cu_3Au were determined from X-ray measurements of Borie.

intensity of the Bragg peaks which are reduced by a factor $e^{-2M'}$. For cubic alloys the expression for $2M'$ is

$$2M' \cong 18{\cdot}9 \frac{m_A}{m_B} (\Delta a_A)^2 \left(\frac{\sin\theta}{\lambda}\right)^2 \quad \text{f.c.c. alloy}$$

$$2M' \cong 30{\cdot}2 \frac{m_A}{m_B} (\Delta a_A)^2 \left(\frac{\sin\theta}{\lambda}\right)^2 \quad \text{b.c.c. alloy} \qquad (79)$$

where m_A and m_B are the atomic fractions of elements A and B, a_{AB} the average or observed X-ray lattice parameter, θ the Bragg angle, λ the wavelength and $(\Delta a_A)^2 = (a_A - a_{AB})^2$ (a_A is the effective lattice parameter of element A in the alloy). If a_A for the A atoms is larger than the average, a_{AB}, then the B atoms must be smaller so that

$$m_A \Delta a_A + m_B \Delta a_B = 0 \tag{80}$$

The factor $e^{-2M'}$ reducing the intensity of the Bragg peaks arises from the fact that the atoms are not at exact lattice points so they are not exactly in phase. In order to measure this reduction in intensity we must know the intensity in the absence of this factor, i.e. if all the atoms were the same size so that $\Delta a_A = \Delta a_B = 0 = 2M'$ and $e^{-2M'} = 1$. We can determine this in several ways, although we are not certain that we understand all the experimental difficulties due to the sparsity of this type of work.

1. If the alloy forms a perfectly ordered structure such as Cu_3Au (Fig. 43) then all the atoms are on the mathematical lattice sites and are in phase. In this case $2M' = 0$ and $e^{-2M'} = 1$.

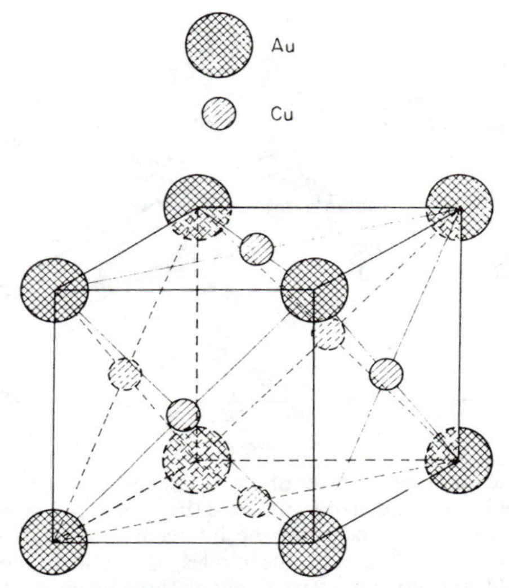

FIG. 43. The ordered Cu_3Au structure. Each copper atom has eight copper and four gold nearest neighbours and each gold atom has twelve copper nearest neighbours.

If the same alloy is disordered many factors are cancelled in the ratio of the intensities of a particular Bragg reflection of the disordered and ordered alloy. Except for minor corrections which can be made (i.e. lattice parameter changes) this intensity ratio is equal to $e^{-2M}e^{-2M'}$(disordered)/e^{-2M}(ordered) where the factors e^{-2M} arise from the thermal vibrations of the atoms (to be discussed on page 158) and $e^{-2M'}$ is that due to size effects in the disordered alloy. In the case of Cu_3Au the factors e^{-2M} for the ordered and disordered alloy were deduced by Borie[10] from other measurements and $e^{-2M'}$ from the X-ray intensity ratio. In this particular disordered alloy the effective size of the gold atom was 4·10 Å and 3·61 Å for copper. The effective size of the gold and copper atoms are shown at three concentrations in Fig. 42 and indicate that the sizes of the atoms change by less than a few percent over the entire alloy range. In fact a crude estimate of the factor $e^{-2M'}$ for an alloy can be made by using the pure metal atomic sizes as an approximation. As an example, in disordered copper-gold (50–50) the 331 peak is reduced by $\sim 30\%$ due to size effects (as compared to 27% due to thermal motion).

The method of Borie for determining $e^{-2M'}$ depends on the presence of a completely ordered phase. Since this is not common in alloys the method has limited applicability. It does have the advantage, though, that the factor e^{-2M}(disordered)/e^{-2M}(ordered) for the ratio of thermal factors is close to unity.

2. A second method applicable to alloys exhibiting a single phase solid solubility over its entire range, like copper-gold, cupronickel, cobalt-nickel, iron-chromium, titanium-hafnium, etc., consists in measuring a particular Bragg reflection for the alloy and for the pure constituents. The ratio, R, of the intensities of the Bragg peak of the alloy to the weighted sum of the same Bragg peaks of the pure elements is for cubic crystals

$$R = \frac{I_{AB}}{m_A I_A + m_B I_B} = \frac{\dfrac{(m_A f_A e^{-M_A} + m_B f_B e^{-M_B})^2 e^{-2M'}(1+\cos^2 2\theta_{AB})}{\mu_{AB} a_{AB}^6 \sin^2\theta_{AB}\cos\theta_{AB}}}{\dfrac{m_A (f_A e^{-M_A})^2(1+\cos^2 2\theta_A)}{\mu_A a_A^6 \sin^2\theta_A \cos\theta_A} + \dfrac{m_B (f_B e^{-M_B})^2(1+\cos^2 2\theta_B)}{\mu_B a_B^6 \sin^2\theta_B \cos\theta_B}} \tag{81}$$

$$\mu_{AB} = \frac{\mu_A \dfrac{\rho_{AB}}{\rho_A} + \dfrac{m_B}{m_A}\dfrac{A_B}{A_A}\mu_B \dfrac{\rho_{AB}}{\rho_B}}{1 + \dfrac{m_B}{m_A}\dfrac{A_B}{A_A}} \tag{82}$$

$$a^6 = (\text{X-ray lattice parameter})^6 \rightarrow a^4c^2 \text{ for h.c.p. metals}$$

where the subscripts A, B and AB refer to the pure elements and the alloy respectively, μ is the linear absorption coefficient, A the atomic weight and ρ the density. The pure elements, of course, do not have a size effect factor. This technique has the advantages that any uncertainty in μ, f, or e^{-M} is partially cancelled in taking the ratio and it can be used for the whole range of alloys. Furthermore, the uncertainty in the temperature factor can be reduced considerably by making the measurements at low temperature. Not much work has been done with this technique which is probably quite good for dilute alloys. Some experimental difficulties concerning sample preparation, methods for measuring the intensity, etc., will be discussed in the next section, page 136.

3. If one knows the temperature factor e^{-2M} approximately from some other measurements like specific heat or elastic constants or else makes the measurement at low temperatures the size factor $e^{-2M'}$ can be determined from absolute measurements (p. 136). Since size effects are quite large for most alloys of concentration $> 20\%$ uncertainties in the atomic scattering factors and absorption coefficients can be kept small. This method has more general applicability than method 2, since it does not require complete solid solubility. In fact, it can also be done for a phase not present in the pure elements.

Short Range Order

The measurement of $e^{-2M'}$ only yields the effective size of the atoms, which is an average of the distances a_1, a_2, a_3 in Fig. 39. In order to determine a_1, a_2 and a_3 we must measure the angular distribution of the intensity removed from the fundamental Bragg peaks due to the size effect factor $e^{-2M'}$ or else measure the short range order. For the most part this intensity appears at angles other than the Bragg peaks and is called *the diffuse scattering*. However no work has been done on the diffuse intensity removed from the Bragg peaks so we shall limit our discussion to *short range order*. If the atoms in an alloy prefer to surround themselves with the opposite species this leads to short range order while if the atoms prefer to surround themselves with their own species this leads to *segregation*.* Short range order and segregation give rise to diffuse scattering which is rather intimately connected with the diffuse scattering due to size effects, so we shall treat them together.

Short range order and segregation are specified in terms of short range order coefficients α_i which are defined as follows. Consider, for example, a f.c.c. cupronickel alloy (70% Cu 30% Ni). Let us count the number of nickel and copper atoms surrounding each nickel atom. Since there are twelve nearest neighbours we might find a series of numbers like 8 Cu, 4 Ni; 7 Cu, 5 Ni; 8 Cu, 4 Ni; 9 Cu, 3 Ni; 8 Cu, 4 Ni; 7 Cu, 5 Ni. etc. If we did this for all the nickel atoms and took the average, we might find it to be 8·1 Cu, 3·9 Ni or 8·1/12 = 0·675 for the probability of finding a copper as nearest neighbour to a nickel. α_1 is then defined as

$$\alpha_1 = 1 - \frac{p_1(\text{Cu})}{m_{\text{Cu}}} = 1 - \frac{0{\cdot}675}{0{\cdot}700} = 0{\cdot}022 \tag{83}$$

where $p_1(\text{Cu})$ is the probability of finding a copper atom as nearest neighbour to a nickel atom and m_{Cu} is the atomic fraction of copper atoms in the sample. α_2 is defined as one minus the probability of finding a copper atom as second nearest neighbour to a nickel or one minus $p_2(\text{Cu})$ divided by the atomic

* Also called clustering.

fraction of copper (0·7). If $p_1(\mathrm{Cu}) = p_2(\mathrm{Cu}) = p_3(\mathrm{Cu}) = p_4(\mathrm{Cu})$, etc., are all equal to $m_{\mathrm{Cu}} = 0{\cdot}7$ then we should have a completely disordered crystal and all the α_i would equal zero. If the probability of finding a copper atom as nearest neighbour to a nickel atom was 0·75 then α_1 would be $\alpha_1 = 1 - 0{\cdot}75/0{\cdot}70 = -0{\cdot}071$. When α_1 is negative the atoms prefer to surround themselves with their opposite species and we have short range order while a positive value of α_1 (segregation or clustering) indicates that the atoms prefer to surround themselves with their own species. The effect of the short range order (or segregation) and the size effect in an alloy is to give rise to X-ray diffuse intensity, I_D (counts/sec), which for conditions of symmetrical reflection (incident beam angle and diffracted angle are the same) is:

$$I_D = N_0 m_A m_B (f_B - f_A)^2 \left(\frac{e^2}{m_0 c^2}\right)^2 \left(\frac{1+\cos^2 2\theta}{2}\right) \frac{I_0}{2\mu}\Omega \times \left[\sum_i \left\{ Z_i \alpha_i \frac{\sin sr_i}{sr_i} - Z_i \beta_i \left(\frac{\sin sr_i}{sr_i} - \cos sr_i\right)\right\}\right] \quad (84)^*$$

where N_0 is the number of atoms /cm^3 calculated from the lattice parameter, Ω the solid angle subtended by the counter, Z_i the number of neighbours in the ith shell ($i = 1$ for nearest neighbours, $i = 2$ for next nearest neighbours, etc., $s = 4\pi \sin\theta/\lambda$, θ = angle of incidence, α_i the short range order parameter for the ith shell, e^2/m_0c^2 the electron scattering amplitude ($0{\cdot}282 \times 10^{-12}$ cm), I_0 the incident monochromatic intensity (c/s) measured with the same detector measuring I_D, μ_{AB} the linear absorption coefficient of the alloy and β_i the size effect parameter,

$$\beta_i = \frac{f_A}{f_B - f_A}\left\{\frac{(m_B/m_A + \alpha_i) f_A}{(f_B)} \frac{r^i_{BB} - r_i}{r_i} - \left(\frac{m_A}{m_B} + \alpha_i\right)\frac{r^i_{AA} - r_i}{r_i}\right\} \quad (85)$$

where r^i_{AA} is the distance between an A atom and an A atom in the ith shell and r_i the same distance calculated from the measured lattice parameter, and similarly for r^i_{BB}. The remaining distance $r^i_{AB} = r^i_{BA}$ can be determined from the experimental lattice parameter since

$$\frac{m_A(1 - p'_i) r^i_{AA}}{r_i} + \frac{(m_A p'_i + m_B p_i) r^i_{AB}}{r_i} + \frac{m_B(1 - p_i) r^i_{BB}}{r_i} = 1 \quad (86)$$

where p_i is the probability of finding an A atom at a distance r_i from a B atom and p'_i the probability of finding a B atom at a distance r_i from an A atom ($\alpha_i = 1 - p_i/m_A$).

The short range order and size effects are not easily separated in eqn. (84), and to compound the difficulty two other contributions are present in the diffuse background, Compton scattering and thermal diffuse scattering. In addition, I_0 is too large to count directly if I_D is large enough to detect above

* Equation 84 only accounts for the size effects on the scattering associated with short range order. The intensity removed from the fundamental Bragg peaks has not been measured to date. It would presumably appear near the fundamental peaks.

cosmic ray background. All three of these problems will be discussed further. In practice, one attempts to fit the experimental data on diffuse scattering by adjusting the β_i and α_i to give the best fit, having first attempted to correct for the Compton scattering and thermal diffuse scattering.

Very little experimental information of this type is available so we are unable to deduce very much about the wave functions of alloys. However some general remarks can be made about the theory. You may recall that in discussing the solutions of the Schrödinger equation for a periodic potential in Chapter II that the basis of the Bloch theorem was the *exact periodicity* of the potential. In a disordered alloy the potential is not identical for each atom nor is it precisely periodic due to the size effect. Both these effects complicate the solutions considerably. If the size effect were small and the potential of the two atoms not very different (Cu-Au alloys are a good example) then we expect the general details of the solution for a perfect lattice would still exist. We might expect the energy gaps to become fuzzy as in a liquid and the Fermi surface to become more diffuse but until either the solutions of the Schrödinger equations for alloys are improved and/or the experimental information is increased considerably we must leave this subject to future consideration (cf. chap. XII).

A fairly crude estimate of the size effect and strain energy introduced by this size effect can be made as follows. Let us assume that the energy required to contract or expand an A atom, when it is introduced into a B lattice, is the same as for an A atom in its own A lattice. This energy is calculable from the compressibility of the A lattice. Let us also assume that the contraction or expansion of the B lattice with the introduction of an A atom can be calculated from the shear modulus of the alloy. This represents the two opposing forces if the atoms are not the same size, for if the A atom is larger it is compressed and exerts an outward force on the B lattice and vice versa. The final size of the A atom is determined by equating the forces. If $\mathcal{R}_{AB}$ is the average size of an atom in the alloy as deduced from the X-ray lattice parameter, $\mathcal{R}_A$ the size of the A atom in the alloy and $\mathcal{R}_0$ the size of the A atom in the pure A lattice then

$$\begin{aligned} \frac{\mathcal{R}_{AB} - \mathcal{R}_A}{\mathcal{R}_A - \mathcal{R}_0} &= \frac{3\kappa_A \mathcal{R}_0}{4G_{AB}\mathcal{R}_{AB}} = \delta_A \qquad \text{for atom } A \\ \frac{\mathcal{R}_{AB} - \mathcal{R}_B}{\mathcal{R}_B - \mathcal{R}_0} &= \frac{3\kappa_B \mathcal{R}_0}{4G_{AB}\mathcal{R}_{AB}} = \delta_B \qquad \text{for atom } B \end{aligned} \tag{87}$$

where κ_A, κ_B are the bulk moduli of pure elements A, B and G_{AB} the shear modulus of the alloy. The total strain energy per mole is

$$\text{Strain energy} = N_0(m_A W_A + m_B W_B)$$

$$W_A = \frac{3\pi\kappa_A \mathcal{R}_0(\mathcal{R}_0 - \mathcal{R}_{AB})^2}{4(1 + \delta_A)} \qquad \text{also for } B \tag{88}$$

where N_0 is Avogadro's number. These expressions should only be used as semi-quantitative guides but are probably good to within a factor of two.

Long Range Order

We now turn to long range order, the last measurement we shall discuss concerning the X-ray determination of atomic positions. An example of this is Cu_3Au (Fig. 43.) When this alloy orders the X-ray diffraction pattern exhibits those additional lines missing according to footnote, p. 121 (h, k, l mixed between odd and even). These are called *superlattice lines.* The structure factor can be calculated from the position of the atoms in the unit cell which are

ordered	Au	0, 0, 0	
Cu_3Au			4 atoms
f.c.c.	Cu	$\frac{1}{2}, 0, \frac{1}{2}; 0, \frac{1}{2}, \frac{1}{2}; \frac{1}{2}, \frac{1}{2}, 0.$	

Another well-known type of ordering occurs in the body-centered cubic lattice and is known as the CsCl structure shown in Fig. 44 for FeCo (50–50).

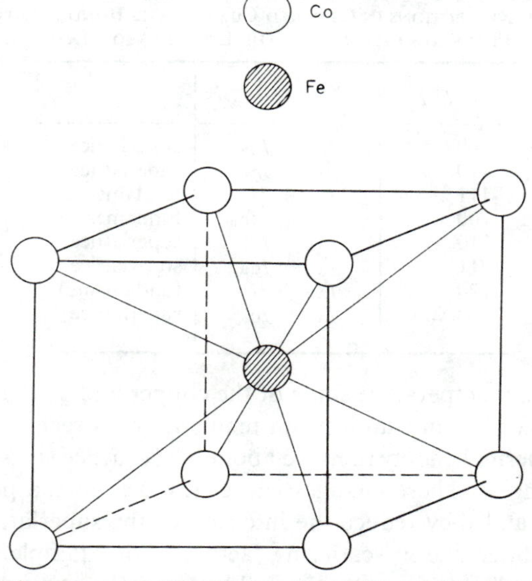

FIG. 44. The ordered iron-cobalt structure. Each atom has eight nearest neighbours of the opposite species.

The positions of the atoms for determining the structure factors are

ordered	Fe	0, 0, 0	
FeCo			2 atoms
b.c.c.	Co	$\frac{1}{2}, \frac{1}{2}, \frac{1}{2}$	

We have no way of guessing which alloys will order nor is there any systematic behaviour amongst the elements as to which do not. *The energy difference between ordered and disordered alloys is generally no more than 0·1 eV per atom, considerably less than the cohesive energy and considerably smaller than can be calculated.* The existence of order has been of interest to the metallurgist but its unsystematic occurrence in the periodic table and its small energy difference has not afforded the theoretician much insight into the nature of the solutions of the Schrödinger equation for the ordered and disordered alloy. However, from the point of view of the arrangement of the atoms in a crystal we do have some understanding of the transition of a crystal from an ordered to a disordered state. We shall consider the case Cu_3Au as an illustration since some measurements are available.

In the perfectly ordered state at $T = 0°K$, Fig. 43, the X-ray diffraction pattern consists of the diffraction lines and respective structure factors listed in Table VII. Except for geometrical factors the superlattice lines are weaker by a factor of about 10 in intensity ($I \propto |F|^2$).

TABLE VII
MILLER INDICES AND STRUCTURE FACTORS OF THE FIRST NINE REFLECTIONS IN ORDERED Cu_3Au. THE FUNDAMENTAL PEAKS ALSO APPEAR IN THE DISORDERED STATE

h, k, l	F	
100	$f_{Au} - f_{Cu}$	superlattice
110	$f_{Au} - f_{Cu}$	superlattice
111	$f_{Au} + 3f_{Cu}$	fundamental
200	$f_{Au} + 3f_{Cu}$	fundamental
210	$f_{Au} - f_{Cu}$	superlattice
211	$f_{Au} - f_{Cu}$	superlattice
220	$f_{Au} + 3f_{Cu}$	fundamental
300, 221	$f_{Au} - f_{Cu}$	superlattice

As we raise the temperature some of the copper and gold atoms will want to exchange places but not until room temperature is reached do the atoms have sufficient thermal energy to move about in the lattice so as to appreciably effect these changes. These changes cause mistakes in the perfect arrangement of Fig. 43 and they reduce the intensity of the superlattice lines which depend on the difference in scattering factors. For example, if 10% of the copper atoms trade places with the gold atoms the structure factor of a superlattice line becomes $F = (0{\cdot}9f_{Au} + 0{\cdot}1f_{Cu}) - (0{\cdot}9f_{Cu} + 0{\cdot}1f_{Au}) = 0{\cdot}8$ $(f_{Au} - f_{Cu})$. The fraction by which the structure factor is reduced from that of perfect order is called the *long range order parameter* denoted by S ($= 0{\cdot}8$ in this case). The intensity is reduced by a factor S^2 so that the interchange of 10% of the gold atoms with copper atoms reduces the intensity 36% $(1 - S^2 = 1 - 0{\cdot}64 = 0{\cdot}36)$. The intensity lost from the superlattice lines appears in the diffuse background but its precise angular dependence depends on the particular arrangement of the gold and copper atoms that have exchanged. Both theoretically and experimentally we know that the mistakes do not occur by a random switching of gold and copper atoms but rather occur in such a fashion as to preserve as many copper-gold nearest neighbours as possible. In the perfectly ordered lattice each gold atom has only copper atoms as nearest neighbours. If we switch one gold and copper atom this particular gold atom will have 3 gold and 9 copper nearest neighbours.* As the crystal disorders further, the tendency would be for one of these three gold atoms to make the next switch rather than a gold atom further away. By so doing, it has increased the number of copper nearest neighbours to 10 for the first gold atom and so produces more copper-gold nearest neighbours. This preference for the atoms to prefer their opposite types gives rise to broad peaks in the diffuse background around Bragg superlattice peaks. As the temperature is increased further the long range order parameter is continually reduced until it reaches $S = 0{\cdot}8$ at $\sim 395°$C. As we heat the crystal further the long range order parameter discontinuously changes to zero since the free energy of the disordered phase becomes equal to the partially ordered phase ($S = 0{\cdot}8$) at $T_c \sim 395°$C, the *critical temperature.* While the superlattice lines disappear for $S = 0$ there is still considerable short range order which contributes to the broad peaks in the diffuse background. Finally, at $T \gg T_c$ all short range order disappears ($\alpha_0 = 1$, $\alpha_i = 0$) and the diffuse scattering (eqn. (84)) peaks in the forward direction (for $i = 0$, $r_0 = 0$ and $\sin sr_0 = 1$), the intensity being given by

$$I_D = N_0 m_A m_B (f_B - f_A)^2 \left(\frac{e^2}{m_0 c^2}\right)^2 \left(\frac{1 + \cos^2 2\theta}{2}\right) \frac{I_0}{2\mu} \Omega \tag{89}$$

limit of eqn. (84) for $\alpha_i = 0$ and $\beta_i = 0$.

Schematic diffraction patterns are given in Fig. 45 for Cu_3Au at $T = 0°$K, $T = 380°$K, $T = 395°$K and $T \cong 800°$K. These patterns are ordinarily superimposed on the patterns due to size effects but the latter have been omitted in Fig. 45.

Table VIII presents a list of known ordering in alloys.

Theory of Ordering

Some theoretical work has been attempted in calculating the variation of S and α_i with temperature.[11] The simplifying assumptions that are made are that only nearest neighbour interactions are considered, that the potential,

* In analogy to spin waves, quantum mechanics tells us that the first excited state is not one in which some particular gold atom exchanges with some particular copper atom but rather a "disorder wave" in which this one mistake spreads out through the entire lattice. However, the potential barrier to atomic diffusion is so high that thermodynamic equilibrium is not established until rather high temperatures are reached, $T \sim 300°$K, so that a classical treatment is adequate.

TABLE VIII

ALLOYS THAT FORM ORDERED STRUCTURES AND THEIR CRYSTAL STRUCTURES.

Alloy	Type	Alloy	Type
FeCo	1	AuCu	3
FeAl	1	CoPt	3
FeV	1	MgIn	3
CuZn	1	MnNi	3
AgZn	1	NiPt	3
AgCd	1	FePd	3
CuPd	1	FePt	3
Ni_3Fe	2	Fe_3Al	4
Ni_3Mn	2	Fe_3Si	4
Ni_3Pt	2	Mg_3Li	4
Cu_3Au	2	Cu_3Al	4
Au_3Cu	2		
$CoPt_3$	2	Ag_3In	5
Fe_3Pt	2	Mn_3Ge	5
$FePd_3$	2	Mg_3Cd	5
Cu_3Zn	2	Cd_3Mg	5
Pt_3Fe	2	Ni_3Sn	5
Mg_3In	2		
Cu_3Pt	2	Cu_2MnIn	6
Pt_3Mn	2	Cu_2MnAl	6
Pt_3Zn	2	Cu_2MnSb	6
Ag_3Pt	2	Ag_2MnAl	6
Pt_3Ag	2	Au_2MnAl	6
		Ni_2MnAl	6

Type 1—FeCo, Fig. 44
Type 2—Cu_3Au, Fig. 43
Type 3—f.c. tetragonal CoPt
Type 4—Cubic Fe_3Al
Type 5 Hexagonal Ag_3In
Type 6 Cubic (Heussler Alloy) Cu_2MnAl Same as type 4 cubic; copper at corner positions, aluminium and manganese alternate in body centre positions.

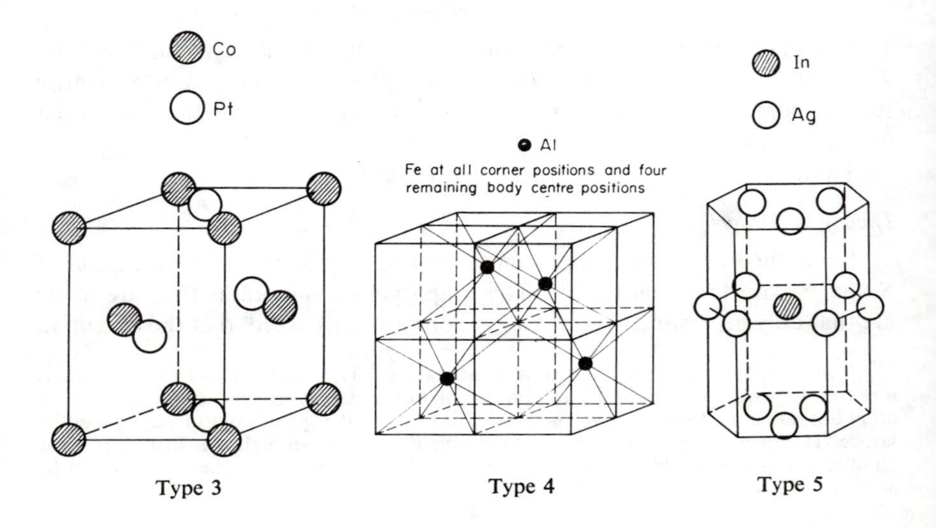

Type 3 Type 4 Type 5

V, between nearest neighbour atoms has some fixed value for specific pairs of species, and that this potential is independent of temperature. We are thus dealing with three potentials, V_{AA}, V_{BB} and V_{AB} representing for example, the copper-copper, gold-gold, and copper-gold potentials respectively. The potential can be reduced to the average potential $V_{AA} + V_{BB} - 2V_{AB}$ which is negative if ordering is preferred and positive if segregation is preferred. The problem is then solved classically by choosing that arrangement of atoms which minimizes the free energy at all temperatures. This phenomenological approach is reasonably successful in explaining the variation of S and α_i with temperature but can not be considered a fundamental approach until

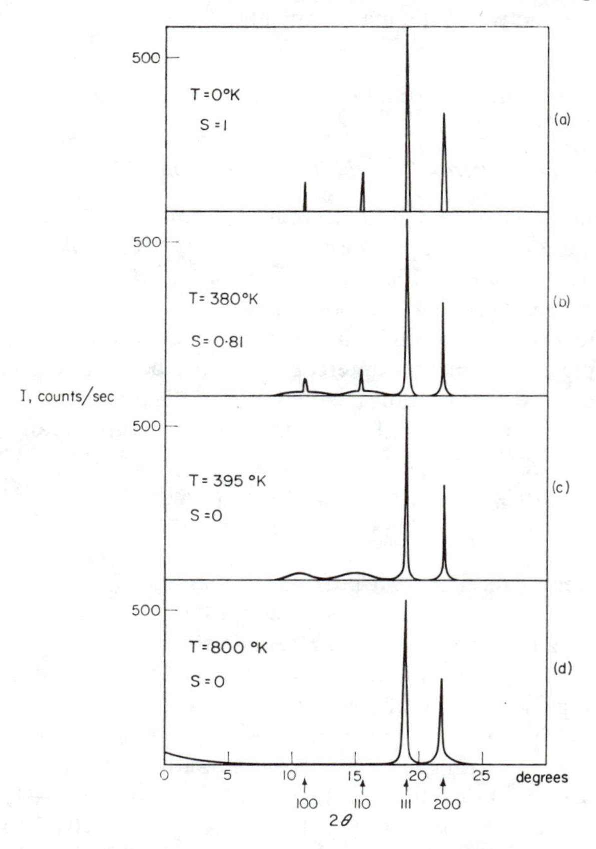

FIG. 45. Schematic X-ray powder patterns for Cu_3Au showing intensity in c/s versus counter angle 2θ in degrees. Pattern *a* is the completely ordered structure at $T = 0°K$ showing 100, 110 superlattice lines. Pattern *b* is at $T = 380°K$ just below the critical temperature showing some short range order around the superlattice lines. Pattern *c* is just above the critical temperature indicating the short range order. Pattern *d* is at high temperature and reveals the disordered Laue monotonic scattering which peaks at $\theta = 0°$, eqn. 89.

the V_{AA}, V_{BB} and V_{AB} can be calculated from the Schrödinger equation, rather than chosen empirically.*

To summarize the experimental X-ray work on atomic positions in metals, there is considerable data on crystal structures, some data on long range order, a lesser amount on short range order and a meagre amount on size effects. The data on crystal structures has not enabled us to make any notable inroads in understanding the eigenfunctions of the Schrödinger equation applied to metals nor have ordering effects been helpful since they are energetically small and occur too sporadically. Insufficient data exists on size effects to say whether it will be very helpful although, in principle, it should be the most instructive of the measurements.

B. Electron Probability Distribution

How to Determine Electron Probability Distribution

At present we have very little accurate information about electron probability distributions in solids. *However, this type of X-ray measurement is the only method yielding direct information about the eigenfunctions of the electrons and as such we shall describe it in some detail.*

If a monochromatic beam of intensity I_0 (counts/sec) impinges on a flat powder sample large enough to intercept the entire beam and always arranged such that the angle of the beam incident on the sample equals $\frac{1}{2}$ the counter angle (symmetrical reflection), Fig. 38, then the structure factor $|F|$ is related to this intensity by

$$|F|^2 = \frac{16\pi R^2 \omega}{Aj}\left(\frac{C}{I_0}\right)\frac{V^2 \mu \sin\theta_B \sin 2\theta_B (1+\cos^2 2\alpha)(m_0^2 c^4/e^4)}{\lambda^3 (1+\cos^2 2\alpha \cos^2 2\theta_B) e^{-2M} e^{-2M'}} \tag{90}$$

where A = area of receiving slit in front of counter (generally $\sim$10–20 mm^2)

R = distance from sample to receiving slit ($\sim$17 cm in Fig. 38)

ω = angular velocity of *counter* in radians/sec

j = multiplicity of peak (the number of ways the indices h, k, l can be arranged both positive and negative. For example, 110, 101, 011, 10 − 1, 1 − 10, 01 − 1, − 1 − 10, − 10 − 1, 0 − 1 − 1, 0 − 11, − 110, − 110, − 101, $j = 12$)

C = total number of counts in a Bragg peak (integrated intensity)

* There is also evidence that second nearest neighbours must be considered in some alloys.

I_0 = Intensity measured by counter in counts/sec by placing counter at $2\theta = 0°$ and removing sample and receiving slit.

μ = linear absorption coefficient of the bulk metal (not powder)

θ = Bragg angle

α = Bragg angle of an ideally imperfect monochromating crystal

$$\left(\frac{e^2}{m_0c^2}\right)^2 = (0{\cdot}282 \times 10^{-12}\ \text{cm})^2$$

e^{-2M} = Debye Waller factor due to thermal motion of atoms (see thermal scattering, eqn. (111))

V = volume of X-ray unit cell = a_0^3 for a cubic crystal

$e^{-2M'}$ = size effect factor for disordered alloys (eqn. (79))

Experimental Difficulties in Determining Electron Probability Distributions

The problems are manifold in applying this expression directly and we shall discuss them.

1. *Extinction*

If one measures the absorption coefficient, μ, by passing a monochromatic beam of X-rays through a polycrystalline foil of a substance of thickness, t, the ratio of the transmitted intensity to the incident intensity is $I/I_0 = e^{-\mu t}$. Ordinarily this attenuation is almost entirely due to true absorption in which the X-ray is annihilated and gives its energy to an electron ejected from an atom. However, if the X-ray is at the Bragg angle both attenuation processes occur in competition, Bragg scattering in which the beam is attenuated in a rather complicated fashion and true absorption which is exponential. While X-rays passing through a polycrystalline foil are primarily attenuated by true absorption since very few grains are at the Bragg angle the measurement of the intensity of X-rays that have undergone Bragg reflection provides the greatest possible exaggeration of this competitive effect. If it were possible to reduce the probability of Bragg scattering to a negligible amount relative to true absorption even when the X-rays were at the Bragg angle then the sample is said to be free of extinction. Equation (90) is based on this assumption, so that μ in the equation is measured by transmission through a polycrystalline foil of the sample. If the absorption due to Bragg scattering is not negligible relative to true absorption then the sample is said to have extinction. Without boring the reader with too many details, we shall say that we are unable to calculate extinction in real crystals nor measure it accurately. However we shall indicate how we can estimate it. In practice the best solution to the problem is to try to obtain a sample free of extinction, but before we say how to do this we must say something of these estimates.

Extinction is traditionally divided into two categories, primary and secondary extinction. This division is not realistic since it is based on an artificial model of a crystal but nonetheless this division represents much of the truth of the situation and has been quite useful. In this model every single crystal grain is assumed to consist of small mosaic blocks (~500–50,000 Å in size) each tilted with respect to the other (~1′ to 1°) so that neighbouring blocks do not diffract the same X-ray, Fig. 46. Within each mosaic block the atoms are assumed to be in a perfect crystalline array. (Of course a metallic grain is not at all like this but contains arrays of dislocations which prevent it from being a perfect crystal.) Primary extinction depends on the size of the mosaic blocks since this determines the range over which the X-ray wave is

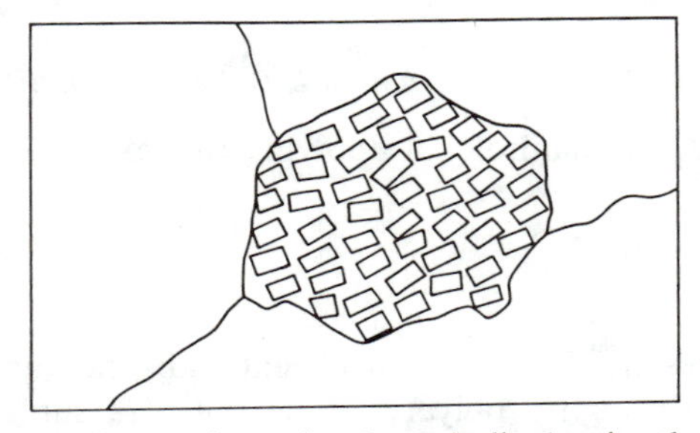

FIG. 46. Four grains of a metal sample schematically showing the mosaic blocks in the centre grain. The mosaic blocks are assumed perfect micro-crystals tilted with respect to each other so that two neighbouring blocks do not satisfy the Bragg conditions simultaneously.

in phase or is coherent. If the blocks get rather large ($t_0 > 10^4$ Å) then the integrated intensity of the X-rays diffracted from this grain is reduced from eqn. (90) by the approximate ratio

$$R_1 = \frac{C_{\text{EXT}}}{C} \cong \frac{\tanh A + (\kappa A)^2/2}{A} \tag{91}$$

$$A \cong \left(\frac{e^2}{m_0 c^2}\right) \frac{\lambda |F|}{V} K t_0 \tag{92}$$

$$\kappa = \frac{\mu V m_0 c^2}{2|F|\lambda e^2 K} = \text{index of absorption} \tag{93}$$

$$^{*}\begin{cases} K = 1 \text{ for parallel polarization} \\ K = |\cos 2\theta_B| \text{ for perpendicular polarization} \end{cases}$$

* For an unpolarized beam calculate the *intensity* for parallel polarization $K = 1$ and perpendical polarization $K = |\cos 2\theta_B|$, and average the two. For a polarized beam weight the two according to their relative amounts.

where C_{EXT} is the integrated intensity of the diffracted beam if extinction is present and C is the integrated intensity if extinction is absent; t_0 is the "effective size" of the mosaic block, $|F|$ the structure factor, V the volume of the unit cell (a_0^3 for cubic metals) and θ_B is the Bragg angle. Unless the linear absorption coefficient is very high, $\mu > 1000$ the term $(\kappa A)^2/2$ is generally small and can be neglected.

If the average angular tilt between the blocks becomes quite small then the absorption of the beam due to Bragg scattering can become quite large for any X-ray, since this increases the scattered intensity of X-rays incident upon the crystal at the Bragg angle. This increased attenuation due to scattering is called secondary extinction and the integrated intensity of the X-rays which have been diffracted is reduced from eqn. (90) by a factor,

$$R_2 = \frac{C_{\text{EXT}}}{C} \cong \frac{2 + \mu D}{2 + \mu D + \{[3QD]/[2\eta\sqrt{(\pi)}]\}}$$

$$Q = \left(\frac{e^2|F|}{m_0 c^2 V}\right)^2 \frac{\lambda^3(1 + \cos^2 2\theta_B \cos^2 2\alpha)}{\sin 2\theta_B(1 + \cos^2 2\alpha)} \tag{94}$$

valid for powders in symmetrical reflection

where D is the size of the grain or the average path length of the X-ray in the grain and η the average angular tilt in radians between the blocks.* In the case of a single crystal whose absorption coefficient is quite high so that $\mu > 2/D$ then

$$R_2' \cong \frac{1}{1 + 0{\cdot}87Q/2\mu\eta\sqrt{(\pi)} - 0{\cdot}08(Q)^2/4\pi\mu^2\eta^2} \tag{95}$$

valid for single crystals in symmetrical reflection

A good deal of confusion about extinction can be eliminated if the reader bears in mind that R_1 and R_2 are merely corrections to eqn. (90) to allow for the fact that due to Bragg diffraction the beam is being absorbed at a significantly greater rate than accounted for by the true absorption coefficient, μ.

The problem in applying the expressions R_1 and R_2 in order to obtain estimates of extinction in powders resides in our lack of knowledge of t_0 in R_1 and D and η in R_2 although an upper limit to D is the size of the individual particles in the powder sample. Nonetheless, experience enables us to make some estimates of η and t_0. For example, if we use Mo$K\alpha$ ($\lambda = 0{\cdot}71$ Å) on 400 mesh iron powder ($\mu \cong 295$/cm) and measure the lowest angle peak (110), the one most susceptible to extinction, then an upper estimate of D is $\sim 3 \times 10^{-3}$ cm (from the mesh size). A typical value of η is 10′ or 1/360 of a radian so that our estimate of R_2 is $\sim 0{\cdot}83$. If we use 5 μ iron powder

* If the mosaic blocks have a gaussian distribution η is the half width at half maximum divided by 1·15.

then a lower limit of R_2 is $\sim 0{\cdot}98$. In the case of R_1 it is known that t_0 is $\sim 10^{-4}$ cm for an annealed powder so that with the same X-ray wavelength $R_1 \cong 0{\cdot}97$. Thus for accuıate intensity measurements in a typical annealed powder extinction effects cannot be neglected.

A clue to a practical approach to this problem can be gained when we consider that the diffraction phenomena as the mosaic block size, t_0, gets small. Diffraction theory tells us that diffraction occurs over an angular range

$$\Delta\theta \cong \lambda/t_0 \cos\theta_B \tag{96}$$

This means that the incident X-ray beam can satisfy the Bragg conditions over an angular range $\Delta\theta$ centred about the Bragg angle θ_B and that the diffracted beam will be broadened by an amount $\Delta\theta_B$. For example, Co $K\alpha$ (1·789 Å) diffracted from the (220) peak of iron with mosaic block size $t_0 \sim 3 \times 10^{-5}$ cm yields a $\Delta\theta_B$ of 4′ of arc. If we examine eqns. (91), (94) for small extinction (R_1 and $R_2 > 0{\cdot}8$) they can be expanded mathematically (assume κA is small) since $\tanh A/A \to 1 - A^2/3 + \ldots$ so that

$$R_1 \cong 1 - \left[\left(\frac{e^2}{m_0c^2}\right)^2 \frac{\lambda^2|F|^2}{V^2}\frac{(1+\cos^2 2\theta_B)}{2}\right)\frac{t_0^2}{3}$$

$$R_2 \cong 1 - \left[\left(\frac{e^2}{m_0c^2}\right)^2 \frac{\lambda^2|F|^2}{V^2}\frac{(1+\cos^2 2\theta_B)}{2}\right]\frac{3\lambda D}{4\sin\theta_B(\cos\theta_B)\eta\sqrt{(\pi)}} \tag{97}$$

averaged over polarization

If t_0 is small then the Bragg conditions are fulfilled over an angular range $\lambda/t_0 \cos\theta$, eqn. (96), but to satisfy our mosaic block model, neighbouring mosaic blocks must be tilted by an angle greater than this lest they both diffract the same X-ray. If the angular tilt between blocks is independent, then the average angular tilt, η, is $\sim(D/t_0)(\lambda/t_0 \cos\theta_B)$ since D/t_0 is the number of mosaic blocks in the path of an X-ray. If we substitute this value of η in R_2, eqn. (97), then both R_1 and R_2 are virtually identical. *Thus we can say that any method that eliminates one form of extinction in powders probably eliminates the other.*

A practical approach to estimating and eliminating extinction in powders is:

(*a*) Select the finest powders available (less than 400 mesh).

(*b*) Measure t_0 by employing relatively long wavelength X-rays at high angles in a high resolution geometry. Use a well annealed powder to determine the geometrical resolution of the instrument.

(*c*) Substitute this value of t_0 in eqn. (91) (for $K = 1$) and estimate the extinction of the most intense peak for the wavelength to be employed. Use an approximate value for the atomic scattering factor f. To minimize extinction a short wavelength should be employed, Mo$K\alpha$ generally being a good choice. If the calculated value of R_1 in eqn. (91) is greater than 0·995 then extinction can probably be neglected. If it is not, try cold working the powder by ball milling to reduce t_0.

2. *Absorption*

The factor μ in eqn. (90) is based on the assumptions that the reflection is symmetrical and the sample is thick enough so that only a negligible fraction of the beam passes through the entire sample without being absorbed. One can correct this latter point by replacing μ by

$$\mu \rightarrow \mu/(1 - e^{-[2(\mu/\rho)(T'/\cos\theta_B)]})$$

where T' is the thickness of the sample in gm/cm^2 (weigh it and divide by the area), ρ is the density in g/cm^3 and θ_B the Bragg angle for each reflection. In order to ensure that the sample is in a symmetrical position obtain a flat polished sample of any material that fluoresces very strongly with the radiation being used (brass with Mo$K\alpha$ is good). Replace the powder specimen with the fluorescent sample such that its surface coincides with the surface of the powder specimen in the specimen holder. By maintaining the usual θ, 2θ relationship between specimen and counter measure the fluorescent intensity as a function of angle particularly at low angles, being certain that the sample intercepts the entire beam. If the intensity is independent of angle (except for Bragg reflections) the reflection is symmetrical but if it is not adjust the angle of the specimen until this condition is fulfilled.

Accurate values of μ are not available in the literature, errors of 5% or so being common. If you are fortunate enough to have a thin foil of uniform cross-section then the absorption can be measured but this is rarely the case. It is in fact somewhat deplorable that accurate measurements of absorption coefficients are not available for X-rays while in the rather "young field" of neutron physics accurate measurements are available. Since the absorption cross-section per atom $\sigma = (\mu/\rho)AN_0$ (A = atomic weight, N_0 = Avogadro's number) is essentially independent of the alloy in which the atom is placed a knowledge of σ for each element would permit us to determine μ for any alloy whose composition is known. But until accurate tables of σ are available the following may guide the reader in this problem.

(*a*) Try to obtain a thin sample of uniform cross-section of the material under study. Determine its thickness T' in g/cm^2 and check its uniformity by measuring the X-ray transmission at several points. Ascertain the true density of the material from the X-ray lattice parameter and its composition and determine μ from the transmission, $I/I_0 = e^{-(\mu/\rho)T'}$.

(*b*) As a second choice obtain foils of the pure materials comprising the alloy and measure μ/ρ. The μ/ρ_{AB} of the alloy is given by eqn. (82.) The density of the alloy, ρ_{AB}, should be determined from the X-ray lattice parameter and its composition.

(*c*) If neither (*a*) nor (*b*) is possible as it might be for a brittle alloy like $Mn_{0\cdot8}Fe_{0\cdot2}$ one of whose constituents is also brittle, determine μ/ρ from foils of the neighbouring elements like titanium, vanadium, iron, cobalt, nickel and copper. Plot $(\mu/\rho)A$ for each element versus $Z^{3\cdot94}$ where Z is the atomic number. This should approximately give a straight line and one can read the interpolated value of $(\mu/\rho)A$ at the appropriate value of $Z^{3\cdot94}$.

(*d*) If method (*b*) or (*c*) is used, analysis should be made of the sample particularly searching for highly absorbing impurities.

(*e*) As an approximation good to about 1% in the vicinity of iron for Mo$K\alpha$ (0·71 Å).

$$\frac{\mu}{\rho} A = 5\cdot52 \times 10^{-3} Z^{3\cdot94}$$

valid for elements up to yttrium.

3. *Geometrical Factors*

In order to secure sufficient diffracted intensity the monochromatic X-ray beam must have an angular divergence of at least $\sim\frac{1}{2}°$ and such factors as θ, λ and R in eqn. (90) may be in error. The following rules will serve as a guide to minimize these errors to about $\frac{1}{2}$%.

(*a*) Use a standard of known lattice parameter like

$$\text{Si } (a_0 = 5\cdot43032 \text{ Å at } 21°\text{C})$$

with a temperature variation $(\Delta a_0/a_0) = 7 \times 10^{-6}/°\text{C}$) to determine the angular scale of the diffractometer and the average wavelength of the beam diffracted from the monochromating crystal.

(*b*) Plot the contour of the monochromatic beam incident upon the power sample by removing the sample, lowering the counter to intercept the direct beam, and scanning with a very narrow slit before the counter. The contour of the beam should be symmetric.

(*c*) Be certain the sample intercepts the entire beam and the centre of the beam passes through the axis of rotation of the sample.

(*d*) If the above conditions are fulfilled R is then the distance from the centre of the sample to the back surface of the receiving slit.

(*e*) Determine the lattice parameter a_0 of the sample on a high resolution diffractometer.

4. *Counter Dead Time*

All counters are limited in their ability to count very rapidly. The dead time is the time it takes a counter and its associated electronics to recover after counting an X-ray and it ranges from about 2 μsec for a scintillation counter to 200 μsec for a Geiger counter. Determine the dead time (see Klug and

Alexander) and maintain the counting rate sufficiently low so that the correction is never more than about 5%.

5. *Thermal Scattering and Size Effect*

The technique for correcting for diffuse scattering due to thermal effects and determining e^{-2M} will be discussed under temperature effects, p. 157.

The size effect parameter has been discussed under *A*. If one is measuring the electron probability distribution in a disordered alloy, then the size effect parameter cannot be determined separately since one must know either f or $e^{-2M'}$ to measure the other. A good estimate of $e^{-2M'}$ can be obtained by assuming the atoms in the alloy to be the same size as in the pure metals (see p. 125). If this estimate indicates the correction to be large then the only accurate method of determining $e^{-2M'}$ is to use neutron diffraction (to be discussed later).

6. *Second Order Contamination*

If monochromatic Mo$K\alpha$ X-rays are used the second order, $\lambda = 0{\cdot}35$ Å, will diffract from the monochromating crystal if the X-ray tube potential is sufficient to excite it. If the potential is kept below 35 kV the second order is eliminated. At 40 kV the second order contamination is $\sim 2\%$.

7. *Absolute Standardization*

Since the counting rate of the incident monochromatic beam is 10 to 100 thousand times greater than the diffracted beam, the accurate measurement of both presents a formidable problem. In order to have reasonable counting rates in the diffracted peaks the incident monochromatic beam must have an intensity I_0 of $\sim 10^6$ c/s or more. Since the dead time error in most circuits is too large to count this accurately one must utilize intermediate steps to determine I_0 in eqn. (90). In general it is wise to limit the counting rates to several thousand c/s in all stages of the experiment. There are several techniques to measure I_0 in spite of this limitation:

(*a*) *Absorbers*. The main beam can be counted by inserting a sufficient number of absorbers (for example, 6–8 absorbers each with a transmission of about 20%). The absorption of each is measured separately by alternately removing and replacing each of them when they are all in the beam. Each measurement must be accurate, for an error of only $\frac{1}{2}\%$ in each foil is multiplied by the number of foils and this is significant. For Mo$K\alpha$ it is best to use Zirconium absorbers.

(*b*) *Intermediate Fluorescent Sample*. Allow the incident beam to pass through a very tiny pinhole so that it can be counted without a large dead time error. Place a sample in the beam that fluoresces strongly and measure the fluorescent intensity at some angle not corresponding to a Bragg reflection in the fluorescent sample. Let the ratio of the counting rates be denoted by β. Remove the pinhole and measure the fluorescent counting rate without altering any other geometry. The incident beam intensity, I_0, is obtained by multiplying this scattered beam intensity by β. It may be necessary to use one absorber of transmission $\sim 20\%$ in conjunction with the fluorescent sample to keep the counting rates within our imposed limitations.

(*c*) *Polystyrene as a Standard.* The total diffuse scattering of polystyrene (Compton plus elastic) can be calculated quite accurately at high values of $\sin\theta/\lambda$ and can be used as a standard. This eliminates the necessity for counting the main beam as in *a* and *b*. By comparing the diffuse scattering from polystyrene against the integrated intensity of a Bragg reflection we can determine $|F|$ which, for cubic crystals, is

$$|F|^2 = \left[\frac{4\pi\omega\mu a_0^6 C \sin\theta \sin 2\theta(1+\cos^2 2\alpha)}{j\lambda^3(1+\cos^2 2\alpha\cos^2 2\theta)e^{-2M}e^{-2M'}}\right]_{\text{SAMPLE}} \times \left[\frac{N_0\sigma(1-e^{-2\mu t/\cos\theta})(1+\cos^2 2\alpha\cos^2 2\theta)}{\mu I_D}\right]_{\text{STANDARD}} \quad (98)$$

where 2θ is the counter angle, C the integrated intensity of the Bragg peak, μ the linear absorption coefficient, α the Bragg angle of the monochromating crystal, I_D the intensity of diffuse scattering from polystyrene in counts/sec at some fixed counter angle 2θ, N_0 the number of carbon-hydrogen pairs per cm^3 in polystyrene (ratio of carbon to hydrogen is 1),

$$\sigma = (\sum f_{ii})^2 + Z - \sum_i \sum_j f_{ij}^2 = \text{total scattering per CH pair}$$

and t the thickness of the polystyrene in cm. Equation (98) assumes the Bragg peak and diffuse scattering are measured with the same slit, scanning in the first case and fixed in the second, and that both powder sample and polystyrene sample are in the symmetrical position. At $\sin\theta/\lambda = 0{\cdot}5$ Å^{-1}, $\sigma = 8{\cdot}252$ per CH pair.

(*d*) *Aluminium powder.* High purity 10μ aluminium powder which is free of extinction with Mo$K\alpha$ radiation can be obtained from Charles Hardy Inc., 420, Lexington Ave., New York. Its preferred orientation can be checked by preparing a thin sample ($\sim 0{\cdot}25$ g/cm^2) and measuring the integrated intensities of the 111, 200, and 220 peaks in symmetrical transmission and symmetrical reflection. If the transmission of the sample $I/I_0 = e^{-\underline{\delta}}$ is determined, then the ratio of a Bragg peak in transmission to reflection is $(\underline{\delta}/\cos\theta)e^{-\underline{\delta}/\cos\theta_B}$ (θ_B is the Bragg angle) provided the sample is free of preferred orientation. If the measured ratios are within a few per cent of the calculated ones then the average of the transmitted and reflected cases are probably better than one per cent. The observed scattering factor (including dispersion) of the aluminium 111 peak with Mo$K\alpha$ radiation is $f^2 = 76{\cdot}2$, $e^{-2M} = 0{\cdot}925$ integrating over the angular range 16–19° 2θ and subtracting background. The absorption coefficient is $\mu = 13{\cdot}54$/cm and $a_0 = 4{\cdot}0496$ Å. The advantage of the aluminium standard is that it enables one to make absolute intensity measurements without monochromatic radiation provided we limit ourselves to observations of the sharp Bragg reflections. To make such a measurement a standard diffractometer is used without the single crystal attachment and we employ Mo$K\alpha$ radiation in the following sequence of steps:

(i) Secure a zirconium and strontium (or yttrium) set of filters of such thickness to give identical transmission at $\lambda = 0{\cdot}8$ Å but thick enough to eliminate the Mo$K\beta$.

(ii) Measure the integrated intensity of the Al 111 peak with the zirconium filter and then the strontium (or yttrium) filter and take the difference.

(iii) Repeat this for the sample under investigation.

(iv) Measure the absorption coefficient of the sample by employing a single crystal in the specimen position set to diffract the Mo$K\alpha$, lowering the tube potential to below 35 kV.

8. *Preferred Orientation and Porosity*

Equation (90) assumes the sample to be a random aggregate of grains so that the relative probability of an X-ray encountering the various planes is given by the multiplicity factor j. Fine, loosely packed metal powders are generally free of preferred orientation but materials like sodium chloride which cleave along preferred crystalline planes evidence preferred orientation even when loosely packed. On the other hand, loosely packed powders introduce a so-called porosity (or packing) effect. Equation (90) assumes the sample to be completely homogeneous such as a solid piece of metal and it can be shown that the equation is still valid as long as the size of the individual powder particles is very small compared to $1/\mu$, the reciprocal of the linear absorption coefficient. When this is the case the X-ray penetrates many particles, but even in the favourable case of 400-mesh iron particles with Mo$K\alpha$ the X-ray penetrates only a few particles. This porosity or lack of homogeneity is virtually impossible to calculate but experimental measurements indicate it to be rather large for loosely packed powders in which the absorption coefficient limits the depth of penetrations to a few particles. It introduces an additional absorption factor so that μ in eqn. (90) is replaced by $\mu\delta$. Fortunately δ can be determined to within $\sim 10\%$ if our sample gives detectable fluorescence with the monochromatic radiation used in our diffraction experiment. A measurement of the ratio of the intensity of fluorescent radiation from our sample and a smoothly polished solid block of the sample material yields for δ

$$\delta \cong \left[1 - \frac{\mu + \mu^*}{2\mu^*}\left(1 - \frac{1}{R_F}\right)\right]^{-1} \tag{99}$$

where μ and μ^* are the absorption coefficients of the sample for the incident monochromatic radiation and the fluorescent radiation (a weighted average of $K\alpha$ and $K\beta$) and R_F is the ratio of fluorescent intensity from the polished sample to the powder specimen. The ratio, R_F should be measured at any angle not near a Bragg angle, but at the same angle for both samples. For Mo$K\alpha$ on loosely packed iron powder δ is about 1·15 but this value can be reduced to unity if the powder sample is pressed in a die, Fig. 47. However,

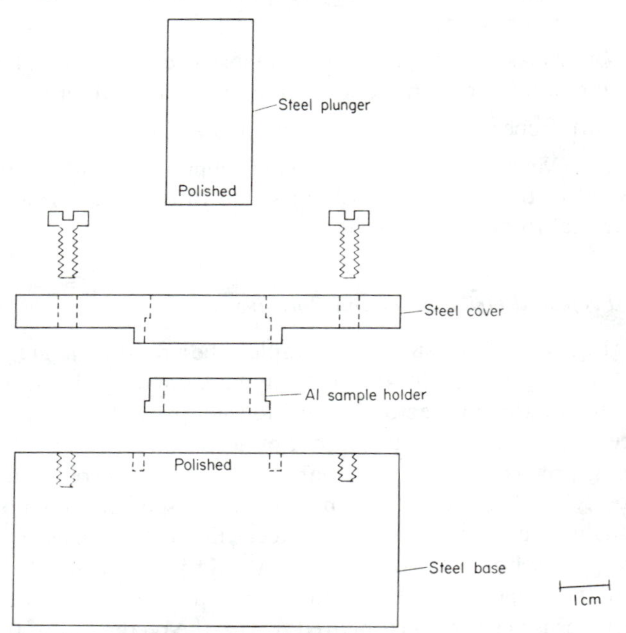

FIG. 47. A die for making powder compacts for X-ray diffraction. The aluminium sample holder which fits the Norelco specimen spinner is placed on the polished steel base and the steel cover screwed on. The sample holder is filled with powder and compressed by the steel plunger in an hydraulic press.

such pressing may cause plastic flow and preferred orientation effects particularly in soft non-cubic metals like zinc. Thus a compromise must be reached and the following precedures are suggested in preparing a sample.

(*a*) If the metal is reasonably hard such as iron determine the minimum pressure necessary to eliminate porosity by measuring the fluorescence. If the sample does not fluoresce add 1% of a powder that does.

(*b*) Check the relative intensities of the diffracted peaks of the pressed sample against that of a loose packed sample assuming the loose packed sample does not exhibit preferred orientation.

(*c*) If the sample can be made thin enough for transmission it means that the absorption coefficient is small enough so that porosity effects can be neglected. Check for preferred orientation as described above for aluminium standardization.

(*d*) If the material is so soft that the exertion of sufficient pressure to eliminate porosity causes preferred orientation, then a loose packed powder must be used and corrected with eqn. (99). In such a case the error in the measurement may be large.

(*e*) A crude estimate of δ (eqn. (99)) can be obtained from a knowledge of the apparent density of the powder specimen, the particle size, d, and the linear absorption coefficient of the bulk material, μ. $\delta \cong (1 - (1 - \alpha)\mu d/2)$ where α is the apparent density divided by the true density.

9. *Dispersion Corrections*

If the energy of the X-ray used in a diffraction experiment is close to the energy required to eject a K or L electron from the atom, then the atomic scattering factor must be modified by the *Hönl* or *dispersion correction*. While the values of these corrections have not been accurately measured, if one uses monochromatic radiation not too close to the excitation energy the correction can be kept small and theoretical values can be used. Dauben and Templeton.[12] have published a partial list of these dispersion corrections which are given as two terms, $\Delta f'$ and $\Delta f''$. As an example of their use consider the structure factor of a f.c.c. element.

$$|F|^2 = |4f|^2 = 16(f_0 + \Delta f')^2 + 16(\Delta f'')^2$$

where f_0 is the atomic scattering factor without the dispersion correction. $\Delta f''$ is only employed if the X-rays have greater energy than is required to eject the K and L electrons. For Mo$K\alpha$ all elements up to zirconium require a small correction. The theoretical values for Mo$K\alpha$ on elements $Z = 1$ to $Z = 38$ is

$$\Delta f' \cong (1{\cdot}23 \times 10^{-7})Z^4(33{\cdot}5 - Z)$$

$$\Delta f'' \cong (2 \times 10^{-6})Z^4$$

valid for Mo$K\alpha$ (0·71 Å) for $Z = 1$ to $Z = 38$.

10. *Single Crystal Measurements*

It is considerably more difficult to eliminate extinction in metal single crystals as compared to powders although it can be reduced considerably by careful cold rolling of the single crystal. Nonetheless, single crystal measurements are important in order to measure the atomic scattering factor at large values of $\sin \theta/\lambda (>0{\cdot}5)$ where the powder peaks are considerably weakened. At these high values of $\sin \theta/\lambda$ extinction effects in many metal single crystals can be eliminated or determined with reasonable accuracy since the atomic scattering factor is much smaller. In addition, the problems of preferred orientation, porosity, and absolute standardization are reduced or eliminated since single crystals diffract a large fraction of the beam ($\sim 1\%$) and they can be made quite homogeneous by electro-polishing the surface.

The experimental arrangement for single crystal measurements is shown in Fig. 48, considerably simplified in comparison to powder measurements. A monochromatic beam is produced by a flat crystal, since focusing is unimportant, and the crystal is arranged to intercept the entire beam and rocked through the Bragg angle with angular velocity ω (radians/sec.). The counter is held fixed at 2θ with its window large enough to intercept the entire diffracted beam. The structure factor is related to the integrated intensity by

$$|F|^2 = \frac{2\mu C\omega_s \sin 2\theta_B V^2(1+\cos^2 2\alpha)(m_0c^2)^2/(e^2)^2}{I_0\lambda^3(1+\cos^2 2\alpha \cos^2 2\theta_B)\, e^{-2M}\, e^{-2M'}} \tag{100}$$

(valid for single crystals in symmetrical reflection.)

The symbols are the same as in eqn. (90) except that ω_s is now the angular velocity of the single crystal. The procedure for single crystal measurements is as follows.

(*a*) After cutting and carefully electro-polishing the crystal approximately parallel to the desired plane measure the rocking curve with a very fine beam ($\sim 0{\cdot}1$ mm) to determine η. This latter step can be accomplished by using very fine pinhole slits to limit the beam divergence to an area $\sim 1/\mu^2$. The half width of the angular range over which the crystal diffracts (rocking curve) can be used as a measure of η. Substitute this value of η into R_2' (eqn. 95) and determine the extinction. If the extinction is too large the crystal can be carefully cold rolled to increase η and reduce the extinction.

(*b*) If one is only interested in the diffracted peaks at large values of $\sin \theta/\lambda$ it is desirable not to cold roll the crystal so these relatively weak outer peaks are sharp. In this case determine R_2' by measuring the absolute intensity of the peak most susceptible to extinction and compare it to the calculated intensity for no extinction. From this value of R_2' determine η and use this value of η to determine the correction, R_2', to be applied to the peaks at large $\sin \theta/\lambda$.

(*c*) If the crystal is not cut parallel to the diffracting plane measure the integrated intensity twice with the crystal rotated 180° azimuthally and average the two. This will give the same intensity as a symmetrically cut crystal. If the crystal is small and of a simple geometric shape (sphere or cylinder) then replace 2μ in eqn. 100 by $e^{-\alpha}/v$ where v is the volume of the crystal in cm^3 and $e^{-\alpha}$ is the average transmission of the beam (see Klug and Alexander), and replace I_0 by G_0 where G_0 is the flux or number of X-rays per cm^2/sec

FIG. 48a, 48b. A possible arrangement of the Norelco diffractometer for single crystal measurements. A monochromating crystal is mounted in the usual powder sample position and selects the desired wavelength. The single crystal specimen is mounted with a motorized drive to rock it through its Bragg diffracting position and the counter is held fixed at twice the Bragg angle. The monochromating crystal and details of the single crystal rocking mechanism are seen in the lower photograph.

striking the crystal. In this case the entire crystal is bathed in the X-ray beam. The problem of the uniformity of the beam flux and absolute standardization can be approximately solved by casting or machining a piece of polystyrene identical to the sample and measuring the total intensity at $\sin\theta/\lambda \cong 0{\cdot}5$.

(*d*) The comments in paragraphs 2, 4, 5, 6 and 9 above are also appropriate to single crystals.

(*e*) If an extremely perfect metal crystal can be obtained then the integrated intensity is related to the structure factor (symmetrical reflection) by

$$\frac{C\omega}{I_0} \cong \frac{8\pi|F|Ke^2\lambda^2e^{-M}(1-2\kappa)}{3V(m_0c^2)\sin 2\theta_B}$$

where κ and K are given in eqn. (93). The perfection of the crystal can be determined by measuring $|F|$ at two widely different wavelengths. $|F|$ should be independent of λ (except for dispersion corrections).

Converting the Structure Factor into an Electron Probability Distribution

Once the structure factor has been determined the atomic scattering factor, f_0, is obtained from it through knowledge of the crystal structure (the crystal structure of most metallic alloys is known).[9] The atomic scattering factor, f_0, is precisely related to the electron eigenfunctions by

$$f_0 = \int \Psi^* \sum_i e^{i\mathbf{s}.\mathbf{r}_i}\Psi \, dx_i \, dy_i \, dz_i \tag{101}$$

where Ψ is the many-electron wave function, r_i is the position of the ith electron relative to its nucleus, $\mathbf{s} = \boldsymbol{\kappa} - \boldsymbol{\kappa}_0$ where $\hbar\boldsymbol{\kappa}$ and $\hbar\boldsymbol{\kappa}_0$ are the momenta of the scattered and incident X rays, and the integral is over the co-ordinates of all the electrons in the polyhedra, Fig. 13.* If the X-ray is scattered by Bragg diffraction then $s = 4\pi \sin\theta/\lambda$ and the direction of $\mathbf{s}$ is normal to the diffracting plane. Eqn. (101) can be immediately simplified by replacing Ψ by the antisymmetric product of one-electron wave functions

$$f_0 = \sum_i \int |\psi_i|^2 e^{i\mathbf{s}\cdot\mathbf{r}_i} dx_i \, dy_i \, dz_i \tag{102}$$

which is just the sum of individual one-electron scattering factors. Equation (102) is further simplified if the ψ_i are spherically symmetric (s wave functions) or they are part of a closed shell (which is also spherically symmetric)

$$f_0 = \sum_i \int |R_i|^2 \frac{\sin sr_i}{sr_i} r_i^2 \, dr_i \tag{103}$$

(independent of the direction of $\mathbf{s}$)

* Strictly speaking eqn. (101) is only applicable to simple elements. In general eqn. (101) relates the structure factor, F, to the electron probability distribution in the unit cell.

where R_i is the radial part of the one electron wave functions. Since the bulk of the electrons on atoms are part of closed shells and even the outer electrons tend to be predominantly s-like in metals we shall limit our present discussion to eqn. (103). It is possible, particularly for the magnetic scattering of neutrons, to observe the deviation of the outer electron wave functions from spherical symmetry and this will be discussed in the section on neutron diffraction. As X-ray techniques improve, such effects may also be seen and in that case the treatment of f_0 in the neutron case is easily adapted to the X-ray case.

If it were possible to measure the structure factor for all h, k, l up to infinity then the number of electrons per unit volume at some point x, y, z in the unit cell would be given by

$$\sum_i |\psi_i|^2 = \frac{1}{V} \sum_h \sum_{k=-\infty}^{\infty} \sum_l F(hkl)\, e^{-2\pi i(hx/a + ky/b + lz/c)} \tag{104}$$

which is just the electron probability distribution that we desire.*
In some cases, crystallographers have actually measured a great many reflections and substituted the structure factors of each into eqn. (104) making some assumptions as to the behaviour of the structure factor as h, k, l becomes larger than can be measured. However, such a procedure is subject to large inaccuracies and is impractical in the case of most metals, particularly for those with simple crystal structures. For one thing, there are not many Bragg reflections for simple metals and for another the reflections at large values of $\sin \theta/\lambda$ (large values of h, k, l) are subject to large errors such as the uncertainty in e^{-2M} due to thermal motion of the atoms. In practice one generally works backwards by comparing the measured atomic scattering factor with a calculated scattering factor.

For many years theoreticians have calculated one-electron free atom wave functions by the Hartree–Fock method but no accurate absolute scattering factor measurements have been made on free atoms to ascertain the accuracy of the X-ray method and/or the accuracy of the Hartree–Fock wave functions. Such a measurement is exceedingly difficult because the scattering from free atoms is so weak, but recently attempts are being made to perform such measurements. In the case of crystals, relatively accurate X-ray measurements have been made recently and theoretical crystal wave functions are just beginning to emerge. Thus, this book has been written at a time when a direct comparison of theoretical and experimental eigenfunctions is on the brink of reality.

* In the case of free atoms for which the atomic scattering factor can be measured at all values of $\sin \theta/\lambda$ rather than only at discrete Bragg reflections, eqn. (104), reduces to a particularly simple form for spherically symmetric probability distributions

$$\Sigma |\psi|^2 = \frac{1}{2\pi r} \int_0^\infty s f_0 \sin(sr)\, ds \tag{105}$$

This form is frequently used as an approximation even in crystals.

In the interim, we can compare scattering factors calculated from free atom eigenfunctions with measured scattering factors in crystals. Let us consider the case of aluminium. Fig. 49 shows the measured scattering factors of f.c.c. aluminium metal[(18)(19)] and the theoretical scattering factor for the free atom of aluminium which is the sum of $1s^2 2s^2 2p^6$ (ten electrons, "neon core") + $3s^2 3p$ (three bonding electrons). There is a significant difference between the measured scattering factor, f_0, and that deduced from the free atom wave functions. One might be tempted to attribute this difference to a change in the aluminium free atom wave functions as the atom becomes part of the metal. However, the $1s$ electrons have an eigenvalue -1559 eV, the $2s$ and $2p$ electrons have eigenvalues of about -73 eV while the cohesive energy of aluminium is only 3 eV per atom, and this cohesive energy is principally due to the overlap of the $3s - 3p$ orbitals which have free atom eigenvalues of a few eV. We thus expect virtually no change in the $1s^2 2s^2 2p^6$ ("neon core") one-electron wave functions when an aluminium atom becomes part of the metal. These ten electrons which comprise the "neon core" of aluminium are responsible for the repulsion in the metal. The free atom scattering factor of these ten electrons is plotted in Fig. 49 as well as the scattering factor for the three outer electrons. As one can see, the three outer electrons do not make any significant contribution to the Bragg scattering. The perplexing thing is that there is no possible arrangement of the three outer electrons which would account for the *negative* difference of the observed and calculated scattering factors. If the X-ray measurements are correct it means the "neon core" wave functions of aluminium would have to be expanded but even a crude calculation shows that it would require at least 50 eV to expand the wave functions sufficiently to account for the difference. Since we are only dealing with a cohesive energy of 3 eV this possibility must be ruled out.

How then do we know whether the Hartree–Fock wave functions are correct ? Actually we do not, and in the case of helium we have sufficiently accurate correlated wave functions to know the scattering factor to high accuracy. If we compare it to the scattering factor obtained from Hartree–Fock wave functions we find errors of several percent. *We must, therefore, conclude that several percent difference exists between theoretical and experimentally determined wave functions and until this is resolved there is little we can say about outer electron crystal wave functions determined from X-ray measurements since theoretical Hartree–Fock wave functions must be used to subtract the core contribution.**

However, there are some aspects of analyzing X-ray scattering factors that

* In the case of perfect single crystals of germanium good agreement has been obtained between the observed scattering factors and those calculated from Hartree–Fock wave functions. In the case of silicon both powder and perfect crystal give f values about 5% lower than Hartree–Fock calculations.

are worth mentioning. The three outer electrons in aluminium contribute so little to the scattering factor of the Bragg peaks that it appears unlikely that we should be able to determine their eigenfunctions from such X-ray measurements. This is true, since any reasonable change in the 3*s*, 3*p* wave functions causes very minor changes in the total atomic scattering factors at the Bragg peaks. For example, if we assume these outer electron wave functions yield a constant charge density (free electron approximation) their atomic scattering factor would only be altered slightly in the region of the Bragg peaks as shown in Fig. 49. For a constant charge density it actually oscillates and passes through zero at each of the Bragg reflections. Since the scattering factor of these outer electrons never contributes more than about 1% to the total

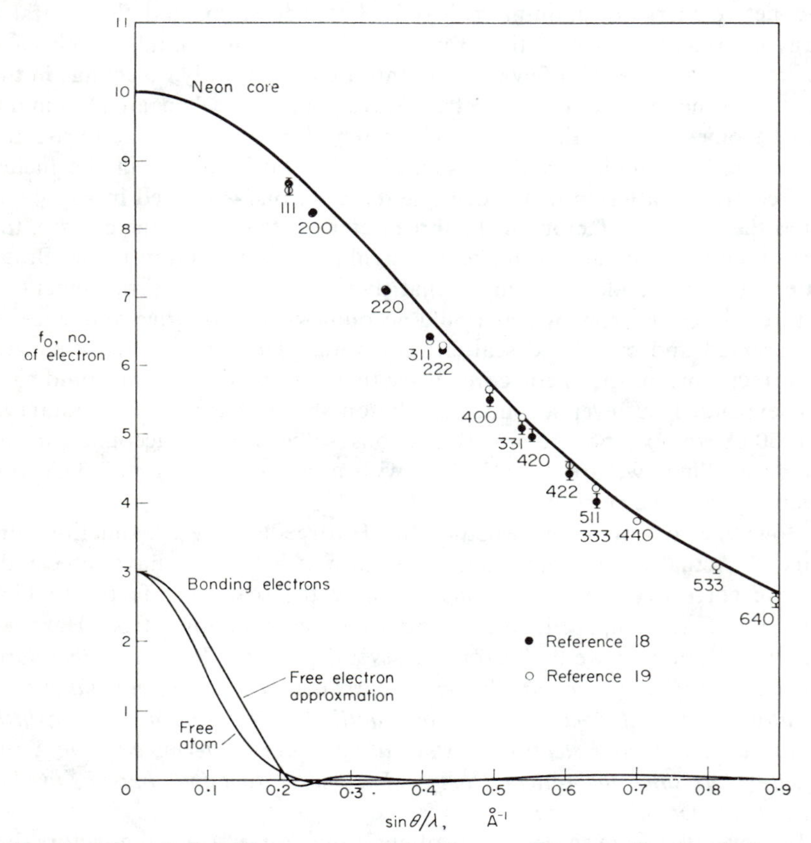

FIG. 49. The calculated free atom scattering factors of Freeman for the ten "neon core" electrons of aluminium and the three outer or bonding electrons. If the three bonding electrons assume the wave functions in the solid appropriate to the free electron theory their atomic scattering factor is shown. The experimental values[18][19] are from two reliable independent sources and indicate the extent of the discrepancy with theory.

scattering factor at the Bragg peaks it would be an exceedingly difficult task to derive very much from the measurements. One might wonder then, why one should bother with X-ray diffraction measurements in solids if they yield very little information about the outer electron wave functions which are responsible for the properties of the solid. The answer is that such is not always the case. Consider the outer electron wave functions of iron in Fig. 21. If we calculate the scattering factor for the $4s$ electrons we should find that they contribute very little to the total scattering at the Bragg reflections just as in

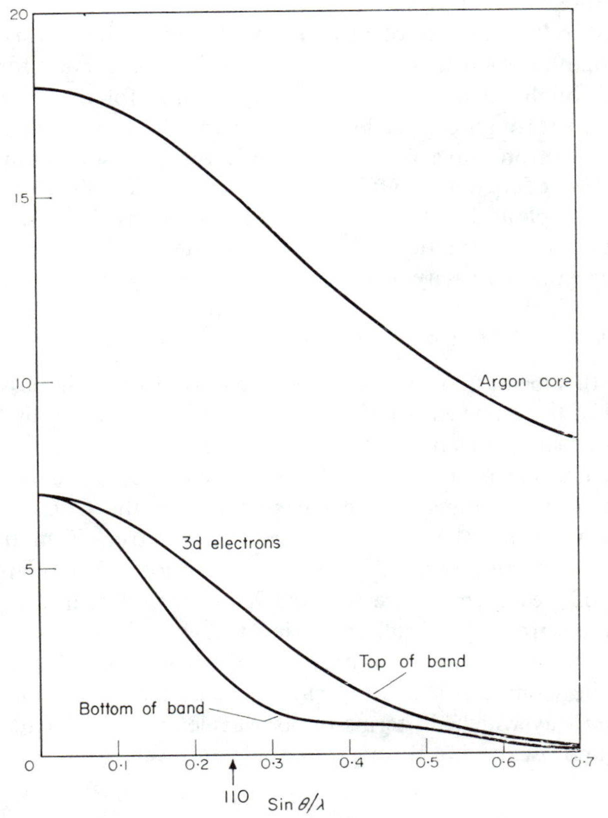

FIG. 50. The free atom atomic scattering factor for the argon core $1s^22s^22p^63s^23p^6$, and the approximately seven $3d$ electrons in b.c.c. iron either assumed all at the bottom of the band or the top, Fig. 21. The position of the 110 peak at $\sin\theta/\lambda = 0{\cdot}249/\text{Å}$ indicates the sizeable effect of bonding on the contribution of the $3d$ electrons to the atomic scattering factor.

the case of aluminium, but Fig. 50 also shows the scattering factor for the $3d$ wave functions at the top and bottom of the band as well as the scattering factor for the "argon core", $1s^22s^22p^63s^23p^6$. If the approximately seven $3d$ electrons in iron had wave functions like those near the top of the band their contributions to the total scattering factor of the 110 Bragg reflection is ~21% or about 40% of the intensity ($I \propto f_0^2$) while their contribution would be about 10% in f_0 or 20% in intensity if their wave functions were like those near the bottom of the band. Thus in the case of $3d$ electrons the expected variations in the wave functions due to binding effects in the solid make significant contributions to the absolute scattering factor. While X-ray measurements have been made on iron it would be premature to attempt to analyze them until the disparity in aluminium is eliminated though such attempts have been made in the literature.

The overall sensitivity of absolute X-ray scattering factor measurements to the bonding electron wave functions in metals is poor for the elements lithium through calcium. The sensitivity is good for the $3d$ wave functions in scandium through copper but poor for the $4s$, $4p$ orbitals. It is poor for the outer electron wave functions in zinc through strontium and only fair for the $4d$ wave functions in yttrium through silver. The $4d$ group of elements is not as favourable as the $3d$ group since the krypton core, $1s^2 2s^2 2p^6 3s^2 3p^6 4s^2 4p^6$, represents a larger fraction of the total scattering factor. For all elements above silver the sensitivity to the bonding electron wave functions is low.

Determination of Electron Probability Distribution from Compton Scattering

Does this mean that the X-ray technique is virtually useless for the determination of the outer electron wave functions in most metals? Fortunately, there is one last measurement that we shall now discuss that can provide this information, in principle, although it requires considerably more work in technique development. The measurement is the Compton or inelastic scattering in which the X-ray displaces the electron from the atom, both X-ray and electron going off in opposite directions. The helpful features in this type of measurement are that the X-ray is preferentially scattered from the outer electrons for small scattering angles since they are not as tightly bound to the atom, and that it occurs at all angles not being limited to Bragg peaks. The Compton scattering can be identified by an energy loss of the scattered X-ray which, in terms of its wavelength, represents a wavelength increase upon scattering through an angle 2θ given by

$$\Delta\lambda \cong 0{\cdot}0243\ \text{Å}\ (1 - \cos 2\theta), \tag{106}$$

The intensity of Compton scattering for a flat specimen intercepting the entire beam in symmetrical reflection is

$$I_c = I_0\left(\frac{e^2}{m_0c^2}\right)^2 \xi\, \frac{(1+\cos^2 2\alpha \cos^2 2\theta)}{2}\, \frac{A}{R^2}\, \frac{N_0}{2\mu}\,(1 - e^{-2\mu t/\cos\theta})(Z - \sum_i \sum_j f_{ij}^2)$$

$$\xi = \left\{\left[1 + \frac{\gamma^2(1-\cos 2\theta)^2}{(1+\cos^2 2\theta)[1+\gamma(1-\cos 2\theta)]}\right]\right\}\left(\frac{1}{1+\gamma(1-\cos 2\theta)}\right)^3$$

$$\gamma = E/m_0c^2 \tag{107}$$

where E is the energy of the incoming X-ray, ($m_0c^2 = 0{\cdot}51$ MeV), I_c is the scattered intensity in counts/sec through a slit of area A at a distance R from the sample, $e^{-\mu t}$ is the transmission of the sample, N_0 the number of atoms per cm^3, α the Bragg angle of the monochromating crystal, I_0 the incident intensity of the monochromatic beam in counts/sec and λ the wavelength of the incident X-ray. The terms

$$\sum_i \sum_j f_{ij}^2$$

represent the *incoherent* one-electron scattering factors

$$f_{ij} \cong \int \psi_i \, e^{i\mathbf{s}\cdot\mathbf{r}} \psi_j \, dx \, dy \, dz \qquad (108)$$

integrated over the atomic polyhedra

where ψ_i and ψ_j are the eigenfunctions of two different states but evaluated for each specific electron. One takes each electron in turn and assumes it has its own wave function, ψ_i, and another electron's wave function, ψ_j, to evaluate eqn. (108). Of course when $i = j$ eqn. (108) gives the ordinary atomic scattering factor that we have dealt with in eqn. (102). Compton scattering functions,

$$\mathscr{F} = \sum_i \sum_j f_{ij}^2,$$

have been evaluated for many free atoms from Hartree–Fock wave functions but not from metallic wave functions. Let us now return to the case of aluminium and plot ther vaious one electron contributions to the Compton scattering factor in eqn. (107)

$$Z - \mathscr{F} = Z - \sum_i \sum_j f_{ij}^2$$

which are calculated from the same free atom wave functions used in determining the scattering factors for Fig. 49. Let us also limit our attention to the region from $\sin\theta/\lambda = 0$ to $\sin\theta/\lambda = 0{\cdot}2$ since this is the region in which the outer electron wave functions make a sizeable contribution. In this region the terms f_{ij} in which $i \neq j$ represent only about 18% of the total Compton scattering and are, in fact, practically a constant ratio of the total in this range. We are left principally with the terms

$$\sum_i \sum_i f_{ii}^2 = 2f_{1s}^2 + 2f_{2s}^2 + 6f_{2p}^2 + 2f_{3s}^2 + f_{3p}^2.$$

The Compton scattering from the neon core, $1s^2 2s^2 2p^6$, is shown in Fig. 51 and is seen to be relatively small, particularly around $\sin\theta/\lambda = 0.1$ whereas the Compton scattering for the outer electrons is rather large in this region. *Thus a measurement of the Compton scattering at small values of* $\sin\theta/\lambda$ *is capable of yielding just the information about outer electron wave functions in metals not suitable for this determination by Bragg diffraction.* Unfortunately, the Compton scattering is very weak particularly since an analyzing crystal must be used to separate the Compton scattering (which has undergone a wavelength increase) from other sources of elastic background scattering. For example, for Co $K\alpha$(1·789 Å) on aluminium at $\sin \theta/\lambda = 0{\cdot}1$, $\Delta\lambda = 0{\cdot}0153$ Å. If one employs a conventional X-ray diffractometer modified to select the Co $K\alpha$, scatters this beam from a sample of aluminium, and attempts to separate the Compton scattering with an analyzing crystal, one would find virtually no intensity! A factor of at least 10 to 100 in the power of the X-ray tube is probably required to perform this experiment satisfactorily. With sufficient time and money this is a feasible endeavour and in the

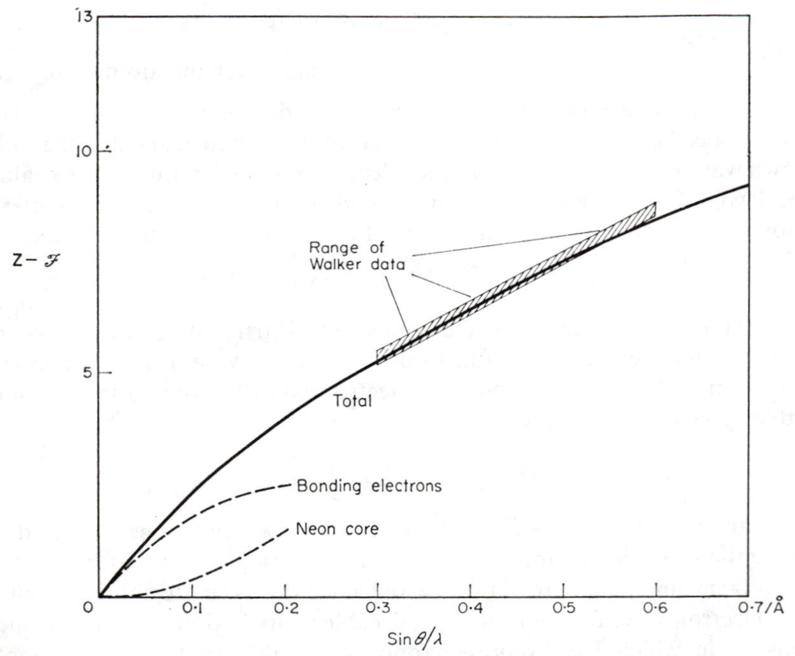

FIG. 51. The calculated free atom Compton scattering factor for aluminium over the range $\sin \theta/\lambda = 0$/Å to 0·7/Å, solid curve. The measurements of Walker give good agreement over the range $\sin \theta/\lambda = 0{\cdot}3$/Å to 0·6/Å, filled in area. The relative contribution of the bonding electrons and "neon core" electrons is indicated for the range $\sin \theta/\lambda = 0$/Å to 0·2/Å, dashed curves.

future we shall see this realized. This technique also enables us to determine the momentum distribution of the electrons which have participated in the Compton scattering. This effectively measures the momentum distribution of the electrons in the band, if the measurements are confined to $\sin \theta/\lambda < 0{\cdot}1$.

A second experimental technique requiring less intensity consists of measuring the total diffuse scattering (without an analyzing crystal) and assuming this to consist of thermal diffuse scattering and Compton scattering only. This is a very tenuous assumption at $\sin\theta \leqslant 0{\cdot}1$ since imperfections in the metal cause scattering at these low angles. The thermal diffuse scattering is separated by making the measurements at low temperatures or making a theoretical estimate. There is no accurate experimental data available for $\sin\theta/\lambda < 0{\cdot}2$ but the data on aluminium above $\sin\theta/\lambda = 0{\cdot}3$ agrees rather well with the free atom calculation,[13] Fig. 51. This method does not enable one to determine the momenta distribution.

A third technique suggested by D. Chipman, which may prove feasible with present diffraction equipment, is to allow the entire polychromatic beam to impinge on the sample and to analyze the scattered beam. One limits one's interest to the characteristic $K\alpha$ line which has undergone Compton scattering and is still distinguishable above a broad background.

C. Thermal Scattering

Thermal Scattering of X-rays

The X-rays which undergo Bragg diffraction by a crystal are elastically scattered in much the same way that a rubber ball bounces off a brick wall. The brick wall is compressed slightly in the neighbourhood of the point of impact and such a trivial amount of energy is transferred to the wall that we call the collision elastic. The momentum of the X-ray and the rubber ball have been changed in direction not magnitude, the crystal and wall absorbing this momentum in the two cases respectively. The momentum change of the X-ray is $s = 4\pi \sin\theta/\lambda$ in a direction normal to the diffracting plane.

In addition to this elastic scattering, X-rays are also capable of exciting or de-exciting one or more of the normal modes of a crystal (phonons) and we call this *thermal scattering.** Of course, both energy and momenta are conserved in thermal scattering. In a typical case, an X-ray of energy 5000 eV ($\lambda = 2{\cdot}42$ Å) may excite a phonon of energy $\sim 0{\cdot}03$ eV so that the energy lost by the X-ray can be considered trivial. On the other hand the X-ray momentum, $h\nu/c = 5000$ eV/(3×10^{10} cm/sec) $= 0{\cdot}167 \times 10^{-6}$ eV/cm/sec is comparable to that of the phonon momentum which for a typical case is $\sim h\nu/v = 0{\cdot}03$ eV/(3×10^5 cm/sec) $= 0{\cdot}10 \times 10^{-6}$ eV/cm/sec, so that the *thermal scattering of X-rays produces significant changes in the direction* (*momentum*) *but essentially no change in X-ray wavelength* (*energy*). In spite of this trivial change in X-ray energy, thermal scattering is still termed inelastic scattering because the crystal is altered internally by the excitation or de-excitation of a phonon in contrast to Bragg scattering in which the crystal recoils in its entirety with no internal changes.

The excitation or de-excitation of a phonon by an X-ray depends on the frequency distribution of the phonons, the temperature of the crystal and the

* By exciting a phonon we mean a phonon is raised to its next higher energy level and vice versa.

mass of the atoms, this latter point determining the momenta. The process is further complicated by the fact that in addition to the X-ray and the phonon a particular Bragg reflection is always involved. If you wish, it can be envisaged as a three body collision comparable to the rubber ball dislodging a small chip as it rebounds from the wall. This may cause the ball to bounce off at some angle other than the angle of incidence but its velocity or kinetic energy is only slightly altered. In the X-ray case a Bragg reflection absorbs the bulk of the momentum,

$$s = 4\pi \sin \theta/\lambda = (2\pi/a_0)(h^2 + k^2 + l^2)^{1/2}$$

(for a cubic crystal), the phonon accounting for the rest. Thus the laws of conservation of energy and momentum are satisfied by requiring the energy loss (or gain) of the X-ray to equal the energy gain (or loss) of the phonon, and the change in momentum of the X-ray to equal the sum of the momenta of the phonon, $h\boldsymbol{\sigma}$, and the crystal **s**. The thermal and Bragg scattering of X-rays are depicted vectorially in Fig. 52.

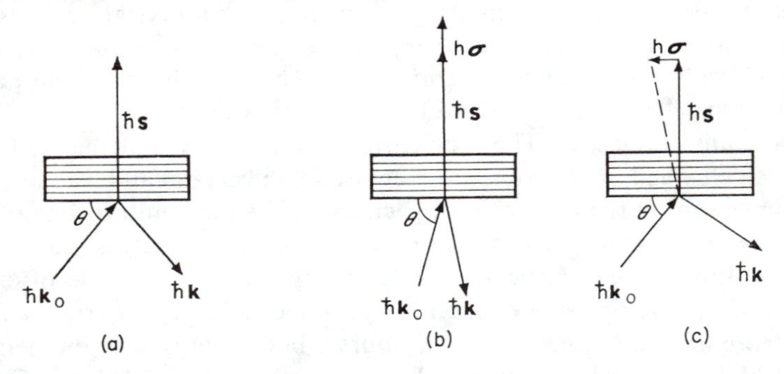

FIG. 52. Vector diagrams showing conservation of momentum in Bragg scattering, diagram (a), longitudinal phonon scattering, diagram (b), and transverse phonon scattering, diagram (c). $\hbar\mathbf{s}$ is the crystal momentum, $h\boldsymbol{\sigma}$ the phonon momentum, and $\hbar\mathbf{k}_0$ and $\hbar\mathbf{k}$ the incident and scattered X-ray momentum.

A certain fraction of the X-rays which are incident at the Bragg angle are *always* thermally scattered so that the intensity of the Bragg peaks is reduced by a factor e^{-2M} called the *Debye–Waller* factor. This factor is related quite simply to the square of the component of the displacement, **u**, of the normal modes in the direction of the scattering vector **s** and averaged over all directions for **u**. For crystals in which the local symmetry around each atom is cubic this average is independent of the direction of s and we have

$$2M = 16\pi^2\overline{u^2}(\sin \theta/\lambda)^2 \tag{109}$$

which turns out to be

$$2M \cong \frac{\hbar}{mN}\frac{(4\pi \sin \theta/\lambda)^2}{3}\left(1 + \frac{kT}{2D}\right)\left[\int_0^\infty \frac{q(\nu)\,\mathrm{d}\nu}{(e^{h\nu/kT} - 1)\nu} + \int_0^\infty \frac{q(\nu)\,\mathrm{d}\nu}{2\nu}\right] \tag{110}$$

where N is the number of atoms, m their mass and the remaining symbols have been defined in eqn. (42). If one compares eqn. (110) with the thermal energy of a crystal, eqn. (42), there is an additional factor $1/\nu$ in eqn. (110) so that the X-rays are less sensitive to the higher frequency modes than the energy or specific heat of a crystal. However, as in the case of energy and specific heat the factor $2M$ is rather insensitive to the distribution of normal modes so that the Debye approximation for $q(\nu)$ is generally quite adequate.* We then have

$$2M \cong \frac{22{,}973T}{A\Theta^2}\left(1 + \frac{kT}{2D}\right)^{\dagger}\left(\frac{\sin\theta}{\lambda}\right)^2\left(\phi(x) + \frac{x}{4}\right)$$

$$\phi(x) + \frac{x}{4} \cong 1 + 0{\cdot}026x^2 \quad \text{for} \quad x < 4 \tag{111}$$

$$\cong \frac{x}{4} + \frac{1{\cdot}62}{x} \qquad \text{for} \quad x > 4$$

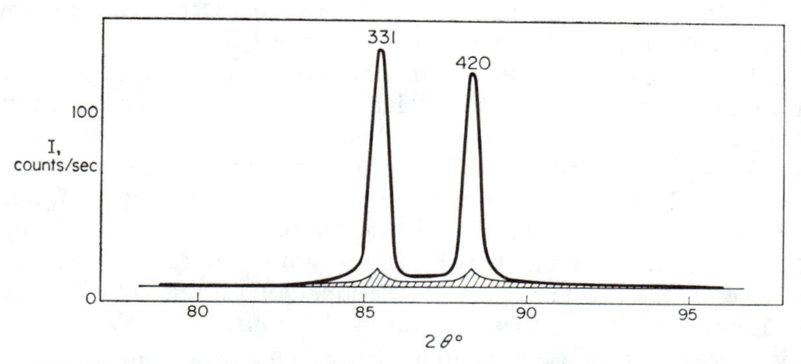

FIG. 53. The intensity in counts/sec versus counter angle showing the 331, 420 powder peaks in lead taken with Cu $K\alpha$ ($\lambda = 1{\cdot}54$ Å) radiation. The shaded area indicates the thermal diffuse intensity ordinarily included in measurements of integrated intensity.

where $x = \Theta/T$, A is the atomic weight, Θ the Debye temperature in degrees Kelvin, λ the wavelength in Å, and T is the temperature in degrees Kelvin. $(1 - e^{-2M})$ represents the fraction of X-rays that are thermally scattered. An appreciable fraction of these, though, are scattered through angles close to the Bragg angle. This is seen in Fig. 53 where the 331, 420 peaks of Pb

* This is not the case for a substance with optical modes clearly separated from the acoustic such as in germanium, silicon, zirconium oxide, etc. In these cases it is essential that the correct $q(\nu)$ be used in eqn. (110) since the Debye Θ values can appear to vary by as much as 20% when determined by the X-ray method (eqn. 111) and by specific heat.

† As an approximation $k/2D$ can be replaced by $(3/2)\,\alpha\gamma$ where α is the coefficient of linear expansion and γ is the *Gruneisen constant*, approximately equal to 2. A partial list of γ values follows:- α Fe, 2·10; Co, 1·95; Ni, 2·00; Cu, 1·96; Al, 2·17; Mn 2·42 K, 1·25; Pt, 2·54; W, 1·62; Ag, 2·40; Pd, 2·23. As a crude approximation γ can be assumed to be equal to 2·0 since the correction is small.

taken with monochromatic Cu $K\alpha$ (1·54 Å) radiation are shown. The thermal diffuse scattering tends to peak at the Bragg angle and any measurement of the integrated intensity of a Bragg peak always includes some of this. Chipman and Paskin[14] have studied this problem both theoretically and experimentally and they find that the intensity ratio, σ, of thermal scattering to Bragg scattering in powders is approximately:

$$\sigma \cong \frac{(0{\cdot}81)}{(N_0)^{1/3}} \frac{\cos\theta\,\Delta(2\theta)2M}{\lambda} \tag{112}$$

$$N_0 = g/a_0^3 \text{ for cubic crystals}$$

where N_0 is the number of atoms per unit volume, g the number of atoms in the unit cell, a_0 the lattice parameter, $2M$ the Debye–Waller factor, and $\Delta(2\theta)$ is the total angular range in radians over which the Bragg peak is integrated to determine the integrated intensity. In general, integration over 2 to 4° in 2θ includes about 10% of the amount lost according to eqn. (111).

In order to determine the factor $2M$, hence the Debye Θ, the following experimental procedure is recommended for powders:

1. For cubic crystals, measure the integrated intensity of a peak at two temperatures, preferably a peak at high $\sin\theta/\lambda$ and one with a high multiplicity, j. Room temperature and liquid nitrogen are recommended, since the anharmonic correction $kT/2D = (3/2)\alpha\gamma T$ in eqn. (111) is only a few percent. Estimate the Debye temperature from the Lindemann melting point formula eqn. (184) and use this value to make the diffuse background correction, eqn. (112). Ascertain that value of Θ which gives the observed ratio of the integrated intensities at the two temperatures (use eqn. (90)). If this value of Θ differs significantly from that estimated from the Lindemann formula use this new value to make the diffuse background correction and continually readjust Θ so as to be internally consistent. If the powder is not cubic, proceed in the same fashion except that $2M$ now depends on the particular h, k, l differing for different crystallographic directions.

In the case of alloys one replaces $|F|^2 e^{-2M}$ in eqn. (90) by $|F_M|^2$ where

$$F_M \cong \sum_i (f_i e^{-M_i}) e^{-2\pi i(hx/a + ky/b + lz/c)} \tag{113}$$

so that each type of atom has its own e^{-M_i} factor. Thus an alloy such as Cu_3Au would have the structure factors

$$F_M = (3f_{Cu}e^{-M_{Cu}} + f_{Au}e^{-M_{Au}})$$

fundamental peaks (both ordered and disordered);

$$F_M = (f_{Cu}e^{-M_{Cu}} - f_{Au}e^{-M_{Au}})$$

superlattice peaks (ordered only).

In effect this permits each element to have its own mean square displacement although in general the mean square displacements for each element in an alloy are about the same, so that $A\Theta^2$ is the same in eqn. (111). If the mean square displacements are the same, the ratio of the Debye Θ's for the elements in an alloy are approximately inversely proportional to the square root of their atomic weights.*

While there has been a scarcity of good work in the determination of Θ by X-rays the several cases that have been done like iron, copper, aluminium and lead give good agreement with the values deduced by specific heat, provided the X-ray measurements are performed below $T \cong \Theta$.

2. If one makes these measurements at $T > \Theta$ the effect of the variation of the elastic constants with temperature is more pronounced. We have already indicated that the anharmonicity in the atomic forces prevents us from strictly describing the thermal excitation of a crystal in terms of independent normal modes but that we do so because the anharmonicity is small. The practical effect, however, is to make it appear that the frequency distribution is slightly shifted to lower frequencies as we raise the temperature. This is demonstrated in the work of Chipman[15] on aluminium showing the decrease of Θ with increasing temperature as determined from X-rays. These results compare favourably with the variation of the elastic constants. A smooth decrease of the order of 20% from $T = 0°K$ to $T \cong 3\Theta$ can be expected for the Debye Θ of most metals.

In performing the measurements at high temperature one must remember to account properly for variations in V, μ, $\sin \theta$, F, etc., in eqn. (90) since the lattice expands with temperature. Surprisingly enough, most of the corrections tend to balance out. It is also important to anneal the crystal at the highest temperature to be measured so that intensity changes due to the onset of extinction can be avoided. However, it may be difficult to determine $2M$ if the extinction is large since the factor e^{-2M} in eqn. (90) assumes no extinction. If one can satisfy oneself that the extinction is less than 20% ($R_1 > 0{\cdot}8$) then one will not be in serious error using e^{-2M}.

High temperature furnaces for use with X-ray diffractometers and adequate to ~1000°C are easily constructed or available commercially (Materials Research Corp., 47, Buena Vista Ave., Yonkers, New York). For liquid nitrogen measurements an attachment built by DeMarco and Chipman is easily duplicated.[16]

The determination of the factor $2M$ provides a value of the Debye Θ and its variation with temperature. However, we must study the details of the diffuse scattering to determine the dispersion relation (ν versus σ) and from this deduce the frequency distribution, $q(\nu)$. A single crystal is required since we must measure the intensity of the thermal diffuse scattering versus

* This is rather a liberal use of the concept of the Debye Θ, since it is a parameter relating to the frequency distribution, and the frequency distribution is the same for all elements in an alloy. This liberal use of Debye Θ is done to maintain the simplicity of eqn. (111).

$\sin\theta/\lambda$ for specific directions in the crystal. The intensity (counts/sec) of diffuse scattering, I_T, for a monatomic cubic crystal of atomic mass, m, in symmetrical reflection intercepting the entire monochromatic beam is

$$I_T = I_0\left(\frac{e^2}{m_0c^2}\right)^2 \frac{f^2e^{-2M}A}{R^2}\,\frac{|\mathbf{s}|^2N_0(1-e^{-2\mu t/\cos\theta})}{m2\mu}\,\frac{(1+\cos^2 2\alpha\cos^2 2\theta)}{2} \times \sum_{\substack{2\text{ transverse, 1 longitudinal wave}}}\frac{E\cos^2\alpha'}{\gamma^2} \qquad (114)$$

where I_0 is the incident intensity in c/s, N_0 the number of atoms per cm^3, E the mean energy of the phonon of frequency ν

$$\left(E = \frac{h\nu}{e^{h\nu/kT}-1}+\frac{h\nu}{2}, \qquad \text{harmonic forces only}\right),$$

α' is the angle between the scattering vector $\mathbf{s}\,(= 4\pi\sin\theta/\lambda)$ and the polarization direction of the phonon and the other terms have been defined before, eqn. (98). The polarization direction of a phonon is the direction of its vibrations. A longitudinal phonon vibrates in a direction along its direction of propagation while the two transverse waves vibrate normal to its propagation direction and 90° apart. The three polarization directions thus form a set of cubic axes. Due to crystal symmetry it is possible to arrange the experiment so that the three waves are approximately separated. For example, consider a f.c.c. crystal whose surface is the 200 plane. The first diagram in Fig. 52 shows the conditions for the 200 Bragg reflection. If the angle of the incident beam is then increased and the counter angle adjusted to equal the incident angle we have the conditions for thermal scattering in diagram 2. The phonon propagation direction (and momentum) is normal to the 200 plane and the angle α' between the polarization direction and the scattering vector $\mathbf{s}$ is zero for the longitudinal mode and 90° for the two transverse modes. Since $\cos^2\alpha'$ appears in eqn. (114) we only observe scattering from the longitudinal phonons. On the other hand, if the incident angle of the X-rays is kept at the Bragg angle and the counter is moved to lower or higher angles we have the conditions in diagram 3 so that only the transverse phonons are seen. By measuring the absolute intensity of thermal diffuse scattering we determine the frequency from eqn. (114) and by determining the phonon momentum from the vector diagrams in Fig. 52 we can determine frequency versus momentum. Since the phonon momentum is the wave number times Planck's constant the *result of thermal diffuse measurements is a determination of frequency versus wave number for longitudinal and transverse normal modes propagating along specific directions in the lattice.* This, of course, is just the dispersion curve and Fig. 54a shows this for Walker's measurements[17] on aluminium in the 100 direction. Such curves are quite typical for all crystal directions. Since the slope of the dispersion curves equals the velocity of the phonon they must join smoothly to the observed velocity of the ultrasonic waves as ν gets very small.

In order to determine $q(\nu)$ several assumptions and laborious calculations must follow. One makes a model of the lattice as in Fig. 23 and adjusts the values of the harmonic spring constants so that the calculated dispersion curves match the observed. Walker connected springs as far as third nearest neighbours in aluminium but concluded that such a central force model would not give the observed dispersion curves. Of course the failure of the central force model could also have been concluded from the deviation from the Cauchy relation; $c_{12}/c_{44} = 2{\cdot}22$ for aluminium (p. 81). With a model that allows for non-central forces but not anharmonicity Walker calculated the effect of the displacement of an atom in various directions on its nearest neighbour, Fig. 54c. The errors are quite large particularly the unrealistic negative displacement in diagram 2 and this may be due to limitations of his model. The calculated frequency distribution is shown in Fig. 54b. The specific heat calculated from this frequency distribution does not give as good an agreement with experiment as one would desire.

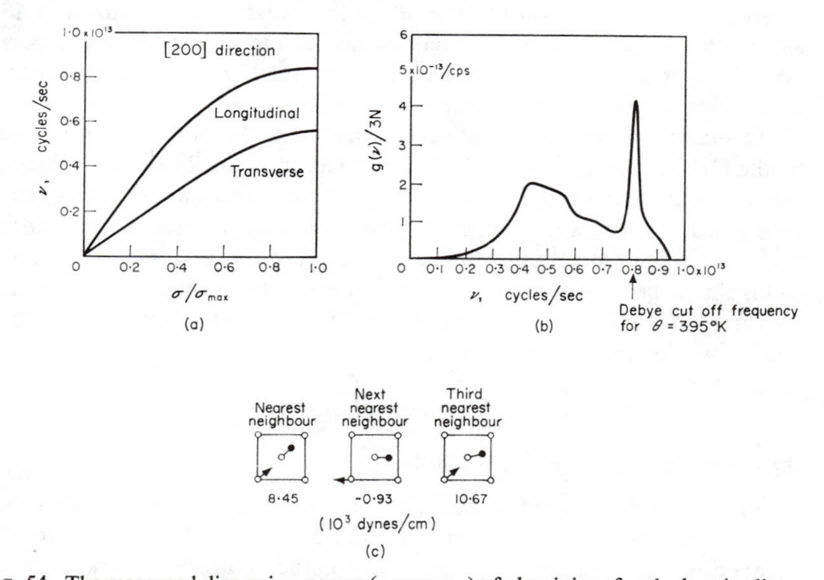

FIG. 54. The measured dispersion curves (ν versus σ) of aluminium for the longitudinal and transverse phonons in the 200 direction, *a*; the frequency spectrum deduced by Walker, *b* (showing the Debye frequency); and the direction and magnitude of the force in dynes/cm on the lower corner atom when the atom at the centre of the face is given unit displacement towards its nearest neighbour, next nearest neighbour, and third nearest neighbour, *c*.

We do not know all the experimental problems in such a measurement but some of them are:

1. The Compton scattering in low Z elements like aluminium comprises a large fraction of the total diffuse scattering ($\sim 75\%$) and should be measured directly by its wavelength change, if possible.

2. The presence of an oxide layer which absorbs more than a few per-cent of the X-rays may cause anomalous results, particularly if its Bragg peaks are at the same angles as the thermal diffuse scattering of the metal.

3. One must be wary of impurities that fluoresce. For example, 0·01 atomic per cent of manganese in aluminium will yield a diffuse scattering as great as the thermal scattering if Cu $K\alpha$ radiation is used.

4. The intensity must be placed on an absolute basis. Presumably this can best be done by using polystyrene (p. 144) as a standard.

It is difficult to pinpoint the weaknesses in the X-ray determination of the frequency distribution. The experimental dispersion curves of Walker have been confirmed by neutron diffraction measurements which are not subject to the same experimental difficulties. This suggests that the model is the source of error and is too naïve. However, any improvement in the model to include anharmonicity increases the complexity of the calculations considerably. In any event, a direct comparison of the forces between the atoms in a metal as calculated from the Schrödinger equation and as determined experimentally from thermal diffuse scattering is not readily forthcoming since both theory and experiment are exceedingly difficult.

In summary then, the integrated intensities of the Bragg peaks are reduced by the Debye–Waller factor, e^{-2M}. A measurement of $2M$ yields values of the Debye Θ and its variation with temperature. The intensity lost from the Bragg peaks appears as thermal diffuse scattering centred about the Bragg peaks. An absolute measurement of the intensity of this diffuse scattering yields the dispersion curve. A model of the crystal is then devised to give the measured dispersion curves and the frequency distribution of the normal modes are deduced from this model.

NEUTRONS

Introduction to Neutron Diffraction

The major difference between neutron and X-ray diffraction is that the true absorption coefficients are several thousand times smaller for neutrons (true absorption means capture by a nucleus for the neutron and fluorescence for the X-ray). In addition the scattering cross-sections are 10 to 100 times smaller for neutrons. In order to obtain appreciable scattering with neutrons the sample must be considerably larger and in most cases the attenuation of a neutron beam in traversing a sample is principally due to scattering not absorption. Except for scattering by the unpaired electrons in a magnetic material the scattering of thermal neutrons by nuclei gives rise to an atomic scattering factor independent of sin θ/λ since the nucleus is so much smaller than the neutron wavelength. Different isotopes frequently have different atomic scattering factors for neutrons (traditionally called *nuclear scattering amplitudes* or *scattering lengths*) even though their identical electronic structure gives rise to identical X-ray atomic scattering factors. Even in the case of elements with a single isotope whose nuclear spin, I, is not zero the nucleus may have different scattering amplitudes in scattering with its spin parallel or antiparallel to the neutron (cf. vanadium, p. 103). For any element the scattering amplitudes of the individual isotopes weighted according to their abundance and according to the quantum mechanical probability of parallel and antiparallel scattering (g on p. 102) represents the average scattering amplitude for the neutrons. This average scattering amplitude is called the *coherent scattering amplitude*, b, and replaces the product of the X-ray atomic, scattering factor, f, and the scattering amplitude per electron, e^2/m_0c^2, in eqn. (77). For very special reasons b can be positive or negative although f is always positive. If the scattering amplitude of any of the isotopes differs from the average this gives rise to an incoherent scattering called *isotope or spin diffuse scattering* which is independent of sin θ/λ and does not contribute to the Bragg scattering. The coherent scattering amplitudes are tabulated in the appendix, the number of cases of negative values of b being rather limited.

Equipment

Experimental Details. Practically all neutron diffractometers have been built to individual specifications since the demand has not been great enough to interest commercial manufacture. In addition, the greater expense (~$50,000) compared to X-ray diffractometers as well as the individual problems of mounting the diffractometer on the neutron source have discouraged commercial interest. However, a working set of 137 blueprints for a spectrometer can be purchased from Cooper Trent Blueprint and Microfilm Corporation, 2701, Wilson Blvd., Arlington, Va., U.S.A., by requesting a complete set of AEC #CAPE-72 taken from TID-4100.*

The neutron source is a reactor creating high energy neutrons (1–2 MeV) by uranium fission. These neutrons make a large number of collisions in a material called a moderator which scatters considerably more than it absorbs, $\sigma(n, n) \gg \sigma(n, \gamma)$. Each scattering event dissipates considerable energy in and occasionally dislodges an atom from the moderator, the neutron losing some energy in each case, until its energy is only 10–20 eV. Below this value the neutron has insufficient energy to dislodge an atom, and excites phonons only. In this way the neutron energy is eventually reduced to thermal energy and it ends up bouncing around in thermal equilibrium with the phonons. A one megawatt reactor creates a steady state flux of about 10^{13} neutrons/cm^2/sec near its uranium core. In this steady state most of the neutrons in

* A German made commercial unit is advertised by American M.A.N. Corp., 500 Fifth Ave., New York 36, N.Y.

excess of those required to sustain the chain reaction are eventually absorbed by the moderator. An extremely small fraction of the neutrons escape through exit ports and a small portion of these are utilized for neutron diffraction. The lifetime of a neutron in a reactor from its birth to its capture is a few milliseconds although a neutron in vacuum has a half life of ~10 minutes decaying into a proton, electron and neutrino.

The neutrons in thermal equilibrium with the moderator have an approximately Maxwellian distribution

$$g_\lambda = \frac{2}{\lambda}\left(\frac{E}{kT}\right)^2 e^{-E/kT} \tag{115}$$

where E is the energy of the neutron whose wavelength is λ ($E = \frac{1}{2}h^2/M\lambda^2$) and g_λ is the fraction of the neutrons whose wavelength is between λ and $\lambda + d\lambda$. For a reactor operating at a temperature of ~100°C the maximum in g occurs at ~1 Å, a most remarkable accident of nature since these neutrons are the right wavelength for diffraction purposes. Unfortunately, the Maxwellian distribution lacks any characteristic wavelength so that the neutron beam must be monochromated by a crystalline reflection. Crystals of lead, beryllium and copper are all useful for this purpose while a crystal of f.c.c. $Co_{0\cdot95}Fe_{0\cdot05}$ can be utilized to obtain polarized neutrons. Monochromatic beams of about 10^8 neutrons/min capable of resolving diffraction peaks 15 minutes apart can be produced.

One's first impression of neutron diffraction is that everything is an order of magnitude larger than in X-ray diffraction, the sample, the source, the counter, the cost and the distance between one's office and one's equipment. A picture of the neutron diffractometer constructed from the AEC #CAPE-72 plans is shown in Fig. 55. In general, at least one mole of sample is required for powder diffraction.

A. Crystal Structure, Atomic Positions and Atomic Sizes (Principally Magnetic)

Determining the Arrangement of Magnetic Moments in a Crystal

Neutron diffraction has been applied successfully to cases in which one of the elements has a very low Z making it insensitive to X-ray scattering. In particular the crystal structures of materials containing hydrogen have been studied by chemists. This advantage of neutron diffraction has rarely evoked metallurgical interest, TiH and ZrH perhaps being the outstanding exceptions. Considerably more interest resides in the magnetic scattering of neutrons and the determination of the crystallographic arrangement of the atomic magnetic moments and their direction in the lattice. The problem of determining the magnitude and crystallographic arrangement of magnetic moments is quite similar to that of determining the crystal structure of a substance by X-rays except for the added complication of the *direction* of the magnetic moment.

FIG. 55. The neutron diffractometer built for Corliss and Hastings at Brookhaven. The large cylinder on the left contains paraffin and cadmium to shield the neutron counter which is about two feet long and two and one half inches in diameter. The neutron beam emerges through the square monitoring counter identified by the 16 bolts around its perifery. A typical angular velocity for the counter is $\sim 2°/\text{hr}$. The height from the ground to the top of the shield is ~ 3 feet. (Brookhaven National Laboratory Photograph).

In crystal structure determinations we try to reduce the analysis to a unit cell in which we specify the position and type of each atom. In magnetic structure determinations we try to reduce the analysis to a *magnetic unit cell* in which we specify the position of each magnetic moment, the type (or magnitude) of each magnetic moment *and the direction of each magnetic moment.* This added complication is compensated by helpful information gained elsewhere. For example, we generally know the positions of the atoms themselves from X-ray work and we know that the magnetic moments are generally due to *d* or *f* electrons so they are centred on the transition or rare earth elements whose positions are generally known. Thus the problem reduces to separating the magnetic scattering contribution to the Bragg peaks from the nuclear scattering contribution.

The structure factor, F_N, for neutrons is

$$|F_N|^2 = |F_{\text{NUCLEAR}} \pm F_{\text{MAGNETIC}}|^2 \tag{116}$$

$$F_{\text{NUCLEAR}} = \sum_j b_j \, e^{2\pi i(hx/a+ky/b+lz/c)} \tag{117}$$

$$F_{\text{MAGNETIC}} = \sum_j p_j \sin \alpha_j \, e^{2\pi i(hx/a+ky/b+lz/c)} \tag{118}$$

where the + and − in eqn. (116) refers to the neutron being scattered with its own magnetic moment antiparallel or parallel, respectively, to the magnetic moment of the atom, α_j is the angle between the neutron scattering vector **s** and the magnetic moment of atom j, and p_j is the magnetic scattering amplitude of the atom j. F_{NUCLEAR} and F_{MAGNETIC} both have the dimensions of length unlike the X-ray case (eqn. 77) in which F is dimensionless. In the X-ray case the factor $e^2/m_0c^2 = 0{\cdot}282 \times 10^{-12}$ cm. which eventually multiplies F in computing intensities, is traditionally kept separate.

In general powder diffraction measurements are done in symmetrical transmission through a plane parallel sample contained, preferably, between two thin vanadium sheets which do not give rise to Bragg peaks themselves since the coherent scattering amplitude of vanadium is almost zero. If the sample intercepts the entire beam, the structure factor is related to the integrated intensity by

$$|F_N|^2 = \frac{8\pi R^2 \omega C \rho V^2 \sin^2 2\theta_B}{A j I_0 \rho' t \lambda^3 e^{-\mu t/\cos\theta} e^{-2M} e^{-2M'}} \tag{119}$$

where ρ is the true density of the bulk material, $\rho' t$ is the thickness of the powder sample in g/cm², $e^{-\mu t}$ is the measured transmission of the sample (at $\theta = 0$) including the sample container, and the remaining symbols have been defined in eqn. (90). By using loose packed powders or even sintered powders the effects of sample porosity and preferred orientation are generally not present since the neutron beam penetrates so deeply. In addition, the scattering amplitudes p and b are so much smaller than in the X-ray case that primary and secondary extinction are essentially non-existent in 400 mesh powders. The absolute standardization or determination of I_0 is easily accomplished by using powdered Al or Pb as an absolute standard since all factors in eqn. (119) are known or can be measured. In general, one combines the factors I_0, A, R^2, ω and λ^3 into one "constant" of the diffractometer. If one has insufficient sample to intercept the entire beam one may fabricate a cylindrical sample which intercepts only a fraction of the beam whereby eqn. (119) is modified by replacing $\sin^2 2\theta$ by $\sin\theta \sin 2\theta/2$, t by the volume v of the sample (ρ' is the apparent density of the loose powder) I_0 by P_0 the intensity/cm^2 and $e^{-\mu t/\cos\theta}$ by an absorption factor tabulated by Bradley.[20]

The various common types of neutron diffraction patterns are illustrated in Fig. 56 for a b.c.c. metal powder. If the metal is non-magnetic, then only the allowed nuclear reflections occur, diagram 1, which are 110, 200, 211, etc., eqn. (117). If the material is ferromagnetic then eqn. (118) tells us that the magnetic contributions add to or subtract from the nuclear contributions since they have the same periodicity. If the neutron beam is unpolarized then one calculates the intensity with the plus sign and minus sign separately in eqn. (116) and averages the two, $|F_N|^2 = |F_{\text{NUCLEAR}}|^2 + |F_{\text{MAGNETIC}}|^2$. This additional magnetic intensity is seen in diagram 2. In a powder sample one has to average over $\sin\alpha_j$ in eqn. (118). For example, if the magnetic moments prefer to point along the [100] direction in one small region of the crystal then they must also point along all six equivalent directions, [010], [−100], [001], [0–10], [00–1], in other regions with equal probability since there can be no preference in a cubic crystal. The average of $\sin^2\alpha_j$ (for the intensity) turns out to be $\frac{2}{3}$ in a cubic material and in fact is independent of the direction preferred by the magnetic moments. It is also possible by application of an external magnetic field to align the magnetic moments in a specific direction relative to the scattering vector **s**. One can alternately arrange $\sin\alpha_j$ to be zero and unity so that the magnetic contribution to the Bragg peaks can be directly ascertained. Diagram 3 shows the diffraction pattern for an antiferromagnetic b.c.c. metal. In this structure the magnetic moments of the equivalent atoms at the body centre position and at the corner positions are equal in magnitude and opposite in directions. For such a case eqn. (118) leads to superlattice lines like 100, 111, 210 etc. which are solely magnetic in origin. The fundamental lines 110, 200, 211, etc. have no magnetic contribution and are identical to the pure nuclear peaks in diagram 1. If the temperature of the material is raised well above its Curie or *Néel temperature*,* the magnetic moments are no longer ordered but point in random directions. In such a case the scattering is demonstrated in diagram 4. Its intensity extrapolated to $\theta = 0$ is given by

$$I = \frac{I_0 A N_0 t \rho' e^{-\mu t/\cos\theta} 2(e^2\gamma)^2 g^2 J(J+1)}{R^2 \rho \cos\theta \, 3(2m_0c^2)^2} \tag{120}$$

valid for symmetrical transmission at $\theta = 0$

* The Néel temperature is the Curie temperature of an antiferromagnetic material, in deference to Professor L. Néel who predicted antiferromagnetism.

where γ is the magnetic moment of the neutron in n.m. This expression permits us to determine $g^2J(J + 1)$, g being the Landé g factor (eqn. 179) and $\hbar\sqrt{[J(J + 1)]}$ the total angular momentum of the atom. The factor $J(J + 1)$ arises if $kT_c \ll E$ where T_c is the Curie or Néel temperature and E is the

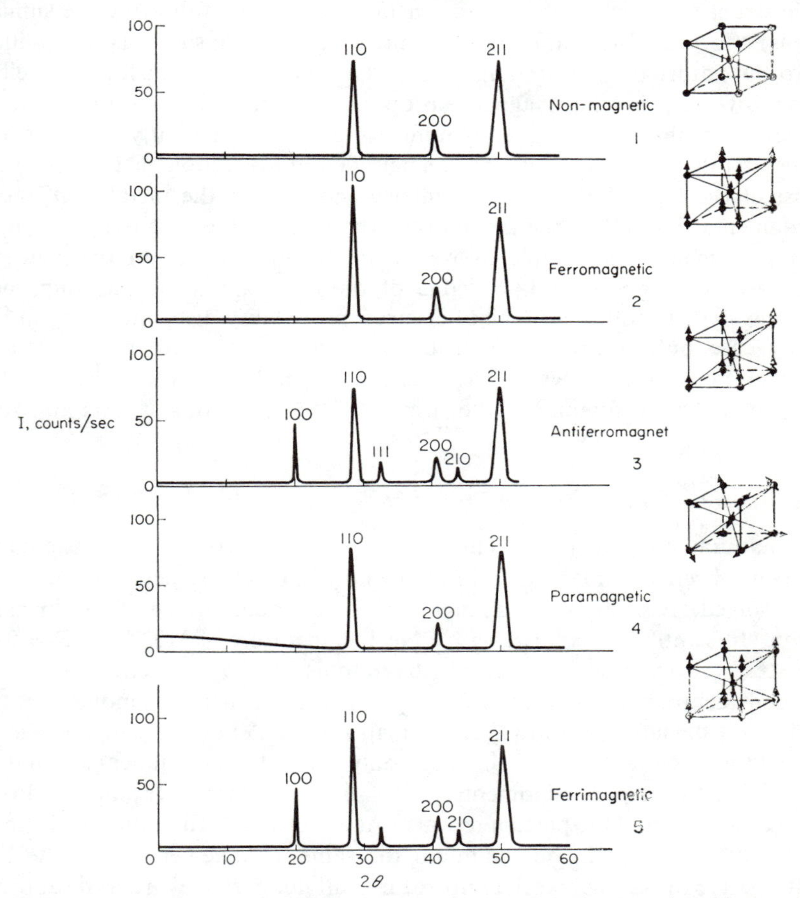

FIG. 56. Typical neutron diffraction patterns for a body-centred cubic metal ($a_0 \cong 2{\cdot}86$ Å and $\lambda \cong 1$ Å). Diagram 1 shows the fundamental peaks 110, 200, 211 from a non magnetic sample. In diagram 2 additional intensity appears at the fundamental reflections due to the ferromagnetic alignment of magnetic moments. In the antiferromagnetic case, diagram 3 magnetic superlattice peaks appear, the fundamental peaks containing no magnetic contribution. If either of these samples are heated to well above their Curie temperatures a magnetic diffuse intensity appears at low angles, diagram 4. In the ferrimagnetic case the magnetic moments are unequal in magnitude but opposite in direction. The sample has a net magnetic moment and both superlattice lines and fundamental lines contain magnetic contributions, diagram 5. Due to the lack of sharp characteristic lines in the neutron distribution from the reactor the powder diffraction peaks broaden significantly with increasing scattering angle.

neutron energy. If $kT_c > E$ then $J(J + 1)$ is replaced by J^2 and intermediate values of T_c yield intermediate values in eqn. (120). In the ferrimagnetic case, diagram 5, there are contributions to both superlattice and fundamental peaks.

Considerable effort has been expended in determining complicated magnetic structures, particularly in non-metals, historically following the similar intense efforts in X-ray diffraction. Similarly, little real insight has been gained as to the nature of the solutions of the Schrödinger equation for those electrons involved in the magnetic properties. One principal observation, however, is that the magnetic moments are predominantly colinear i e. either parallel or antiparallel to a single definite direction in the unit cell Most of the exceptions occur, peculiarly enough, for the metals. Of more fundamental interest to the metallurgist though, are the magnetic structures of the elements and simple alloys. While the determination of magnetic structures presupposes a knowledge of the magnetic scattering amplitude and this depends on the unpaired electron probability distribution, as in the X-ray case only approximate probability distributions are necessary and Hartree–Fock $3d$ or $4f$ wave functions can be employed as the case may be.

We shall now summarize the results for the transition metals and rare earths.

The Magnetic Structure of the Transition Metals

Titanium and vanadium have no unpaired electrons. A considerable effort has been expended on chromium whose magnetic structure is only partially solved. Each atom has a magnetic moment of about 0·4 of a Bohr magneton. It is antiferromagnetic with a Neel temperature of 313°K. At $T = 0$°K the structure is shown in Fig. 57 (diagram 1).* The magnetic moment at the body centre position is almost antiparallel to the magnetic moment at the corner, the deviation from 180° being small at point A but gradually increasing so that there is a rather rapid change in only a few atomic distances at point B. At point C the magnetic moments are just out of phase with point A a dozen unit cells apart, and the pattern repeats. At about 112°K the spins all abruptly rotate 90° (diagram 2) but maintain the same arrangement along the line ABC. At 313°K, the Néel temperature, all long range order disappears leaving a short range order which can be seen for 100° or so above the Néel temperature. (There is some evidence that the magnetic moment of each chromium atom becomes zero at temperatures somewhat above the Néel point.)

We can understand the origin of this peculiar magnetic structure in terms of our vector model of magnetism (page 101) by considering a line of magnetic moments with a nearest neighbour interaction, $W_{EX} = -2J_1\mathbf{S}\cdot\mathbf{S}$ favouring antiferromagnetism (J_1 negative). At $T = 0$°K this leads to the ordered

* Private communication, J. Hastings.

arrangement in diagram 3, Fig. 57. If we now add a second nearest neighbour interaction $|J_2| < |J_1|$, also favouring an antiferromagnetic arrangement the line of atoms must find some compromise. Each atom cannot arrange itself so that both its nearest and next nearest neighbour is antiparallel to itself,

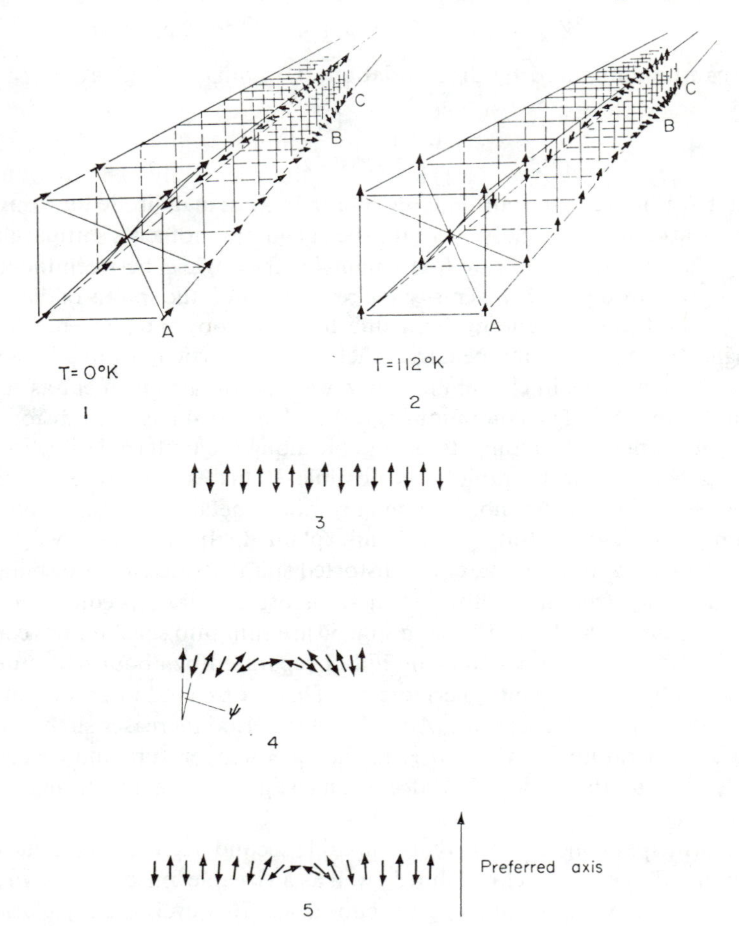

FIG. 57. A plausible arrangement of magnetic moments in chromium metal suggested by the neutron diffraction patterns. The magnetic moment at the corners of the cube are almost opposite to those at the body centre and form a distorted spiral arrangement along the line ABC. Diagram 1 is the structure from $T = 0°K$ to $T = 112°K$ at which temperature the magnetic moments abruptly rotate 90° with relatively no change in the spiral arrangement along ABC. Diagrams 3, 4 demonstrate the arrangement of magnetic moments of a linear chain if both nearest and next nearest neighbour interactions are antiferromagnetic with the ratios $|J_1|/|J_2|$ greater and less than four respectively. If an anistropy energy favouring a preferred axis is added to the case in diagram 4 we obtain the arrangement in diagram 5 illustrative of pure chromium in diagrams 1 and 2.

since they are nearest neighbours to themselves. The compromise is seen in diagram 4 in which the spins form a continuous spiral. By making the crude approximation that $\mathbf{S}\cdot\mathbf{S} \cong S^2 \cos\varphi$ the magnetic energy of each atom in the chain of diagram 4 is

$$W_{EX} = -4J_1S^2 \cos\varphi + 4J_2S^2 \cos 2\varphi$$

which can be minimized by differentiation, $\mathrm{d}W_{EX}/\mathrm{d}\varphi = 0$, giving

$$\cos\varphi = \frac{1}{4}\frac{J_1}{J_2} \tag{121}$$

so that if J_1 is more than four times greater than J_2 in magnitude the minimum energy is attained by rigidly aligning the magnetic moments antiparallel at $T = 0°$ K as in diagram 3. For J_1 less than $4J_2$ the angle φ between the neighbours is given by eqn. (121). This crude result predicts the spirals of diagram 4 but if we add another energy term due to anisotropy, i.e., an energy term favouring the magnetic moments parallel to some particular direction in the crystal (the cube axis in chromium), then we have an arrangement as seen in diagram 5, Fig. 57. The continuous spiral of diagram 4 has been distorted to keep as many magnetic moments as possible along the preferred direction, the spiralling being done as rapidly as possible. This essentially explains the arrangement of magnetic moments in chromium metal. The transition from diagram 1 to diagram 2 at 112°K is unexplained. It involves a very small energy change but does not alter the distorted spiral arrangement of magnetic moments. The structure factor for such distorted spirals predicts that the superlattice lines 100, 111, 210 etc. in Fig. 56 are split into several lines centred about the 100, 111, 210 positions in 2θ but separated by about half a degree or so depending on the spiral periodicity. This is observed in chromium and the periodicity is ~14 unit cells. Actually, this period increases slightly as we raise the temperature from $T = 0$°K to the Néel temperature and this is presumably due to the ratio J_2/J_1 decreasing slightly with increasing lattice parameter.

The principal conclusion is that a sizeable second nearest neighbour magnetic interaction exists in chromium as well as a sizeable anisotropy favouring magnetic moment alignment along the cube axes. However, the calculation of J_2 and J_1 from the solutions of the Schrödinger equation is still beyond our ken. One can venture to guess, though, that the magnetic moments are principally in the e_g orbitals, Fig. 17, and that the anisotropy is due to spin orbit coupling (page 18).

Manganese crystallizes in four allotropic modifications denoted α, β, γ and δ in order of stability as the temperature is increased. α-Mn and β-Mn are quite complicated crystallographically but γ-Mn and δ-Mn form f.c.c. and b.c.c. structures respectively. α-Mn is antiferromagnetic since additional neutron diffraction lines appear below its Néel temperature, 95°K, but its exact magnetic structure is not known. Crystallographically there are four kinds of atoms in α-Mn each with a different nearest neighbour configuration. These four atoms appear to have different magnetic moments ranging from zero to ~3 Bohr magnetons with no evidence for spirals, although single crystals may be required to decide this. The evidence from both neutron diffraction and nuclear resonance (chap. XI) is that β-Mn has no unpaired electrons. γ-Mn is antiferromagnetic with a Néel temperature of 580°K and a magnetic moment per atom of ~2·5 Bohr magnetons. Above the Néel temperature it is f.c.c. but distorts to face centred tetragonal, $c/a = 0{\cdot}95$, below this temperature. Its magnetic structure is seen in Fig. 58, consisting of an alternation of magnetic moments in 200 planes in the c direction, with the magnetic moments normal to these planes. This distortion arises from the peculiarity in f.c.c. structures that of the twelve nearest neighbours of an atom one third of them are nearest neighbours to each other so it is impossible to arrange all nearest neighbours antiparallel even if favoured by the magnetic exchange energy. γ-Mn compromises with the magnetic structure in

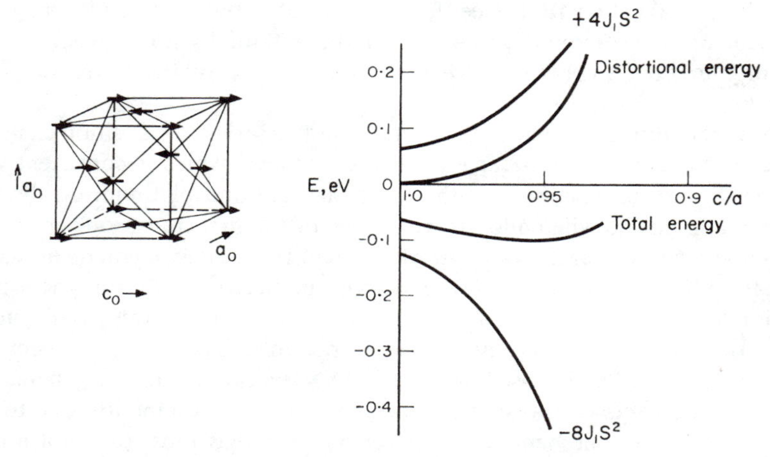

FIG. 58. The antiferromagnetic arrangement of magnetic moments in γ-Mn relative to the face centred tetragonal unit cell, $c/a = 0{\cdot}95$. The tetragonal distortion increases the relative separation of parallel magnetic moments to antiparallel magnetic moments. The contributions of the magnetic energy of parallel and antiparallel nearest neighbours and of the distortional energy is shown as a function of the c/a ratio. Above the Néel temperature the crystal structure is f.c.c.

Fig. 58 in which eight of the twelve nearest neighbours are antiparallel and four are parallel, but distorts its cubic structure so that the energetically unfavourable four parallel neighbours are further away than the eight antiparallel ones. Since it requires about 0·1 eV per atom to distort the structure from cubic to $c/a = 0{\cdot}95$, it shows that the magnetic exchange energy or J_1 depends strongly on distance. The curve in Fig. 58 schematically indicates the various contributions to the total energy of γ-Mn as a function of c/a ratio showing the minimum at $c/a = 0{\cdot}95$. The second nearest neighbour interaction is probably significantly smaller than in chromium since the ratio of second nearest neighbour distance to nearest neighbour distance in a f.c.c. metal is about 22% greater than in a b.c.c. metal. δ Mn cannot be stabilized at room temperature so we do not know its magnetic structure, but Mn-Cr (50-50) forms a disordered b.c.c. alloy which can be stabilized. It is antiferromagnetic although single crystals would be required to determine the existence of spirals. It has an average magnetic moment per atom of 0·7 Bohr magnetons,[21] which suggests by extrapolation that pure b.c.c. Mn is antiferromagnetic with a magnetic moment per atom of about one Bohr magneton.

α-Fe, h.c.p. and f.c.c. Co, and f.c.c. Ni are known to be ferromagnetic from magnetization measurements and this has been supported by neutron diffraction work. While it appears impossible to retain f.c.c. Fe at low temperatures by quenching, alloying with Mn stabilizes this phase. Neutron diffraction reveals these alloys to be antiferromagnetic with an arrangement of magnetic moments identical to γ-Mn, Fig. 58. Extrapolation to pure iron suggests that it is antiferromagnetic with a Néel temperature of about 80°K and a magnetic moment of $\sim 0{\cdot}5$ Bohr magnetons. There is essentially no noticeable tetragonality in f.c.c. Fe (as in γ-Mn) presumably due to the lower magnetic energy.

Neutron diffraction evidence for the $4d$ and $5d$ series of transition elements indicates the absence of sizeable magnetic moments which is consistent with magnetization measurements although palladium and platinum do develop magnetic moments when alloyed with some of the first transition series.

So far we have not indicated to what extent the atomic magnetic moments arise from the intrinsic magnetic moments of the electrons and from the orbital motion of the electrons (see page 6). It is not easy to deduce this from neutron diffraction but we rather rely, whenever possible, on other more sensitive measurements to be discussed in Chapter IX. We can say that, in general, the magnetic moments of transition metal atoms are predominantly due to the intrinsic magnetic moments of the electrons and that most (but not all) of the orbital magnetic moment is "quenched". Suppose, for example, that the magnetic moment of a transition metal atom in a cubic crystal were entirely due to two unpaired $3d$ electrons in the e_g orbitals, Fig. 17, the total $m_s = 1$. Since these orbitals are comprised of the $m_l = 0$ and equal amounts of $m_l = +2$ and $m_l = -2$ orbitals the total orbital angular momentum is zero. Thus the total magnetic moment of two electrons in the e_g orbitals is due solely to the intrinsic magnetic moment of the two electrons. These two electrons have chosen these e_g orbitals due to the crystalline field i.e. the Coulomb potential of the neighbouring atoms. On the other hand, if the Coulomb potential of the neighbours was removed as it would be for the free atom, then the two electrons would select the $m_l = -2$ and $m_l = -1$ orbitals due to spin orbit coupling (page 18). In the metal there is a competition between the two effects, the electrostatic crystalline field forcing the electrons into orbitals that generally have a total orbital angular momentum equal to zero and the magnetic spin orbit coupling generally preferring orbitals with total orbital angular momentum not equal to zero. If the crystalline field is the larger of the two the orbitals tend to be *quenched* while if the spin orbit coupling is larger the orbitals are not quenched. In the transition metals the crystalline field wins about 80–90 % of the battle so that the magnetic moment of the atoms is primarily "spin only".

Magnetic Structure of the Rare Earth Metals

From magnetization measurements the rare earth metals are known to exhibit ferromagnetism at very low temperatures and except for gadolinium undergo a transition to a "sort of antiferromagnetism" at higher temperatures. Their Néel temperatures are all lower than room temperature. The first clue to the behaviour of the rare earth metals comes from an inspection of the $4f$ wave functions in Fig. 18. Since they are so much closer to the nucleus relative to the $5d$ and $6s$ bonding orbitals the crystalline field effects are much weaker and the spin orbit coupling is much stronger. This latter point is due to the larger apparent positive charge of the nucleus. As a result the electrons prefer the same orbitals as in the free atoms although some weak crystalline field effects sometimes cause a partial "quench" of the orbitals at very low temperatures. The second clue has been intuitively gathered and is that the nearest neighbour interaction, J_1, favours ferromagnetism but that the second nearest neighbour interaction, J_2, is appreciable and favours antiferromagnetism. At low temperatures the ratio of $|J_2|/|J_1|$ favours ferromagnetism, Fig. 59,

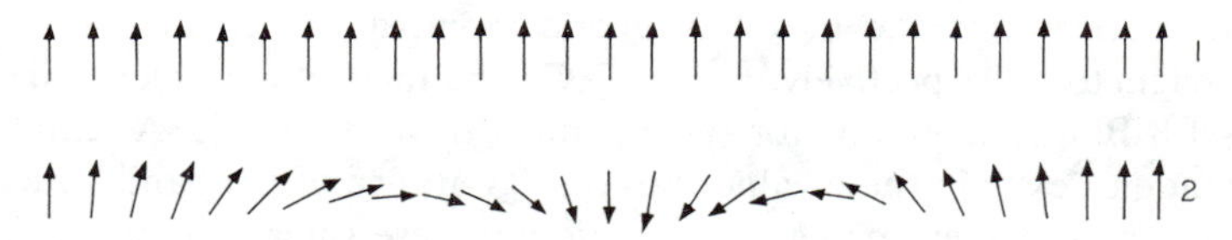

FIG. 59. The arrangement of magnetic moments in a linear chain illustrative of the rare earth metals in which the nearest neighbour magnetic interaction is ferromagnetic and the next nearest neighbour interaction antiferromagnetic. Diagrams 1 and 2 represent the cases for $|\mathbf{J}_1|/|\mathbf{J}_2|$ greater and less than 4 respectively.

diagram 1, but the ratio $|J_2|/|J_1|$ increases with lattice parameter (or temperature) so that a spiralling arrangement is preferred at higher temperatures, diagram 2. The measurement of the magnetization of a sample with the magnetic moment arrangement in diagram 2 leads us to believe a "sort of antiferromagnetism" exists because there are an equal number of magnetic moments pointing in opposite directions. In fact, the rare earths are primarily ferromagnets with relatively strong antiferromagnetic second nearest neighbour interactions. The spiralling arrangement leads to complicated neutron diffraction patterns that require single crystals for analysis and this has only been done in a few cases. Except for the partial quenching at low temperatures in some rare earths the magnetic moments are the same as in the free atoms so that there is generally a significant contribution of the $4f$ orbital motion to the magnetic moment.

Ordered Alloys

If a non-magnetic alloy has an ordered structure like CuZn the structure factor leads to superlattice lines whose integrated intensity is proportional to $S^2(b_{Cu} - b_{Zn})^2$ where S is the long range order parameter. Since the nuclear scattering amplitudes b_{Cu} and b_{Zn} are known it is possible to determine S by measuring the intensity on an absolute basis. This provides a decided advantage over the X-ray analysis since the X-ray scattering factors are not known accurately. As such it is possible to combine neutron diffraction on a partially ordered alloy to determine S and then X-ray diffraction to determine the difference in the atomic scattering factors of the two elements. If the alloy is ferromagnetic it is possible to determine S by application of an external magnetic field which eliminates the magnetic scattering. One can then reorient the external magnetic field and measure the magnetic scattering to determine the *difference* in the magnetic moments of the two elements. An independent measurement of the saturation magnetization of the alloy determines the *sum* of the magnetic moments and this combination of the sum and difference can provide the value of the individual moments. There are a limited number of ordered transition metal alloys where this is possible such as Ni_3Fe, Ni_3Mn and FeCo. In Ni_3Fe and Ni_3Mn, the Fe and Mn atoms have magnetic moments of $\sim$2·8 and $\sim$3·2 Bohr magnetons, respectively, aligned ferromagnetically to the nickel magnetic moments which are 0·6 and 0·3 Bohr magnetons, respectively.[22] In FeCo the magnetic moments are $\sim$2·8 and $\sim$1·8 Bohr magnetons for the Fe and Co, respectively. Attempts have been made to explain these values but our ignorance of the band structure in transition metals prevents us from predicting these values any more than we can predict the values of the pure elements, nor is there sufficient experimental data to provide the necessary clues for a theoretical understanding.

Size Effects

An opportunity is presented, in the case of neutron studies of disordered alloys, to determine the size effect factor $e^{-2M'}$ since all other factors except e^{-2M} in eqn. (119) are known or measurable. By making measurements at two temperatures such as room temperature and liquid nitrogen temperature e^{-2M} and $e^{-2M'}$ can be determined since the size effect factor is probably independent of temperature over this range. There has been little work in this direction, in spite of its decided advantage over X-ray methods.

Extinction Effects

It is frequently desirable to employ single crystals for neutron measurements since the ratio of background intensity to diffracted intensity is reduced. For a single crystal slab in symmetrical transmission, intercepting the entire beam we have

$$|F_N|^2 = \frac{C\omega_s V^2 \sin 2\theta_B \cos \theta_B}{I_0 \lambda^3 t e^{-\mu t/\cos\theta} e^{-2M} e^{-2M'}} \tag{122}$$

where e^{-ut} is the transmission of the crystal with the counter at $\theta = 0$, t is the thickness in cm and the other symbols have been defined in eqn. (100) and eqn. (90). Except for very perfect crystals, such as pure silicon and germanium, primary extinction is generally negligible but secondary extinction is not. By replacing $\frac{1}{2}\mu$ by $D/\sin\theta$ (D is crystal thickness) eqn. (95) can be used to estimate extinction except that Q for neutrons is

$$Q_N = \frac{|F_N|^2 \lambda^3}{V^2 \sin 2\theta} \tag{123}$$

In an extreme case such as an iron crystal 1 mm thick with $\lambda = 1$ Å and $2\eta\sqrt{\pi} = 10'$, R_2' in eqn. (95) is $\sim 0{\cdot}55$. Thus one must always be cautious of secondary extinction effects in single crystal measurements. The comments on pages 136ff are also appropriate to neutron diffraction.

Diffuse Scattering

Measurements of the diffuse scattering of neutrons are sometimes difficult because the background scattering can be quite high. Since the absorption cross-sections are small relative to the scattering cross-sections the attenuation of the neutron beam is principally due to scattering and as such appears as an isotropic background. In the case of ferromagnetic disordered alloys, however, this background is easily subtracted by applying an appropriate external magnetic field to alternately align the magnetic moments parallel and perpendicular to the scattering vector. In both these cases the background scattering is about the same so that one directly measures the diffuse magnetic scattering as the difference in the two intensities. With a few minor modifications eqn. (84) for the diffuse scattering of X-rays in symmetrical reflection can be adapted to the diffuse scattering of neutrons in symmetrical transmission. We replace $1/2\mu$ by $\rho' t e^{-\mu t/\cos\theta}/\rho \cos\theta$ where $\rho' t$ is the thickness in g/cm^2, $e^{-\mu t}$ the transmission of the sample at $\theta = 0$ and ρ the density of the bulk material; and we replace the difference in the atomic scattering factors $(f_B - f_A)^2$ by $[(b_B \pm p_B \sin\alpha) - (b_A \pm p_A \sin\alpha)]^2$ where the choice of plus or minus depends on the polarization of the neutron beam. For an unpolarized beam we have $(b_B - b_A)^2 + (p_B - p_A)^2 \sin^2\alpha$. By application of a suitable external magnetic field the magnetic scattering amplitudes of iron and chromium were determined from the diffuse scattering of polycrystalline FeCr alloys. The magnetic moment of the iron atom varies only a small amount as chromium is added to b.c.c. iron. The chromium magnetic moment is small and parallel to the iron magnetic moments.

Virtually no work has been reported on the use of eqn. (84) to determine the diffuse scattering due to size effects. The large background present in neutron diffraction powder patterns would make this difficult although this troublesome background could be diminished with single crystals. A small amount of work has been done on short range order measurements but this work has not contributed significantly to our understanding of the phenomenon. Perhaps the most interesting diffuse scattering measurements have been those on α-Fe above its ferromagnetic Curie temperature. We have already seen in Fig. 32 that there is appreciable magnetic specific heat well above the Curie temperature of 1040°K and this must reflect the continued short range order above this temperature. The probability curve in Fig. 32 at $T = 1140$°K was deduced from such neutron diffuse scattering measurements.(23)

Summary

The principal points to remember about neutron diffraction measurements of atomic and magnetic moment positions are:

1. The presence of spirals in magnetic transition metals and rare earths suggests relatively strong second nearest neighbour interactions in terms of the Heisenberg vector model.

2. Since neutron diffraction is readily placed on an absolute basis it is well suited to measuring size effects in alloys, $e^{-2M'}$, and Debye–Waller factors in elements and ordered alloys, e^{-2M}.

3. Diffuse scattering measurements confirm the existence of the short range order of magnetic moments above the Curie temperature.

B. Unpaired Electron Probability Distributions

Determination of the Unpaired Electron Probability Distribution

The magnetic scattering amplitude is given by

$$
\begin{aligned}
p_j &\cong \frac{e^2\gamma^2}{2m_0c^2}\,(2\mathbf{S}\,f_S + \mathbf{L}f_L) \\
&= \frac{e^2\gamma}{2m_0c^2}\,gJ, \text{ at } \sin\theta/\lambda = 0
\end{aligned} \tag{124}
$$

where **S** and **L** are the projections of the total spin (total m_s) and orbital (total m_l) angular momentum on J (see page 18), γ is the magnetic moment of the neutron in *n.m.*, $e^2\gamma/m_0c^2 = 0{\cdot}539 \times 10^{-12}$ cm and g is the Landé g factor, eqn. (179). f_S is the form factor for the scattering of neutrons by the unpaired electron spins and f_L is the form factor for scattering of neutrons by the orbital motion of the electrons. In the former case the neutrons interact with the magnetic fields created by the intrinsic magnetic moments of the electrons while in the latter case the neutrons interact with the magnetic fields created by the negatively charged electrons whirling around the nucleus. The form factor due to magnetic spins depends on the electron probability distribution in a fashion similar to the X-ray case. In the one-electron approximation it is

$$f_S = \sum_i \int (\uparrow|\psi|_i^2 - \downarrow|\psi|_i^2) e^{i\mathbf{s}\cdot\mathbf{r}_i}\, dx\, dy\, dz \tag{125}$$

where $\uparrow|\psi|_i$ is the wave function for the ith electron with $m_s = +\frac{1}{2}$ and $\downarrow|\psi|_i$ is the wave function for the ith electron with $m_s = -\frac{1}{2}$. The form factor due to the orbital motion of the electrons is considerably more complicated,

$$f_L \cong 2 \sum_i \int (\uparrow|\psi|_i^2 - \downarrow|\psi|_i^2) \left(\frac{e^{i\mathbf{s}\cdot\mathbf{r}_i}(i\mathbf{s}\cdot\mathbf{r}_i - 1) + 1}{(i\mathbf{s}\cdot\mathbf{r}_i)^2} \right) dx\, dy\, dz \tag{126}$$

If the unpaired electron spins are spherically symmetric then eqn. (125) reduces to

$$f_S = \sum_i \int (\uparrow|R|_i^2 - \downarrow|R|_i^2) \frac{\sin sr_i}{sr_i}\, dr_i \tag{127}$$

Since $L = 0$ for a spherically symmetric unpaired electron probability distribution and the orbital part of the magnetic scattering amplitude vanishes, only a crude "order of magnitude" simplification of eqn. (126) is obtained by averaging over the angle between **s** and $\mathbf{r}_i$

$$f_L \cong 2 \sum_i \int (\uparrow|R|_i^2 - \downarrow|R|_i^2) \frac{(1 - \cos sr_i)}{(sr_i)^2}\, dr_i \tag{128}$$

This simplification is useful for "rough" but simple estimates of orbita contribution.

As in the case of X-rays one generally "works backwards" by fitting the measured magnetic scattering amplitudes to some theoretical wave functions. However the problems of the orbital contribution and lack of spherical symmetry in neutron diffraction generally necessitate the use of the more complicated relationships (eqns. (125) and (126)). The experimental and theoretical approaches are perhaps best illustrated with a specific problem or two. Let us first consider nickel metal.

Neutron Diffraction of Nickel

From magnetization measurements nickel metal is known to have a magnetic moment per atom of 0·606 Bohr magnetons and a Curie temperature $\sim$633°K. From gyromagnetic measurements to be discussed in Chapter IX, $g = 1{\cdot}835$ so that $L \cong 0{\cdot}056$ and $S \cong 0{\cdot}275$ in eqn. (124) and the orbital contribution is significant. Since $b = 1 \times 10^{-12}$ cm and $p \cong 0{\cdot}11 \times 10^{-12}$ cm for the 111 Bragg reflection of nickel metal the square of the structure factor is $|F_N|^2 = 16(1 \pm 0{\cdot}11)^2 \times 10^{-24}$ cm^2, the plus and minus sign referring to the two polarization directions of the neutron. The ratio of the intensities for antiparallel to parallel polarization is 1·53. On the other hand, the square of the structure factor for unpolarized neutrons is $|F_N|^2 = 16 \times 10^{-24}$ cm^2 + $0{\cdot}194 \times 10^{-24}$ cm^2 so that the magnetic contribution is a mere 1%. The very great enhancement in the case of polarized neutrons has enabled us to attack problems such as the unpaired electron probability distribution in nickel, not possible for unpolarized neutrons. Since the use of polarized neutrons requires the magnetic moments in a sample to be aligned in some definite direction this limits the technique to ferromagnetic and ferrimagnetic substances for which an external field can cause such alignment and to a few special cases of antiferromagnets in which the crystal creates a preferred direction of alignment.

The best technique for polarizing neutrons is to use the 111 reflection of a magnetized $Co_{0{\cdot}95}Fe_{0{\cdot}05}$ f.c.c. single crystal. In this case the nuclear scattering amplitude, b, and the magnetic scattering amplitude, p, are about equal in magnitude. If the crystal is magnetized so that $\sin \alpha = 1$ the structure factors are $16\,(b - p) \cong 0$ and $16(b + p) = 32b$ for the two directions of polarization. Thus only those neutrons with their magnetic moments antiparallel to the direction of magnetization of the $Co_{0{\cdot}95}Fe_{0{\cdot}05}$ crystal are diffracted supplying us both with a monochromatic and polarized neutron beam. These crystals are easily grown from the melt if pure materials are used, and a recrystallized alumina crucible is quite satisfactory for containing the molten metal. A permanent magnet can be used to magnetize the crystal machining the crystal to fit snugly between the pole pieces. Because of the fringing field of the magnets there is a tendency for the neutrons to depolarize since they always try to remain parallel to the field. This is particularly true when the magnetic fields of the polarizing crystal and sample are antiparallel. However, an Fe collimator shielding the neutron path generally suffices to prevent this. The experiment consists of measuring the ratio of the integrated intensities of the Bragg reflections for the neutron, polarized parallel and antiparallel to the direction of magnetization of the sample. Rather than reverse the magnetic field of the sample the polarization of the neutron beam is reversed by passing it through a radio frequency field whose frequency matches the Larmor precessional frequency.[24] The ratio of the integrated intensities directly gives $(b + p)^2/(b - p)^2$ so that we can determine p from the known values of b. The nickel sample is in the form of a thin single crystal with a broad rocking curve, to eliminate extinction. A simple technique employed by Shull and Nathans[25] and illustrated for the 110 reflection in iron enabled them to determine the secondary extinction. In Fig. 60 is plotted the ratio $(b + p)^2/(b - p)^2$ as two crystals are slowly rocked through their angular range of mosaic blocks, η. Extinction is a maximum at the centre of the rocking curve since the probability for Bragg diffraction is a maximum. It also is a maximum for the antiparallel polarization, $(b + p)^2$, since the total scattering amplitude is greatest. If the ratio $(b + p)^2/(b - p)^2$ is independent of crystal angle then extinction has been eliminated. Fig. 60

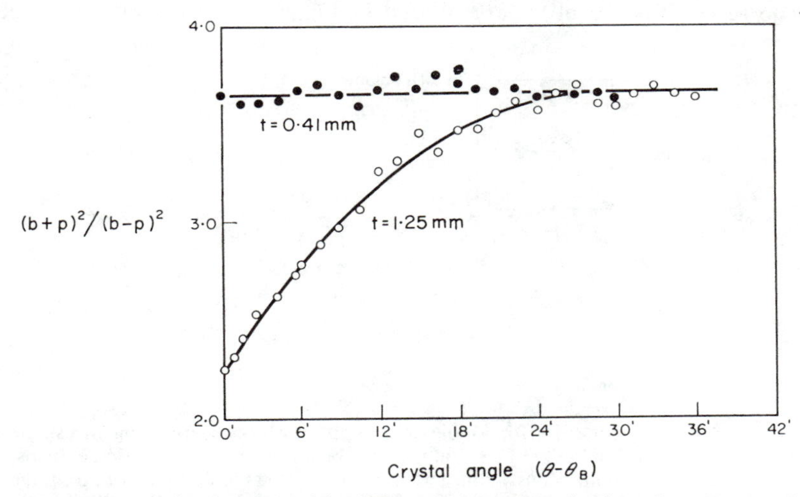

FIG. 60. The ratio of the intensities diffracted by the 110 reflection of two Fe crystals with the neutron magnetic moment antiparallel and parallel to the magnetic field, as a function of the angle the crystal is rotated from the Bragg angle. The thin crystal $t = 0{\cdot}41$ mm appears devoid of extinction. The value of η, the average angular tilt between the coherent regions (mosaic blocks), is about half a degree.

shows this to be the case for a 0·41 mm crystal but not a 1·25 mm crystal. A second technique for estimating extinction consists in applying a suitable magnetic field to eliminate the magnetic scattering and measuring the integrated intensity of the nuclear scattering contribution to the Bragg peak. Since this can be done on an absolute basis, the extinction can be determined.

One experimental problem that frequently arises in neutron diffraction of single crystals and occasionally in X-ray diffraction of crystals with small absorption coefficients is termed *umweganregung*. Translated from the German this refers to neutrons arriving at their destination via a detour. For example, if an iron crystal is aligned for the 310 reflection it is possible for the neutron beam to diffract from the 110 plane provided the azimuthal angle around the

[310] direction is just right. It is further possible for the beam diffracted from the 110 plane then to diffract from the 200 plane before emerging from the crystal. The net effect is that this "detour" adds more intensity to our 310 peak, since double Bragg diffraction from a cubic crystal "appears" to have emerged from a single diffraction process. A necessary but not sufficient condition is that the indices of the diffracting plane are the sum or the difference of two Bragg planes (310 = 110 + 200, add h, k, l separately) This double Bragg scattering may seem unlikely, but it is illustrated in Fig. 61. One must study the integrated intensity versus azimuthal angle around the axis normal to the diffracting plane (arrow in Fig. 61) to be certain one has chosen a region free of umweganregung.

The results for nickel[22] are plotted in Fig. 62 as well as the magnetic

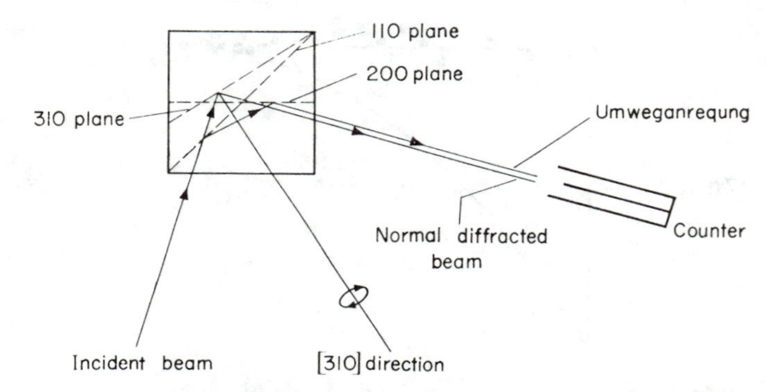

FIG. 61. A possible condition for double Bragg scattering (umweganregung) in a single crystal of body centred cubic Fe. The incident beam of monochromatic neutrons, $\lambda = 1{\cdot}28$ Å, can simultaneously diffract from the 310 plane, $\theta = 45°$, or from the 110 plane, $\theta = 18{\cdot}4°$, and 200 plane, $\theta = 26{\cdot}6°$, in succession. Rotating the crystal around the axis normal to the 310 plane destroys the Bragg conditions for the 110 plane but not the 310 plane.

scattering amplitudes based on the Hartree–Fock $3d$ radial wave functions eqns. (127) and (128). At values of $\sin \theta/\lambda > 0{\cdot}7$/Å the orbital contribution comprises at least half the scattering. However, this does not explain the fact that the Bragg reflections 400, 331 do not lie on a smooth curve. This must come from the angular parts of the unpaired wave functions. *In addition the calculated scattering factor lies lower than the observed and to date the Hartree–Fock method is unable to account for this difference.*

Neutron Magnetic Scattering of Iron

To illustrate more clearly the lack of a smooth form factor curve let us look at the neutron diffraction results for iron in Fig. 63 since iron has only

3% orbital contribution which can be neglected for our purposes. We now make the assumption that the unpaired electron probability distribution is either entirely in the e_g orbitals or t_{2g} orbitals (Fig. 17) and substitute these orbital wave functions into eqn. (125) together with the Hartree–Fock radial wave function to determine f_S. The e_g and t_{2g} orbitals have a specific orientation relative to our cubic axes so that f_S is calculated for specific values of

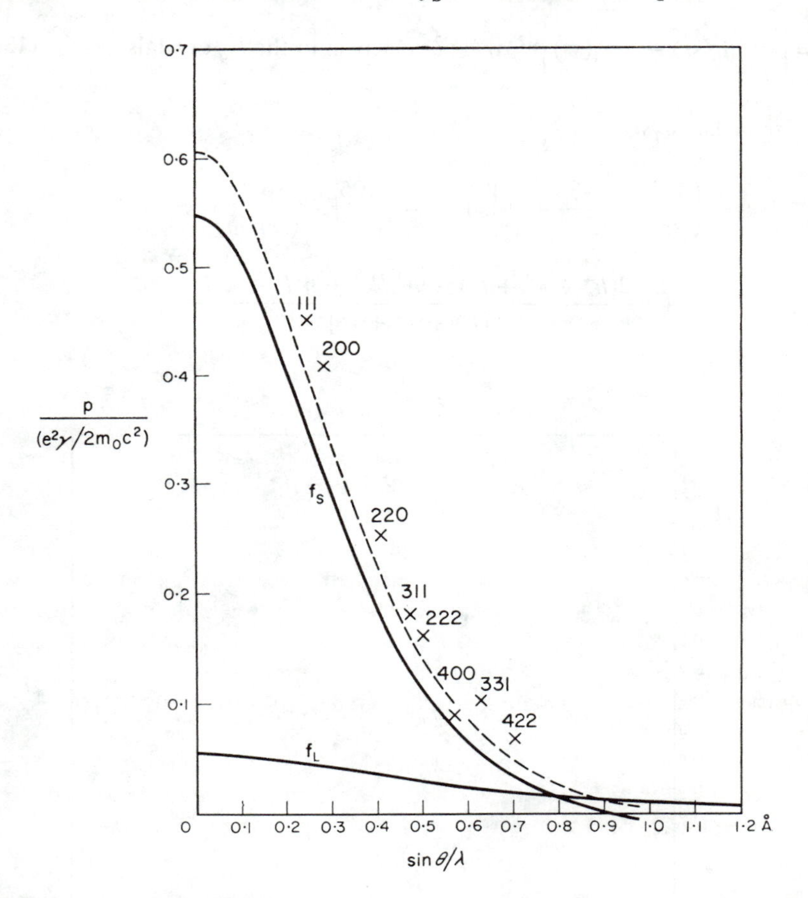

FIG. 62. The calculated spin and orbital contributions (f_S and f_L) to the magnetic scattering in f.c.c. nickel metal, solid curves, and their sum, dashed curve. The magnetic scattering amplitude, p, is plotted in units $e^2\gamma/2m_0c^2$ ($= 0{\cdot}27 \times 10^{-12}$ cm) so that the value at $\sin\theta/\lambda = 0$ is given by gJ. The measured values of Nathans and Shull are included (crosses), evidencing marked deviation from spherical symmetry in the reversal in intensity of the 400, 331 peaks.

h, k, l. After first integrating eqn. (125) over the orbital parts of the wave functions we are left with

$$f_S = \int R^2 \left\{ j_0(sr) + \frac{A}{2} j_4(sr) \right\} r^2 \, dr \quad \text{per electron in the } e_g \text{ orbitals} \tag{129}$$

$$f_S = \int R^2 \left\{ j_0(sr) - \frac{A}{3} j_4(sr) \right\} r^2 \, dr \quad \text{per electron in the } t_{2g} \text{ orbitals} \tag{130}$$

$$j_0(sr) = \sin sr/sr$$

$$j_4(sr) = \left\{ \frac{105}{(sr)^5} - \frac{45}{(sr)^3} + \frac{1}{sr} \right\} \sin sr - \left\{ \frac{105}{(sr)^4} - \frac{10}{(sr)^2} \right\} \cos sr$$

$$A = \frac{3(h^4 + k^4 + l^4) - 9(h^2k^2 + h^2l^2 + k^2l^2)}{[h^2 + k^2 + l^2]^2}$$

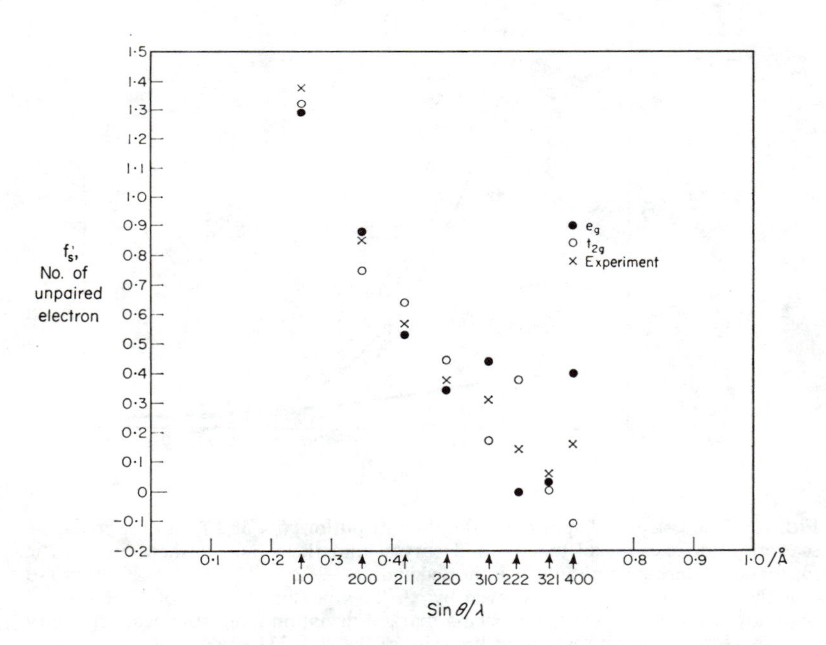

FIG. 63. The spin only magnetic form factor, f_S, for b.c.c. iron calculated on the basis that the 2·2 unpaired electrons per iron atom are either entirely in e_g or t_{2g} orbitals. Since these orbitals are oriented relative to the crystal axes the form factor is only calculated for the Bragg reflections. The data of Shull[26] suggests a slight preference for the e_g orbitals.

where h, k, l are the Miller indices for the particular reflection and R is the radial part of the wave function.*

The procedure, then, is to evaluate numerically the integrals $\int R^2 j_n(sr) r^2 \, dr$ (for $n = 0, 2, 4$) from the Hartree–Fock radial wave functions and determine the appropriate coefficients for the particular h, k, l in eqns. (129)–(133).[27] Fig. 63 shows the theoretical calculations for b.c.c. iron based on the Hartree–Fock radial wave functions near the top of the $3d$ band, Fig. 21. It is readily seen by comparison to Fig. 50 that after renormalization to 2·2 at $\sin \theta/\lambda = 0$ /Å the experimental points lie significantly closer to the form factor calculated from the wave functions at the top of the band rather than near the bottom. On the other hand, if the unpaired electrons in iron are either entirely e_g or t_{2g} then we expect a significant difference in the scattering factors, particularly 400 and 222! The experimental points of Shull[26] in Fig. 63 suggest that the unpaired electrons in iron are at the top of the band and that approximately half are in the e_g *orbitals* and half in the t_{2g} *orbitals*.

While a similar analysis could be made for nickel the orbital contribution complicates matters seriously and we leave this for the interested reader to pursue further. The preference in nickel is for t_{2g} orbitals though, while in the case of f.c.c. cobalt the preference is for the e_g orbitals.

The principal conclusions to be drawn are:

1. Based on the best available wave functions the unpaired electrons are near the top of the band in iron, nickel, and cobalt.

2. They slightly favour the e_g orbitals and "spin only" in b.c.c. iron, and slightly favour the t_{2g} orbitals but with a sizeable orbital contribution in nickel.

* In the case of hexagonal crystals the various orbital wave functions do not combine to form a new orbital as in cubic crystals but the hydrogenic orbitals themselves have the crystal symmetry if the c axis of the crystal becomes the z axis in Fig. 1. The solutions are

$$fs = \int R^2 \left\{ j_0(sr) + \frac{5}{7}\frac{(B-2)}{(B+1)} j_2(sr) + \frac{9}{28}\frac{(3B^2 - 24B + 8)}{(B+1)^2} j_4(sr) \right\} r^2 \, dr$$

per electron in the $m_l = 0$ orbital eqn. (131)

$$fs = \int R^2 \left\{ j_0(sr) - \frac{5}{7}\frac{(B-2)}{(B+1)} j_2(sr) + \frac{3}{56}\frac{(3B^2 - 24B + 8)}{(B+1)^2} j_4(sr) \right\} r^2 \, dr$$

per electron in the $m_l = \pm 1$ orbital eqn. (132)

$$fs = \int R^2 \left\{ j_0(sr) + \frac{5}{14}\frac{(B-2)}{(B+1)} j_2(sr) - \frac{6}{28}\frac{(3B^2 - 24B + 8)}{(B+1)^2} j_4(sr) \right\} r^2 \, dr$$

per electron in the $m_l = \pm 2$ orbital eqn. (133)

$$j_2(sr) = \left(\frac{3}{(sr)^3} - \frac{1}{sr}\right) \sin sr - \frac{3}{(sr)^2} \cos sr$$

$$B = \frac{4}{3}\left(\frac{c_0}{a_0}\right)^2 \left(\frac{h+k}{l}\right)^2$$

where c_0 and a_0 are the lattice parameters and h, k, l the Miller indices.

3. Polarized neutrons are required for such accurate measurements.

4. These measurements have provided the most accurate wave functions in solids and qualitative agreement has been attained with theoretical calculations. However, theory has not yet provided a calculation of the orbitals involved in the partial quenching of these metals. In general the differences between observed and calculated scattering factors are of the order of 10%.

C. Thermal Scattering of Neutrons

Thermal Scattering of Neutrons. The energies and momenta of tnermal neutrons and phonons are both comparable, unlike X-rays and phonons for which only the momenta are comparable. In an inelastic collision the neutron loses or gains an appreciable fraction of its energy and momentum so that one can measure both the change in wavelength (energy change) and change in direction (momentum change). If a single phonon is excited or de-excited by the neutron, the wavelength change of the neutron determines the energy and frequency of the phonon, $E = h\nu$, and its momentum change, $\hbar(\mathbf{k} - \mathbf{k}_0)$, determines the wave number of the phonon. If a single crystal is used the

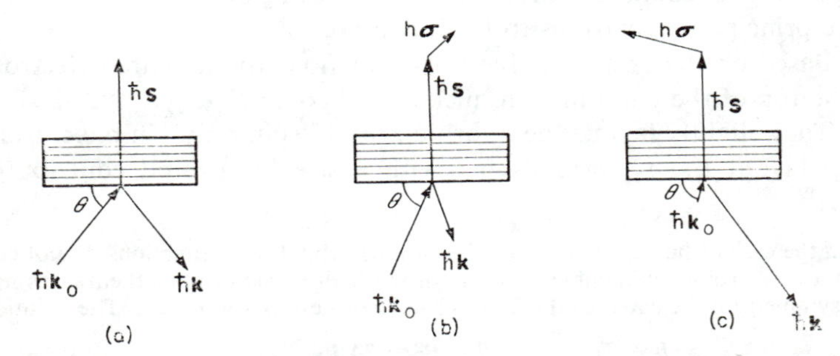

FIG. 64. Vector diagrams of momentum conservation in Bragg scattering of neutrons *a*, phonon absorption *c*, and phonon creation *b*. The wavelength of the neutron is given by $|k| = 2\pi/\lambda$. $\hbar\mathbf{s}$ is the crystal momentum, $\hbar\sigma$ the phonon momentum and $\hbar\mathbf{k}_0$ and $\hbar\mathbf{k}$ the incident and scattered neutron momentum.

separation into transverse and longitudinal contributions can be accomplished as in the X-ray case, and the crystal momentum $\hbar\mathbf{s}$ can be subtracted from the change in neutron momentum. We thus determine the dispersion curves (frequency versus wave number) of the phonons just as in the X-ray case. The principal difference is that we determine the frequency of the phonon from the change in neutron wavelength rather than from the absolute intensity as in the X-ray case. Some examples of inelastic scattering of neutrons are illustrated in the vector diagrams of Fig. 64. In *a* the neutron is elastically scattered by a Bragg plane. In *c* a low energy neutron (long wavelength) absorbs energy from a phonon while in *b* the neutron has excited a phonon in the crystal thereby reducing its own energy. Fig. 65 shows the dispersion curve of aluminium in the [100] direction as determined by Walker with X-rays, dashed curve, and the measured neutron points of Brockhouse.[(28)] While the X-ray data is somewhat better, the development of higher intensity neutron reactors may in the future improve the neutron data relative to the X-ray data. In any event, one must generally resort to the same analysis of the dispersion curve for both neutrons and X-rays to determine the frequency spectra. A case, though, where the frequency distribution is determined directly is discussed on p. 192.

Experimental Techniques. Two essentially different experimental techniques are employed in neutron inelastic scattering. In one the monoenergetic neutrons selected by Bragg diffraction are allowed to impinge on the crystal to be studied and the wavelength distribution of the neutrons scattered in various directions is analyzed by a third crystal. The intensities diffracted by the third crystal are rather weak and the experiment is rather difficult with ordinary reactors. A reactor of ten megawatt power is desirable for such

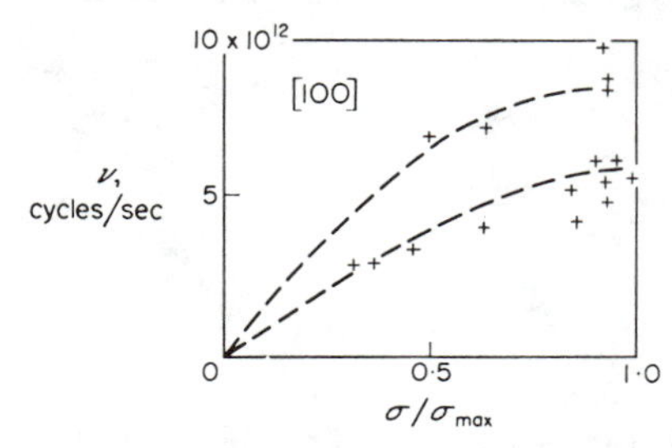

FIG. 65. A comparison of the dispersion curve (phonon frequency versus wave number) obtained for the [200] direction of Al by neutrons (crosses) and X-rays (dashed) lines, corresponding to the data of Walker in Fig. 54).

measurements ($\sim 10^{14}$ neutrons/cm^2/sec near the centre). In the second technique, extremely low energy neutrons are produced by filtering through polycrystalline beryllium.* At room temperatures these neutrons will principally absorb phonon energy in scattering from a crystal, their final energy being so much greater than their initial energy that one can assume the neutrons to have zero initial energy. This permits one to utilize all the filtered neutrons even though they are not monochromatic but arise from the long wavelength tail of the Maxwellian distribution in eqn. (115). The filtered beam is allowed to impinge on the crystal to be studied and the energy distribution of the scattered neutrons is determined by a *time of flight* arrangement. A rotating shutter is placed just after the sample and a neutron counter is placed

* All neutrons of wavelength greater than 3·9 Å pass through the beryllium essentially unhindered, since the absorption and thermal scattering cross-sections are extremely small. All neutrons of shorter wavelength undergo Bragg diffraction and are removed from the collimated beam. The average energy of the filtered neutrons is $\sim$0·005 eV.

at a reasonable distance from the shutter (~ 10 feet). The shutter permits a spurt of neutrons to pass and the distribution in time required for the scattered neutrons to reach the counter is measured by an electronic timing device. A knowledge of the distance and the time determines the velocity; and the wavelength of the neutrons is determined from $\lambda = h/Mv$ where M is the neutron mass and v its velocity.*

Thermal Scattering from Si *and* Al. Examples of the type of data obtained by the two techniques are shown in Fig. 66, diagram 1 obtained with the

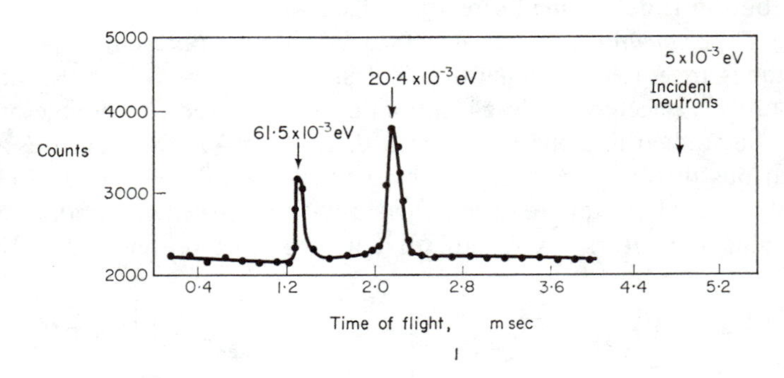

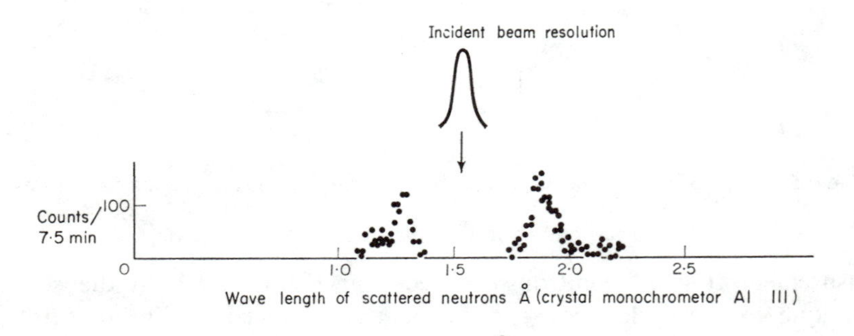

FIG. 66. The type of data taken in neutron inelastic scattering measurements by the beryllium filter technique, diagram 1 for silicon; and the three crystal technique, diagram 2 for aluminium. In diagram 1 the intensity in neutron counts is plotted against their time of flight and the incident filtered beam, 5×10^{-3} eV, is indicated at $\sim 4{\cdot}8$ milliseconds. The two distinct groups of neutrons have gained energy by absorption of optical and acoustic phonons respectively. In diagram 2 the positions of the incident beam and the two neutron groups which have gained and lost energy, respectively, are shown.

* In a variation of this technique Brockhouse employed a set of matched beryllium–lead filters so that their difference isolated neutrons between 3·9 Å and 5·7 Å. This gives one a more homogeneous incident beam. The longest wavelength which can undergo Bragg diffraction is called the cut off wavelength. It is equal to twice the largest interplanar spacing giving rise to Bragg diffraction. In lead, for example,

$$\lambda = 2a_0/\sqrt{(h^2 + k^2 + l^2)} = 2a_0/\sqrt{3} = 5{\cdot}7 \text{ Å}.$$

beryllium filtered neutron technique on silicon[29], and diagram 2 obtained by the three crystal technique on aluminium.[22] In diagram 1 we see two distinct groups of neutrons at 20·4 and 61·5 × 10^{-3}eV each having absorbed phonon energy of ~15 and ~56 × 10^{-3} eV, respectively. The main disadvantage to the filtered beam technique is that most substances absorb the long wave length neutrons rather strongly. In diagram 2 we see two groups of neutrons at about 1·3 Å and 1·85 Å, respectively, having absorbed and given up energy to the phonons which accounts for the wavelength difference from the incident beam at 1·5 Å.

The results for aluminium have already been discussed in regard to the X-ray measurements of Walker and, since the neutron results agree with the X-ray results within experimental error, no further discussion is necessary.

When we analyze the silicon results (diamond structure) we see something new in the dispersion curve, Fig. 67. In addition to the phonons whose

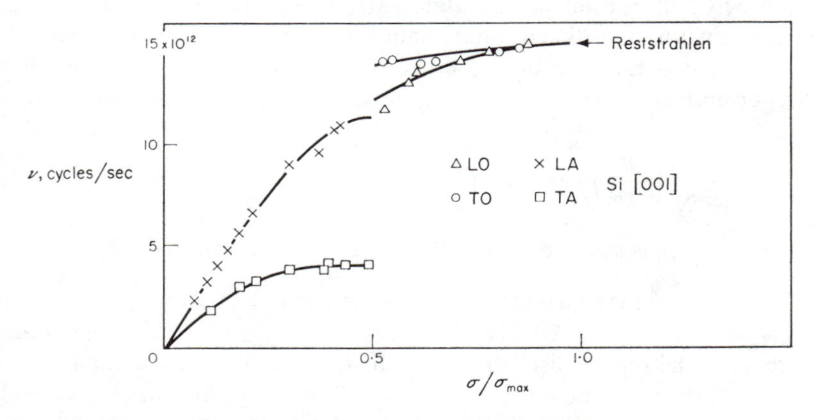

FIG. 67. The measured dispersion curve (phonon frequency versus wave number) for the optical and acoustical longitudinal and transverse modes, labelled *LO*, *LA*, *TO* and *TA*, respectively, in the [001] direction in silicon. The "reststrahlen" frequency occurs at ~15 × 10^{12} cps.

frequency are zero at zero wave number we find a group of phonons separated by a frequency gap whose frequency is relatively independent of wave number. These latter phonons in the upper part of Fig. 67 are called optical modes as compared to the ordinary or "acoustical" phonons in the lower curves. The optical phonons occur in all elemental crystals in which there is more than one type of atom, i.e. the surroundings of each atom are not identical. They are identical in f.c.c., b.c.c. and h.c.p. elements but not in the diamond structure, Fig. 16, in which half the atoms are rotated 90° with respect to the other half. In the diamond structure a line along the 111 direction (body diagonal) finds the atoms spaced alternately at distances $a_0\sqrt{(3)}/4$ and $3a_0\sqrt{(3)}/4$ whereas they are equally spaced at $a_0\sqrt{(3)}/2$ in a b.c.c. metal. *This unequal spacing leads to unequal spring constants.* If we return to our simple one dimensional chain of atoms (page 72) and alternate the spring constants, κ_1 and κ_2, between the atoms we find the frequency versus wave number (ν versus σ) is given by

$$\nu = \frac{1}{2\pi}\left[\frac{2(\kappa_1+\kappa_2) \pm \sqrt{[4(\kappa_1+\kappa_2)^2 - 16\kappa_1\kappa_2 \sin^2 \pi\sigma a_0]}}{2m}\right]^{1/2}$$

plus sign—optical modes $(1/2a_0) \lesssim \sigma \lesssim (1/a_0)$ (134)

minus sign—acoustical modes $0 \lesssim \sigma \lesssim (1/2a_0)$

leading to the dispersion curve of Fig. 68 diagram 1. a_0 is the unit cell size or repeat distance, twice the distance between atoms.

Optical normal modes also occur in any ordered alloy or chemical compound in which the masses are different. If we keep the spring constants the same in the one dimensional chain and alternate the masses m and M ($M > m$) we obtain the dispersion curve for an ordered one dimensional alloy, given by

$$\nu = \frac{1}{2\pi}\sqrt{\left(\frac{\kappa}{Mm}\right)}\,[M + m \pm \sqrt{(M^2 + m^2 + 2Mm \cos 2\pi a_0\sigma)}]^{1/2} \qquad (135)$$

plus sign—optical modes $(1/2a_0) \lesssim \sigma \lesssim (1/a_0)$

minus sign—acoustical modes $0 \leq \sigma \lesssim (1/2a_0)$

where a_0 is the repeat distance (twice the distance between atoms). This is plotted as the solid lines in Fig. 68, diagram 2, for the particular case $M = 2m$. Lastly we can simulate a disordered alloy by arranging the two masses at random but keeping the lattice spacing and spring constants the same. This leads to the dashed dispersion curves and frequency distribution[30] of diagram 2 in Fig. 68. The gap between the optical modes and acoustical modes has disappeared but in both the ordered and disordered cases the total number of normal modes is the same.

A clear picture demonstrating the onset of the frequency gap at $\sigma = 1/2a_0$ or $\lambda = 2a_0$ is not easily made but the upper sketch in diagram 1 of Fig. 68 indicates the nature of the displacements at $\sigma = 1/2a_0$. At the lower frequency, ν_1, the atoms are vibrating in opposite directions against the weaker spring κ_2, while at the upper frequency, ν_2, the atoms are vibrating in opposite directions against the stronger spring, κ_1. For $\kappa_1 = \kappa_2$ there is no preference. It is just that it is impossible to set up a standing wave of the type labelled ν_0 in the upper sketch of diagram 1, Fig. 68, when $\kappa_1 \neq \kappa_2$. At long wavelengths, $\sigma \to 0$, the acoustical modes are identical to the sound vibrations.

At the high frequency limit, $\sigma = 1/a_0$, the optical modes are called the "*reststrahlen*". In the case of silicon the "reststrahlen" occur at $15{\cdot}4 \times 10^{12}$ cps, Fig. 67.*

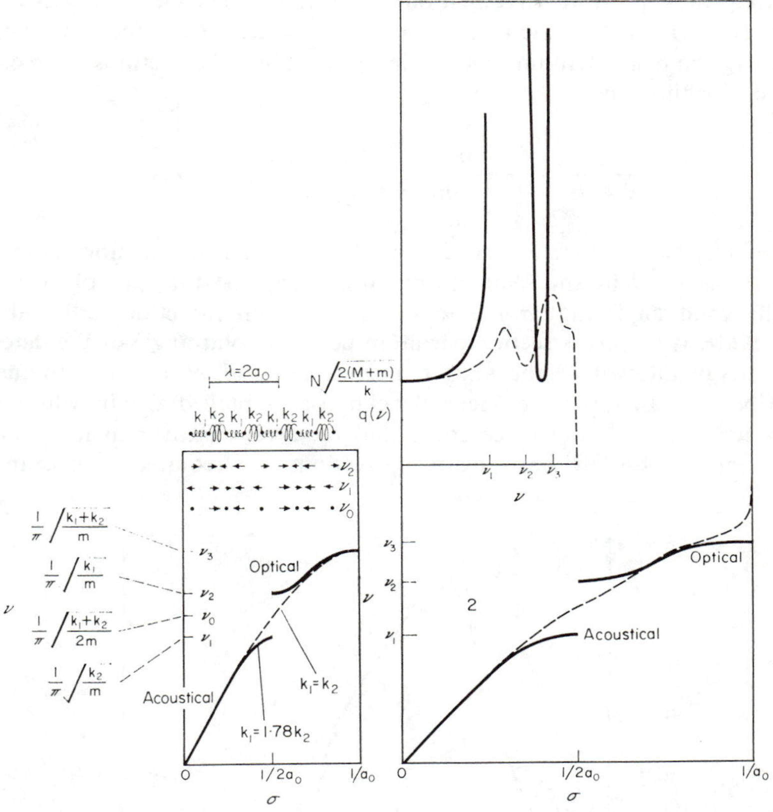

FIG. 68. The dispersion curve (phonon frequency versus wave number) for the linear chain with alternating spring constants, κ_1 and κ_2, connected to equal masses m spaced $a_0/2$ apart, where a_0 is the repeat distance, solid curve diagram 1. If the spring constants are identical, $\kappa = (\kappa_1 + \kappa_2)/2$, we have the dashed curve in diagram 1. If the spring constants are identical and the masses are unequal in the linear chain we obtain the dispersion curves and frequency distributions of diagram 2; solid curves are for an ordered arrangement and dashed curves for a disordered arrangement. The critical frequencies are $\nu_1 = 1/2\pi\sqrt{(2\kappa/M)}$, $\nu_2 = 1/2\pi\sqrt{(2\kappa/m)}$, $\nu_3 = 1/2\pi\sqrt{(2\kappa(M+m)/Mm)}$ where the ratio M/m has been taken to be two for the curves in the figure. The additional phonons above the optical limit, appearing in the frequency distribution of the disordered arrangement are due to small groupings of light mass atoms. In both ordered and disordered states a_0 is the unit cell repeat distance for the *ordered state*, equal to twice the nearest neighbour distance.

* While the effect is weak in silicon a strong absorption of photons of the same frequency as the optical modes of non-metals is the historical reason these normal modes are called "optical". They are, however, only lattice vibrations. (cf. chap. VII for a fuller discussion).

In all the cases mentioned so far the measured dispersion curves can be converted to a frequency distribution through the use of a model and elaborate calculations. In some special cases, though, it is possible to measure the frequency spectrum directly if one can arrange the coherent cross-section to be zero. In such a case there is no Bragg scattering and the conservation of energy and momentum can occur in virtually all directions. The cross-section for this process is

$$\frac{d^2\sigma}{d\Omega dE} = \frac{\sigma\hbar\, e^{-2M}|k|(\mathbf{k}_0 - \mathbf{k})^2}{16\pi^2 M|k_0|(E - E_0)^2(e^{h\nu/kT} - 1)} q(\nu) \tag{136}$$

where $\hbar\mathbf{k}_0$ and $\hbar\mathbf{k}$ are the incident and scattered neutron momentum, E_0 and E the incident and scattered neutron energy, M the mass of the atom, Ω the solid angle and σ the total scattering cross-section (all diffuse). $d^2\sigma/d\Omega dE$ is the cross-section per atom per unit solid angle of the detector per energy interval of the scattered neutrons, and $q(\nu)$ is the frequency distribution. In this experiment Brockhouse employed the beryllium-lead filter technique,[(31)] but since eqn. (136) requires a measurement of intensities on an absolute basis many corrections are required. For example,

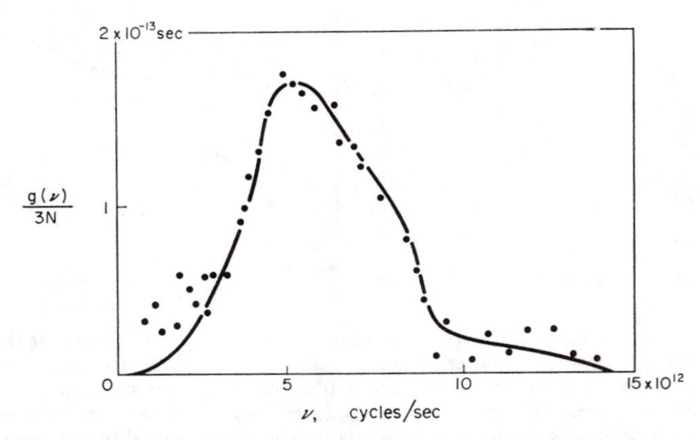

FIG. 69. The frequency distribution, $q(\nu)$, divided by $3N$ versus frequency as determined by neutrons for the f.c.c. disordered alloy $Mn_{0\cdot42}Co_{0\cdot58}$. The additional phonons at high frequencies may be due to short range order while those at low frequency may be due to magnetic scattering.

due to the distribution of energies in the scattered beam one must know the relative absorption of the shutter as a function of wavelength, the relative counter sensitivity, relative air absorption, absorption and rescattering in the sample etc. These corrections as well as those for the absorption of more than one phonon can be made reasonably well and the results for vanadium have been shown in Fig. 29. This technique has also been applied to a disordered alloy $Mn_{0.42}Co_{0.58}$ whose weighted average or coherent cross-section is zero since b is negative for manganese and positive for cobalt. The results are shown in Fig. 69.[31] The excess number of points at low frequency may be a result of extraneous magnetic scattering while the points at high frequency may be the influence of optical modes due to short range order. Referring back to Fig. 68 for the ordered and disordered diatomic linear chain, we can expect the distinct optical modes to begin to disappear as we disorder the crystal giving rise to frequencies in the forbidden gap between the acoustical and optical modes. The precise manner in which this occurs may be quite interesting but to date we have neither theoretical nor experimental evidence of the nature of this transition. However, as we said it may explain the high frequency phonons in Fig. 69 or as the author states it may be an experimental error.

A proper theoretical calculation which allows for size effects, and variations in spring constants may be rather formidable. Fortunately, it is possible to make an ordered alloy with no coherent scattering to study this effect. It is also possible by varying the ratio of the isotopes to obtain samples of nickel, and lithium with no coherent scattering.

It would be very interesting to study the inelastic scattering of vanadium with X-rays to gain a better appreciation for the problems in converting the

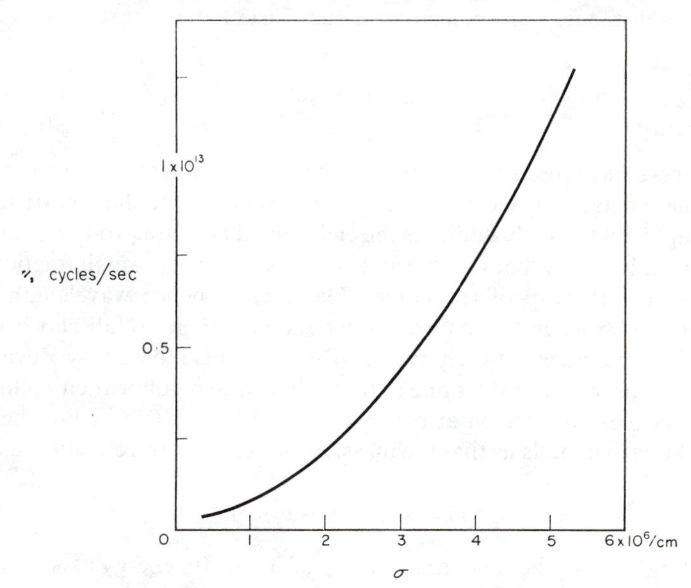

FIG. 70. The dispersion curve (frequency versus wave number) of the spin waves in f.c.c. $Co_{0.92}Fe_{0.08}$ measured with neutrons. These measurements of Sinclair and Brockhouse[32] provide the first unambiguous confirmation of the quadratic dispersion law for spin waves ($\nu \propto \sigma^2$).

measured dispersion curves to frequency distributions, since $q(\nu)$ has already been measured directly with neutrons. In summary, then, the inelastic scattering of neutrons is a rather powerful tool for the study of lattice vibrations. It has some decided advantages over the X-ray technique in cases where the coherent cross-section is zero and for substances with optical modes but in other cases the two techniques are comparable. Unfortunately the time required for either an X-ray or neutron determination of $q(\nu)$ is rather prohibitive, ~ two man years, so that higher intensity sources are needed.

A second source of inelastic scattering unique to neutron scattering arises from the excitation or absorption of energy from one of the magnetic excitations or spin waves. This experiment has been done by Sinclair and Brockhouse for f.c.c. $Co_{0.92}$ $Fe_{0.08}$ and the results of their measurements are shown in Fig. 70. They have verified the quadratic dispersion law for spin waves ($\nu \propto \sigma^2$, eqn. (57)) and have deduced a value of $J_1 \cong 1{\cdot}6 \times 10^{-2}$ eV. The measurements were performed at room temperature still within the temperature range of validity of spin wave theory ($T/T_c < 0{\cdot}2$; $T_c \cong 1300°$K). However, as we shall see in chap. X this value of J_1 for cobalt disagrees with other measurements.

This technique has not been extensively exploited but is probably the only experimental technique capable of measuring the energy levels of magnetic substances, particularly near the Curie temperature.

ELECTRONS

Introduction

So far we have discussed the use of photons and neutrons for the study of the intimate details of electronic structure of solids. The third particle extensively employed for such studies is the electron. These are produced copiously by heating a metal so that the thermal energy ejects a very small fraction of the electrons near the top of the band. The energy, hence wavelength, of the electrons can be accurately controlled by allowing them to fall between plates with a fixed voltage between them. They can also be easily deviated by magnetic fields which enables one to focus them and regulate their collimation.

The mass of an electron at rest is $m_0 = 9{\cdot}1084 \times 10^{-28}$ g but the special theory of relativity tells us that the mass, m, increases with velocity v according to

$$m = m_0/\sqrt{(1 - v^2/c^2)} \tag{137}$$

the limiting velocity being c, the velocity of light. Its energy as a function of velocity is

$$E = m_0c^2\left\{\frac{1}{\sqrt{[1 - (v/c)^2]}} - 1\right\} \tag{138}$$

which reduces to the simple $E = m_0v^2/2$ by expanding the square root for low velocities ($v^2/c^2 \ll 1$). Its wavelength in Å is

$$\lambda = \frac{12\cdot225}{(1 + 9\cdot76E \times 10^{-7})^{1/2}(E)^{1/2}} \tag{139}$$

where E is in eV.

We have already indicated that photons interact with a solid causing the electrons in the solid to execute enforced vibrations due to the vibrating electric field of the photon. Neutrons interact with the nuclei in solids through the nuclear forces and with the magnetic field of the unpaired electrons through its own magnetic moment. The electron, however, interacts with the electric fields of the electrons and nuclei through their Coulomb potential. The atomic scattering factor for electrons depends on the difference between the attractive nuclear scattering and the repulsive electron scattering and is given by

$$f_E(\theta) = \frac{me^2}{2h^2}\frac{\lambda^2}{\sin^2\theta}(Z - f)$$
$$f_E(0) = \langle r^2\rangle/3a_1 \quad \text{valid for } \theta = 0 \tag{140}$$

where f is the X-ray scattering factor for this particular group of electrons, $\langle r^2\rangle$ is the expectation value of r^2 for the electron distribution and a_1 is the Bohr radius ($a_1 = 0\cdot529 \times 10^{-8}$ cm). The X-ray atomic scattering factors are $\sim 10^{-12}$ to 10^{-11} cm in magnitude and the neutron scattering amplitudes are 10^{-13} to 10^{-12} cm in magnitude whereas the electron scattering factor is $\sim 10^{-8}$ cm or a factor of 10^3 to 10^4 times larger than in the case of X-rays. This accounts for the major difficulty in electron diffraction for it gives rise to very large extinction effects. On the other hand, the absorption coefficients for electrons commonly used for diffraction purposes ($E \sim 50$ keV, $\lambda \cong 0\cdot05$ Å) is only about one hundred times greater than typical X-ray absorption coefficients so that we again find scattering processes primarily accounting for the attenuation of the beam just as we did for neutrons.*

* There are several effects which may occur when an electron of 50 keV energy enters a solid.

1. It may be elastically scattered (coherent scattering)
2. It may be entirely absorbed by an atom by ejecting an inner electron of an atom.
3. It may suffer an inelastic collision by ejecting an electron, the two electrons departing in opposite directions.
4. It may undergo a large deviation if it comes close to a nucleus and, if so, will emit a photon to absorb the momentum (*Brems-strahlung*).
5. It may excite or absorb a phonon.
6. It may be scattered by the unpaired spins on a magnetic atom.

All these processes have a high degree of probability except the last. This has about the same magnitude as neutron magnetic scattering so is about 10^6 smaller than processes 1–5.

A and B. Atomic Positions and Electron Probability Distributions.

Electron Distribution in Argon Gas

Since the techniques for electron diffraction differ greatly from the techniques employed in X-ray and neutron diffraction we shall describe the problems encountered in an illustrative case, the measurement of the electron distribution on argon atoms by Bartell and Brockway. We have combined the sections A and B on atomic positions and electron probability distributions since it is convenient to do so. The section C on thermal inelastic scattering is summarily eliminated since such experiments on solids are more easily performed with X-rays or neutrons. The experiment, furthermore, would be exceedingly difficult since we could not measure the $\sim 0{\cdot}03$ eV energy change of a 40 keV electron nor easily separate the thermal diffuse scattering from the background scattering.

The following sketch reveals the components of the Bartell and Brockway equipment, Fig. 71. Electrons in the range of 40 keV, $\lambda \sim 0{\cdot}06$ Å were chosen as a compromise since lower energy electrons are absorbed more strongly and higher energy electrons compress the diffraction pattern into a very small angular range. There are technical problems of voltage stabilization at much higher voltages but they are adequately controlled at less than ~ 100 keV. The necessity for good voltage stabilization is to ensure good monochromatization hence sharp diffraction patterns.

The intensity scattered by a volume of gas atoms is given by

$$I = I_0 N_0 \Omega \frac{e^{-\alpha}}{\tau} \left(\frac{me^2}{2h^2}\right)^2 \frac{\lambda^4}{\sin^4 \theta} \{(Z - f)^2 + S\}^* \qquad (141)$$

where Ω is the solid angle subtended by the counter, I_0 the number of incident electrons per sec striking the sample, I the number of scattered electrons per sec, N_0 the number of gas atoms per cm, S is the incoherent or Compton scattering function given approximately in Table VIIIa and τ is the relativistic correction,

$$\tau = \left[1 - \left(\frac{v}{c}\right)^2\right]\left[1 - \left(\frac{v}{c}\sin\theta\right)^2 + \frac{vZ}{137c}\sin\theta\right].$$

There is no Debye–Waller factor for a gas since the electrons always remain centred about their nucleus and one is not concerned with the relative positions of atoms.

* Since the electrons interact with the Coulomb potentials through their own electronic charge and not their spin, there is practically no tendency for polarization which accounts for the absence of a polarization factor in eqn. (141). There are, of course, strong spin dependences in the Coulomb interaction due to the Pauli principle but this is only significant for electrons with similar wave functions. The wave function of the incoming electron is quite unlike the wave function of any atomic electrons.

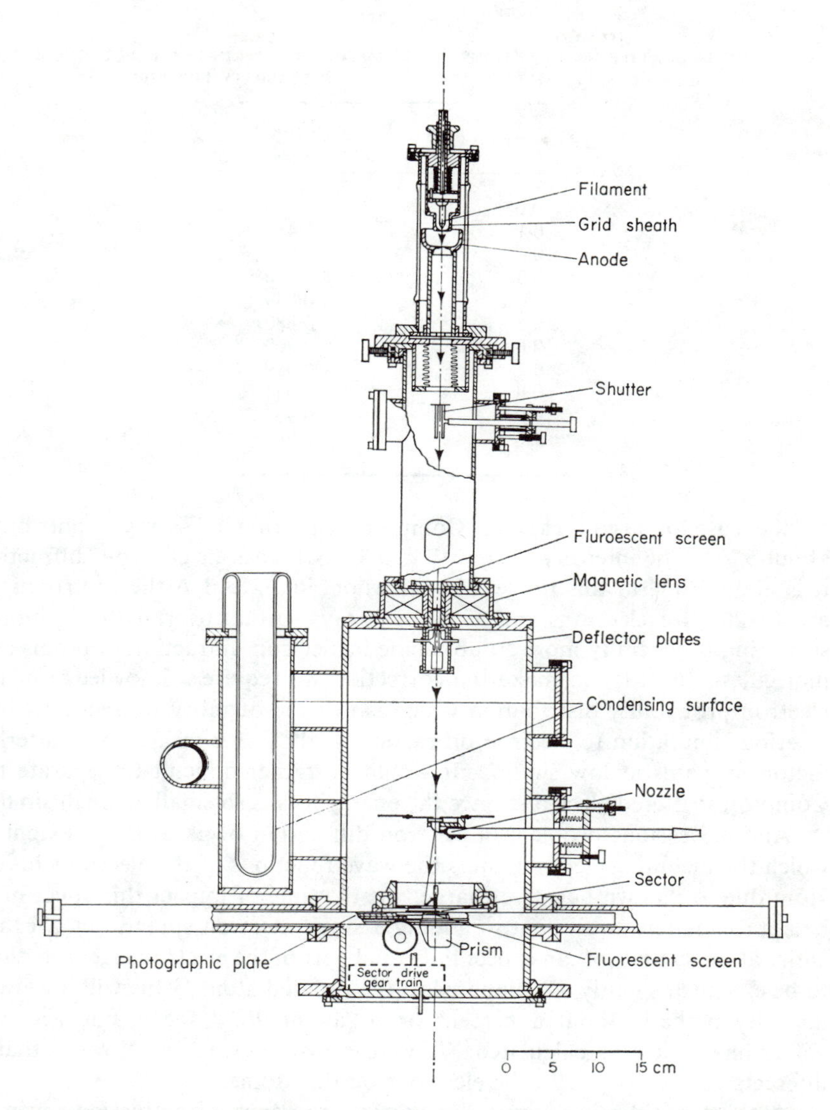

FIG. 71. The essential components of an electron diffractometer used by Brockway and Bartell[33] for the diffraction of argon atoms. The electrons are produced at the filament and accelerated toward the anode. The shutter can be used to stop the beam. The magnetic lens and deflector plates are used to collimate the beam. The argon gas is introduced through the nozzle and is immediately swept toward the liquid nitrogen cooled condensing surface where the gas solidifies and is pumped out by a vacuum pump. The fluorescent screens are used for focusing purposes to observe visually the position of the electron beam. The arrows trace out a typical electron path.

TABLE VIIIa

APPROXIMATE SCATTERING FUNCTIONS FOR COMPTON SCATTERING OF ELECTRONS BASED ON CLASSICAL FERMI–THOMAS ELECTRON PROBABILITY DISTRIBUTION

$\frac{2\cdot22}{Z^{2/3}} \sin \theta/\lambda$	S/Z
0·05	0·320
0·1	0·486
0·2	0·674
0·3	0·776
0·4	0·839
0·5	0·880
0·6	0·909
0·7	0·929
0·8	0·944
0·9	0·955
1·0	0·963

In the case of argon gas the Compton scattering of X-rays contributes about 5% to the intensity at $\sin \theta/\lambda < 0\cdot3/\text{Å}$ whereas in electron diffraction it is about 30% in this range. At values of $\sin \theta/\lambda > 1/\text{Å}$ the contributions are 15–20% for electrons and 35% for X-rays so that overall the Compton scattering is generally more troublesome in electron diffraction of atoms and molecules. In order to make the correction we require a knowledge of the electron probability distribution which is just the quantity we seek! This is a serious limitation to electron diffraction methods for measuring scattering factors in gases at low $\sin \theta/\lambda$. It would be rather difficult to separate the Compton scattered electrons since the energy loss is so small at small $\sin \theta/\lambda$.

Another serious problem in electron diffraction of gases is the extent to which the incoming electron alters the wave functions of the electrons on the atom due to its own negative charge. At the time of impact this effect must be approximately equivalent to adding a single electron spread out over the entire atom so that we can expect the radial part of the electron eigenfunctions to be extended slightly as compared to the isolated atom. This will lower the intensity perhaps about a percent or so at $\sin \theta/\lambda \cong 0\cdot5/\text{Å}$ but accurate values have not been calculated. The great advantage to the X-ray is that it interacts very weakly with the electrons on the atoms.

A third problem concerns the possible presence of extinction, even in free atoms, since the scattering factors are so large. One can estimate this in two ways. Firstly, use the X-ray expression eqn. (91) for primary extinction in a crystal but use the size of the atom in its crystalline form as an estimate of t_0, the region of coherence. This sort of estimate indicates there is negligible extinction in argon gas. Another method is to evaluate the average electrostatic potential felt by the incoming electron as it passes through the atom (or crystal). On the average the potential is attractive since any penetration of the incoming electron into the core electron probability distribution increases the relative contribution of positive nuclear potential to negative electron potential. For high energy electrons ($E > 10$ keV) this average potential, called the *inner potential*, is in eV approximately

$$\phi \cong +\frac{4{\cdot}32Z}{R^3} \tag{142}$$

where R is the radius of the atom in the solid. These potentials range from 10–15 eV. The condition for negligible extinction is that $\rho^2/3 \ll 1$ where ρ is

$$\rho = \frac{\phi}{E}\frac{2\pi R}{\lambda} \tag{143}$$

E is the electron energy and λ its wavelength. For an argon atom $\rho \cong 0{\cdot}04$ which yields a negligible extinction correction $R_1 \cong 1 - \rho^2/3 = 5 \times 10^{-4}$*

Due to the factor $1/\sin^4 \theta$ in eqn. (141) the intensity falls off quite rapidly with increasing angle. While λ is quite small for 40 keV electrons and one can make measurements to large $\sin \theta/\lambda \sim 3/\text{Å}$ ($\theta \cong 17°$), the variation in intensity of about 50,000 over this angular range presents special problems. Since the intensities are recorded on films and since accurate intensity measurements cannot be made over such wide ranges on one film a rotating sectored disk is placed in the scattered beam. It is shaped so as to reduce the intensity of the low angle reflections relative to the high angle. The disk is accurately calibrated and is adjusted *such* that the ratio of intensity on the film does not exceed about 10 for the various peaks. In order to minimize the variations in the density of optically sensitive material in the photographic plate Brockway arranged to spin the film during microphotometering to average out these variations. The problems in converting photographic density on a film to electron intensity are manifold and are discussed in the reference.[33] It will be a vast improvement in technique if photomultipliers can be adapted to accurate counting of individual electrons. Unfortunately the attenuation of electrons in all solid materials is so great that conventional counters which require windows are useless. It may be possible to seal an electron sensitive fluorescent crystal like stilbene in the vacuum system but this awaits future developments.

Since no windows can be used to contain the gas sample one may wonder how the experiment is arranged. Due to the high absorption and scattering even of the gas one cannot allow the argon gas to fill the entire volume from the electron source to detector, Fig. 71. The gas is introduced through the

* The index of refraction, n, of electrons is always greater than unity and is given by

$$n^2 - 1 = \left\{1 + \frac{E}{2m_0c^2}\right\}\frac{\phi}{E} \tag{144}$$

For 40 keV electrons $n - 1 \cong 10^{-4}$ for most typical crystals.

nozzle in Fig. 71 with such velocity that it passes through the electron beam and continues on to be pumped out by the diffusion pump as fast as it is introduced.

In spite of the large Compton scattering correction and the inability to perform the measurement on an absolute basis Brockway finds good agreement between the experimental scattering factor and the scattering factor computed from the Hartree–Fock wave functions. Fig. 72 illustrates the relative intensity as a function of $s = 4\pi \sin \theta/\lambda$ compared to calculated intensities. The agreement over a range of 10,000 in intensity is remarkably good.

Electron Diffraction of Molecules, Surfaces, etc.

A considerable effort has been made in the study of the arrangement of atoms in gaseous molecules because the scattering power of electrons is so much greater than X-rays and is particularly suited for molecules containing light atoms like hydrogen, carbon and oxygen. These measurements have been concerned principally with the distances and angles between the atoms in the molecules with little emphasis on the electron distribution. In general, these measurements have yielded good results in those cases where the interatomic distance can be checked against the moment of inertia. For example, the Cl_2

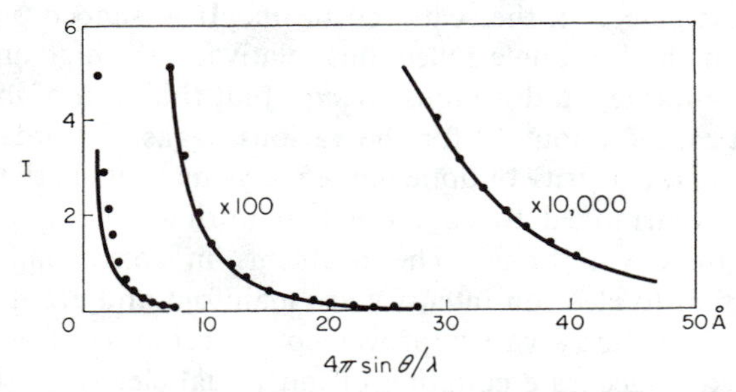

FIG. 72. The relative intensity, I, of the electron beam as a function of $s = 4\pi \sin \theta/\lambda$ showing the experimental observations of Brockway and Bartell[34] on argon gas and the theoretical curve (solid line) matched at $s = 42$ and calculated from the Hartree–Fock electron probability distribution.

molecule is like a dumb-bell with a distance $d = 2{\cdot}00 \pm 0{\cdot}02$ Å between the Cl nuclei determined by the interference maxima in the electron diffraction pattern. This dumb-bell can rotate about an axis normal to the line joining the Cl atoms and has a moment of inertia, I, about this axis of $2M(d/2)^2$ where M is the mass of the chlorine atom. The Cl_2 molecule does in fact rotate about this axis and quantum mechanics tells us that the eigenvalues are $E = \kappa(\kappa + 1)h^2/8\pi^2 I$; $\kappa = 0, 1, 2, \ldots$. The optical lines emitted by the Cl_2 molecule when it makes a transition from one rotational eigenvalue to another have been measured and yield $d = 1{\cdot}983$ Å in good agreement with electron diffraction measurements.

If we wish to use electrons for the study of atomic positions or electron probability distributions in solids the problem of extinction is quite serious. Consider aluminium for example, a metal easily obtained extinction free for X-rays and neutrons when used in powder form. According to eqn. (91) we should require a coherent region size, t_0, less than 10 Å, a size of 100 Å giving considerable extinction. From X-ray line *broading* measurements on cold worked aluminium powder the coherent region size is ~ 1000 Å according to eqn. (96). Since extinction effects are virtually always present in solids and since absolute intensity measurements are virtually impossible with electrons it seems unlikely that electron diffraction will be useful for the bulk of the measurements discussed in the sections on X-ray and neutron diffraction.

On the other hand these particular characteristics of electron diffraction can be turned to advantage in many practical metallurgical problems. They include

1. Studies of polished metal surfaces.
2. Studies of oxidation of surfaces.
3. Studies of thin films such as obtained in electroplating, evaporating, epitaxy etc.
4. Photography of moving dislocations by electron microscopy.
5. Complex crystal structures when only very small single crystals are available.

In addition, the ability to focus electrons enables one to make a chemical analysis over the surface of a substance by measuring the fluorescent X-rays. The resolution is ~ 1 (micron)^2 in surface area.

We shall discuss separately the above items.

1. Since the polishing of metals is done on bulk samples, electron diffraction studies of polished surfaces must be done in reflection. Due to the high absorption coefficients of most substances samples ordinarily investigated in reflection are deliberately given rough surfaces so that the electron beam might be diffracted *in transmission* by small irregularities jutting out of the surface. However, polished metal surfaces are indeed smooth so that electron diffraction is truly performed in reflection. Various investigations of polished metal surfaces reveal the usual polycrystalline lines to be replaced by "haloes" somewhat suggestive of liquid like or amorphous structures. The precise arrangement of atoms in these structures has not been solved and more work is required but it would be of interest to ascertain whether metastable stıuctures can be produced. The energy of such a structure cannot be very different from the stable crystal structure and it might provide some further evidence of the relative stability of various structures. Furthermore, other physical properties of these amorphous structures like resistivity etc. are of interest since we base much of the electron theory of metals on their regular periodicity (use of Bloch waves, Brillouin zones etc.)

2. The precise nature of the oxide structure of a metal is of interest from the practical problems of providing surface protection of a metal. Some oxides, like the one on stainless steel, are quite stable and impede further oxidation. While the arrangements of atoms in oxide surfaces have been studied by electron diffraction, the precise manner in which the oxygen diffuses into the metal, the relative stability of various oxides and the thickness of such oxides are not known.

To illustrate the complexity of the problem, four iron oxide crystal structures have been observed by electron diffraction on an oxidized surface of Fe. The presence of any one or several oxides depends on heat treatment, atmospheric environment etc., in an as yet irreproducible fashion. These oxide films have been studied in both reflection and transmission. In the latter case a method for dissolving the metal away from the oxide has been employed.

3. The structure of evaporated and electroplated metallic films has obvious technical importance. Some electron diffraction work on *epitaxy* has been reported. The tendency for the first layers of an evaporated film to have some crystallographic alignment with the surface on which the film condenses (called the substrate) is termed epitaxy. For example, copper and nickel will orient their cube faces parallel to the cube faces of NaCl (along which it cleaves). The first layers of metal atoms condensing on the surface will generally increase or decrease their interatomic distance to try to match the lattice spacing of the substrate. As more layers are added, the lattice spacing returns to its normal value but the preferred crystallographic orientation continues. However, if the film exceeds about 1000 Å it begins to lose its preferred orientation and becomes polycrystalline. Most of the studies have used rock salt or mica as a substrate with little conclusion as to the precise atomic distribution or electron probability distribution near the boundary. Here again it may be possible to prepare metastable crystalline films by evaporation on to an appropriate substrate and to study their physical properties (electron probability distribution, resistivity, optical properties etc.). This may lead to some better theoretical understanding of the relative stability of various crystal structures. This field requires considerably more work before any ideas emerge.

4. An extremely fascinating and somewhat accidental discovery are the moving pictures of dislocations which can be taken by combining electron microscopy and electron diffraction. In electron microscopy a narrow monochromatic parallel beam of electrons passes through a thin film of the material or a thin plastic replica of the surface of the material (the plastic is poured on the surface, allowed to harden and then peeled off) and the variations in electron absorption over the surface bring out details of the contour of the thin film or surface. After the parallel electron beam traverses the film or replica it is expanded with a suitable magnetic field and recorded on film for visual observation. Since monochromatic electrons are used they are subject to the

usual diffraction effects particularly if the film is crystalline. Hirsch, Horne, and Whelan[35] passed the electron beams through extremely thin foils of aluminium ~ 1000 Å thick (actually at the edge of a hole in the foil eaten through by acid) and observed noticeable variations in the intensity of the electron beam in the neighbourhood of a *dislocation*. If the aluminium foil is a single crystal properly oriented for Bragg diffraction in the region traversed by the electron beam, the presence of a dislocation in this region interrupts the perfect periodicity of the crystal and alters the intensity of the diffracted beam. The contour of the dislocation is revealed quite clearly. Since the aluminium foil diffracts so strongly only a fraction of a second is required for a picture and it was possible to employ a motion picture camera to follow the dislocations as some stress was placed on the foil.

The subject of dislocations and plastic flow is not covered in this book. It is primarily a geometrical problem depending on the atomic arrangement in a crystal rather than a problem of the forces between atoms and their electron probability distribution. The use of the Hirsch technique has become quite popular, the principal problem being one of sample preparation.

5. There are many cases of complicated crystal structures for which it is quite difficult to obtain the single crystals required for structure analysis. In these cases it may be possible to select a macroscopic crystal approximately *one micron* in size, too small for X-ray diffraction. There is some problem in manipulating a crystal this size but nevertheless it should be quite feasible.

We shall conclude this section by outlining some additional advantages and practical experimental difficulties in electron diffraction. As far as equipment is concerned, most work to date has been performed on "home-made" units but in recent years several good commercial diffractometers have become available so that a newcomer might best be advised to purchase a commercial unit.

a. Voltage stabilization hence wavelength stabilization is only good to about one part in one thousand which means that lattice parameter measurements are a factor of ten to one hundred poorer than in X-ray determinations.

b. Absorption coefficients are not known particularly for elements heavier than aluminium.

c. For very thin crystals (< 100 Å) or for crystals where the absorption coefficient limits the penetration of the electron beam to less than 100 Å, the diffraction pattern begins to appear two-dimensional i.e. the diffracting crystal is effectively so thin compared to its surface dimensions that the diffraction pattern is representative of a two-dimensional crystal. In this case the restriction that the diffracted ray be in the plane formed by the incident ray and normal to the diffracting plane is relaxed and the diffracted beam is broadened out of this plane. Neither two- nor three-dimensional theory can generally be applied to these results so that analysis is difficult.

d. For most work a vacuum of 10^{-5} mm Hg is adequate.

e. Transmission specimens are more desirable due to the small diffracting angles (at 40 keV the 110 peak in iron is at $2\theta = 1{\cdot}7°$). In reflection the beam must strike a rather extended part of the specimen with a loss in resolution. In many cases the samples can be prepared by evaporation onto a water soluble crystal like sodium-chloride and then floating the specimen onto a wire mesh, the electron beam being smaller than the mesh distance.

f. For details of the first few layers of atoms on the surface of a substance slow electron diffraction has been used in the range 10–200 eV (~4–1 Å). The diffracting angles are now quite reasonable but the absorption coefficient is so high as to limit the depth of penetration to a few atomic layers. The diffraction pattern is now distinctly two-dimensional and is produced by directing the electron beam normally to the surface. The detector is placed at some angle θ to the incident beam and the diffraction maxima occur according to $n\lambda = D \sin \theta$, where D is the spacing of the rows of atoms along the surface, and n is the order (Fig. 73). In practice the crystal is rotated in the

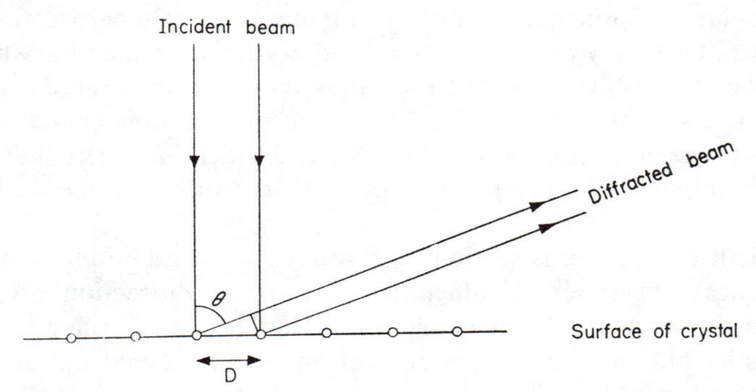

FIG. 73. A schematic representation of a two-dimensional diffraction pattern observed in slow electron diffraction. The condition for the diffraction maxima is that $n\lambda = D \sin \theta$ where D is the spacing between *rows of atoms* on the surface.

plane of its surface (angle φ) and for given values of θ and φ the voltage, hence wavelength, is varied until maxima appear. Since the beam only penetrates a few atom layers it is essential to remove surface contaminants. This can be done by vaporizing some of the surface with the electron beam before taking the diffraction pattern. For recent work on this problem see the work of Germer and Hartman on "The Detection of Oxygen on the Surface of Nickel" (*J. App. Phys.* **31**, 2085, (1960)).

Problems

1. Giving due consideration to the material covered in chapters IV, V and VI describe in some detail the experimental procedure in solving the following problems. Include the method of sample preparation, the type of diffraction to be employed, and the maior difficulties one might encounter.

(*a*) Short range order (or segregation) of an alloy $Cu_{0\cdot 98}Co_{0\cdot 02}$ quenched from 1300°K.

(*b*) The crystal structure and lattice parameter of the chromium plate on the bumper of an inexpensive automobile.

(*c*) The long range order and magnetic structure of Fe-V (50-50).

(*d*) The crystal structure of some microscopic inclusions in niobium metal.

(*e*) The homogeneity of a piece of gold beaters skin.

(*f*) The limit of solubility of gadolinium in f.c.c. copper at 300°C.

(*g*) Whether the 3*d* electrons in vanadium are principally in the e_g or t_{2g} orbitals.

(*h*) The coefficient of expansion of γFe (f.c.c.) as a function of temperature from 910°C to 1200°C.

2. Determine whether there is a positive or negative deviation from Vegard's law in as many iron rich alloys as you can find in Pearson's book.[9] Is the sign of the deviation related to the position of the second element in the periodic table? Can you find any physical property that correlates with the sign of the deviation?

3. (For the mathematically minded)

Plot the dispersion curves for the ordered linear chain with equal spring constants, κ_1, and with $M/m = 1\cdot 0$, $1\cdot 1$ and $1\cdot 5$ respectively, (m is the same for all three cases). Determine the frequency distribution from the slope of the dispersion curve and from the relationship $g(\nu) = N_0 a_0\, d\nu|d\sigma$ where N_0 is the number of unit cells and a_0 the repeat distance. So that they can all be plotted on the same graph use twice the nearest neighbour distance as the repeat distance in all cases even though the nearest neighbour distance is the repeat distance for $M/m = 1\cdot 0$. Evaluate the very low temperature specific heat for all three cases remembering that in one dimension the number of normal modes equals the number of atoms.

$$\left[\text{Answer } C = \frac{2RT}{\theta}\sqrt{\left(\frac{M}{m}+1\right)} \qquad \text{where } \theta = \frac{h\nu^*}{k} = \frac{h}{2\pi k}\sqrt{\left(\frac{2\kappa_1}{m}\right)}\right]$$

4. Write a report on field emission microscopy. In particular, discuss some possible metallurgical applications.

Summary

The interaction of X-rays, neutrons and electrons with atoms and solids is sufficiently different so that all three diffraction techniques complement one other. While all three techniques are capable of determining atomic arrangements, X-ray diffraction is most suitable for determining the total electron probability distribution on each atom, neutron diffraction for magnetic studies, and electron diffraction for thin film and surface studies. All in all the three techniques have provided some of the most important information about the structure of metals.

CHAPTER VII

SPECTROSCOPY OF THE SOLID

Introduction

In chapter VI we have placed some emphasis on the use of X-ray and neutron diffraction in determining the eigenfunctions of electrons in metals. In this chapter we shall discuss the techniques aimed at determining the energy eigenvalues of the electrons in metals. The principal tool for these determinations are photons varying in wavelength from a fraction of an Angstrom in the X-ray region to many thousands of Angstroms in the visible and infrared region. In order to discuss this subject we shall ıeview the field of the interaction of photons with solids since photons have been the most successful probes of the solid state (if you wish, they have thrown some light on the subject).

In order to learn anything about a substance you must disturb it in some way. Even photographing an individual requires photons to be deflected from that person and this causes the electrons that are struck by the photons to vibrate a little. Of course this is close to ideal in that the photons produce a negligible disturbance. To gain even more information about this individual, like an ulcer resulting from overwork, X-rays might be employed and these cause a more permanent although generally reparable damage. The guiding principle is to disturb the object a negligible amount in gaining information about its structural details. This is not always possible in solid state physics and, in fact, proves to be the major problem in determining eigenvalues in solids.

How Photons Interact with Solids

Photons consist of an electric field and a magnetic field both vibrating in a direction normal to their direction of propagation and with frequency ν. The energy of a photon is $h\nu$, its momentum $h\nu/c$ and its wavelength $\lambda = c/\nu$. When a photon comes close to an electron its electric field interacts with the electric charge on the electron causing it to vibrate with frequency ν. This interaction can cause the photon to be either (1) elastically or (2) inelastically scattered or (3) completely absorbed and we shall discuss these three processes separately. In the main this chapter deals with (3) absorption (and emission) since the other two processes have already been discussed in chapter VI. The absorption and emission processes enable one to glean information about the energy levels in a solid.

1. Elastic Scattering

The only true elastic scattering occurs when the momentum of the photon is absorbed by the crystal as a whole. Examples of this, which have already been discussed, are Bragg scattering, short range order diffuse scattering and size effect diffuse scattering all of which leave the *internal* state of a crystal unaltered. Another example of elastic scattering is the reflection of photons from a mirror surface, at an angle below the critical angle when the index of refraction is less than one, and at all angles when the index of refraction is greater than one. For photons in the X-ray region (0·1 Å to ~100 Å) the index of refraction of solids is less than one and the reflecting power (fraction of photons reflected) of a mirror surface is given as a function of the angle of incidence θ, as

$$
\begin{aligned}
R &= \frac{(\theta - p)^2 + q^2}{(\theta + p)^2 + q^2} \\
p^2 &= \tfrac{1}{2}\{\sqrt{[(\theta^2 - 2\alpha)^2 + 4\beta^2]} + \theta^2 - 2\alpha\} \\
q^2 &= \tfrac{1}{2}\{\sqrt{[(\theta^2 - 2\alpha)^2 + 4\beta^2]} - \theta^2 + 2\alpha\} \qquad (145) \\
\beta &= \lambda\mu/4\pi \\
\alpha &= \frac{\lambda^2 e^2}{2\pi m_0 c^2} \sum_i N_i(Z_i + \Delta f_i)
\end{aligned}
$$

where λ is the wavelength, μ the linear absorption coefficient, and N_i is the number of i atoms per cm^3 having an atomic number Z_i and a Hönl (dispersion) correction Δf_i. The index of refraction is $1 - (\alpha + i\beta)$ and for substances of small absorption the critical angle below which total reflection occurs ($R = 1$) is equal to $\sqrt{2\alpha}$.

For photons of longer wavelength ($\lambda \geq \sim 100$ Å) particularly in the optical regions the index of refraction of solids is greater than one and is $n - ik$. The reflecting power at 180° is

$$R = \frac{(n-1)^2 + k^2}{(n+1)^2 + k^2} \qquad (146)$$

and is always less than one. n and k are the real and imaginary parts of the index of refraction.

In all cases of elastic scattering the electrons in the solid are caused to vibrate the negligibly small distance of about 10^{-5} Å or so (except near an absorption edge) and at a frequency corresponding to that of the incident photon. This is quite close to the ideal case of negligible disturbance and the wave functions obtained from Bragg scattering are virtually identical to the undisturbed eigenfunctions. Fig. 74, diagram 1, illustrates the regions commonly covered in elastic scattering.

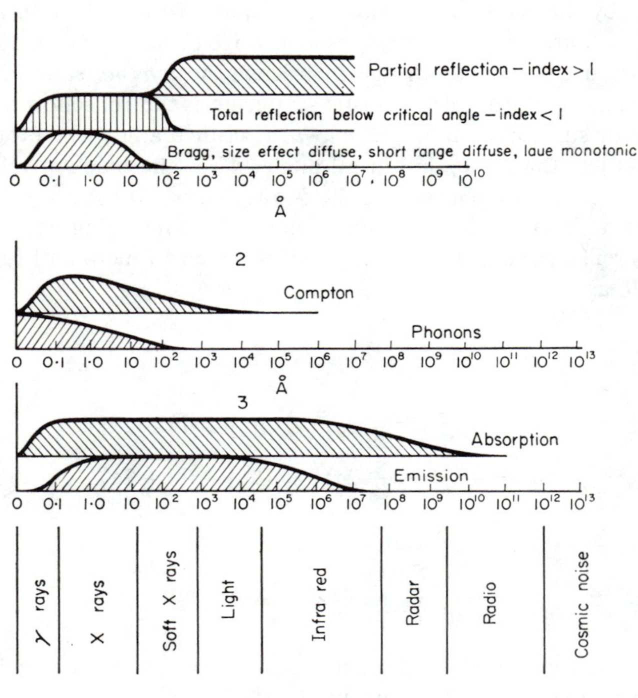

FIG. 74

Diagram 1

The wavelength ranges over which various types of elastic scattering of photons by solids commonly occur. Partial and total reflection from surfaces occur respectively for the index of refraction greater than and less than 1.

Diagram 2

The wavelength ranges over which inelastic scattering of photons occurs. In Compton scattering a portion of the photon energy is lost in ejecting an electron from an atom while in phonon scattering the photon loses or gains the energy of the phonon it excites or de-excites.

Diagram 3

The wavelength ranges over which total absorption or emission of photons occurs in solids. In absorption the entire photon energy is transferred to an electron or phonon. In emission the entire energy of an electronic transition is transferred to a photon.

2. *Inelastic Scattering*

If the photon absorbs or gives up a fraction of its energy (less than one), the scattering is termed inelastic. This can be done by exciting a phonon in which case the momentum is shared by the lattice and the phonon; or the photon can lose energy by ejecting an electron (Compton effect) the momentum loss of the photon being absorbed by the electron. At long wavelengths the total thermal inelastic scattering decreases inversely as the square of the wavelength. The Compton scattering also decreases inversely as the square of the wavelength since the lower energy photons can only eject less tightly bound electrons. If B is the binding energy of an electron in eV then a photon of energy $h\nu$ (in eV) scattered 180° will transfer an energy E (in eV) to an electron ($h\nu > B$).

$$E = \left(\frac{h\nu - B/2}{500}\right)^2 \tag{147}$$

Since the Compton scattering ejects the electron the atom involved is highly disturbed in the measurement. The Compton scattering expressions traditionally assume that only the wave function and eigenvalue of the *scattered electron* are altered in the process. This is generally a reasonable approximation but not nearly as ideal a situation as in Bragg scattering. The regions commonly covered for these processes are shown in Fig. 74 diagram 2.

3. *True Absorption and Emission*

If a photon is completely anihilated so that its entire energy, $h\nu$, is absorbed by an electron (which is generally ejected from its atomic position), the process is termed true absorption. After a short period of time some other nearby electron will enter the "hole" left by the initially ejected electron and a photon will be emitted, equal in energy to the energy difference. This second photon is separate in all respects from the first and this latter process is called emission.

In absorption all the energy and momentum of the photon is transferred to the electron. If the electron is in some initial state like a $1s$ electron in iron metal then the photon must have sufficient energy to essentially remove it from the atom since all other allowed states like $2s$, $2p$, $3s$, $3p$, are filled as well as all states from the bottom of the band to the top. The first available energy level is at the top of the band in iron and 7111·2 eV is required for this process. As the energy of the incoming photon is decreased fewer electrons can be ejected until we reach the visible light region ($\sim$ several eV) where only electrons within a volt or so of the top of the band are affected. While the process of true absorption in a metal occurs for all photon energies, emission only occurs at energies less than the energy absorbed by an electron dropping from the top of the band to a hole in the $1s$ level (we neglect the rare case where we create two holes in the 1s shell). Emission also occurs as a result of the thermal excitation of electrons near the top of the band as they fall back into the empty lower energy states. This process gives rise to radiation of visible light by solids at high temperatures (cf. problem 4, chap. III). The ranges ordinarily covered in emission and absorption are shown in Fig. 74, diagram 3.

It is convenient to discuss the experimental determination of eigenvalues separately for the X-ray region ($\lambda < 1000$ Å) and the optical region ($\lambda > 1000$ Å).

X-ray Region—Absorption and Emission

Absorption of X-rays in Argon

Since the absorption and emission of photons represent energy differences of atoms in different states these processes have been used to determine eigenvalues. Let us examine this in more detail in a good illustrative case[(36)] i.e. the free atom of argon $1s^2 2s^2 2p^6 3s^2 3p^6$. The absorption coefficient in the vicinity of the absorption edge is shown in Fig. 75. At long wavelengths

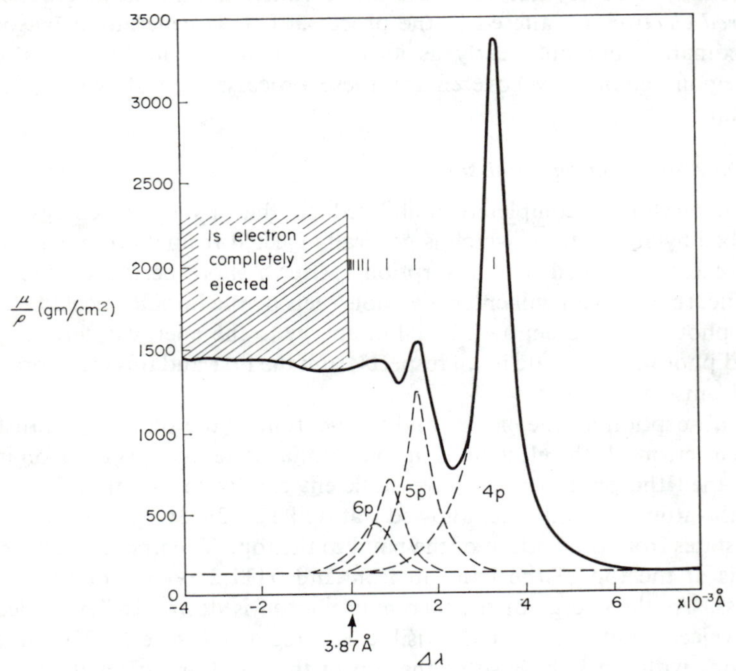

FIG. 75. The mass absorption coefficient of argon gas as a function of wavelength near the critical absorption limit at 3·87 Å. The 1s electron is successively excited to 4p, 5p etc. levels. At wavelengths shorter than 3·87 Å the ejected 1s electron completely escapes from the atom.

the photon does not have sufficient energy to remove one of the 1*s* electrons whose eigenvalue is ~3190 eV but as the wavelength is increased until the photon energy is close to the eigenvalue a sudden increase in absorption occurs. This increase is associated with the excitation of one of the 1*s* electrons to the first unoccupied level it is permitted to fill, the 4*p* (see p. 15). While the 4*s* may be the lowest unoccupied energy level the conservation of angular momentum requires the angular momentum quantum number of the electron to change by one unit since the photon carries with it one unit of angular momentum relative to the argon nucleus. As the photon energy is increased the successively higher states 5*p*, 6*p*, etc. are occupied by the ejected electron until finally the X-ray energy is sufficient to completely remove the 1*s* electron from the atom leaving it ionized. By definition this point is called the critical absorption edge. The positions of the levels are identified in Fig. 75, the higher levels 7*p*, 8*p* to ∞p being too close to resolve. The shape of the curve of the absorption coefficient versus wavelength is governed by the expression for the linear absorption coefficient

$$\mu = \frac{2N_0 e^2}{m_0 c} \sum_{\nu_{if}=0}^{\nu_{if} \leq \nu} g_{if} \frac{\Gamma/2}{(\nu_{if} - \nu)^2 + \Gamma^2/4} \tag{148}$$

where N_0 is the number of atoms per cm^3, ν the frequency of the incoming photon, ν_{if} the frequency difference between the initial state designated i and the allowed final state designated f (i.e. the energy difference divided by h, $E/h = \nu$), g_{if} is the so called *oscillator strength* which is a number expressing the relative probability that the atom will jump from state i to f when irradiated with photons of frequency ν and is given by

$$g_{if} = \frac{4\pi m}{\hbar} \nu_{if} \left| \int \psi_f \mathbf{r} \psi_i \, \mathrm{d}x \, \mathrm{d}y \, \mathrm{d}z \right|^2 \tag{149}$$

Γ is the so called "damping constant" and is approximately equal to the natural energy width of the final state divided by h, generally ~1 eV/h, and ψ_f and ψ_i are the wave functions of the final and intial states of the atom. If we plot eqn. (148) we find it to be a rather sharp function around the resonant frequency ν_{if} for each transition and to match the individual curves in Fig. 75 for $\Gamma \cong 0{\cdot}58$ eV/h. The summation in eqn. (148) gives the total linear absorption coefficient which is the sum over all final states which can be excited by the photon, $\nu_{if} \leq \nu$. The index of refraction of the gas, $n + ik$, is related to μ in that

$$\mu = 4\pi nk/\lambda$$

$$n^2 - k^2 = \varepsilon = \text{dielectric constant} = 1 + \frac{e^2}{\pi m_0} \sum_{\nu_{if}=0}^{\nu_{if} \leq \nu} \frac{g_{if}}{(\nu_{if} - \nu)^2 + 2i\nu\Gamma} \tag{150}$$

The eigenvalue of the 1*s* electron is the sum of its potential and kinetic energy when it is in its ground state in the argon atom and the levels are occupied as shown in Table IX column 2. The first transition 1*s* → 4*p*

TABLE IX

THE OCCUPATION OF THE VARIOUS LEVELS IN AN ARGON ATOM IN ITS GROUND STATE, EXCITED STATE AND IONIZED STATE

Ground State		First Transition		Ionized State	
Level	Number of Electrons	Level	Number of Electrons	Level	Number of Electrons
1*s*	2	1*s*	1	1*s*	1
2*s*	2	2*s*	2	2*s*	2
2*p*	6	2*p*	6	2*p*	6
3*s*	2	3*s*	2	3*s*	2
3*p*	6	3*p*	6	3*p*	6
		4*p*	1	escaped	1

and the ionized states have the levels occupied as shown in Table IX, columns 4 and 6. If the eigenvalues of all the electrons except the one removed remained the same in the ground state and the state marked first transition or ionized state then the energy absorbed from the photon by the atom would equal the eigenvalue of the 1*s* electron and we would have a simple method of measuring eigenvalues. However, the "hole" left by the 1*s* electron changes the Coulomb potential felt by all the other electrons on the atom and this new potential will yield different wave functions and eigenvalues when inserted into the Schrödinger equation. Thus the energy of the absorbed photon is only equal to the energy difference between the ground state and excited state of the *entire atom*. Nonetheless, we may still ask how close the critical absorption energy is to the eigenvalue of the 1*s* electron. The answer is that in the case of elements in the neighbourhood of argon and the first transition group the difference between the Hartree–Fock calculated eigenvalues and the experimental energies required to eject an electron is a minimum of tens of volts *for all electrons*. (Table X for Fe)

TABLE X

THE OBSERVED AND CALCULATED ONE ELECTRON ENERGIES FOR Fe IN eV

Level	calculated[37]			observed
	Hartree–Fock free atom			X-ray absorption edge in metal
	Fe $3d^6 4s^2$	Fe $3d^8$	Fe^+ $3d^7$	
K_I 1*s*	−7080·9	−7072·2	−7083·4	−7111·2
L_I 2*s*	−864·8	−855·0	−866·4	−841·0
L_{II} 2*p*	−742·2	−732·6	−744·0	−720·4
M_I 3*s*	−112·4	−103·0	−114·7	−93·3
M_{II} 3*p*	−74·0	−65·02	−76·06	−53·3
M_{IV} 3*d*	−17·3	−5·69	−18·78	~1·9

Since this difference is somewhat larger than the error in the Hartree–Fock calculations due to the one electron approximation (which omits correlation), one is led to the following guide. *Experimental determination of one electron eigenvalues are in error by at least several eV since either the initial or final states of the system are excited and all electron eigenvalues are altered.* Actually the experimental difference is smaller than one might have thought and a theorem called *Koopman's Theorem* relates to the extent to which the measured values agree with the eigenvalues. When they accidentally do (and they generally do not) physicists say that Koopman's theorem is satisfied.

Emission of X-rays from Argon

In emission studies one records the energy of photons emitted when the atom returns to its ground state from an excited state. Consider the ionized argon atom in Table IX as the starting point. Several things happen. An electron from the $2p$ shell may fall into the $1s$ shell emitting a photon called a $K\alpha$ X-ray. At a subsequent time a $3s$ electron may fill the vacancy now left in the $2p$ shell and the emitted photon is called an L X-ray. Finally a free electron wandering in the neighbourhood may be attracted to the ionized atom, since it is positively charged, and fill the vacancy in the $3s$ shell by emitting an M X-ray.

After a $1s$ electron has been ejected from the $1s$ shell it takes about 8×10^{-15} sec for the $2p$ electron to drop into the vacancy and emit the $K\alpha$ X-ray. This can be determined from the so-called measured natural width of the $K\alpha$ X-ray ($\sim\frac{1}{2}$eV) and the uncertainty principle ($\Delta t = h/\Delta E \cong 8 \times 10^{-15}$ sec). A similar time is required for the emission of the L X-ray but the time required for M emission depends on the availability of free electrons in the neighbourhood.

It is also possible for the singly ionized atom to return to its ground state by a $3p$ electron filling the vacancy rather than a $2p$ electron. This gives rise to a $K\beta$ X-ray and happens about $\frac{1}{5}$ as often as the $K\alpha$ process. This is followed by the atom returning to its ground state by capturing a wandering electron attracted to its positive charge.

By tradition the X-ray spectroscopists have given special designations to the various electron shells and the various photons emitted when a higher energy electron falls into a vacant lower state. Fig. 76 indicates some of the designations.

If the argon gas is solidified there is a decided change, and the sharp final states are blurred into a band of energies. Even though the cohesive energy of solid argon arises principally from the weak closed shell van der Waals forces nonetheless the neighbouring atoms perturb the sharp p levels and broaden them. This is rather definitive evidence of the existence of bands even though we are unable to calculate μ from eqn. (148) because we do not know the wave functions of the unfilled eigenvalues in the band required in eqn. (149).

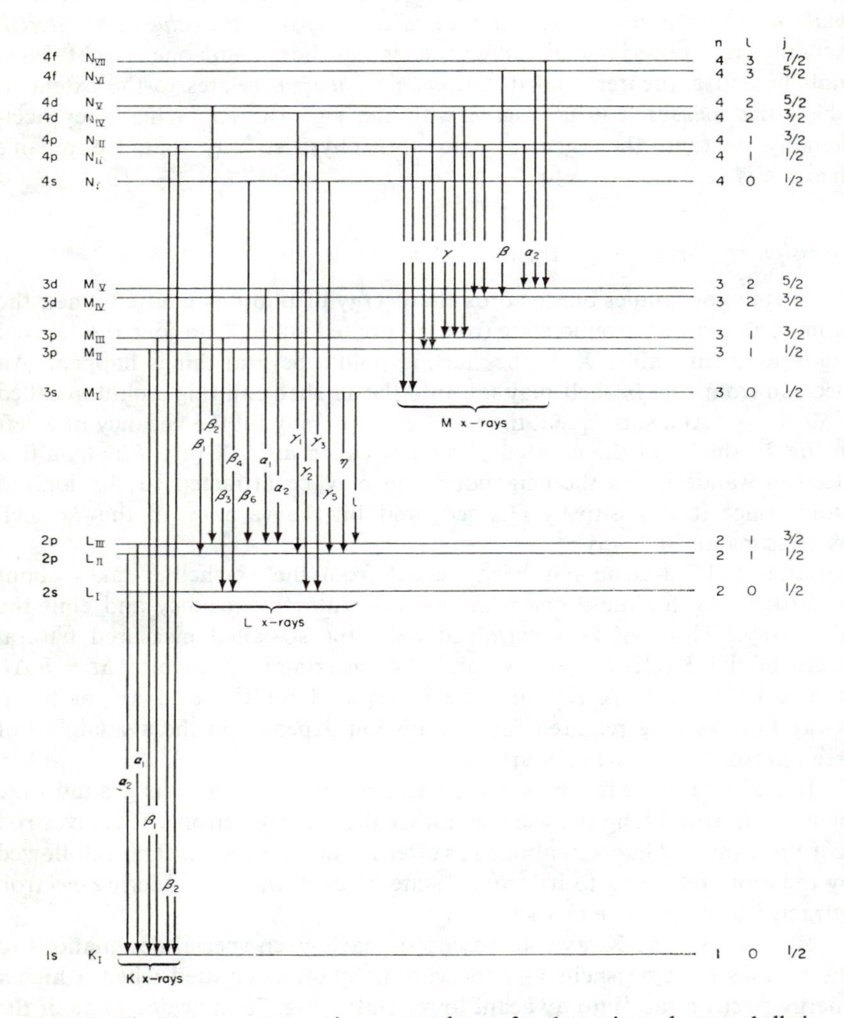

FIG. 76. The X-ray spectroscopists nomenclature for the various electron shells in an atom and the nomenclature of the X-ray lines emitted when an electron makes a transition from an upper level to a vacancy in a lower level. n and l are the quantum numbers and j represents the level for which m_s adds or subtracts from m_l. For example, $l = 3$, $m_l = 3$, $m_s = +\frac{1}{2}$, $j = \frac{7}{2}$; $l = 3$, $m_l = 3$, $m_s = -\frac{1}{2}$, $j = \frac{5}{2}$.

X-ray Emission from Iron, Cobalt, Nickel, Copper, Zinc, Gallium and Germanium

Let us now look at the absorption and emission lines of the metals[38] iron, cobalt, nickel, copper, zinc, gallium and germanium in the neighbourhood of the K absorption edge, Fig. 77. The lines designated K_β are emitted

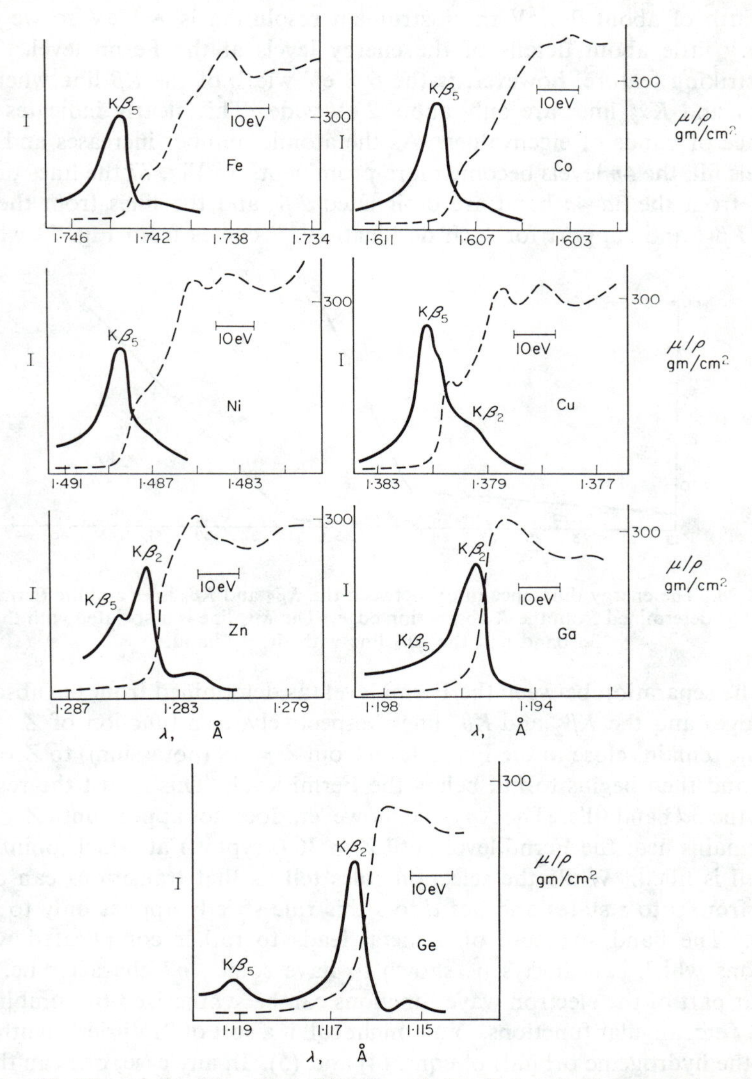

FIG. 77. The relative intensities of the K_β emission lines near the absorption edges of iron, cobalt, nickel, copper, zinc, gallium and germanium, solid curves, and the approximate mass absorption coefficients of these elements, dashed curves, as a function of wavelength in Å.

by the metal when an electron within a few eV of the Fermi level falls into the vacant $1s$ level. In iron the absorption edge occurs around 1·744 Å and tells us the minimum energy required to excite a $1s$ electron to the first available state above the Fermi level, or at least the first available state on an atom with a $1s$ electron removed. Although the $K\beta$ lines have a natural half-width of about 0·2 eV the instrument resolution is $\sim$1 eV so we can say very little about details of the energy levels at the Fermi level. The most striking feature, however, is the 6–8 eV width of the $K\beta$ line whereas the $K\alpha_1$ and $K\alpha_2$ lines are only about 2 eV wide. This clearly indicates the existence of bands of eigenvalues. As the atomic number increases and the $3d$ levels fill, the $4p$ levels become more prominent. In Fig. 77 the lines originating from the $3d$–$4s$ band are designated $K\beta_5$ and the lines from the $4p$ band, $K\beta_2$, and support for such designation[39] comes from Fig. 78 which

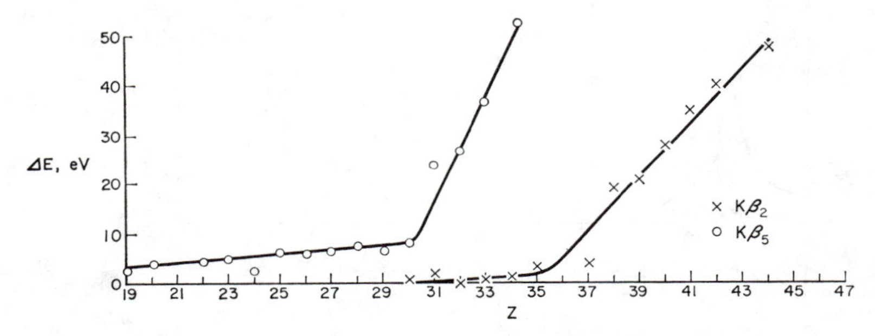

FIG. 78. The energy difference in eV between the $K\beta_2$ and $K\beta_5$ lines and the Fermi level as determined from the K absorption edge. The $K\beta_5$ line is associated with the $3d$ band and the $K\beta_2$ line with the $4p$ band.

plots the separation between the Fermi level (as determined from the absorption edge) and the $K\beta_2$ and $K\beta_5$ lines, respectively, as a function of Z. The $K\beta_5$ line remains close to the Fermi level from $Z = 19$ (potassium) to $Z = 30$ (zinc) and then begins to fall below the Fermi level. This is just the region where the $3d$ band fills. The $K\beta_2$ line, however, does not appear until $Z \cong 29$ and remains near the Fermi level until $Z \cong 36$ (krypton) at which point the $4p$ shell is filled. While the selection rules tell us that transitions can only occur from p to s states and not d to s this rule strictly applies only to free atoms. The band structure of a metal leads to rather complicated wave functions which can always be shown to have some "p" character i.e. the angular part of the electron wave functions can be synthesized by combining s, p, d, f etc. angular functions. You might call it a sort of "Fourier" synthesis using the hydrogenic orbitals of eqns. (4) and (5). In any case, one can think of the $K\beta_5$ line as a transtition from the "p" character of the $3d$–$4s$ band to the vacant $1s$ level.

As one can see in germanium the $K\beta_5$ line has moved about 26 eV below the Fermi surface. In all the cases the total band width, $3d$–$4s$ or $4p$ as the case may be, is about 6–10 eV. If we measure the width of the $L\alpha$ line[40] of copper as shown in Fig. 79 we see that it is only about 4 eV wide as compared to 6 eV for the $K\beta_5$, Fig. 77. However, this line involves the transition $3d \rightarrow 2p$ rather than $4p \rightarrow 1s$. Thus the $K\beta_5$ line gives us the width of the "p" character in the $3d$–$4s$ band ($\sim$6 eV) and the $L\alpha_1$ line gives us the width of the d character in the $3d$–$4s$ band. That the $3d$ part of the band is narrower has in fact already been discussed on p. 58.

If we now investigate the experimental width of the M emission band[41] of copper this involves transition from the $4s$ part of the $3d$–$4s$ band to the vacant $3p$ levels. This is plotted in Fig. 79, diagram 2, and is seen to be

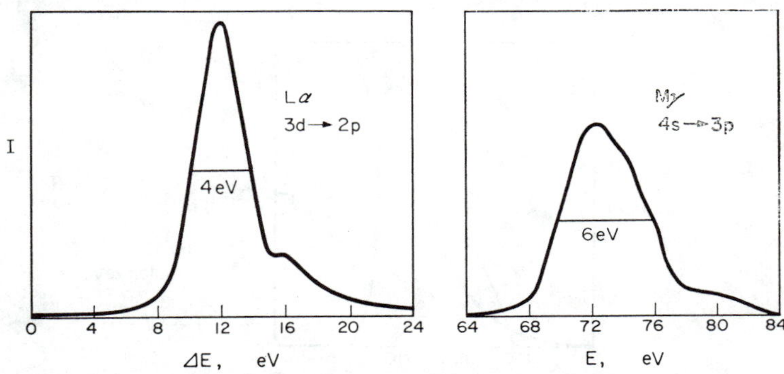

FIG. 79. The $L\alpha$ ($3d$–$2p$ transition) and the $M\gamma$ ($4s$–$3p$) emission lines of copper showing the $\sim$4 eV width of the $3d$ band and the 6 eV width of the $4s$ band. The energy of the $L\alpha$ line is $\sim$940 eV.

$\sim$6 eV wide, again confirming our picture of the electronic structure of copper on p. 58.

Determining the Density of States in Beryllium

While X-ray emission lines give us an indication of the widths of the bands in metals such information falls far short of that desired. We should really like to know the energy levels within the band or more specifically the density of states in the band. This is rather difficult to calculate and, as it turns out, is also rather difficult to determine experimentally. In addition to technical problems the difficulty arises because the intensity versus energy of the X-ray emission lines depends on the product of the density of states and the transition probability. The number of photons emitted per sec is proportional to

$$I \propto \nu \frac{\mathrm{d}n}{\mathrm{d}E} \left| \int \psi_f^* \frac{\partial}{\partial x} \psi_i \, \mathrm{d}x \, \mathrm{d}y \, \mathrm{d}z \right|^2 \tag{151}$$

where $\mathrm{d}n/\mathrm{d}E$ is the number of energy levels per atom per unit energy interval, ν the photon frequency, ψ_f^* the wave function of the atom in its final state, ψ_i the wave function in its initial state. It is impossible in eqn. (151) to separate the density of states from the transition probability which involves a knowledge of the wave functions. *Since we do not know the outer electron wave functions in a metal and cannot calculate the transition probability we must combine this with the failure of Koopman's theorem to form the two major obstacles in determining eigenvalues of the outer electrons in metals.*

Nonetheless, let us examine the more favourable case of the $K\alpha$ emission lines of beryllium which involve transitions from the "$2p$" character of the $2s$–$2p$ band in beryllium to the vacancy in the $1s$ shell. The calculated density of states curve of Herring and Hill[(2)] has been given in Fig. 15 and the experimental emission line is given in Fig. 80. The agreement is quite good

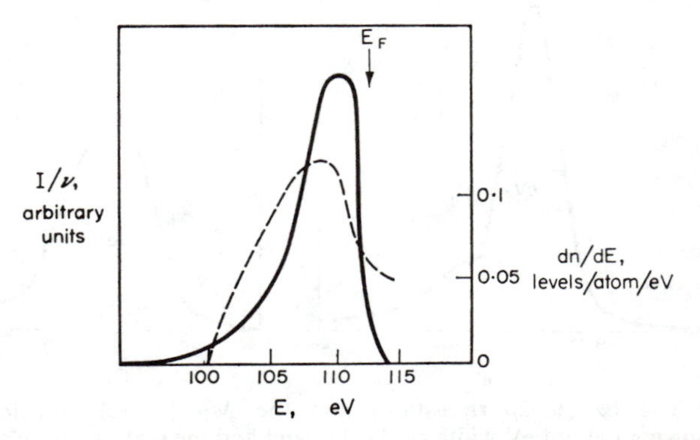

Fig. 80. The calculated density of states of beryllium (dashed curve) and the intensity of the $K\alpha$ emission line of beryllium, $2p \rightarrow 1s$. If the transition probability is independent of energy the curve suggests that the $2p$ mixture in the $2p$–$2s$ band increases as we approach the Fermi surface from below.

for the band width but only fair for the density of states. Of course this assumes that the transition probability is a slowly varying function of energy through the band which means that the "p" character of the wave functions does not vary appreciably over the band width so that the transition probability is independent of energy. Most probably the "p" character increases near the top of the band.

Perhaps the main conclusion to be drawn is that neither theory nor experiment are sufficiently accurate to expect any more than an accidental correlation of the density of states in metals. However, there is reasonable hope in the near future to expect theory and experiment to agree at least on band widths.

Before discussing the experimental techniques of emission or absorption spectroscopy there are several effects to be mentioned in X-ray spectroscopy.

(*a*) *Fluorescent Yield.* If an electron is ejected from the $1s$ shell this is followed by a $K\alpha$ or $K\beta$ line only in about one case in three. This fraction is called the *fluorescent yield.* In the other two out of three cases the atom returns to its ground state by a co-operative regrouping of the remaining electrons to fill in all the lowest energy states, the energy so gained being transferred to an outer electron which is ejected from the atom. This now leaves the atom minus two outer electrons, but these vacancies are filled from electrons in the band.

(*b*) *Structure of X-ray Absorption Edge*. It has been suggested that the structure on the high energy side of an absorption edge as in Fig. 77 is due to the ejected electron undergoing Bragg scattering. The energy of the ejected electron is just the energy difference of the photon and the energy required to excite the electron to the Fermi level. Converting this energy, E, to wavelength and then to the conditions for 180° Bragg scattering, $\sin\theta = 1$, we have for a b.c.c. crystal like lithium.

$$\lambda = \frac{h}{\sqrt{(2mE)}} = 2d = \frac{2a_0}{\sqrt{(h^2+k^2+l^2)}}$$
$$E = 3{\cdot}14(h^2+k^2+l^2)\text{ eV} \tag{152}$$

The absorption on the high energy side of the K absorption edge of lithium ($\lambda = 226{\cdot}6$ Å) is shown in Fig. 81 with the energies of the Bragg peaks

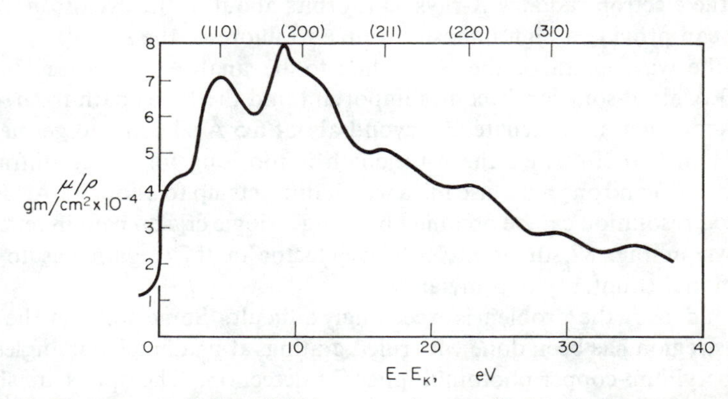

FIG. 81. The mass coefficient of lithium metal around the K absorption edge. The Bragg peaks of b.c.c. lithium metal appropriate to the ejected electrons of energy $E - E_K$ are designated.

indicated; while there is some correlation in lithium metal this is not always the case and requires more study.

(*c*) *Satellites*. It is sometimes possible to simultaneously eject a 1*s* and 2*s* electron from an atom leaving two vacancies. If this is followed by a 2*p* electron dropping into the 1*s* shell we find the $K\alpha$ line so emitted to be shifted relative to a $K\alpha$ line emitted from an atom without a 2*s* vacancy. This new line which is very weak is called a *satellite*. In the first series of transition metals these satellites are shifted about 20 eV to higher energies. This indicates the extent of the disturbance caused by the 2*s* vacancy.

Experimental Technique in the X-ray Region

Experimental Techniques in X-ray Emission. In order to obtain fine details of the energy bands it is desirable in studying the emission spectra of solids to be capable of measuring energies to $\frac{1}{10}$ eV. Even though the $K\alpha$ lines of a substance like iron have a natural width of ~ 2 eV due to their short life time the $K\beta_5$ line is much longer lived so that its natural width is a few tenths of an eV. If we wish to measure the iron $K\beta_5$ line with this resolution a crystal with a natural X-ray rocking curve of about 5 sec is required. With the advent of high purity silicon and germanium this is now feasible, particularly in high order. The experimental arrangement requires two crystals in the so-called 1, +1 position shown in Fig. 82, diagram 1.

In order to excite the atoms it is customary to bombard them with electrons but this gives rise to a large *Brems-strahlung** background. Since the $K\beta_5$ line is rather weak it is probably better to excite the electrons with a high intensity source of X-rays such as can be obtained from a fluorescent tube. In this case the *Brems-strahlung* is absent. High intensity sources of X-rays are also becoming available in high energy electron cyclotrons ($\sim$BeV) in which the electron radiates X-rays as it orbits about in the cyclotron. These X-rays can provide a high intensity source to fluoresce the sample.

As the wavelength of the X-ray line to be analyzed increases beyond about 3 Å air absorption becomes important and the X-ray path from source to detector must be evacuated. Beyond about 6·5 Å silicon and germanium are no longer useful since the wavelength is too long for Bragg diffraction ($\lambda > 2a_0/\sqrt{3}$) and one must use mica which diffracts up to about 18 Å. In this case good resolution can be obtained by using a single crystal bent in reflection as shown in Fig. 82, diagram 2. The detector in this region has to be a proportional counter with a mylar window.

Beyond 18 Å the problem is exceedingly difficult. Some work in the 20 Å to 200 Å region has been done with ruled gratings at small incident angles and with a beryllium-copper photomultiplier for detection. The optics are shown in Fig. 82, diagram 3. In this range everything is highly absorbing and a thin oxide coat of 10 to 100 Å on the sample can cause great difficulty in studying, say, the M spectra of iron in the region 250 Å. Due to the long wavelength, high resolution is possible ($\sim$0·01 eV) but considerable effort is required to work out the experimental problems of the sample surface. A technique of bombarding the metal surface with high energy electrons or inert gas ions to evaporate the oxide layer, checking this by electron diffraction employing relatively slow electrons $\sim$1000 eV, then measuring the X-ray emission lines all in a vacuum of 10^{-8} to 10^{-9} mm of mercury appears requisite to good measurements. It is difficult at the present time to know whether good work

* Brems-strahlung is the X-radiation emitted by electrons as they are decelerated in a target.

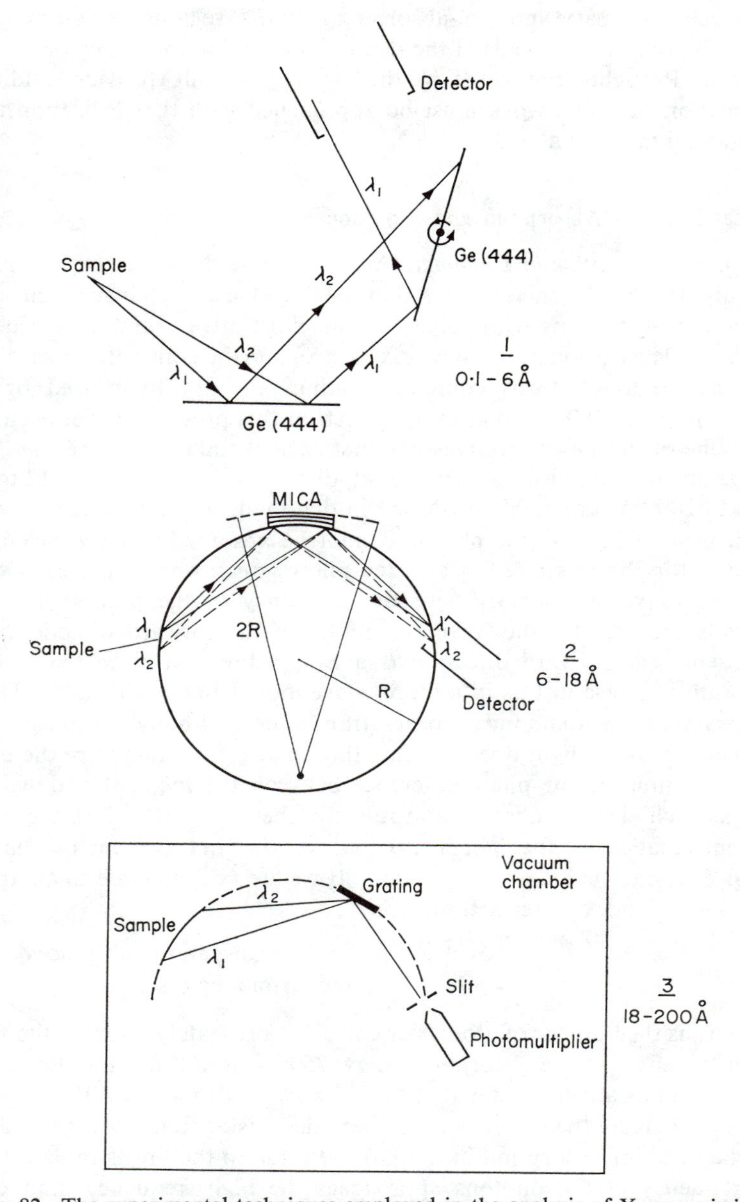

FIG. 82. The experimental techniques employed in the analysis of X-ray emission lines in various wavelength ranges. *Diagram 1* (0·1 to 6 Å) Double crystal (Ge 444) spectrometer, the second crystal rotating to analyze the beam selected by the first crystal. *Diagram 2* (6–18 Å) A mica crystal bent to twice the radius, *R*, of the focal circle. *Diagram 3* (18–200 Å) A ruled grating at grazing incidence with sample, grating and receiving slit on the focal circle.

in the relatively easy and non-absorbing *K* and *L* regions, for elements like iron, will provide the details of the band structure that the *M* region will only confirm. Probably the work in the *M* region will provide additional information but such work must be approached with the realization that a painstaking task lies ahead.

Optical Region—Absorption and Emission

Optical Absorption in Diamond. As we increase the wavelength to greater than about 1000 Å (energies less than ~12 eV) many changes occur in the interaction of photons with solids. As an illustrative example consider the diamond. The carbon atoms have six electrons in the configuration $1s^2 2s 2p^3$, the outer four electrons completely filling the sp^3 hybridized orbitals. Since it requires 280 eV to eject a $1s$ electron this process cannot occur. To excite one of the sp^3 electrons to the first excited state across the *energy gap* (first zone filled) requires ~6 eV so that all photons in the range ~12 to 6 eV (1000 to 2000 Å) are highly absorbed in diamond. If we increase the wavelength beyond 2000 Å the photon cannot be absorbed. The wavelength is too long for Bragg scattering and the photons can only cause each of the electrons to vibrate slightly with the frequency of the photon and these electrons then scatter the wave coherently in the forward direction. In a strict sense the forward direction *is* a Bragg direction since the scattered waves are in phase just as in a more conventional Bragg direction. The net effect is that the diamond appears transparent, although it must be remembered that the light does not pass through as if nothing were there since there is a time lag or phase-difference between the incident and scattered wave as each electron absorbs and re-emits the wave. It is just this phase-difference that causes the light energy to slow down in traversing the diamond and gives rise to its index of refraction. If we are not too close to an absorption edge the index of refraction, n, is

$$n^2 \cong 1 + \frac{N_0 e^2 \lambda_k^2}{\pi m_0 c^2} \quad \text{valid for transparent substances in the visible region} \tag{153}$$

where N_0 is the number of atoms per cm^3. Unfortunately, even in the visible region the effect of the absorption below 2500 Å is still felt and this formula is only good as a rough estimate ($N_0 e^2 \lambda^2 / \pi m_0 c^2$ ~ to about 10%).

As we increase the wavelength further, (decreasing frequency) we suddenly find the onset of absorption in the infra-red due to the "reststrahlen", when the frequency of the photons approaches the high frequency limit of the optical phonons (p. 191). At the high frequency limit of the phonon dispersion curve, Fig. 68, the nearest neighbour atoms are vibrating approximately 180° out of phase. The relative motion of the outer electrons on the neighbouring atoms induces electric fields and these can interact strongly with the electric vector of a photon of similar frequency. Thus the photon can actually be absorbed by creating a phonon. In the diamond this occurs at $\nu \cong 62 \times 10^{12}$ cps or a wavelength of about 48,000 Å*. As the wavelength is increased further the absorption due to lattice vibrations diminishes since the lower frequency phonons do not create such strong electric fields. Finally, in the long wavelength radio-wave region the diamond is fairly transparent again.

The Colour and Band Structure of Copper and Silver. In the case of a metal, however, there are always possible excitations no matter how low the photon energy since the unfilled levels above the Fermi level are quite closely spaced. In addition the vibrating electric field of the photon can create an a.c. current in the Fermi level electrons. These two effects are not clearly separated although there are theoretical attempts to do so by applying the free electron theory to the electrons at the Fermi level. In the case of metals we have both a real and imaginary part of the index of refraction, $n + ik$,

$$n^2 - k^2 = 1 - \frac{n_0 e^2}{\pi m^*} \frac{1}{\nu^2 + \gamma^2}$$

$$nk = \frac{n_0 e^2}{4\pi^2 m^* \nu} \frac{2\pi\gamma}{\nu^2 + \gamma^2} \tag{154}$$

$$\mu = 4\pi nk/\lambda = \text{absorption coefficient}$$

where n_0 is the number of electrons per cm^3, γ is approximately 2π times the reciprocal of the time between electron collisions when the electron normally flows in the metal under an applied voltage and ν is the photon frequency. $(n_0 e^2/2\pi m^* \gamma) = \sigma_0$ is just the conductivity of a metal (a typical value of γ is $\sim 6 \times 10^{13}$/sec) and the reflecting power is given by eqn. (156).† These expressions are simplified further for the cases:

* Actually in diamond, silicon and germanium the "reststrahlen" absorption is rather weak since the nearest neighbour atoms are a mirror image of each other (known as a centre of symmetry). The strong absorption of photons requires that an electric dipole moment be created by the vibrating atoms and this cannot occur for mirror image atoms since they are identical. On the other hand a substance like MgO can absorb very strongly since the transfer of some outer bonding electrons from the Mg to the O leaves the Mg positively charged and the O negatively charged. The observed weak absorption in diamond presumably arises from a small electric quadrupole moment.

† For scattering angles other than 180° the reflecting power is

$$R_s = \frac{a^2 + b^2 - 2a\cos\theta + \cos^2\theta}{a^2 + b^2 + 2a\cos\theta + \cos^2\theta}$$

$$R_p = R_s \frac{a^2 + b^2 - 2a\sin\theta\tan\theta + \sin^2\theta\tan^2\theta}{a^2 + b^2 + 2a\sin\theta\tan\theta + \sin^2\theta\tan^2\theta}$$

$$a^2 = \tfrac{1}{2}\{[(n^2 - k^2 - \sin^2\theta)^2 + 4n^2k^2]^{1/2} + n^2 - k^2 - \sin^2\theta\}$$

$$b^2 = \tfrac{1}{2}\{[(n^2 - k^2 - \sin^2\theta)^2 + 4n^2k^2]^{1/2} - n^2 + k^2 + \sin^2\theta\} \tag{156}$$

where the subscript s and p refer to the reflecting powers for the parallel and perpendicular planes of polarization.

1. In the infra-red where $\nu \ll \gamma$ we obtain

$$R \cong 1 - 2\left(\frac{\nu}{\sigma_0}\right)^{1/2} \tag{155}$$

which is approximately satisfied when one compares the reflecting power of a metal and its resistivity, ρ $(= 1/\sigma_0)$.

2. The wavelength, λ_0, at which the real part of the index of refraction changes from less than one to greater than one or when the substance becomes totally reflecting for $n > 1$ $(\nu \gg \gamma)$ is given by

$$\lambda_0 = 2\pi\left(\frac{m_0 c^2}{4\pi n_0 e^2}\right)^{1/2} \tag{157}$$

The agreement with eqn. (157) is good only for the alkali metals, for which λ_0 varies from about 1500 Å for lithium to 3800 Å for cesium. But in any case it is expected that λ_0 varies from about 500 Å to 3000 Å for every metal.

In the free electron theory the absorption coefficient μ, varies from about 10^4/cm in the ultra-violet to about 10^7/cm in the infra-red. Thus the penetration depth is quite appreciable in the ultra-violet but the photon only penetrates a few atomic layers in the infra-red.

That the outer electron eigenvalues of the metallic bands are important in their interaction with light is quite apparent from the different colours of metals. Most metals, even the silvery ones, have distinctive colours that can be detected by the naked eye. This is particularly true of copper and gold. Let us examine the optical properties of two metals with vastly dissimilar colours, copper and silver, to understand this effect in more detail. Fig. 83 gives the wavelength dependence of the real and imaginary parts of the index of refraction[42], the reflecting power at $\theta = \pi/2$, and the absorption coefficient. The reflecting power of silver is quite high over the entire visible range 4000 Å–7000 Å giving it its characteristic whitish or metallic colour. On the other hand, the reflectivity of copper favours the red and is responsible for the reddish colour, the blue end of the spectrum being strongly absorbed. The absorption coefficients of both copper and silver show peaks at ~5000 Å (2·5 eV) and 2500 Å (5 eV) respectively and while the origin of these peaks is not known it is reasonable to assume that they are due to the excitation of an electron from the *d* band to the Fermi level ~2·2 eV in copper and 5 eV in silver (see pp. 58, 377). If this is the case one should expect this excitation to be followed by an emission line, above 5000 Å in copper and 2500 Å in silver when an electron at the Fermi level drops down to fill this vacancy. This has not been established (in fact the interaction of photons with metals over the range 2000–9000 Å has lacked much experimental attention). The real and imaginary parts of the index of refraction in silver and copper also reflect the strong absorption at 2500 Å and 5000 Å.

In the infra-red and at longer wavelengths all metals behave classically in that the photons set up a current in the metal as if an a.c. electric field were applied. The very high absorption coefficients limit this current to within

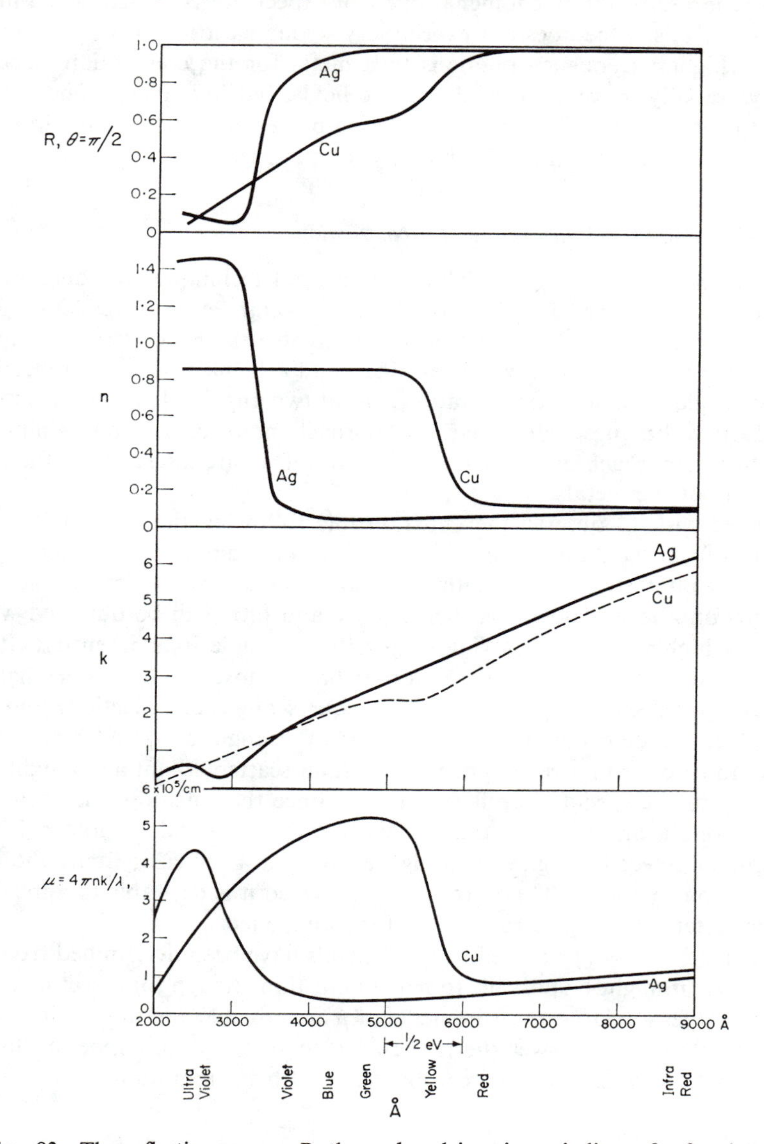

FIG. 83. The reflecting power, R, the real and imaginary indices of refraction n and k; and the linear absorption coefficient μ of copper and silver as a function of wavelength from 2000 to 9000 Å. The colour of copper is due to the absorption at ~5000 Å.

about 10 to 100 Å of the surface of a good conductor and this interaction is limited to the electrons at the Fermi surface. While observations have not been reported of the phenomena, one can expect "reststrahlen" even in the case of metals. One does not expect very strong electric fields to be created by the highest frequency phonons in a metal for the conduction electrons always rapidly move to neutralize these fields, but in principle one may be able to detect some effects in the reflecting power of a metal in the infra-red providing it lacks a centre of symmetry (cf. p. 223).

Experimental Techniques in the Optical Region

Experimental Techniques. The experimental techniques for determining n and k are reviewed in *Methods of Experimental Physics*, Academic Press **6B** (1959) pages 254–262, most of which involve the general reflectivity expression, eqn. (157). One can determine n and k by measuring the reflectivity at two angles or determine the ratio R_s/R_p at two angles. This involves rather standard techniques. However, an approach more closely resembling the soft X-ray approach may prove more fruitful if we are interested in the band structure of the metals.

If we wish to observe the emission line following the excitation of the electron from the $3d$ band to above the Fermi surface we require an intense source of photons of wavelength less than $\sim$5000 Å with virtually no contamination above 5000 Å. A mercury arc and filter can be obtained which yields a high ratio of ultra-violet intensity to visible light intensity. If the copper sample is polished to a mirror finish most of the visible light is specularly reflected (angle of incidence equals angle of reflection) and one can place the detector at an angle outside the angle of specular reflection. This can be done in a velvet lined box to limit scattered light and a light pipe can absorb the specularly reflected light. Since the emission line should be quite broad a prism can be used to analyze the light and a photomultiplier or film to detect the light. The high absorption coefficient limits the light penetration to about 1000 Å from the surface so it is desirable to anneal the copper after polishing to remove cold working effects.

While the absorption coefficient of metals have been determined from the reflectivity it might be possible to make films thin enough for direct measurements. At least relative measurements of μ are possible and absorption edges can be detected. To cover the range 2000 to 10,000 Å in copper a film 0·1 microns thick is required and can be prepared by evaporation.

Problems

1. Secure pieces of cobalt, iron, nickel, vanadium, titanium, manganese, silver, tin and germanium. Give them all a light polishing to remove the oxide but do not produce a mirror finish. Look at them under an incandescent light. Are any two indistinguishable?

Does this suggest that the curves in Fig. 83, hence the band structures, differ? At what wavelength in the visible region do you estimate the absorption maxima occurs in gold? Why is graphite black?

2. From the material covered in Chapters IV to VII and from the *Handbook of Chemistry and Physics* estimate and plot the linear absorption coefficient of copper metal (on log–log paper) for each of the electron shells ($1s^2$; $2s^2 2p^6$; $3s^2 3p^6$; $3d4s$ band) from 0·1 Å to 100,000 Å. Sum these up to give the total.

3. (For the mathematically minded).

Estimate the K oscillator strength of the $1s$ electron in copper by assuming the initial wave function to be hydrogenic with Z_{eff} adjusted to give the observed energy (K absorption edge) of the $1s$ electron and the final wave function to be that of a free electron at the Fermi surface in copper. Replace the Wigner–Seitz call by a sphere and first average over direction of the momentum of the free electron. The total K oscillator strength is obtained by summing over all final states and is related to the linear absorption coefficient, μ_k, at the K absorption edge (λ_k) by $g_k = \Sigma_f g_{if} \cong mc^2\mu_k/5{\cdot}67 N_0 \lambda_k e^2$ where N_0 is the number of atoms per cm^3. The Hönl dispersion corrections are given by $\Delta f' \cong \{g_k \ln|(\lambda/\lambda_k)^2 - 1|\}/(\lambda/\lambda_k)^2$; $\Delta f'' \cong mc^2\mu/2e^2\lambda$. Calculate g_k and the Hönl dispersion correction at $\lambda = 1{\cdot}54$ Å. Ans.: $g_{if} = 1{\cdot}26 \times 10^{-3}$; $g_k = 1{\cdot}49$; $\Delta f = -1{\cdot}70$.

Summary

In summary, the entire range of photon wavelengths from 0·1 Å to the infra-red has given us some information about the eigenvalues in the metallic bands but no measurement has unequivocally provided us with a reasonably accurate density-of-states curve. The entire range of wavelengths is beset with great experimental difficulty, yet no other technique offers a remote chance of an experimental determination of the eigenvalues in metallic bands. In addition, the initial or final state of the system in such measurements is necessarily an excited state and absolute errors of several eV can generally be expected. *On the other hand, the density of states may still retain its same shape even though shifted by several eV*, but this remains to be seen.

CHAPTER VIII

TRANSPORT PROPERTIES

Introduction

In the preceding two chapters we have discussed experimental methods aimed at determining eigenfunctions and eigenvalues of the outer electrons in solids. In this chapter we shall discuss experiments involving only those outer electrons very close to the Fermi level. These experiments have given us rather detailed information of the momentum of the electrons at the Fermi surface. It is indeed, unfortunate, that these experiments are limited to the eigenvalues within a few hundredths of an eV of the Fermi level for these levels play a negligible role in the cohesion of the metal, that role being principally fulfilled by the electrons at the bottom of the band.

In Chapter II, p. 34, we have pointed out the existence of regions or paths in a metal in which the potential is practically constant and that the solution to the Schrödinger equation yields a constant electron probability distribution for these regions. Since electrons move so readily in a metal it is reasonable to suggest that the electrons at the Fermi level spend most of their time in these regions of constant potential. Of course, the wave functions of these electrons must join smoothly from the outer regions of the atoms to the regions inside the core but the wave functions of these electrons are fairly small inside the core. Thus we say that the current in a metal is carried primarily along paths between the atoms, Fig. 9. A more sophisticated question asks whether the current is carried by a few isolated electrons hurrying along at about 1/100 the velocity of light, or by the entire electron community moving along at a snail's pace. For example, if one ampere flows in a copper wire of 1 mm radius then $6{\cdot}28 \times 10^{18}$ electrons/sec pass any point along the length of the wire. If this current was carried by the outer 4*s* electron of each copper atom then the entire outer electron community would have to move with a net velocity of $\sim 2{\cdot}3 \times 10^{-3}$ cm/sec. On the other hand, if we use the free electron approximation, the electrons at the Fermi level have a velocity of $\sim 10^8$ cm/sec and if they actually move through the crystal with this velocity then only one in $\sim 5 \times 10^{10}$ is required to carry one ampere current. Which of these two extremes is correct, all the outer electrons moving along at $2{\cdot}3 \times 10^{-3}$ cm/sec or only one in a hundred billion moving at $\sim 10^8$ cm/sec? Probably the former, since all outer electrons in copper are identical and one cannot single out isolated electrons for this current carrying mission. Furthermore, it takes about 1 eV to squeeze an electron between two neutral atoms. A conceptual picture of electrical conduction in a metal then, is a mass co-operative motion of the electrons along the paths of their normal probability distributions with each electron maintaining about the same distance from its neighbour. If we introduce thermal motion of the atoms into this picture it becomes more difficult for this co-operative electron movement since some of the electrons will be deviated or scattered by atoms that get into its straight line path. This causes a drag or resistance. However, the many electron picture we have just given is not theoretically tractible and we *must* use the one electron wave functions and, in the case of a crystal, the one electron Bloch waves. As such, the electron wave functions become wave functions of the entire crystal and we can no longer legitimately inquire about the intimate details of electron motion since quantum mechanics only tells us the expectation values of momentum, velocity etc., for the electrons over the *entire crystal.*

In this Bloch wave treatment the eigenvalues form an energy band which is filled to the Fermi level. When an external electric field is applied to a metal the electrons are accelerated in the direction of the field and thus increase their energy. Since all the eigenvalues are filled up to the Fermi level only the electrons at the Fermi level can increase their energy and occupy higher unfilled energy levels so that conduction is limited to these few electrons We are thus back to the picture of only a few electrons carrying the current! Actually this is a common misconception. One must remember that each of the one-electron wave functions, from the bottom of the band to the top, are only mathematical functions which are solutions of the Schrödinger equation, and that it is not correct to identify each electron with a specific one of these wave functions. To determine the wave function for an electron one must make the proper antisymmetrized combination and this requires an equal amount of *every one-electron wave function with identical m_s value.* By the time this is done every electron with the same m_s is identical and we have a much more realistic picture of the electrons in the band. Therefore, we must remember that whenever we speak of the electrons at the bottom of the band or at the Fermi level etc., *we strictly mean the one-electron wave functions.* The extent of the difference can be seen in Fig. 14 representing the probability distribution for two electrons in a one-dimensional box. If we identify each electron *separately* (both with same m_s) with the two lowest energy wave functions their probability distributions would be independent and would be P_0 and P_2, respectively. *The total wave function is not antisymmetric.* On the other hand, a properly antisymmetrized wave function leads to a probability distribution that depends on the position of both electrons. If electron 1 is at $x = 1{\cdot}75$ Å in Fig. 14 then electron 2 has a probability distribution given by P_1. The two electrons are then indistinguishable. The difference between P_1 and P_2 for the electron probability distribution of electron 2, when electron 1 is at 1·75 Å, is like night and day.

Types of Measurements

There is a wide variety of measurements that are made on these conduction electrons and we shall first briefly enumerate and then describe them in more detail.

1. *Resistivity*—The resistivity is a measure of the extent to which the electrons are deviated or scattered from their straight line paths as they are accelerated by an external electric field.

2. *Magnetoresistance*—In this measurement the resistivity is measured as a function of the magnitude and direction of an externally applied magnetic field. The interaction of the moving negative electric charge of the electron with the applied magnetic field causes further deviations of the electron's path and generally increases the resistance.

3. *de Haas van Alphen Effect*—In this measurement the external magnetic field is varied and the magnetization or susceptibility of the metal is measured for each value. Oscillations are found in the susceptibility versus magnetic field curve which are due to the *quantized* orbital motion of the electrons. The susceptibility is diamagnetic since the orbital motion of the electrons creates a magnetic field opposed to the applied field, having its classical analog in the Lenz law.

4. *Cyclotron Resonance*—If an electron is moving in the x direction and a magnetic field is applied in the z-direction the force exerted on the electron is always at right angles to the magnetic field direction and its propagation direction. This leads to spiral paths whose frequency for completing one orbit is governed by the same equation as in the cyclotron ($\nu = eH/2\pi m^*c$). If we apply an external electromagnetic field at this frequency a resonance condition exists. Since the electro magnetic field is in the microwave region it penetrates only 10–100 Å and only the conduction electrons near the surface of the metal are affected. This is but a crude "classical picture" of the process, for in reality the application of an external magnetic field creates a new set of eigenvalues and we directly measure electron transitions between these energy levels by absorption of photons of appropriate energy.

5. *Anomalous Skin Effect*—The absorption of electromagnetic radiation in the microwave region is measured as a function of crystal direction. Only the electrons at the Fermi surface and near the surface of the metal are affected, and the absorption depends on the eigenvalues near the Fermi level of those electrons travelling in different crystallographic directions.

6. *Hall Effect*—If a magnetic field is applied along the z-direction of a conductor carrying current along the x-direction the electrons experience a force which deviates them in the y-direction. The Hall effect is a measurement of the current (or voltage) in the y-direction.

7. *Thermoelectric Power*—When one end of a metal wire is heated the electrons at the Fermi level are excited to higher energy states. This creates

a potential difference between the two ends of the wire. The thermoelectric power is the potential difference per degree and is generally measured relative to some other metal.

8. *Superconductivity*—Below a certain critical temperature many metals exhibit zero resistivity and are called superconductors. The electrons at the Fermi level enter a new lower energy state separated from their first excited state by an energy gap. They are not scattered because this new state is characterized by a special pairing of electrons of opposite m_s and opposite momentum and any scattering would destroy this state. At low temperatures there is insufficient energy available to scatter, for the electrons would have to be excited above the energy gap to do so.

While thermal conductivity is also a transport property we have left its discussion to the chapter on Thermodynamics (chap. IX).

We shall now discuss each of these eight measurements:

1. *Resistivity*

This is the simplest transport property to measure. The specific resistivity ρ in ohm-cm for a wire of uniform cross sectional area A is

$$\rho = AR/l \tag{158}$$

where R is the measured resistance, in ohms, over the length, l. The common unit of resistivity is ohm-cm and is generally of order of magnitude $\sim 10^{-6}$ ohm-cm for a metal at room temperature although it may vary by a factor of several thousand, from absolute zero to the melting point.

Fig. 84 is a schematic diagram of a general purpose circuit capable of measuring resistance with good accuracy. Neither current measurement nor current control is required since the current passes in series between the bridge and the specimen. Only short pulses of current pass through the sample which minimizes increasing the temperature, hence resistance, of a specimen during the measurement. For accurate work careful temperature control is necessary since a temperature gradient in the specimen will produce extraneous thermoelectric potentials. It is preferable to immerse the sample in a stirred non-conducting liquid bath as a thermal reservoir. Isopentane, 110°K to 300°K, various oils, 300°K to 600°K, and salt baths like LiBr, 550°K to 1500°K, are generally suitable. From 4°K to 77°K (liquid helium to liquid nitrogen boiling temperatures) the sample can be placed in an evacuated cryostat and allowed to heat up slowly from 4°K. At high temperatures the potential leads should be of the same material as the sample since alloying can change the resistivity. It is also advisable to average the separate measurements, with the current flowing in opposite directions, to minimize spurious effects.

The major problems in resistivity measurements arise from sample preparation. If one is measuring the resistivity of a very pure material at low temperature one may not wish to solder or spot-weld leads to it for fear of contamination. Current and potential leads of some soft metal like lead or indium, can be joined to the sample with "knife edge" pressure-contacts. If the substance is exceedingly brittle and cannot be fabricated into a sample of uniform cross-sectional area, one can make a mold of the specimen and cast a piece of lead, to duplicate the shape of the specimen. A "shape factor" can be obtained from the ratio of the cast lead sample to a lead sample of uniform cross-section, and the resistivity of the brittle sample obtained from this shape factor. Brittle samples frequently suffer from internal cracks and imperfections, and several samples should be checked for reproducibility. One might also check its measured density (weigh it in water and air) against the calculated X-ray lattice parameter density.

Irregularly shaped samples can also be measured by placing them inside a coil and measuring the a.c. inductance. The a.c. electromagnetic field is absorbed by the sample due to its resistivity but the effect is small at low

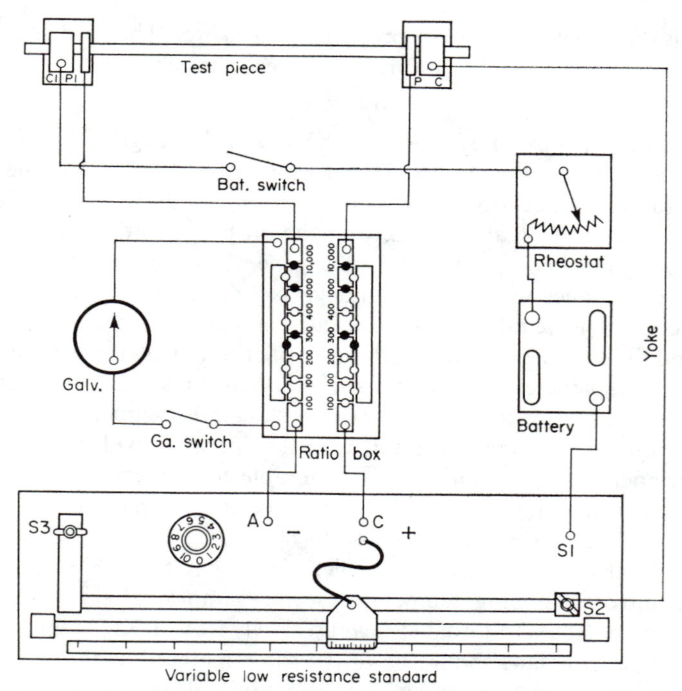

FIG. 84. The Kelvin bridge method of measuring resistivity. The potential drop across the specimen (test piece) is balanced against the potential drop across the standard variable resistance, the current passing in series between the two. The battery switch and galvanometer switch are connected to a double contact key so that the battery switch is closed first and opened last. Pulses are kept short to avoid heating the specimen.

frequencies.* At high frequencies the electromagnetic field only penetrates the surface and the surface resistivity may differ from the bulk resistivity.

Depending on the complexity of the problem the theory of conductivity uses Bloch waves or the more simplified plane waves of the free electron theory. The scattering of the electrons by phonons is treated similarly to the thermal scattering of X-rays or neutrons. Except at very low temperatures the inelastic scattering involves the electron, the phonon and the lattice as a whole through a Bragg reflection. This is called an umklapp or U process. At low temperatures the lattice is not involved and such an inelastic scattering process is called an N process. As in the case of X-rays and neutrons the thermal scattering of conduction electrons depends on the frequency distribution of phonons but is not nearly as sensitive. The Debye approximation is quite adequate for these purposes. But unlike X-rays and neutrons one cannot specifically measure the scattering of individual conduction electrons. The flow of current in a wire represents an overall drift velocity of the electrons as they are scattered in all directions. If both phonon and lattice are involved (umklapp process) the electron is scattered through large angles, since the lattice absorbs considerable momentum. However, it is still under the influence of the electric field and its path will be altered in that direction. At most temperatures the electrons are scattered in all directions and only a small net flow in the direction of the electric field is observed as a curient. It is, of course, the excitation of phonons by the conduction electrons, that converts current into heat.

While it is extremely difficult to calculate the thermal part of the resistivity it will suffice to give the following expression which provides a reasonable description of the temperature dependence of most elements (except for very low temperatures where electron-electron collisions may be important):

$$\rho_T = \frac{BT^5}{\Theta^5}\mathscr{J}\left(\frac{\Theta}{T}\right)$$

$$\mathscr{J}\left(\frac{\Theta}{T}\right) = \int_0^{\theta/T} \frac{x^5\,dx}{(e^x-1)(1-e^{-x})} \tag{159}$$

$$\begin{cases} \rho_T \cong BT/4\Theta & \text{valid for } T > 1{\cdot}5\Theta \\ \rho_T \cong 124{\cdot}4BT^5/\Theta^5 & \text{valid for } T \ll \Theta \end{cases}$$

$$\mathscr{J}\left(\frac{\Theta}{T}\right) = \frac{1}{4}\left(\frac{\Theta}{T}\right)^4 - \frac{1}{72}\left(\frac{\Theta}{T}\right)^6 \qquad \text{valid for } \Theta/T < 0{\cdot}8$$

where Θ is the Debye temperature and B is a constant which differs for each material. Table XI tabulates the values of $\mathscr{J}(\Theta/T)$.

* J. E. Zimmerman, R.S.I. **32,** 402 (1961).

TABLE XI

VALUES OF $\mathscr{J}(\Theta/T)$ APPEARING IN THE DEBYE APPROXIMATION FOR THE THERMAL PART OF THE ELECTRICAL RESISTIVITY. SOME TYPICAL VALUES OF B IN MICRO OHM-CM ARE ALSO GIVEN.

Θ/T	$\mathscr{J}(\Theta/T)$	Θ/T	$\mathscr{J}(\Theta/T)$
∞	124·43	4	29·488
20	124·42	3	12·771
13	123·14	2	3·2293
10	116·38	1·5	1·1199
8	101·48	1·2	0·47907
6	70·873	1	0·23662
5	50·263	0·8	0·0988

Element	B micro ohm-cm	Element	B micro ohm-cm
Fe	59	Ti	273
Ru	48	Zr	167
Os	32	Hf	92
Co	34	V	99
Rh	27	Nb	52
Ir	21	Ta	43
Ni	30	Cr	76
Pd	44	Mo	31
Pt	34	W	25
Cu	7.3	Mg	18
Ag	4·6	Zn	18
Au	4·9	Cd	16
Li	39	La	112
Na	8·7	Al	15
K	8·5	Pb	23

In general the T^5 dependence at very low temperatures is not experimentally confirmed but at high temperatures most substances are linear in T. The calculation of the value of B is rather difficult but it is approximately inversely proportional to the Debye Θ, the mass, the number of atoms per cm^3, the momentum of the electrons at the Fermi surface and the energy of the electrons at the Fermi level. Typical values for B are given in Table XI. Only in the case of the alkali metals can we calculate B with reasonable accuracy. A resistivity of one to ten micro ohm-cm (10^{-6} ohm-cm) corresponds to a mean free path between collisions of several hundred Å and a mean time between collisions of 10^{-14} to 10^{-15} sec.

There are contributions to the resistivity other than those due to lattice vibrations. If a second element is present in a metal either as an impurity or as an alloying element this destroys the perfect periodicity in the potential and gives rise to elastic scattering, similar to the X-ray or neutron scattering of disordered alloys. This scattering is temperature independent since the potentials are generally independent of thermal motion and permits us to make the assumption called *Matthiessen's rule* that the thermal part of the resistivity, ρ_T, and impurity or alloy resistivity,* denoted ρ_i, is additive. For small impurity concentrations the resistivity, ρ_i, is proportional to the concentration and typical values are 0·1 to several micro ohm-cm per atom per cent impurity. The impurity resistivity is zero at zero concentrations but even the purest metals like copper and aluminium contain some impurities so that their resistivity at $T \cong 0°K$ are approximately 10^{-3} micro ohm-cm, compared to several micro ohm-cm at room temperature. If we estimate $\frac{1}{2}$ micro ohm-cm per atomic percent impurity, this gives the purity as 99·9995% which is in approximate agreement with other measurements and indicates the presently attainable limit of purity in these metals. At higher concentration the impurity or alloy resistivity is given by

$$\rho_i \cong \bar{\rho}_i x(1 - x)/100 \qquad (160)$$

where $\bar{\rho}_i$ is the resistivity increase for one atomic per cent impurity and x is is the atomic percent of the alloy. This holds approximately for disordered solid solutions but is not valid if short range or long range order is present for then the resistivity decreases with increased ordering. Even though the resistivity is rather sensitive to ordering, due to theoretical difficulty we do not have a general expression relating the degree of order to the resistivity change.

A third contribution to the resistivity arises from the scattering of the conduction electrons by atoms with unpaired electrons, such as the rare earth metals and some transition metals. One cannot offer the magnetic scattering of neutrons as an analogy for in that case the interaction is magnetic. While there is also a magnetic interaction between the conduction electron's magnetic moment and the magnetic field created by the unpaired electrons, this interaction is negligible compared to the electrostatic interaction arising from the Pauli principle. If the conduction electron m_s value is parallel to the unpaired $3d$ or $4f$ electrons then its wave function is different than if it is antiparallel. If its wave function is different, the Coulomb interaction with the other electrons is different so that, effectively, the electrostatic potential is different.†

When the spins are parallel the antisymmetrized wave function effectively keeps the electrons further apart and reduces their Coulomb repulsion. This

* Frequently called the residual resistivity

† For example, neglecting all the paired electrons, if we have one unpaired $3d$ electron wave function and one $4s$ electron wave function, the proper antisymmetrized product of one electron wave functions, as a function of the electron positions, r_1 and r_2 is $\psi_{3d}(r_1)\psi_{4s}(r_2) + \psi_{3d}(r_2)\psi_{4s}(r_1)$ for antiparallel m_s values and $\psi_{3d}(r_1)\psi_{4s}(r_2) - \psi_{3d}(r_2)\psi_{4s}(r_1)$ for parallel m_s values, where the subscript one and two refer to the individual electrons. The electron probability distribution is entirely different for the two cases, as are the energy and potential.

can be seen in Fig. 14. At very low temperatures the magnetic moments in a ferromagnetic or antiferromagnetic metal are all aligned and the potential is perfectly periodic for both conduction electrons, $m_s = +\frac{1}{2}$, $m_s = -\frac{1}{2}$. When the potential is periodic there is no scattering and the magnetic resistivity is zero. As we raise the temperature some of the spins are misaligned due to the spin waves and these scatter the electrons. The magnetic resistivity continues to increase up to the Curie temperature and remains constant above this temperature. While we are unable to calculate the magnetic resistivity from $T = 0°K$ to the Curie temperature a simple theory yields an expression for the constant part of the magnetic resistivity, ρ_M, above the Curie or Néel temperature.

$$\rho_M \cong CS(S+1) \quad \text{valid above the Curie or Néel temperature} \tag{161}$$

where S is the spin (total m_s) of the $3d$ or $4f$ electrons and C is a constant $\cong 30 \times 10^{-6}$ ohm-cm for transition metals and $\cong 8 \times 10^{-6}$ ohm-cm for rare earth metals, both determined experimentally. It is reasonable to assume that the magnetic resistivity is independent of the thermal or impurity resistivity as long as the atomic magnetic moments on each atom change only their direction and not their magnitude with temperature. This approximation is extremely good for the rare earths and moderately good for the transition metals (chromium and α Mn excepted).

We finally have that the total resistivity of a metal, including all contributions, is

$$\rho \cong \rho_T + \rho_i + \rho_M \tag{162}$$

An example of a metal showing all three contributions to the resistivity is $FeNi_3$ (Fig. 85). This alloy forms an ordered structure of the Cu_3Au type

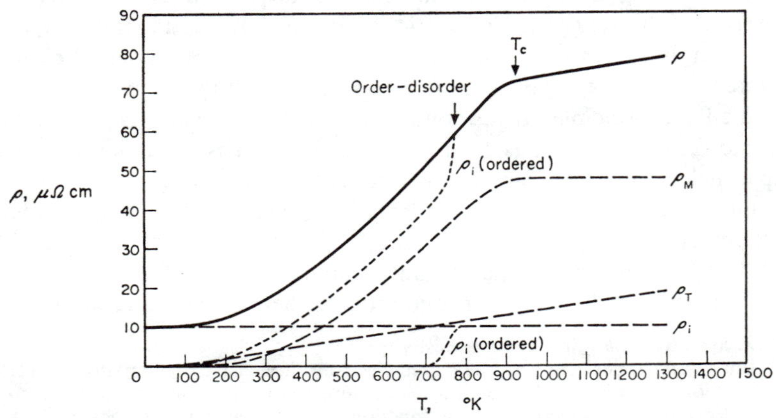

FIG. 85. The observed resistivity of disordered $FeNi_3$ (solid line) and ordered $FeNi_3$ (dotted line) and its magnetic, thermal and alloy components labelled ρ_M, ρ_T and ρ_i, respectively. The Curie point is at 930°K and the order–disorder critical temperature at 780°K.

(Fig. 43) with an order-disorder transformation temperature ~780°K. It is ferromagnetic in both ordered and disordered states with a Curie temperature of ~930°K. By slow cooling or quenching from above 780°K it is possible to measure the low temperature resistivity in either the ordered or disordered state but above 780°K it is only possible to measure the disordered state. The following details of the analysis of Fig. 85 illustrate the techniques in separating the various components of ρ in eqn. (162).

(*a*) At low temperature the ordered state has essentially zero resistivity indicating that ρ_T, ρ_i and ρ_M are all ~0.

(*b*) At low temperatures the disordered state has a resistivity $\rho_i \cong 10$ micro ohm-cm which according to eqn. (160) yields $\rho_i \sim 0{\cdot}5$ micro ohm-cm per percent iron in nickel.

(*c*) At temperatures above the Curie temperature, where both ρ_i and ρ_M are constant, the thermal resistivity is linear in T ($\rho_T = BT/4\Theta$) so that the slope $d\rho/dT$ gives $B/4\Theta$. From specific heat we know Θ to be ~400°K so that B is 22·8 micro ohm-cm. From eqn. (159) we can now plot ρ_T for all temperatures and this is shown in Fig. 85. (If the magnetic and order-disorder contributions are absent in a metal, one can determine $B/4\Theta$ from the high temperature slope and Θ from the value of the resistivity at low temperatures, having first subtracted the constant value ρ_i, determined from measurements at $T \cong 4$°K.)

(*d*) The magnetic resistivity can be obtained by subtracting the sum of ρ_i and ρ_T from the observed resistivity of the disordered state and is shown in Fig. 85. It rises rapidly with temperature and becomes constant above the Curie temperature.

The curves ρ_M, ρ_T and ρ_i in Fig. 85 are typical of the type of magnetic, thermal, and impurity or alloy resistivity curves encountered in metals. In most cases these results are independent of the direction of current flow in the metal, particularly in cubic metals, so that single crystals are not required. In non-cubic metals there are some differences reported at room temperature but these may be due to variations of the phonon vibrational amplitudes with crystal direction. In cubic metals the mean square displacements of these amplitudes, averaged along any crystal direction, is the same. Thus, any electron (or X-ray or neutron) traversing a cubic crystal experiences the same mean square displacement independent of crystal orientation.

Before discussing some of the theoretical consequences of these results it might be mentioned that rather noticeable changes in resistivity occur with phase changes, either allotropic changes in crystal structure or changes from solid to liquid (iron is a rare exception). For most metals the resistivity increases by a factor 1·5 to 2·0 upon melting although there are odd cases of complicated crystal structures, like bismuth and antimony, for which melting is accompanied by a resistivity decrease. Considering the fact that liquids lack long range order and that Bloch waves are no longer solutions to the

Schrödinger equation it appears surprising that the resistivity of the liquid is not much larger. Until we have an adequate electron theory for liquids, though, this shall remain an interesting oddity (cf. chap XII).

What, then, can we deduce about the electron structure of metals from resistivity measurements Actually, very little, for the theory is quite complex. A calculation of resistivity requires a knowledge of the Bloch wave functions and the potentials on the atoms, but as we have pointed out in chapter II, we do not know these. Nor can we use the resistivity measurements to determine them for resistivity is relatively insensitive to the choice of potential or wave function. Thus it is relatively easy to calculate the resistivity to within a factor of 2 or 3 but exceedingly difficult to within 10%. One must remember that resistivity is only a measure of a net flow of electrons in a metal, the intimate details of the individual scattering events being averaged out. In spite of these comments, though, certain qualitative remarks about the current carriers in a metal can be made.

(*a*) The relatively large values of B (Table XI) for the transition metals and rare earths compared to the noble metals, alkali metals and group II elements, indicate that the Bloch waves at the Fermi level are more concentrated around these atoms. This suggests that there is considerable mixing of *d*- and *s*-wave functions so that the momentum at the Fermi level, p_F, is effectively larger, (the greater the slope of the wave function the higher the momentum). It is almost obvious that as the electron probability distribution is more concentrated around the atoms and the probability of an electron being between the atoms is decreased it becomes more difficult for an electron to move through the lattice.

(*b*) The rather large value of magnetic resistivity in all magnetic metals indicates a strong interaction between the current-carrying electrons and the unpaired *d* or *f* electrons. This is further evidence that they are possibly responsible for the magnetic exchange energies J_1, J_2, etc. (p. 61). We have already said that the 4*s* electrons on an atom energetically favour parallel alignment to their own unpaired 3*d* electrons by perhaps $\frac{1}{2}$ to 1 eV. Depending on the magnitude and shape of these 4*s* wave functions, as they join on to the nearest neighbours, they can favour parallel or antiparallel alignment of the unpaired 3*d* electrons on the neighbouring atom, hence ferromagnetism or antiferromagnetism. This effect is sometimes called the *Kasuya or Yosida mechanism* and is depicted in Fig. 86. The unpaired 3*d* wave functions have their magnetic moments pointing upwards and the 4*s* wave functions with their m_s parallel (solid line) and antiparallel (dashed line) to the 3*d* are shown. The approximate potential energy favouring parallel versus antiparallel alignment is shown in the upper part of the curve. If the nearest neighbour midpoint is at A, then ferromagnetism is preferred since the 4*s* wave function is larger and favours parallel alignments with the neighbouring unpaired 3*d* function while at B antiferromagnetism is preferred since the 4*s* wave function is larger and can align itself *parallel* to the reversed 3*d* wave function. This mechanism is still qualitative since it is not readily amenable to calculation.

In summary, then, resistivity is a useful tool for evidencing changes in atomic and electronic structure but does not readily provide quantitative values of eigenvalues or eigenfunctions. Due to its ease of measurement it it should be a standard tool in every metallurgical laboratory.

2. *Magnetoresistance*

The magnetoresistance is the change in resistance of a substance with the application of a magnetic field divided by the resistance in a zero field. When the magnetic field is applied in a direction normal to the current flow it is termed transverse magnetoresistance while a parallel magnetic field is called longitudinal magnetoresistance.

If one is dealing with a magnetic metal the application of an external magnetic field increases the alignment of magnetic moments in opposition to the thermal vibrations which decreases the alignment. Since kT is the average thermal energy of an atom and an external magnetic field H lowers the energy of an atom by μH (μ is the magnetic moment) one effectively lowers the temperature of the magnetic moments by $\Delta T \sim \mu H/k$. If T is below the Curie point then the magnetic resistivity is decreased to its value

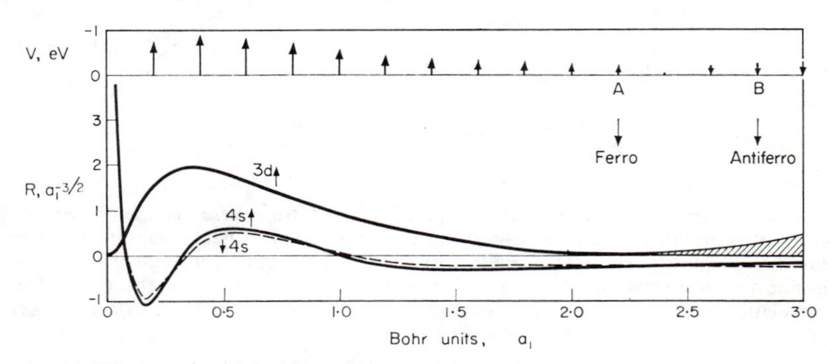

FIG. 86. The unpaired 3*d* wave radial wave function for iron metal and approximate 4*s* wave functions with spin parallel and antiparallel to the 3*d* function. The direction and magnitude of the exchange potential, V, is shown schematically, evidencing the reversal of sign at about 2·4 Bohr units ($a_1 = 0{\cdot}529 \times 10^{-8}$ cm) beyond which the ↓4*s* wave function has greater magnitude.

at $\sim T - \Delta T \cong T - \mu H/k$. For example, in the case of $FeNi_3$, $\Delta T \cong 0{\cdot}7$°C for $H = 10{,}000$ gauss so that ρ_M decreases by $\sim 0{\cdot}06$ micro ohm-cm at $T \cong 600$°K. The magnetic magnetoresistance is thus -3×10^{-3} for 10,000 gauss. Above the Curie point we expect no change since ρ_M is independent of T.

This is a rather crude estimate of the magnetic magnetoresistance in high fields but it suffices to give us a conceptual and semi-quantitative insight into the effect. At low magnetic fields there are changes in resistivity which are apparently associated with magnetostriction, i.e. the elongation or contraction of a metal depending on the direction of the magnetic moments. This change in resistivity is not understood but is presumably due to changes in the Fermi surface. An example of these two effects is shown in Fig. 87,[44] diagram 1, in which the longitudinal and transverse magnetoresistances of nickel at room temperature are plotted. The resistivity in the low field region (<4000 gauss) is governed by magnetostrictive effects while at higher fields the decrease in magnetoresistance due to alignment of the magnetic moments is predominent. Nickel decreases its length as it is magnetized and diagram 2 shows the change in longitudinal magnetoresistance as the length decreases to $\Delta l/l \cong -35 \times 10^{-6}$ at magnetic saturation. Magnetostriction arises from spin-orbit coupling which creates a slight change in the orbital wave functions relative to the spin direction. In nickel this decreases the atomic distance in the direction of the magnetic moments. By extrapolating to zero field, Smit has found that the longitudinal resistance in a ferromagnet is greater than the

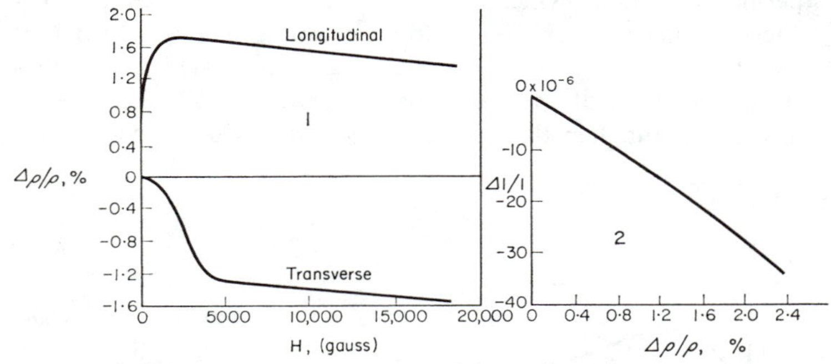

FIG. 87. The room temperature longitudinal and transverse magnetoresistance $\Delta\rho/\rho$ of nickel metal as a function of an externally applied field (diagram 1). The low field region is subject to magnetostrictive effects while the high field region is due to the increased alignment of the magnetic moments by the field. The correlation of longitudinal magnetoresistance and magnetostriction, $\Delta l/l$, in nickel is shown in diagram 2.

transverse resistance and he calls this the *orientation effect*. The extrapolation to zero field provides a measurement of the resistance of a single magnetic domain sample. The orientation effect decreases with increasing temperature and at 20°K is about $\frac{1}{2}\%$ for iron and cobalt and 3% for nickel. By preparing iron-nickel alloys and cobalt-nickel alloys he finds the orientation effect reaches a maximum of $\sim 20\%$ at alloys with an average value of $gJ = 1$. The origin of this effect is not understood but is clearly separable from the magnetostrictive effect.

In the case of the impurity or alloy resistivity, ρ_i, very little change is expected in the magnetoresistance since the application of the external magnetic field does not alter the potentials of the atoms giving rise to ρ_i. Likewise, virtually no change is expected in the frequency spectrum of lattice vibrations so that ρ_T is unaltered at most temperatures. *The real interest in magneto-resistance, in both magnetic and non-magnetic metals, arises in the very low temperature resistivity, relatively independent of its origin, as long as the resistivity is small* ($<0{\cdot}005$ *micro ohm-cm*). In particular, transverse magneto-resistance measurements made on single crystals evidence pronounced dependences on magnetic field direction relative to the crystal axes, especially in pure metals. Furthermore, the variation of magnetoresistance with magnetic field strength may be quite different for the magnetic field in different crystallographic directions. In some directions it may increase with the field, then fall off and saturate at some constant value, while in other directions it may continue to rise even at the highest fields attainable.

It is quite essential that single crystals be used for the qualitative understanding of magnetoresistance in low resistivity metals. The measurement itself is not very difficult for one can generally arrange to keep the temperature of a wire at 4°K and rotate it in a transverse magnetic field. There may be some problems in obtaining single crystal wires or rods in a variety of orientations. However, a good deal of information can be gained by employing a single crystal wire whose axis is in the [110] direction in a cubic crystal or is along the a_0 axis in a hexagonal crystal. In the former the transverse magnetic field can be oriented in the three principal cubic directions [100], [110] and [111] by rotating the wire about its axis while in the latter one can obtain the normals to the three principal hexagonal planes, 100, 001, and 101. If the metal to be studied has a low melting point a single crystal may be grown in a quartz capillary by appropriate seeding. For high melting point metals a cylindrical specimen can be machined from a sample and etched to remove the strained surface.

The clue to the conceptual understanding of the interesting magneto-resistance effects at low resistivity comes from an understanding of the solutions of the Schrödinger equation for an electron in a magnetic field. For our purpose let us consider the case of the free electron ($V = 0$) in the three dimensional cubic box since the wave functions of the electrons at the Fermi level provide a reasonable approximation to the wave functions of the current carriers in a metal. In the absence of a magnetic field the eigenvalues are given by eqn. (20). If we apply a magnetic field in the z-direction, the classical solutions give rise to electron paths which spiral around the z-axis, but the quantum mechanical limitation that the wave functions of these spiral orbits be single valued gives rise to quantized orbits. The solutions are unaltered in the z-direction (the direction of the applied field) but the solutions in the x- and y-directions are altered by the replacement:

$$\frac{\hbar^2}{2m}\left(\frac{n_x^2\pi^2}{L^2}+\frac{n_y^2\pi^2}{L^2}\right)=\frac{\hbar^2}{2m}(k_x^2+k_y^2)\rightarrow\frac{(n+\frac{1}{2})ehH}{2\pi mc} \tag{163}$$

where n is a new quantum number. The new eigenvalues are

$$E = \frac{(n + \frac{1}{2})ehH}{2\pi mc} + \frac{n_z^2 \hbar^2}{8mL^2}; \quad \begin{matrix} n = 0, 1, 2 \text{ etc} \\ n_z = 1, 2, 3 \text{ etc} \end{matrix} \tag{164}$$

The net effect, as seen in Fig. 88, diagram 1, is that all the energy levels between $(n + \frac{1}{2})ehH/2\pi mc$ and $(n + 1 + \frac{1}{2})ehH/2\pi mc$ that existed in zero field ($H = 0$) *coalesce* to form a new single level at $(n + 1)ehH/2\pi mc$ *but only for those electrons whose component of momentum in the z-direction is zero.* The second term in eqn. (164) due to electron momentum in the z-direction still yields closely spaced energy levels. For all levels below the Fermi level this coalescence of levels does not alter their average energy but at the Fermi

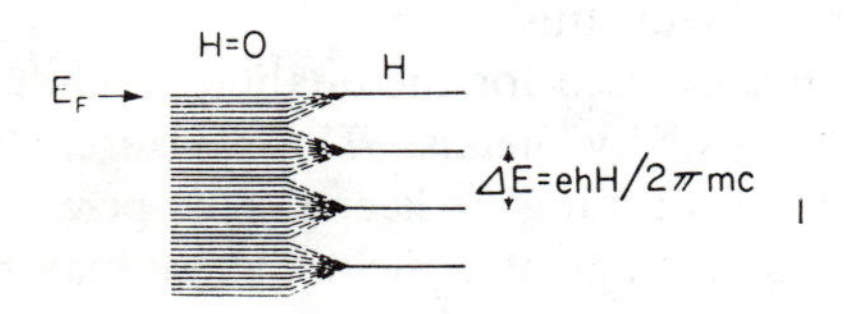

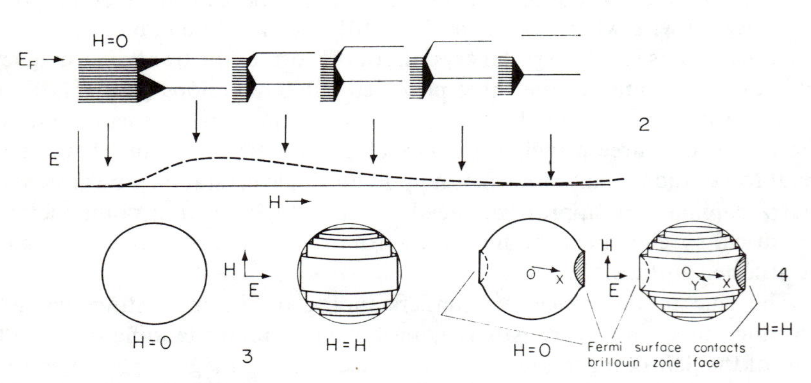

FIG. 88. The effect of an external magnetic field in the z-direction on the energy levels of free electrons with zero component of momentum in the z-direction. In diagram 1 the levels over an energy range $ehH/2\pi mc$ coalesce to form a single level. In diagram 2 the oscillatory total energy is shown as a function of magnetic field strength. In diagram 3 the spherical free electron Fermi surface is altered in a magnetic field to form a series of concentric cylinders while in diagram 4 a Fermi surface is schematically shown making contact with a Brillouin zone face (shaded area). Application of a magnetic field does not alter the energy levels of the electrons on the shaded contact areas since they are occupying states well below the Fermi level.

level the coalescence raises the average energy. A plot of the oscillatory total energy of the crystal versus magnetic field strength is shown in Fig. 88, diagram 2. The spacing between energy levels is $ehH/2\pi mc$. Alteration of the Fermi surface is seen in diagram 3, Fig. 88. In the absence of an external magnetic field the Fermi surface is a sphere of radius $r\ (= k_x = k_y = k_z)_{E=E_F}$

but the magnetic field in the z-direction creates a series of concentric cylinders centred around the z-direction whose areas, $\mathscr{A}_0$, are given by

$$\mathscr{A}_0 = \pi r^2 = \pi(k_x^2 + k_y^2)_{E=E_F} = 2\pi(n + \tfrac{1}{2})eH/\hbar c \qquad (165)$$

How does this new distribution of energy levels effect the resistivity of a metal? In the absence of an external magnetic field the external electric field accelerates the electrons in the direction of the electric field and they successively occupy the closely spaced higher energy levels above the Fermi level until they are scattered. If we apply a magnetic field in the z-direction those electrons with components of momenta in the x- or y-direction require considerably more energy to be excited to the next level. For example, the free electron theory applied to copper metal gives the spacing between energy levels at the Fermi level as at most $\sim 10^{-20}$ eV whereas the application of a magnetic field of 10,000 gauss increases the spacing to $\sim 1{\cdot}2 \times 10^{-4}$ eV $= ehH/2\pi mc$. Thus, unless it can absorb this much energy from the external electric field the electron will not be accelerated in the x- or y-direction. Actually, the spacing of the energy levels given by eqn. (164) is valid only in the presence of an external magnetic field *and in the absence of an external electric field.* When the electric and magnetic fields are both applied there is a slight mixing or hybridization of adjacent levels just as there is a mixing of wave functions in a metallic energy band. The electrons can thus be mostly in one state and partly in the neighbouring state but the important thing is that they can now be accelerated in the direction of the electric field E. However, as it moves in the direction of E it spirals around the magnetic field direction, z, so that it moves about one of the concentric cylinders in Fig. 88, diagram 3, until it gains sufficient energy to move to the next higher concentric cylinder. As it moves on the surface of the cylinder its wave numbers k_x and k_y are constantly changing, since its direction is changing, but $(k_x{}^2 + k_y{}^2)^{\frac{1}{2}}$, the radius of the cylinder remains constant until it is excited to the next higher energy cylinder. Actually, nothing very interesting happens if one has closed orbits as in diagram 3 for there is very little change of resistance with applied magnetic field. However, if the electron is on a distorted cylinder (diagram 4, Fig. 88) resulting from the Fermi surface touching one of the Brillouin zone faces, then it cannot be accelerated to the next higher level since the energy gap forbids the existence of energy levels in the contact area (shaded areas in diagram 4). Of course, any electrons whose wave numbers lay on the surface of contact (shaded areas in diagram 4) before the magnetic field was applied, have an energy well below the Fermi level. However, an electron with wave number OX in diagram 4 ($H = 0$), having a component of momentum in the direction of E, could carry the current. But this same electron in a magnetic field ($H = H$) would proceed to spiral on the Fermi surface and eventually reach the area of contact with the Brillouin zone and would be forbidden to accelerate because of the energy gap. As the magnetic field is increased, more and more electrons from directions further removed such as OY are affected and the magnetoresistance increases as H^2. If the spiral path of the electron on the Fermi surface runs into a Brillouin zone face this is called an *open orbit* since the electron cannot complete the cycle, and these can be observed in transverse magnetoresistence measurements by seeking orientations of the crystal relative to the magnetic field that yield an H^2 dependence. Closed orbits are relatively independent of H and this is called *saturation*. Some measurements on a gold crystal,[45] whose axis is in the 110 direction, are shown in Fig. 89, diagram 1. If the crystal wire were exactly along the 110

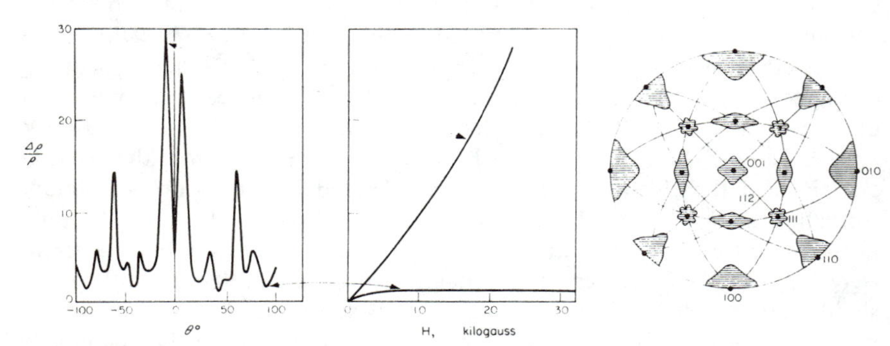

FIG. 89. The transverse magnetoresistance of a gold single crystal wire at 4·2°K in a field of 23,500 gauss. The axis of the wire is in the [110] direction and the angle θ, is the angle of rotation in the magnetic field, diagram 1. In diagram 2 the field dependence of the magnetoresistance at a peak and at a minimum is shown, the H^2 dependence indicating an open orbit. Diagram 3 is a stereographic projection for gold showing the H^2 regions (shaded and solid lines).

direction the curve would be symmetrical. The high points in the curve show an H^2 independence while the low points saturate, diagram 2. In Fig. 89, diagram 3, there is plotted a stereograph projection of gold indicating the H^2 regions (shaded and black lines). The Fermi surface of gold, deduced from these measurements, is similar to that of copper, Fig. 90.

Such measurements offer great promise and the field is rather untouched since only a few substances have been measured. The necessity for low resistivity comes from the requirement that the levels be sharp. The higher the resistivity the shorter the time, τ, between electron collisions. From the uncertainty principle $\Delta E = h/\Delta t = h/\tau$. For $\Delta E \cong 10^{-5}$ eV, $\tau \cong 4 \times 10^{-10}$ sec. so that a resistivity of about 10^{-3} micro-ohm cm is required.

3. *de Haas van Alphen Effect*

The de Haas van Alphen Effect directly measures the oscillatory energy absorption of a metal, Fig. 88, diagram 2, as the magnetic field is varied. Since the energy of the metal is raised by application of a magnetic field conservation of energy requires the energy gain by the metal to be absorbed from the magnetic field. It does this by virtue of its diamagnetism and decreases the strength of the magnetic field. Classically this means that the spiral paths of the negatively charged electrons are in a direction which create a magnetic field opposed to the applied magnetic field (like Lenz law). The oscillatory behaviour of the energy is proportional to the quantum number n in eqn. (164) and is determined in the de Haas van Alphen effect by measuring the diamagnetic susceptibility of the sample. The period $\Delta n = 1$ is thus

$$\Delta n = 1 = \mathscr{A}_0^* \hbar c / 2\pi e H \tag{166}$$

which, if plotted as a function of $1/H$, is

$$\Delta\left(\frac{1}{H}\right) = \frac{2\pi e}{\hbar c \mathscr{A}_0^*} \tag{167}$$

where $\mathscr{A}_0^*$ is the extremal area. The extremal area is the cross-sectional area of the Fermi surface in a direction normal to z, (the direction of the applied field) for those wave numbers whose component in the z-direction is zero. The reason for this is that the energy levels for the components of momenta

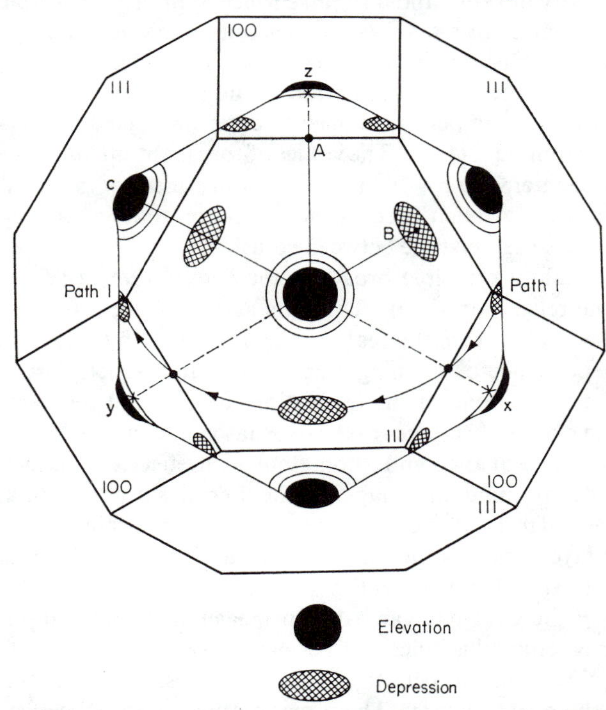

FIG. 90. The Fermi surface of copper deduced by Pippard and showing the contact areas at the 111 Brillouin zone faces. The diameter of the contact area is about 18% of the diameter of the body.

in the z-direction are still closely spaced so that the oscillations are too small to be seen. If thermal oscillations are introduced the energy levels around the Fermi level are spread out and the Fermi surface of Fig. 88, diagram 3, is blurred. The amplitudes of the energy oscillations diminish. An analysis of this effect shows that the susceptibility in a real metal is given as the second derivative of the free energy:

$$\chi = -\mathrm{d}^2F/\mathrm{d}H^2$$

$$F \propto TH^{3/2}\, e^{-(2\pi^2 Tm^* c/eH\hbar)} \cos(\hbar c \mathscr{A}_0^*/eH) \tag{168}$$

$$m^* = \frac{\hbar^2\, \mathrm{d}\mathscr{A}_0^*}{2\pi\, \mathrm{d}E} = \hbar^2/(\mathrm{d}^2E/\mathrm{d}k^2)_{E=E_F}$$

where m^* is the so-called *effective mass* and tells us the reciprocal of the second derivative of the energy versus wave number at the Fermi level. If the energy varies as k^2, as in the free electron theory, then $m^* = m_0$, the electron mass. As the power of k^2 decreases from two, m^* increases and offers evidence for an increasing density of states and a Fermi surface approaching a zone boundary. This use of the effective mass, however, is only a mathematical convenience, the gravitational mass of the outer electrons in a metal always remaining m_0. By measuring the oscillatory period of the diamagnetic susceptibility and the value of χ at two temperatures one can determine the average value of $\mathscr{A}_0^*$ and $\mathrm{d}\mathscr{A}_0^*/\mathrm{d}E$ from eqn. (168). The values of the quantum number n eqn. (164) ordinarily encountered are $\sim 10^3$ to 10^4. A third piece of information yielded by the de Haas van Alphen effect is the relaxation time, τ, of the electrons at the Fermi surface (i.e. the time between collisions). We have already indicated that the uncertainty principle broadens the energy levels ($\Delta E = h/\Delta t = h/\tau$ where τ is the relaxation time). This acts as an apparent temperature effect and can be separated from the real temperature effect by measuring the magnetic field dependence of the magnitude of the diamagnetic oscillations, and observing the broadening of the oscillations at constant temperature.

One of the major advantages of the de Haas van Alphen effect, as compared to the cyclotron resonance and anomalous skin effect techniques (to be discussed) for determining the shape of the Fermi surface, is that the latter techniques use high frequency electromagnetic fields which penetrate only a thin surface layer whereas the diamagnetic susceptibility of the de Haas van Alphen effect arises from the entire sample.

In making the de Haas van Alphen measurement it is desirable to use very high magnetic fields since the periodic variation is only a few gauss in 10,000. Liquid helium temperature (4°K) is also desirable to minimize thermal blurring of the energy levels. These measurements have been pioneered by D. Shoenberg(46) who has used magnetic fields up to 10,000 gauss by charging a 2000 microfarad condenser to 1700 volts and then discharging it through a solenoid. The pulse lasts about a millisecond and produces field changes $dH/dt \sim 5 \times 10^6$ gauss/sec. The solenoid does not overheat in these short times. The sample is placed inside one of two balanced coils of about 10^3 turns which give a signal of ~ 10 millivolts. Only the variation in the signal from the coil is followed on an oscilloscope since only the periodic part of the susceptibility is required. Single crystal samples of dimensions ~ 1 mm are used. They must be of high purity so that τ is large ($eH\tau/2mc \gg 1$).

The details of the Fermi surface of gold and copper, as measured by the anomalous skin effect (to be discussed), have been generally confirmed by de Haas van Alphen measurements including the small area oscillations around the 111 necks (Fig. 90), and the measured effective masses, m^*, in various directions agree with cyclotron resonance measurements (to be discussed). de Haas van Alphen oscillations have been seen in many metals including chromium and molybdenum but the details of the Fermi surface have not been worked out for these transition metals. There seems to be little doubt that the Fermi surface of most elements will be worked out in a few years. With the construction of machines capable of yielding very high magnetic fields ($\sim$250,000 gauss) we should be able to make measurements on less pure samples since we can tolerate shorter relaxation times.

4. *Cyclotron Resonance*

The energy levels given by eqn. (164) for the free electrons in a magnetic field are altered slightly when dealing with real metals. In such cases we replace m in eqn. (164) by the effective mass m^*, eqn. (168). The cyclotron resonance experiment is a measure of the absorption of electromagnetic radiation of frequency $\nu = E/h = eH/2\pi m^* c$ which excites electrons with $k_z = 0$ to the next higher energy level. For a field of 10,000 gauss and $m^* = m$ the resonant frequency is $\nu \cong 28 \times 10^9$ cps well into the microwave region ($\lambda \cong 1$ cm). Consequently the cyclotron resonance measurement determines m^* in an extremal plane $\mathscr{A}_0^*$ for electrons with $k_z = 0$. By varying the magnetic field direction relative to the crystal axes the shape of the Fermi surface can be deduced from the variation of m^*. For example, if the extremal plane includes a point of contact with a Brillouin zone face then $m^* = \infty$ and the resonance disappears.

In performing the experiment a very pure metal crystal ($\nu\tau \gg 1$, so that the energy levels are not blurred) is cut parallel to some crystallographic plane and electropolished, and possibly annealed, to remove cold work in the surface layer. The magnetic field is applied parallel to the surface and the crystal becomes one end wall of a cylindrical microwave resonant cavity tuned to some fixed frequency, commonly $24{\cdot}47 \times 10^9$ cps. The magnetic field is varied and the absorption of the microwaves is measured until a resonance absorption is reached. Considerable background noise can be eliminated by amplitude modulation of the magnetic field at some frequency that is not a harmonic of the 60 or 50 cycles used for power. The microwave detector is "locked in" at the modulating frequency. The measurements are made at liquid helium-temperatures, to minimize thermal blurring of the energy levels, and one can arrange to rotate the microwave cavity so that the field is always in the plane of the crystal surface. Sample preparation is a major problem since the surface must be smooth, and free of strains and irregularities in addition to the demands of high purity (τ long). While many cyclotron resonance observations have been made with the magnetic field other than parallel to the surface we shall consider only the parallel condition in giving a semi-classical picture of cyclotron resonance.

If we examine Fig. 91 we see the spiral path of an electron with no component of momentum in the z-direction ($\hbar k_z = 0$, z is the direction of the applied

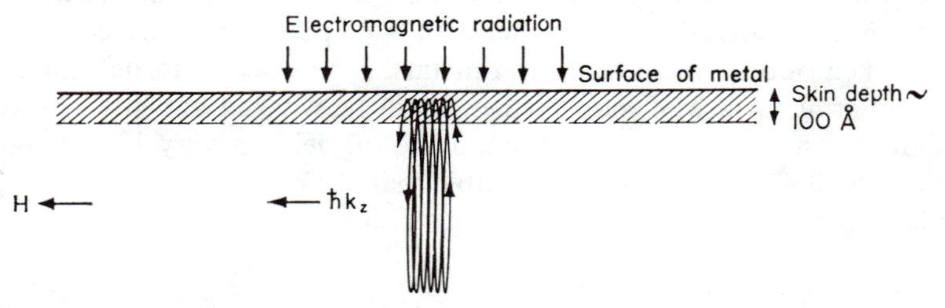

FIG. 91. The classical picture of the condition for cyclotron resonance for electrons with little or no momentum in the magnetic field direction along the surface of the metal. The electromagnetic radiation only penetrates the skin depth, ~100 Å.

magnetic field along the surface). Since the electromagnetic radiation only penetrates a short distance into the metal (~100 Å) the electron only experiences the influence of the electromagnetic field when it comes into this shallow surface layer called the skin depth. The clue to the ability to detect the absorption of the electromagnetic radiation is that the radiation is coherent or in phase with the electron's orbit. If the electromagnetic radiation consisted of completely independent but monochromatic photons, such as the monochromatic X-rays diffracted from a crystal, it would be virtually impossible to detect their absorption. Thus the electron in resonance experiences the same phase of the electromagnetic wave each time it traverses the skin depth. If the electron spirals in its orbit many times before it makes a collision ($\nu\tau \gg 1$ where τ is the relaxation time between collisions) then the resonance level is "sharp". As the resistivity increases, so that τ decreases, the levels become blurred and this is the reason why very low resistivity is required ($< 0{\cdot}005$ micro-ohm cm.). It is particularly difficult to realize this condition on the surface and rather careful electropolishing is required to remove strains.

A typical set of data is shown in Fig. 92[(47)] for the magnetic field 10° from the [100] direction and along the 110 prepared crystalline surface of copper. The excitation to many higher quantum levels are seen. Two extremal orbits are observed, one exhibiting three harmonics, the other nine. The effective masses for copper in various crystalline directions were found to vary from 0·5 to 5 times the free electron mass with the bulk of the data around 1·1 to 1·3. Discontinuities were observed at the areas where the Fermi surface contacted the 111 Brillouin zone faces. The two extremal paths observed in Fig. 92 are of the type labelled path 1 in Fig. 90, the Fermi surface of copper deduced by Pippard[48] from the anomalous skin effect. The cyclotron resonance experiments confirm this picture of the Fermi surface. It is not a straight forward matter to deduce the Fermi surface from a set of cyclotron resonance measurements and a certain amount of trial and error is required.

5. *Anomalous Skin Effect*

Ordinary resistivity measurements represent a drift of the electrons averaged

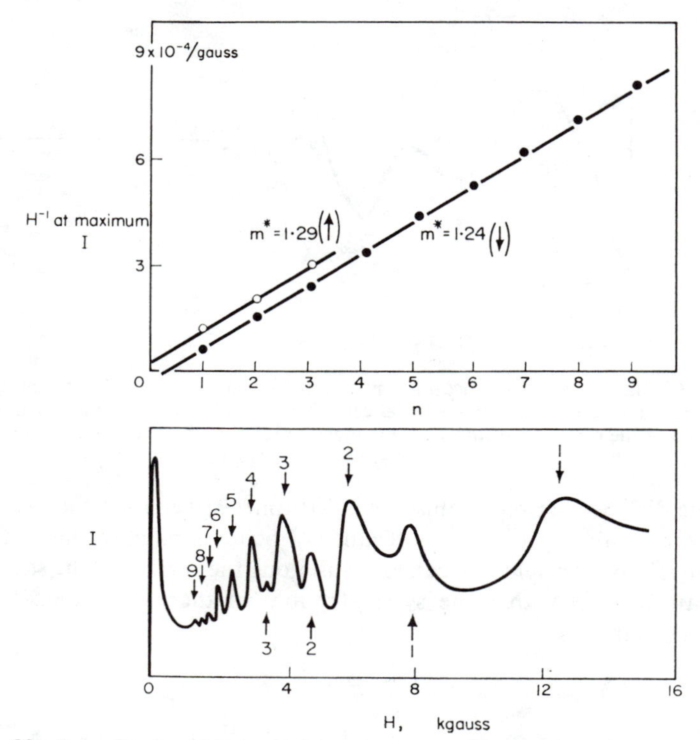

FIG. 92. Intensity in arbitrary units versus magnetic field in kilogauss for the cyctron resonance signal in copper. Two periods with slightly different effective masses are resolved and the excitation to successively higher quantized levels are numbered 1 to 9 and 1 to 3 for the two cases. The values of $1/H$ at the positions of the peaks are plotted versus the quantum number (upper curve).

over the entire Fermi surface, since most electrons in a metal have components of momenta in the direction of the applied electric field. By employing high frequency electromagnetic radiation which is absorbed within the first 100 Å of the single crystal surface of a metal like copper one can limit the applied electric field (now a.c.) to electrons travelling within this thin surface layer. If the copper is very pure and kept at 4°K, the mean free path of the electrons is ~ 0·1 mm so that the resistance of the metal single crystal due to the applied a.c. electromagnetic field is limited to electrons travelling in a narrow range of directions corresponding to a very small segment of the surface area of the Fermi surface. Each electron must make its collision in this thin surface layer path *A*, Fig. 93, or else it does not contribute to the surface resistance as in path *B*, Fig. 93, in which the electron is elastically scattered from the surface.

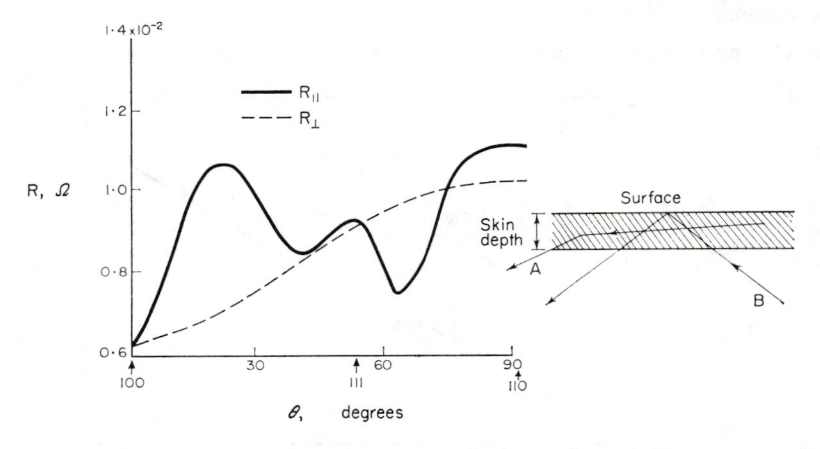

FIG. 93. The surface resistance of copper resolved into $R_\parallel$ and $R_\perp$ as measured by Pippard in the anomalous skin effect. The crystal is cut parallel to the 110 planes and the direction of the electromagnetic radiation is varied around the [110] axis.

Since the distance between collisions is 10^4 times larger than the skin depth, the electron path is within thirty seconds of being parallel to the surface.

If the microwave field of frequency ν is along the surface of the sample and makes an angle ϕ with some symmetry axis of the crystal then the total surface resistance is

$$R = R_\parallel \cos^2 \phi + R_\perp \sin^2 \phi$$

$$R_\parallel = \left\{\frac{2\pi^5 \nu^2}{3\sqrt{(3)}e^2} \int |\rho_\perp| \, dk_\perp\right\}^{1/3} \tag{169}$$

$$R_\perp = \left\{\frac{2\pi^5 \nu^2}{3\sqrt{(3)}e^2} \int |\rho_\parallel| \, dk_\parallel\right\}^{1/3}$$

where $\rho_\perp$ and $\rho_\parallel$ are the radii of curvature of the Fermi surface in directions perpendicular and parallel to the plane defined by the specimen surface and $\int dk$ is the integration over the Fermi surface, (k = wave number). Directions like [100], [110], [111] define symmetry axes since the Fermi surface is symmetrical about these directions. Fig. 93 shows the measurements of Pippard[(48)] resolved into $R_\parallel$ and $R_\perp$ for a copper crystal whose surface is parallel to 110. The electromagnetic field direction is varied along the surface, the [100] and [111] directions indicated in the figure. It is not a simple matter to determine the shape of the Fermi surface from these measurements which only give the average value of the curvature. However, Pippard was able to deduce the Fermi surface in Fig. 90 from these measurements by a trial and error method and subsequent other measurements (de Haas van Alphen, cyclotron resonance, magnetoresistance, ultrasonic attenuation) have confirmed this Fermi surface for copper. The number of electrons which can participate in the surface resistance is obtained from the surface area of the Fermi surface containing all electrons whose directions are within the small angular range approximately given by the ratio of the skin depth to the electron mean free path. The radius of curvature is this surface area divided by this angular factor.

As in cyclotron resonance measurements sample preparation is important. The surface must be free of strains so that the skin depth has a low resistance. The sample is made part of a microwave cavity and the "Q" of the cavity measured.

Qualitatively we can say that any marked variation in the anomalous skin effect over crystal direction is evidence for deviations from a spherical Fermi surface. If the Fermi surface is spherical, or the mean free path short as in most metals, the surface resistance is given by the classical expression eqn. (155), independent of crystal orientation in cubic crystals.

6. *Hall Effect*

In the Hall effect a thin metal strip several mm wide in the y direction and several cm long in the x-direction is placed in a magnetic field, H, in the z-direction and current caused to flow in the x-direction. If potential leads are placed across the strip and adjusted in the x-direction so that there is no potential between them when the current is flowing and $H = 0$ then the potential across the strip in the y-direction, when the magnetic field is applied, is called the Hall voltage. It ranges from 0·1 to tens of microvolts in typical cases so that stray potentials due to thermal effects create difficulty in these measurements. No standard measurement technique has been devised which is applicable to all problems but the circuit by S. Foner and K. Tauer in Fig. 94 is capable of giving satisfactory measurements down to 0·01 μV. The sample is machined with dog ears (a, b, c Fig. 94) for soldering the potential leads and the micro-pot voltage divider R_1 is adjusted so that the potential is zero at the photo-galvanometer with current flowing and $H = 0$. If the dog ears a and b were absolutely parallel the voltage divider would be unnecessary but this condition is virtually impossible to attain. When the magnetic field is applied the Wenner bridge is balanced with the split photo-cell galvanometer. In the split cell photo-galvanometer the light beam is optically split, directed into two photomultipliers and the unbalanced signal measured. The Hall potential is measured with the four combinations of reversal of current direction and magnetic field direction and the average determined. This corrects all stray potentials except one called the Ettingshausen effect in which a temperature difference is created between c and b by the action of the current and magnetic field and this produces an additional

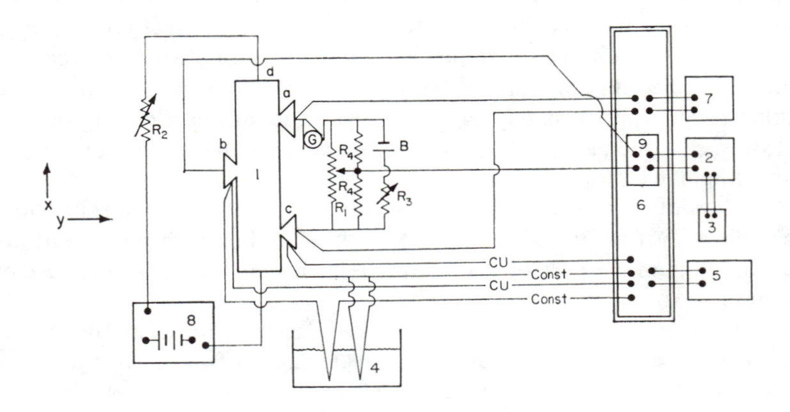

FIG. 94. A schematic diagram for measuring Hall potentials with a Wenner bridge. R_3, R_4, B and G are introduced to reduce the parallel current flow in R_1 to zero since R_1 is small enough to enable one to match the ~13 ohm impedance of the Wenner bridge. R_1 is adjusted so that the potential drop from d to b equals that from d to the centre tap of R_1 with the current flowing and $H = 0$.

1. Sample ~ 0·001 ohms
2. Wenner Potentiometer
3. Split Cell Photo-galvanometer
4. Ice Bath for Cu-Constantin Thermocouple
5. Portable Potentiometer for Temperature Measurement
6. Oil Immersed Switches
7. Kelvin Bridge for Resistivity Measurement
8. Battery and Current Reversing Switch
9. Current Reversing Switch

R_1 100 ohm helipot
R_2 1 ohm
R_3 Decade Box
R_4 10 ohm precision
B Car Battery
G Galvanometer and Switch

potential which adds to the Hall potential. It is generally a small correction and can be made by measuring the temperature difference between b and c and utilizing the known thermoelectric power of the sample. If it is not known it can be measured.

The Hall potential V_H is generally given as

$$V_H = R_0 IH/t \tag{170}$$

where t is the thickness of the sample, I the current flowing in the $+x$-direction, H the magnetic field in the $+z$-direction, V_H the potential measured in the $+y$-direction and R_0 a constant of the metal called the Hall coefficient. R_0 is negative when we have electrons moving in the $+x$-direction. The results of any measurement are generally given in terms of values of R_0 in units of 10^{-13} V cm/amp gauss. For example, copper gives $-5{\cdot}5$ in these units.

The interpretation of the Hall potential is divided into its two component parts, the ordinary Hall effect present in all metals and the anomalous Hall effect present only in ferromagnetic metals. For the ordinary Hall coefficient the free electron theory yields a rather simple temperature-independent result for R_0 which is frequently employed and seems to give reasonable results on occasions. For a metal with N_0 atoms per cm^3.

$$R_0 = -\frac{1}{n_0 ecN_0} \tag{171}$$

where n_0 is the number of current-carrying electrons per atom. For example, measurements of R_0 for copper, silver, gold, lithium and sodium give values of n_0 between 0·75 and 1·5 and for Al a value of $n_0 \cong 3$. Table XII lists the values of R_0 for many elements as well as the values of n_0 deduced from eqn. (171). All the numbers in the first column of Table XII are reasonably close to the number of outer electrons in each metal but if we apply eqn. (171) to beryllium zinc and cadmium we get the wrong sign for R_0, suggesting that the current flowing through the wire is positively charged. It is rather absurd to suggest that the current is carried by other than electrons. How, then, do we interpret the positive sign for R_0?

In Fig. 95 we see two schematic methods whereby an electron is transferred from one end of a metal to the other, simulating an electric current. In diagram 1 an electron is introduced into the neutral metal at point A and replaces the electron which has left its parent atom and hopped down the length of wire to point B. Consequently one electron has travelled the length of the wire. In diagram 2 we see a series of successive stages. An electron leaves the wire at point B leaving a positive ion behind, and another electron is simultaneously introduced into the wire at A. Suppose for some reason this latter electron is rather sluggish and does not move too readily through

TABLE XII

ROOM TEMPERATURE VALUES OF R_0 AND R_1 IN UNITS 10^{-3} V OHM/AMPS GAUSS. n_0 IS THE NUMBER OF CURRENT CARRIERS PER ATOM BASED ON THE FREE ELECTRON MODEL

Metal	R_0	n_0 (sign)	Metal	R_0	n_0 (sign)	Metal	R_0	n_0 (sign)
Li	−17·0	−0·8	Ti	~ −1·3	−8·3	Ta	+9·7	+1·2
Na	−24·5	−1·0	V	+8·20	+1·1	W	+11·3	+0·8
K	−42·0	−1·2	Cr	+36·3	+0·2	Re	+31·50	+0·3
Rb	−59·2	−1·0	Mn	+8·44	+0·9	Ir	+3·3	+2·7
Cs	−78	−0·9	Fe	+2·45	+3·0	Pt	~ −2·2	−4·3
Cu	−5·5	−1·3	Co	−13·3	−0·5	Th	−12·0	−1·8
Ag	−8·4	−1·3	Ni	−5·6	−1·2	U	+4·1	+3·2
Au	−7·2	−1·5	Sn	−0·54	−40	La	−8·0	−2·9
Be	+24·4	+0·2	Pb	+0·95	+20	Ce	+18·1	+1·2
Mg	−11·3	−1·4	Zr	+13·7	+1·1	Pr	+7·09	+3·0
Ca	−17·8	−1·6	Nb	+8·0	+2·7	Nd	+9·71	+2·1
Zn	+3·3	+2·9	Mo	+18·0	+0·5	Sm	−2·2	−9·5
Cd	+6·0	+2·2	Ru	+22·0	+0·4	Gd	−9·5	−2·1
Hg(−60°C)	−7·84	−2·0	Rh	+5·0	+1·7	Dy	−13·0	−1·5
Al	−4·0	−2·6	Pd	−6·75	−1·4	Er	−3·41	−5·5
Ga	−6·3	−1·9	As	+1350	+0·032	Tm	−18	−1·1
			Sb	+2000	+0·01	Yb	+37·7	+0·7
			Bi	~ −1000	−0·02	Lu	+5·35	+3·5
			In	−65·5	−25	Y	−7·70	−2·7
			Tl	+1·97	+9			

Metal	R_1(295°K)
Fe	+790
Co	+250
Ni	−750
Gd	~ −111,000

the wire. While the positive ion at the left is waiting for this electron to travel the length of the wire and neutralize it, the electron on the neighbouring atom makes a jump. This is followed by a succession of neighbouring jumps

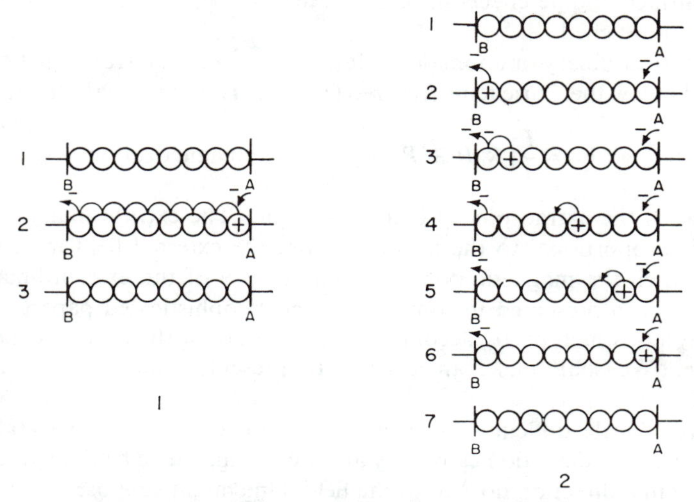

FIG. 95. Diagram 1, normal electron conduction. An electron introduced at *A* replaces an electron which hops down the wire and leaves at *B*. Diagram 2, hole conduction. An electron is introduced at *A* and one simultaneously leaves at *B*. A series of successive neighbourly electron jumps moves the hole down to *A* (pictures 3 to 6).

ending with the last figure. Effectively an electron has been introduced at *A* and an electron has left the wire at *B* the same as in diagram 1. However, in diagram 2 the positively charged hole has moved down the wire and this gives the opposite sign to the Hall coefficient, R_0.

The reason for the sluggishness of electrons in travelling large distances through the metal can be seen within the framework of the Bloch one electron picture if we assume the Fermi surface to be very close to the Brillouin zone indicating an almost filled band. If we apply an electric field to the metal the electrons are accelerated increasing their momentum. When their momentum corresponds to the momentum of the Brillouin zone they can no longer be accelerated and effectively run into a stone wall. On the other hand, electrons which travel only a short distance in the metal are not accelerated as much and do not run into the Brillouin zone. This is the case in diagram 2 in which no electron ever travels more than one atomic distance as compared to diagram 1 in which it has travelled much further and gained more momentum.

Since the sign of the Hall coefficient is reversed for the two effects the latter case is termed "*hole*" *conduction* even though it is always a result of electron movement. We are unable to predict the sign or magnitude of the ordinary Hall coefficient in most metals since it involves a knowledge of the Fermi surface. Very few measurements have been made in single crystals, particularly in low resistivity crystals, to observe details of the shape of the Fermi surface. Some effects have been seen in copper but the field remains virtually untouched.

The extraordinary or anomalous Hall effect occurs in ferromagnetic metals near and below the Curie temperature. The total Hall potential is then written

$$V_H = \frac{I}{t}\,[R_0 H + R_1 \mathscr{M}] \qquad \mathscr{M} = \text{magnetization} \tag{172}$$

where R_1 is the extraordinary Hall coefficient. The extraordinary Hall potential is proportional to the magnetization, the external field merely being used to align the magnetic domains. The theory of the extraordinary Hall effect has been presented in some of the most sophisticated papers on solid state physics, which partly accounts for the failure of theory and experiment to agree. We shall make an attempt to present some of the conceptual aspects.

There is little difficulty in visualizing the ordinary negative Hall effect classically as the deviation caused by an external magnetic field on an electron travelling in a direction normal to the field. One might suppose that a similar effect should occur in the case of a ferromagnet except that an additional magnetic field due to the unpaired $4f$ or $3d$ electrons would be present. An outer s electron in a metal like iron or gadolinium would experience an average field of $4\pi\mathscr{M} \cong 15$ to 25 kgauss due to the magnetic moments of the atoms. If this value is substituted for $\mathscr{M}$ in eqn. (172) we can determine R_1 from the measured Hall voltage. Since $\mathscr{M}$ reaches an approximately constant value when the magnetic domains are aligned we can separate R_0 and R_1 by studying the Hall voltage versus the externally applied field H. We might thus expect R_1 to be approximately the same as R_0 for $H \cong 15{,}000$ gauss but we find it to be 100–1000 times larger and sometimes of opposite sign. In addition, it is strongly temperature-dependent whereas R_0 is approximately independent of temperature. Therefore, there is a real difference between the two. It is believed that the anomalous Hall effect does not arise from deviations of the electrons by the externally applied magnetic field or the internal magnetic field of the unpaired electrons but by a skew scattering process i.e. one which scatters the conduction electrons more in the $+y$-direction than in the $-y$-direction with H and $\mathscr{M}$ in the $+z$-direction and I in the $+x$-direction. The only reasonable mechanism suggested so far, which qualitatively explains the results, is spin-orbit coupling (p. 18).

We have already pointed out that a p or d electron prefers to have its own spin, m_s, antiparallel to its orbital moment, m_l, due to the magnetic field created by the apparent motion of the nucleus and other electrons relative to itself. The resistivity of a specimen of iron at $T = 0°K$ is essentially zero since the crystal is perfectly periodic. Even though the conduction electrons may experience a magnetic field due to the unpaired electrons the Hall coefficient R_1 is zero because the magnetic field is periodic. (In a crude way this is similar to a long wavelength neutron traversing a beryllium filter without being scattered, p. 187.) If we raise the temperature slightly some phonons and spin waves are excited and the electrons can be scattered. However, spin-orbit coupling allows the electrons to pass through a region of lower energy if they scatter to one side of the atom so that their own orbital motion around the atom is opposed to their own spin, m_s. Thus electrons with $m_s = -\frac{1}{2}$ will scatter to the left and those with $m_s = +\frac{1}{2}$ will scatter to the right.* However, this is not enough to explain the effect since half the conduction electrons have $m_s = +\frac{1}{2}$ and half $m_s = -\frac{1}{2}$ so that there will be just as much scattering to the right as to the left and the Hall potential will be zero. Since only the *d* or *p* part of the conduction electron wave functions experience spin orbit coupling we must require that the electrons with m_s, parallel to the magnetic moments, have a different ratio, say in iron, of 3*d* to 4*s* character than those which are antiparallel, the one with more *d* character being scattered more. Consequently we can end up with more scattering to the right than to the left or vice versa. It is essential, though, that the substance be ferromagnetic so that the exchange interaction between the current carriers and the unpaired electrons can alter the ratio of *d* to *s* character for the conduction electrons with spin parallel to the unpaired electrons compared to those with spin antiparallel. This is the reason why R_1 is temperature dependent. It depends on the resistivity to provide the scattering and the resistivity is temperature-dependent. R_1 can be much larger than R_0 since the spin-orbit coupling produces apparent fields at the outer electrons of 100,000 to 1,000,000 gauss.

The anomalous Hall effect has been measured in polycrystalline iron, cobalt, nickel and gadolinium and it is interesting to compare the signs of R_1 and R_0 for these elements given in Table XIII,

TABLE XIII

SIGNS OF R_0 AND R_1 IN FERROMAGNETIC METALS

	Iron	Cobalt	Nickel	Gadolinium
R_0	+	—	—	—
R_1	+	(0–290°K) — (> 290°K) +	—	—

The minus sign for R_0 stands for electron conduction and the plus sign for hole conduction, while the minus sign for R_1 indicates that the current

* Try hopping on one leg around a one foot circle on the ground. It is easier to move about the circle counter clockwise if the left leg is used for hopping.

carriers with m_s parallel to the unpaired $3d$ or $4f$ electrons have a smaller ratio of d to s character and vice versa. Our electron theory of metals cannot predict the sign of R_1, especially the change in sign in cobalt. Furthermore it is difficult to understand how R_0 and R_1 can have opposite signs.

Theoretical attempts have been made to estimate the dependence of R_1 on the resistivity ρ. It is found experimentally that R_1 varies $\sim a + b\rho^2$ where theory predicts only $b\rho^2$. It is rather difficult to estimate the magnitude and temperature-dependence of R_1 since one must know the wave functions of the current carriers and their temperature dependence.

What have we learned from the Hall effect? Since the band theory of metals does not yield sufficiently accurate details of the Fermi surface we cannot calculate the ordinary Hall effect. We may be able to learn something of the Fermi surface from Hall effect measurements but this probably requires high purity single crystals at low temperatures. As for the extraordinary Hall effect in ferromagnets we are still not certain of the spin-orbit coupling mechanism nor can we predict the magnitude or proper temperature dependance for R_1. In short we have learned very little from Hall effect measurements about the wave functions, eigenvalues or momenta of the electrons at the Fermi surface in metals.

7. *Thermoelectric Power*

A measurement extremely sensitive to slight changes in the composition and state of a metal is thermoelectric power but like resistivity its sensitivity exceeds our theoretical capacity to predict or even understand the results. Of course, there is a practical interest in thermocouples for temperature measurement but beyond this there is a sparcity of data on thermoelectric power.

If one end of a metal wire is heated the electrons at the Fermi level are excited to higher levels and a free energy difference is created relative to the cold end. This free energy difference is not always in the direction favouring the flow of electrons from the hot end of the metal to the cold end. If we attempted to measure the potential difference by connecting a high impedance voltmeter across the ends of the wire we would realize that the voltmeter leads will have come to thermal equilibrium with the ends and we would only measure the difference in potential between the wire and the voltmeter leads. As a general rule, then, we measure differences in potential although absolute measurements are possible.

The electric field, E, along a wire with a thermal gradient, is generally written

$$E = Q \, \mathrm{d}T/\mathrm{d}l \tag{173}$$

where $\mathrm{d}T/\mathrm{d}l$ is the thermal gradient along the wire (deg/cm) and Q is the constant of proportionality called the thermoelectric power (units of V/deg) and commonly listed in experimental tables. Q has been placed on an absolute basis by the measurement of a metallic standard against a superconducting metal for which the thermoelectric power can be shown to be zero. All subsequent measurements were placed on an absolute basis relative to the standard. In general Q is approximately a microvolt per degree.

The general theoretical expression for Q is

$$Q = -\frac{\pi^2}{3}\frac{k^2T}{e}\left[\frac{\partial \ln \Lambda + \partial \ln \mathfrak{G}}{\partial E}\right]_{E=E_F} \tag{174}$$

where Λ is the electron mean free path and $\mathfrak{G}$ the surface area of the Fermi surface. Ordinarily we expect the potential to favour current flow from the hot section of the wire to the cold (Q-negative) since the mean free path of the electrons generally increases with energy (they are more difficult to scatter) as does the surface area of the Fermi surface. Thus both terms in the brackets in eqn. (174) are positive and Q is negative. However, if the Fermi surface contacts the Brillouin zone the surface area will decrease with increasing energy since the contact area does not contain any higher energy states and must not be included.* The second term in the bracket of eqn. (174) is then negative and the sign of Q may be either negative or positive, depending on the relative magnitude of the two terms. The reason a Fermi surface area decreasing with energy causes a reversal in the current flow is that the number of available energy states is decreasing and a given number of electrons would have to seek higher energy states in order to be accommodated. Since we are considering a drift velocity the electrons are continually scattered to occupy all states on the Fermi surface. These two conditions are schematically shown for a spherical Fermi surface inside a cubic Brillouin zone. In diagram 1, Fig. 96, the Fermi surface does not contact the Brillouin zone and the hot metal has the electrons at the Fermi surface thermally excited to wave numbers $\sim k_2$. Since the energy is merely the surface area ($E = \hbar^2k^2/2m$) the hot metal is at higher energy than the cold and the electrons will move to lower energies by flowing from hot to cold if the metals are brought into contact (Q negative). In diagram 2 the Fermi surface contacts the Brillouin zone and the contact area contains occupied levels below the Fermi surface. In this case the energy of the available surface area is greater for the cold metal and the electrons at the Fermi level will flow from cold to hot. This does not mean that the total energy of the hot metal is less than the cold. It is just that more of the electrons in the hot metal occupy states on the Fermi surface below the Fermi level (contact area on Brillouin zone) and are not available for conduction.

* Imagine a rubber balloon expanding inside a cubic box with rigid walls and dimension a. The surface area of the balloon is $4\pi r^2$ until $r \geq a/2$ whereupon the surface area decreases if you do not include the area of contact. Eventually the balloon is a cube in which case the surface area is zero and the Brillouin zone is filled.

Copper, silver and gold all have positive values of Q which is consistent with the picture that the Fermi surface makes contact at the 111 Brillouin zone faces. However, the positive sign in lithium is rather surprising since

TABLE XIV

ABSOLUTE THERMOELECTRIC POWER AT 0°C IN μV/°C

Li	+ 13·2	Cu	+ 1·5	Ni	− 16·8	Mg	− 1·3
Na	− 5·4	Ag	+ 1·3	Pd	− 6·2	Cd	+ 1·8
K	− 12·5	Au	+ 1·7	Pt	− 4	Zn	+ 1·8
				Al	− 1·6		
				Pb	− 1·2		

it is "occasionally" believed that the Fermi surface does not make contact in lithium. Some representative values of absolute thermoelectric power are given in Table XIV. It is too early to conclude anything about the relationship of the values of Q in Table XIV to the shape of the Fermi surface. The temperature dependence as well as the impurity or alloy variation is quite erratic. For example, the Q of platinum and copper actually change sign at low temperature.

Equation (174) is only valid at temperatures $T \geq \Theta$. At low temperatures a new effect becomes significant, called *phonon drag*. There will always be a flow of phonons from the hot to the cold end of a metal wire and these phonons can drag the electrons along with them. Conceptually one can see nons can drag the electrons along with them. Conceptually one can see this by imagining a single phonon travelling along a wire. The atoms are all successively displaced in the direction of the wave and since the current-carrying electrons are attached to the atoms they must go along with them. The effect is to cause an added negative contribution to Q of an amount approximately given by

$$Q_L \cong -\frac{k}{e}\frac{1}{n_0}\frac{4\pi^4}{5}\left(\frac{T}{\Theta}\right)^3 \qquad T < \Theta$$

$$Q_L \cong -\frac{k}{e}\frac{1}{n_0} = -\frac{86}{n_0}\frac{\mu\text{V}}{\text{deg}} \qquad T > \Theta \tag{175}$$

contribution of phonon drag to thermoelectric power

where n_0 = number of conduction electrons per atom. This effect may explain why Q changes sign in copper at low temperatures.

While we have limited our discussions to thermoelectric power there are many variations of the thermoelectric effect all of which are beyond our capacity to calculate with any precision. An effect just opposite to the one we've discussed (i.e. the Seebeck effect) is to place an external electric potential across a junction of two metals and thus create a temperature difference

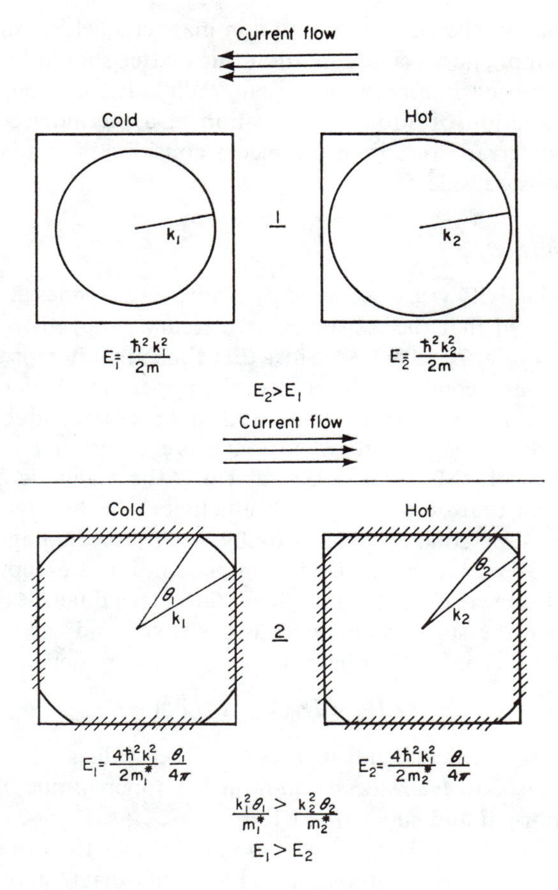

FIG. 96. Diagram 1. A spherical Fermi surface at two temperatures within a cubic Brillouin zone. The energy of the electron at the Fermi level in the hot metal is greater than the cold and causes electron flow from hot to cold when the two are brought into contact.

In diagram 2 the Fermi surface makes contact with the Brillouin zone over a larger area in the hot metal than the cold. The total energy of the electrons at the Fermi level is greater in the cold than the hot so that the electron flow at the Fermi level is from cold to hot.

between the two. This is known as the Peltier effect and some effort has even been expended in using this effect to build a refrigerator. Just as in the Seebeck effect one would have to know rather precise details of the Fermi surface to estimate the magnitude and signs of the effect.

In summary, we cannot make any definite conclusions about the Fermi surface from thermoelectric power. Certainly more measurements are needed, probably on single crystals and in magnetic fields. Since Q is rather sensitive to sample purity and condition this matter should be carefully controlled and reported in any measurement. While it has been suggested that the precise variation of alloy concentration at a thermoelectric junction is responsible for erratic results, more recent considerations tend to minimize this as a serious consideration.

8. *Superconductivity*

Superconductivity was discovered by Kamerlingh Onnes at Leiden in 1911 when he observed that the resistance of mercury dropped to zero below its *critical temperature*, $T_c = 4{\cdot}2°K$. Since that time any attempt to measure the resistivity of a superconductor has failed, the upper limit being set at $\sim 10^{-18}$ micro ohm-cm. In 1913 Onnes discovered that a superconductor returned to its normal state if a certain value of current was exceeded in a wire. It has since been pointed out by Meissner (1933) that it is the magnetic field associated with the current that destroys superconductivity. In the *Meissner effect* it is shown that a superconductor is virtually a perfect diamagnet in that it prevents any external magnetic field from penetrating it except for a thin surface layer. However, if the magnitude of the external field exceeds a critical value, H_c, then the superconductivity is destroyed and the metal returns to its normal state. This critical field has a temperature dependence

$$H_c = H_0(1 - T^2/T_c^2) \tag{176}$$

where H_0 is the critical field at $T = 0°K$. At values of H greater than $(\frac{2}{3})H_c$ there exists an *intermediate state* in the superconductor consisting of domains of normal and superconducting states.

In the region $T = 0°K$ to T_c the specific heat in the normal and superconducting state are quite different. The specific heat of the normal state below T_c can be measured by applying a magnetic field greater than H_c and destroying the superconducting state. Fig. 97 for vanadium[(49)] is illustrative of the specific heat curves in the two states. In the normal state the electronic specific heat is given as γT where γ is the electronic specific heat coefficient but in the superconducting state the temperature dependence is given as

$$C_s = a\, e^{-bT_c/T} \tag{177}$$

Furthermore, the entropy increase of the normal state is $\sim 10^{-3}R$.

In 1950 the *isotope effect* was discovered which showed that the superconducting critical temperature for various isotopes of lead was inversely proportional to the square root of the mass, suggesting that the critical temperature is proportional to the Debye temperature. By using high frequency radiation which limited the depth of penetration it was discovered about the same time that the resistivity of a thin layer at the surface of a superconductor was normal. In addition, one observed a decrease in thermal conductivity in the superconducting state suggesting that the superconducting electrons no longer carried any energy. Other experimental observations include the fact that the thermoelectric power was zero in the superconducting state and in some cases such as vanadium superconductivity was found in thin amorphous evaporated films.

A list of superconducting elements is given in Table XV. Superconductivity is absent in the alkali metals, alkaline earths, noble metals, semi-metals, semi-conductors and ferromagnetic and antiferromagnetic elements. Peculiarly enough there is a considerable number of superconducting compounds like NbN, $T_c = 14{\cdot}7°K$; SnSb, $T_c = 3{\cdot}9°K$; TaSi, $T_c = 4{\cdot}4°K$;

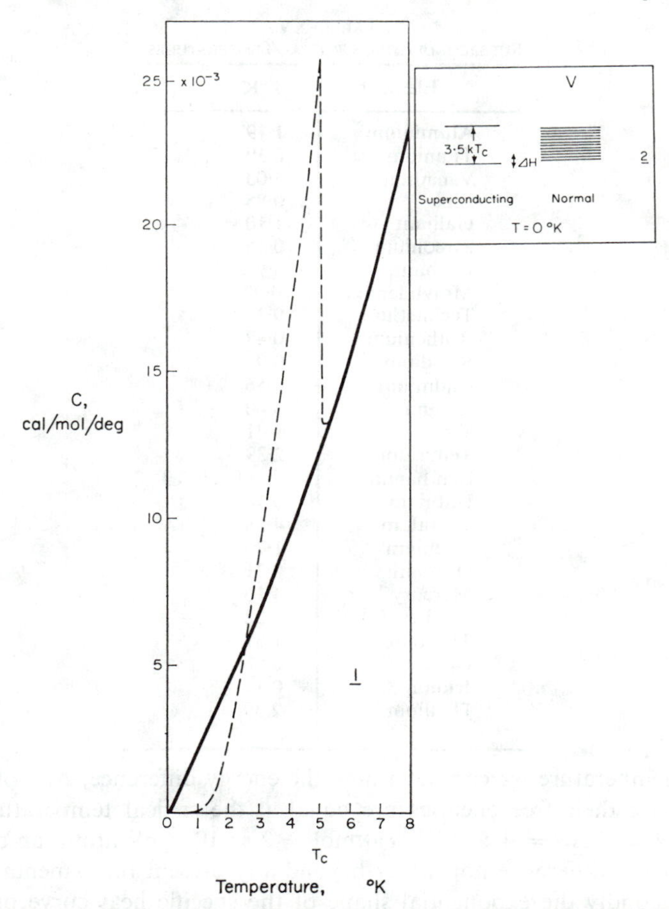

FIG. 97. The specific heat of vanadium in the normal state (solid line) and superconducting state (dashed curve). Diagram 2 is a schematic drawing of the energy levels at $T = 0°K$ in the superconducting and the normal state.

MoC, $T_c = 8°K$ and Nb_3Sn, $T_c = 18°K$ the last one noted for its high critical field, $H_0 > 100$ kilogauss in the cold worked state.

Ever since superconductivity was discovered it has intrigued the theoretician and in 1957 Bardeen, Cooper and Schrieffer produced a theory (commonly called *BCS-theory*) which removed the exotic nature of superconductivity and has given us a physical picture which is almost within the realm of popular understanding. But before we discuss the theory let us see what deductions can be made from the experimental observations.

Firstly the small entropy difference suggests that while the superconducting state is more ordered it is probably confined to the small fraction of electrons within kT_c of the Fermi surface. From the entropy difference and

TABLE XV
SUPERCONDUCTING CRITICAL TEMPERATURES

Element	T°K
Aluminium	1·197
Titanium	0·39
Vanadium	5·03
Zinc	0·85
Gallium	1·10
Zirconium	0·55
Niobium	8·9
Molybdenum	0·92
Technetium	9·3
Ruthenium	0·47
Rhodium	0·9
Cadmium	0·56
Indium	3·40
Tin	3·41
Tellurium	2·39
Lanthanum	α, 4·9; β, 6·3
Hafnium	0·165
Tantalum	4·38
Rhenium	1·70
Osmium	0·71
Mercury	4·16
Lead	7·22
Thorium	1·37
Uranium	0·68
Iridium	0·14
Thallium	2·39

critical temperature we can determine the energy difference, ΔH, of the two states since their free energies are equal at the critical temperature. This gives $\Delta H = T_c \Delta S = 4 \times 10^{-3}$ cal/mol $= 2 \times 10^{-7}$ eV/atom, an extremely small energy difference hopelessly beyond any present fundamental calculation. Secondly the exponential shape of the specific heat curve provides a clue that there is actually an energy gap $\sim kT_c$ in the superconducting state since the thermal excitation from one state to a higher state behaves exponentially at low temperatures (problem 5, p. 92). Thirdly the unusual diamagnetic interaction with the external magnetic field indicates that the quantum mechanical state of lowest energy requires the electrons with $m_s = +\frac{1}{2}$ to have identical energy to those with $m_s = -\frac{1}{2}$ since the magnetic field lowers the energy of the electrons whose intrinsic magnetic moment is parallel to the field and raises the energy of those which are antiparallel. The exclusion of the magnetic field from the superconductor (perfect diamagnetism) presumably arises from the orbital diamagnetic motion of the electrons in the thin surface layer. In fact the energy required for the magnetic field to penetrate the space occupied by the superconducting electrons is approximately kT_c so that H_0, the critical field, is approximately kT_c/μ_B where μ_B is the magnetic moment of the electron. Fourthly, the isotope effect indicates that the superconducting electrons are strongly coupled to the lattice vibrations and due to the low values of T_c suggests that the zero point oscillations are *required* for the effect.

The physical basis of the BCS-theory pieces this reasoning together. It proposes that the electrons at the Fermi surface with opposite m_s values and opposite momenta can lower their energy by vibrating in phase with the zero point oscillations. This does not merely involve nearest neighbour electron pairs but rather a collective grouping of electrons extending over many thousands of angstroms. Somehow in this paired state their Coulomb repulsion must diminish by riding rigidly along with the lattice vibrations. Consequently they show no resistivity for none of them can singly scatter without breaking up the group arrangement.*

The energy gap in the BCS-theory turns out to be $\sim 3{\cdot}5kT_c$ at $T = 0°K$. By irradiating a superconductor with photons in this energy region (the infrared) an absorption edge appears quite sharply at the energy gap and has confirmed the BCS-prediction. The BCS-theory also predicts the correct temperature dependence of the specific heat.

In the BCS-theory a wave function for the outer electrons is written down which is characterized by the pairing of electrons with opposite m_s† and opposite momenta, and the variational principle is used to minimize the energy. However, the theory is still not able to correlate the presence or absence of superconductivity with the electronic structure of the metal although it is obvious that such correlation exists since the absence of superconductivity coincides with the rows in the periodic table. In addition, the existence of

* Imagine a marching column of soldiers each tied to his nearest neighbours with taut ropes. In the absence of the ropes a stone on the ground might cause one of the soldiers to stumble and in effect slow down the group. But when the soldiers are tied together the neighbours might keep the clumsy one moving along if he trips until the time he regains his balance. In this case there would be no slow down of the group movement.

† An attempt to find this pairing of m_s values was unsuccessfully sought in neutron diffraction measurements by searching for superlattice lines that might indicate the electrons to be in a regular anti-ferromagnetic array. However, the conduction electrons cannot generally be seen in Bragg diffraction (see p. 152).

superconductivity in compounds which are not typically metallic is inexplicable. We might also add that since the energy gap ($\sim 3{\cdot}5kT_c$) is considerably greater than the energy difference of the two states ($\Delta H \cong 10^{-3}kT_c$) that the normal and superconducting states represent a delicate balance of energy. Since the total energy of the electrons in the superconducting state is lower than the normal state, the virial theorem tells us that the electrons in the superconducting state have greater kinetic energy than those in the normal state and lower potential energy. This means greater curvature of the electron wave functions in the superconducting state, hence greater localization. On the average the superconducting electrons probably keep out of each others way more effectively. The energy levels are shown schematically in Fig. 91, diagram 2.

We can say then, that superconductivity is understood well enough so that it is no longer considered a major theoretical threat. However, the intimate details of the superconducting state in terms of the electronic structure of the substance is not known. This latter fact is not surprising since we know so little about the electronic structure of the elements anyway. It does support the point though that long range interactions between electrons are important.

(Note added in proof.) Recent observations indicate the isotope effect to be absent or diminished in transition metal superconductors suggesting another mechanism coupling the electrons with opposite momenta (perhaps spin-orbit coupling?).

Problems

1. (*a*) Estimate and plot the resistance, power and visible light output for a 110 V, 100 W incandescent lamp as a function of time in milliseconds.

(*b*) Design an inexpensive bimetallic strip that regulates the current in a circuit not to exceed 1 amp.

2. Look up the changes in resistivity with pressure for copper, lead, iron and aluminium (A.N. Gerritsen *Handbuch der Physik*, **XIX**, 192, (1956) Springer Verlag, Berlin). Assume this to be due to a change in the Debye Θ caused by a reduction in the average amplitude of vibration. Devise an independent experiment to corroborate this change in Debye Θ.

Compare the change in resistivity of pure copper due to the addition of 1% vacancies, 1% interstitials, and 1% gold. How can the first two samples be prepared?

3. (For the mathematically minded) Calculate the average angle of scattering of a free electron at the Fermi level in copper (*N*-processes) with the Debye distribution of phonons at 5°K, Hint; set up the equations of conservation of energy and momentum and average over all possible phonon directions Ans. $\sim 0{\cdot}1$ sec. of arc.

4. Make a literature survey to ascertain which of the eight transport properties discussed in this chapter have been measured for titanium, lead, beta brass, aluminium and tungsten. See if you can make a qualitative analysis of the data in terms of the electronic structure of the metals.

Summary

Measurements of the transport properties of the electrons at the Fermi surface in metals are generally difficult to interpret. Only in the case of very pure metals at low temperatures (for which the electron mean free is long) are transport properties capable of yielding very specific information such as the shape of the Fermi surface and the effective mass. This must be considered the most important contribution of transport measurements in understanding the electronic structure of metals.

As general tools resistivity, thermoelectric power, and superconductivity can be of practical importance to the metallurgist even though one cannot calculate their magnitudes for any specific metal. The de Haas van Alphen effect, cyclotron resonance and the anomalous skin effect are highly specialized measurements which have yielded detailed information about the Fermi surface but magnetoresistance and Hall effect are only understood qualitatively.

CHAPTER IX

THERMODYNAMICS AND COHESION

Introduction

In the field of thermodynamics we shall discuss measurements which alter the phonon distribution and/or volume of a metal and measurements of cohesion in alloys. These include four distinct types of measurements:

1. Thermal energy is introduced into a sample at various temperatures (specific heat and heat content).
2. The transport of phonons is measured (thermal conduction or ultrasonic attenuation).
3. The heat of formation and free energy changes accompanying alloying are measured.
4. The volume of a metal is decreased by application of an external pressure.

1. Specific Heat and Heat Content

Experimental Details of Specific Heat and Heat Content Measurements

Since the specific heat is simply the derivative of the heat content, one can measure either and derive the other. It turns out in practice, though, that for most cases (but not all) one measures specific heat from $T = 0°K$ to room temperature and heat content from room temperature to the boiling point. The technical problems favour the use of different apparatus for specific heat measurements over the five temperature ranges 0–1·2°K; 1·2–4·2°K; 4·2–20°K; 20–300°K; 300–1500°K. 4·2°K and 20°K are the boiling points of liquid helium and liquid hydrogen at one atmosphere pressure. If one reduces the pressure above its surface liquid helium boils more rapidly and this reduces the temperature of the remaining liquid since the latent heat of vaporization is being removed. By controlling the rate of evaporation the temperature of the liquid helium can be reduced to 0·8°K but due to certain practical limitations the lower limit is 1·2°K. Below 1°K one can use liquid He^3 which boils at a lower temperature and enables one to reach ~0·35°K. A paramagnetic salt like cerium magnesium nitrate used in a strong magnetic field enables one to reach temperatures below 0·1°K. The cerium atoms have a magnetic moment due to their single unpaired $4f$ electron ($L = 3$, $S = \frac{1}{2}$, $J = \frac{5}{2}$) but are too far apart to be coupled either ferromagnetically or antiferromagnetically. Initially the cerium atoms are partially aligned by application of a magnetic field. This raises the temperature of the sample since the magnetic energy of the sample is lowered and the thermal energy must be raised to conserve the total energy. This increased thermal energy is removed by making thermal contact between the paramagnetic salt and liquid helium with a solid lead pipe. If the magnetic field is suddenly removed, the lead becomes superconducting and thermally isolates the paramagnetic salt from the liquid helium (a superconductor is a poor thermal conductor). The magnetic moments of the cerium atoms then return to their random orientation of higher magnetic energy by absorbing thermal energy from the lattice and reducing its temperature. This technique is called *adiabatic demagnetization.* The final temperature of the cerium magnesium nitrate can be determined from its magnetic susceptibility, χ, by measuring its magnetic moment per unit volume, M, in a magnetic field H.

$$\chi = \frac{M}{H} = \frac{N_0 g^2 J(J+1)\mu_B^2}{3kT}$$

$$g = 1 + \frac{J(J+1) + S(S+1) - L(L+1)}{2J(J+1)} = \text{Landé } g \text{ factor} \qquad (179)$$

where N_0 is the number of cerium atoms per cm^3, χ is the susceptibility per unit volume, T the temperature, and μ_B the Bohr magneton ($0{\cdot}927 \times 10^{-20}$ ergs/gauss). Since everything is known and χ is measured we can determine T. However the spin-orbit coupling in the cerium magnesium nitrate and the lack of cubic symmetry in this crystal gives rise to some anisotropy in χ so that eqn. (179) is modified slightly. It is thus more practical to measure χ at the boiling point of liquid helium (which is at the accepted standard temperature of 4·2143°K, or liquid hydrogen at 20·390°K) and take the ratios of the χ to determine other temperatures since the measured χ is still inversely proportional to T.

Once a thermal reservoir is available for cooling a sample to a low temperature a thermometer is required to measure the temperature. For temperatures lower than 1°K the susceptibility of the cerium magnesium nitrate is used but if one does not plan to go below 1°K single crystals of manganous ammonium sulphate have the advantage of having an isotropic susceptibility ($S = 5/2$, $L = 0$ for the Mn atoms). Another popular thermometer at these temperatures is a commercial Allen Bradley Co. carbon resistor which follows a general expression for the resistance R,

$$\ln R + K/\ln R = A + B/T \qquad (180)$$

valid from 0·1°K to ~6°K. The constants K, A and B can be determined by calibration against the vapour pressure—temperature scale (1–4°K) of liquid helium,[50] commonly accepted as a standard. Table XVI can be used as a

TABLE XVI

A GUIDE TO THERMAL RESERVOIRS, THERMOMETERS, AND METHODS OF CALIBRATION IN VARIOUS TEMPERATURE RANGES

Temperature range	Reservoirs to cool sample	Thermometers		Thermometer calibration
< 1°K	1. Single crystal cerium magnesium nitrate	Carbon resistor	→	Susceptibility of cerium magnesium nitrate
		or		
	2. He^3 (to ~0·35°K)	Susceptibility of cerium magnesium nitrate	→	Vapour pressure of He^4, 1–4°K
1°K–4·214°K	Liquid He^4	Carbon resistor	→	Vapour pressure of He^4
		or		
		Vapour pressure of He^4	→	Secondary Standard
		or		
		Susceptibility of manganous ammonium sulphate	→	Vapour pressure of He^4
4·214°K–10°K	Liquid He^4	Susceptibility of manganous ammonium sulphate	→	Boiling points of He^4 and H_2
		or		
		He gas thermometer	→	Boiling points of He^4 and H_2
10°K–20·4°K	1. Liquid H_2	Susceptibility of manganous ammonium sulphate	→	Boiling point of H_2
	2. Liquid He^4	or		
		Carbon resistor	→	He gas thermometer
		or		
		Vapour pressure of H_2	→	Secondary Standard
		or		
		He gas thermometer	→	Boiling points of He^4 and H_2
20·4°K–300°K	1. Liquid H_2	Pt resistance	→	Tertiary Standard see p. 115, ref. (50)
		or		
	2. Liquid N_2 (77°K–300°K)	Cu-constantin thermocouple	→	Pt resistance thermometer checked against H_2 boiling point
300°K–1900°K	Room temperature	Pt-Pt Rh thermocouple	→	Secondary standard
> 1900°K	Room temperature	Optical pyrometer	→	Melting point of Au 1063·0°C

guide to the thermal reservoirs and thermometers used in various temperature ranges.

After we have a reservoir to cool the sample to as low a temperature as is desired and a thermometer in contact with the sample, we must then isolate the sample thermally from the reservoir. This can either be accomplished by employing a superconductor and an external magnetic field (if the temperature range is appropriate) or by employing a mechanical metal switch, somewhat in the form of a pair of ice tongs, to make and break physical contact with the reservoir. Once the sample is isolated from the reservoir a known amount of heat can be introduced with a small electric pulse passed through a fine wire wrapped around the specimen and making good thermal contact. During this time the temperature of the sample is measured at frequent intervals and a heating curve is obtained like the one in Fig. 98. The gradual change in

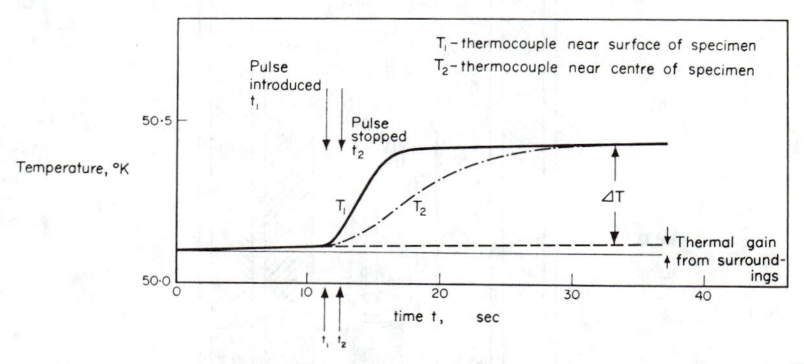

FIG. 98. A representative heating curve in specific heat measurements. The pulse is introduced at time t_1 and terminated at time t_2. The temperature of the sample is recorded by a thermocouple near the surface T_1 and near the centre T_2. Thermal equilibrium is reached when $T_1 = T_2$. The gradual rise in temperature is due to heat absorbed from the surroundings due to imperfect insulation.

temperature before the pulse is introduced is due to heat gain (or loss) to the surroundings since no insulator is perfect. After the heat pulse is introduced and stopped the temperature of the sample continues to rise until the thermal energy spreads uniformly over the sample. It is generally advisable to have several thermocouples distributed throughout the sample in order to ascertain the time at which the various parts of the sample are in thermal equilibrium. For example the curve T_1 in Fig. 98 might be that of a thermocouple near the surface of the specimen and T_2 one closer to the centre since more time is required for the heat pulse which has been introduced at the surface to reach the centre. The specific heat is given by

$$C = \Delta Q/\Delta T \tag{181}$$

where ΔQ is the energy absorbed by the sample from the electric pulse.

ΔQ should always be small enough so that C does not change by more than $\sim 3\%$ over the temperature rise ΔT. Since the specimen must be contained in a sample holder one must subtract the heat absorbed by the sample holder and this is determined in a separate measurement without the specimen (of course one tries to minimize the mass of the sample holder). In all specific heat measurements it is difficult to determine ΔT if the sample has poor thermal conductivity, such as in alloys, for then a long time is required for equilibrium to be reached. There is always a certain amount of heat loss (or gain) to the surroundings by radiation (photons). This depends on the fourth power of the absolute temperature difference between the sample and surroundings and is

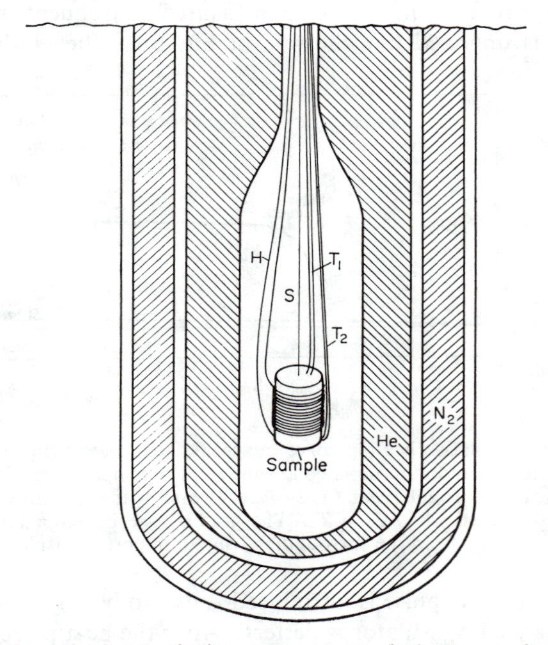

FIG. 99. A schematic sketch of a low temperature calorimeter employed in specific heat measurements. The sample is surrounded by a heating coil H and suspended by a poor conductor S, T_1 and T_2 are thermocouples. The volume around the sample is ordinarily evacuated except in the initial lowering of the temperature when an exchange gas is introduced that carries heat from the sample to the liquid helium by convection.

generally small at low temperatures. A schematic arrangement of a calorimeter is shown in Fig. 99. An exchange gas such as helium is introduced initially to bring the sample down to 4°K and the volume around the sample is then evacuated.

At temperatures above room temperature measurements are susceptible to large radiation losses which are difficult to rectify. As such, published specific heat data above 800°K are in variance by as much as 10% and very little data has been published above 1300°K. Recent attempts have been made to reduce the radiation losses in metals by rapid pulsed heating. If one introduces a large square shaped current pulse of short duration into a wire whose resistance versus temperature is accurately known, one can follow the potential drop during the pulse on an oscilloscope, and measure the specific heat in a time short compared to the time required for appreciable radiation loss from the surface. A recent application of this technique[51] to iron employed a square current pulse of ~40 millisecond duration. The current pulse was turned on and off with mercury switches, and a large capacity resistor in series with the iron wire assured a constant current. The range of temperatures was room temperature to 1300°K, a separate experiment determining the resistance versus temperature of the specimen. In this experiment the specific heat per mole of wire of uniform cross-section is given by

$$C = I^2 R \Delta t / mJ\Delta T \tag{182}$$

where I is the current, in amps, R the resistance, in ohms, at a particular temperature, $\Delta T/\Delta t$ the slope of the oscilloscope trace at any temperature converted from a resistivity-time to a temperature-time scale, J the mechanical equivalent of heat (4·185 joules/cal), and m the mass of the wire in g. Since the heat is introduced into the wire through an electric current the time required for thermal equilibrium is $\ll 10^{-12}$ sec enabling the pulse to be quite short. Each pulse raised the sample temperature about 50°C, the sample being in a vacuum furnace whose temperature is in the range to be measured. The method appears promising but more measurements are required to assess its limitations.

For most substances above room temperature the specific heat is linear in temperature and the drop method for measuring heat content is probably quite adequate. The sample is heated to some temperature, T, and dropped into a reservoir at room temperature. Sample and reservoir share this extra heat with a subsequent temperature rise of the reservoir. A knowledge of the masses of the sample and reservoir and the specific heat of the reservoir enables one to determine the heat content (enthalpy) of the sample from room temperature to T. A schematic diagram of this type of apparatus is shown in Fig. 100. In this specific diagram the reservoir consists of an enclosed container of ice and water with a layer of mercury at the bottom, connected to an external mercury column. As the ice melts following the introduction of the sample the volume of the ice-water mixture decreases and the mercury level rises in the calorimeter and drops in the mercury column. The change in the ice-water volume determines the mass of ice which has been converted to water hence the total heat absorbed by the ice water mixture (latent heat of fusion of ice is 79·71 cal/g). With careful work an accuracy of 0·1% is possible.

Analysis of the Specific Heat of Iron

Specific heat and/or heat content measurements yield considerable information and we shall consider three particular cases to illustrate the procedure for extracting this information. The first is one of the most complicated cases encountered, the specific heat of iron metal, the second the low temperature specific heat of cobalt and the third the low temperature specific heat of rhenium. Fig. 101 shows the specific heat of iron from $T = 0°K$ to the melting temperature at 1808°K and reveals several striking features which are understood. There is a lambda type anomaly at $T = 1043°K$ which is known from magnetization measurements to be the

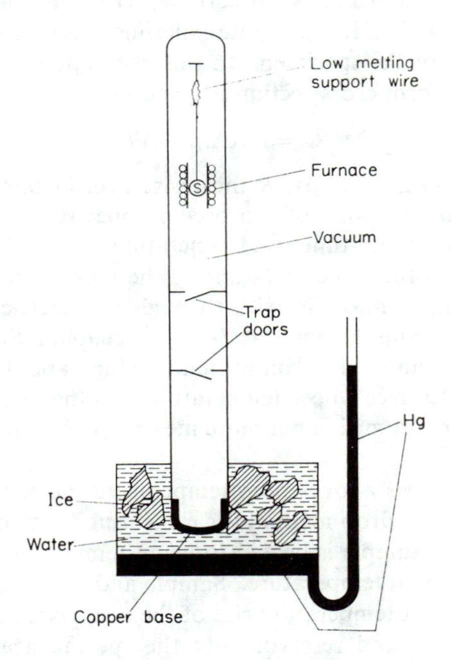

FIG. 100. A type of drop calorimeter employed in heat content measurements above room temperature. The sample, *S*, is heated in a furnace to a temperature *T* and then dropped through trap doors which shield the calorimeter from the furance. The ice and water mixture completely fills the volume above the mercury.

ferromagnetic Curie temperature. There is a discontinuity at 1183°K and another discontinuity at 1673°K both of which are known from X-ray data to be allotropic transformations, the low temperature b.c.c. structure converting to f.c.c. at 1183°K and then back again to b.c.c. at 1673°K. Iron remains b.c.c. until the melting temperature, 1808°K. Two pieces of information not contained in the specific heat curves of Fig. 101 are the latent heats of the transformations at 1183°K and 1673°K which are 210 and $\sim$ 230 cal/mol, respectively.

At very low temperatures (1–4°K) the lattice, electronic and magnetic specific heats of α iron are given by eqns. (51), (56) and (60) respectively. As the magnetic specific heat is negligible compared to the lattice and electronic specific heats in this temperature range it is possible to separate the lattice and electronic contributions since the former varies as T^3 and the latter as T. It is common procedure to plot C/T versus T^2 which gives a straight line whose slope is $234R/\Theta^3$ and whose intercept at $T = 0°K$ is γ. From the measured specific heat of α iron in this temperature range we determine $\Theta = 432°K$ and $\gamma = 12 \times 10^{-4}$ cal/mol/deg^2. Combining this information

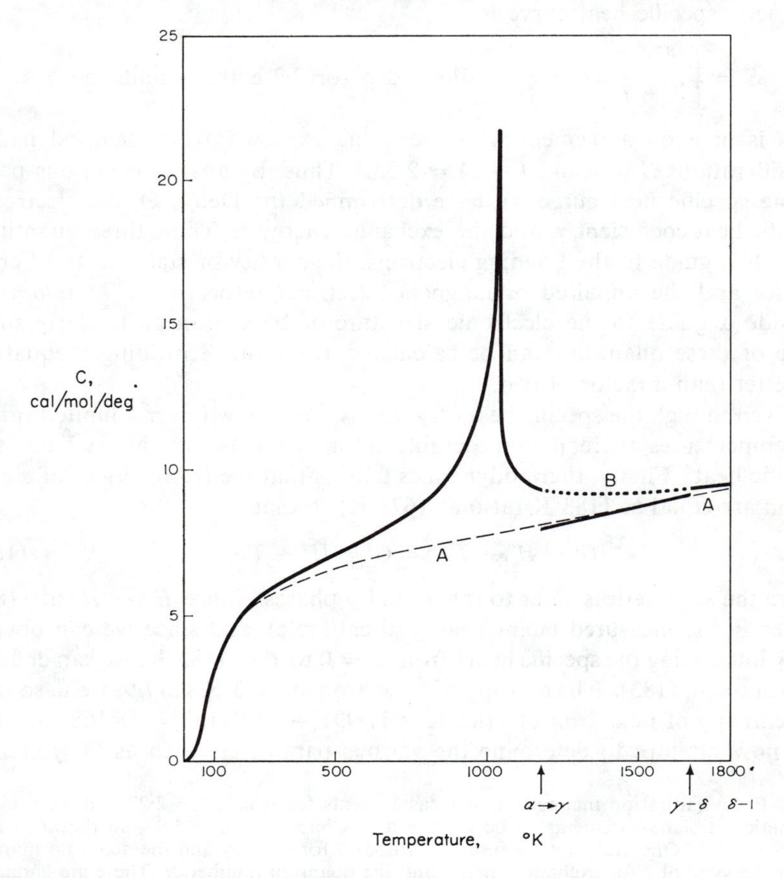

FIG. 101. The measured specific heat of iron from $T = 0°K$ to 1808°K (solid curves). Curve A is the estimated non-magnetic specific heat of b.c.c. iron and curve B is the estimated total specific heat of b.c.c. iron in the range over which f.c.c. iron is stable, 1183°K to 1673°K. The discontinuities occur at the α-γ transition temperatures.

with an average linear coefficient of expansion ($\alpha = 15 \times 10^{-6}$/deg) and using 2 for the Gruneisen constant we can determine the lattice and electronic specific heats of iron from eqn. (43) (tabulated in appendix I) and eqn. (56). This sum is plotted in Fig. 101 as curve *A*. If α iron was stable from 1183°K to 1673°K its specific heat in this range would probably be given by curve *B* since the two points must join smoothly. The difference between the total specific heat and curve *A* is the magnetic specific heat and this is plotted in Fig. 32. The total area under the magnetic specific heat curve is the total magnetic energy and is ~2000 cal/mole. This provides a direct measure of the exchange energy J (p. 61) since the total magnetic energy is $W_{EX} = RzJS^2/k = 2000$ cal/mol so that $J = 0{\cdot}010$ eV. The total entropy under the magnetic specific heat curve is

$$S = \int_0^{1800} \frac{C}{T}\,dT = 2{\cdot}2 \text{ cal/mol/deg (or 2·2 entropy units, e.u.).}$$

This is in good agreement with the value expected from thermodynamic considerations $S = R \ln(2J + 1) = 2{\cdot}35$.* Thus, by analyzing various parts of the specific heat curve we have determined the Debye Θ, the electronic specific heat coefficient γ, and the exchange energy J. These three quantities serve as a guide to the bonding electrons, the density of states at the Fermi surface and the unpaired or magnetic electrons respectively. *In toto* they provide a guide to the electronic structure of b.c.c. iron particularly since none of these quantities can be calculated from the Schrödinger equation to better than a factor of two.

Even though the specific heat of γ iron is only known over a limited range of temperatures there is considerable information to be gained from the specific heat. Firstly, thermodynamics tells us that the free energies of α and γ iron are equal at 1183°K (also at 1673°K) so that

$$G^{\alpha} = H^{\alpha} - TS^{\alpha} = G^{\gamma} = H^{\gamma} - TS^{\gamma} \tag{183}$$

where the superscripts refer to the α- and γ-phases. Since $H^{\gamma} - H^{\alpha}$ at 1183° is merely the measured latent heat (210 cal/mole), and since we can obtain S^{α} by integrating the specific heat† from $T = 0$ to $T = 1183$°K, we can deduce S^{γ} from eqn. (183). The entropy of b.c.c. iron at 1183°K is 17·991 e.u. so that the entropy of f.c.c. iron at 1183°K is $17{\cdot}991 + 210/1183 = 18{\cdot}168$ e.u. We can now attempt to determine the various parameters such as Θ, γ etc. of

* From saturation magnetization measurements for iron, $gJ = 2{\cdot}22$ and from gyromagnetic ratio measurements to be discussed in Chapter X, $g = 1{\cdot}93$, so that $J = 1{\cdot}15$ and $S = 1{\cdot}07$. One must not confuse the symbol S for entropy and the quantum number S nor the symbol J for exchange energy and the quantum number J. These are common notations.

† This can be done because the entropy of a pure element or an ordered structure is zero at $T = 0$°K. This is sometimes called the *Nernst Theorem*. The enthalpy, though, is not zero at $T = 0$°K but is the cohesive energy denoted H_0.

f.c.c. iron in order to see if we can independently determine its entropy at 1183°K. From the specific heat of γ iron in the range 1183°K to 1673°K we can determine the electronic specific heat coefficient since the lattice specific heat has reached its high temperature limit $3R(1 + 3\alpha\gamma T)$, eqn. (46), and there is no evidence for any magnetic specific heat above 1183°K. Of course, we cannot determine Θ in this temperature range since eqn. (46) is essentially independent of Θ. From the coefficient of linear expansion ($\alpha = 28 \times 10^{-6}$/deg) and using two for the value of the Gruneisen constant we obtain $\gamma \cong 8 \times 10^{-4}$ cal/mol/deg^2 from the measured specific heat in Fig. 101. The Debye Θ of f.c.c. iron can be determined reasonably well with the Lindemann formula which relates the melting temperature, T_m, to the Debye Θ in degrees Kelvin,

$$\Theta^* \cong 137 T_m^{1/2} \rho^{1/3} / M^{5/6} \tag{184}$$

where T_m is the melting temperature in °K, ρ the density in g/cm^3 and M the atomic weight. An inspection of the iron-platinum, iron-manganese and iron-nickel phase diagrams indicates that the melting points of b.c.c. iron amd f.c.c. iron are quite close and since the densities are also very close we expect Θ to be similar. Whether we use the Debye Θ of b.c.c. iron 432°K, or the value obtained from the Lindemann formula, 410°K, the total entropy of f.c.c. iron at 1183°K due to lattice vibrations (see appendix I) and electronic excitation ($S = \gamma T$) is $\sim$16 e.u. at least 2 e.u. short of the 18·168 e.u. required. *The only additional source of entropy appears to be magnetic in origin.*

A considerable amount of effort has been expended in determining the magnetic structure of f.c.c. iron and it is still not known with any accuracy. We can say, though, that the linear specific heat of f.c.c. iron from $T =$ 1183°K to 1673°K suggests that 1183°K must be well above any Curie or Néel temperature. Since the total magnetic energy is approximately RT_c and since 500°K is a reasonable upper limit to a critical temperature for f.c.c. iron we can set 1000 cal/mol as the upper limit for the magnetic energy.† From our estimate of the Debye Θ, from the determination of γ for f.c.c. iron and from the upper limit set for the magnetic energy we can determine the enthalpy of f.c.c. iron at 1183. Thus,

$$H^\gamma - H^\alpha = 210 \text{ cal/mole} = \left[\int_0^{1183} (C^\gamma - C^\alpha)\, \mathrm{d}T \right] + H_0^\gamma - H_0^\alpha \tag{185}$$

We are left only with the difference in cohesive energy between b.c.c. and f.c.c. iron ($H_0^\gamma - H_0^\alpha$) which varies from 800 cal/mol to 1800 cal/mol depending

* For b.c.c. Ti and Zr the constant is 108.

† The best guess is that f.c.c. iron is antiferromagnetic at $T = 0$°K with $gJ = 0{\cdot}5$ and that it is thermally excited to a high temperature magnetic structure with $gJ = 2{\cdot}8$ by a Schottky type excitation (p. 92). The estimated magnetic energy is $\sim$500 cal/mol. (For further details see Kaufman, Clougherty and Weiss, *Acta Met.*, 1963.)

on whether the magnetic energy is 100 cal/mol or zero respectively. This is one of the most important results of this analysis. The best experimental estimate of the difference in cohesive energy of α and γ iron at 0°K is ~1300 cal/mol corresponding to ~0·05 eV per atom and is a factor of *10 to 100 smaller than the error in cohesive energy calculations for transition metals.* The principal conclusion is that we must rely on this type of analysis to obtain estimates of the differences in cohesive energies of allotropic transformations because these differences are much smaller than the errors in our calculations.

Low Temperature Specific Heat of Cobalt

In most elements and alloys magnetic effects are not present and the separation into lattice and electronic specific heats is considerably simplified. There are, however, two additional sources of specific heat that are limited to elements with nuclear spin $I > \frac{1}{2}$ and to very low temperatures. The first can be seen in Fig. 102, the low temperature specific heat of cobalt metal.[52]

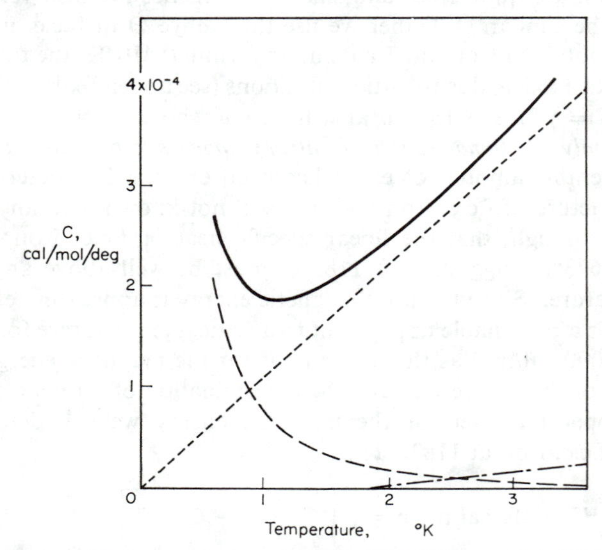

FIG. 102. The total low temperature specific heat of cobalt metal (solid curve). The separated lattice (dot-dashed) electronic (dotted) and nuclear (dashed) specific heats are shown. The nuclear specific heat yields a value of H_{eff} = 200 kilogauss.

At the very lowest temperatures the specific heat appears to increase with decreasing temperature. This arises from the hyperfine interaction of the 1*s*, 2*s*, 3*s* and 4*s* electrons with the cobalt nucleus, which has a magnetic moment $\mu = +4{\cdot}6389$ nuclear magnetons and a nuclear spin, $I = 7/2$. Due to the unpaired 3*d* electrons in ferromagnetic cobalt metal the 1*s*, 2*s*,

3*s* and 4*s* wave functions are different depending on whether their m_s is parallel or antiparallel to the unpaired 3*d* electrons. This arises from the Pauli principle and antisymmetrization. *s*-wave functions have a finite value at the nucleus and create an effective magnetic field at the nucleus, denoted H_{eff}, arising from the electron's magnetic moment. The energy of this hyperfine interaction is $\sim \mu H_{eff}$ ($\sim 10^{-5}$ eV) and would align completely the nuclei in a ferromagnetic array at $T = 0°K$. However the interaction is so weak that at temperatures above about 0·1°K the nuclear spins are almost completely disordered and the specific heat varies as $1/T^2$. It is per mole,

$$C = R[(I + 1)/3I][(\mu H_{eff})^2/(kT)^2 \tag{186}$$

In Fig. 102 the specific heat of cobalt has been separated into lattice, electronic and nuclear specific heats with T^3, T and $1/T^2$ temperature dependences, respectively. The nuclear specific heat leads to a value of $H_{eff} = 200{,}000$ gauss without specifying whether it is parallel or antiparallel to the direction of ferromagnetism. While the application of an external magnetic field of 10,000 gauss would increase or decrease the nuclear specific heat by a measurable 1% depending on whether H_{eff} is parallel or antiparallel, the *antiparallel* determination has been made in nuclear resonance work, Chapter XI. Hartree–Fock calculations on the free atom of cobalt yield antiparallel contributions from the 1*s* and 2*s* electrons and parallel contributions from the 3*s* and 4*s* electron with a total *parallel* contribution. To explain this discrepancy it has been suggested that in a metal the 4*s* contribution reverses its sign for reasons we shall discuss on, p. 342 Chapter XI. The nuclear specific heat in terbium, and α Mn have also been observed yielding $H_{eff} = 6$ million gauss and 90,000 gauss, respectively.

Nuclear Quadrupolar Contribution to Specific Heat

The second low temperature source of specific heat arises from the interaction of the nuclear quadrupole moment with the electric field gradients created by the neighbouring atoms. As a general rule the effect is small down to 0·1°K but has been observed[53][54] in rhenium, gallium and probably zinc. The high temperature "tail" of the specific heat is given as

$$C = bR\left(\frac{e^2qQ}{kT}\right)^2 \quad \text{valid for } kT \gg e^2qQ$$

$$b = \tfrac{1}{16}, I = 3/2; \quad \tfrac{7}{200}, I = 5/2; \quad \tfrac{3}{112}, I = 7/2; \quad \tfrac{11}{480}, I = 9/2. \tag{187}$$

where eq is the electric field gradient and eQ is the electric quadrupole moment of the nucleus. The specific heat of h.c.p. rhenium yields an observed value of $eq = 0{\cdot}14 \times 10^{16}$ e.s.u. Any nonspherical or noncubic electron probability distribution around the nucleus will give rise to such an electric field gradient. Thus, s wave functions or closed shells or orbitals with cubic symmetry like the e_g or t_{2g} probability distiibutions in Fig. 17 do not contribute. While we cannot calculate the outer electron wave functions in metals with sufficient accuracy to determine eq the results would suggest that zinc, for example, has appreciable $4p$ character in its outer electron wave function. The large deviation of the c/a ratio from the ideal ratio in zinc certainly supports this. It would appear that such measurements of the electric field gradients in noncubic elements and in alloys would guide us in understanding the type of wave functions present in the solid.

Let us review the various contributions to the specific heat of the elements and see what conclusions can be drawn about the outer bonding electrons. The next table summarizes these measurements and their temperature dependences. In no case can we calculate the specific heat exactly (from the Schrödinger equation) and all the expressions are approximate. However they are probably good enough for most metallurgical applications.

Information Gained from Specific Heat Measurements

(*a*) *Lattice Specific Heat*

Table XVIII lists the approximate Debye Θ values of the elements obtained mostly from specific heat measurements. These values offer further evidence of the similarity of the outer electron structure in the solid for elements in the same column of Table II. In fact if the Morse type potential curve is quite similar for elements in the same column of the periodic table then $\sqrt{(M)}\Theta$ should be constant since the spring constant between atoms is proportional to $M\Theta^2$. $\sqrt{(M)}\Theta$ is given in Table XVIII grouped according to elements with identical crystal structures and occupying the same columns of Table II. Except for lithium in the alkaline metals, beryllium in the alkaline earths and carbon (diamond) in the semiconductors the deviations are generally less than 15% from the average. Since the repulsive forces in lithium, beryllium and carbon come from closed s shells rather than p shells the higher Debye Θ values may reflect the difference. Cobalt is somewhat low and may be a result of its approximately 2 unpaired $3d$ electrons not participating in the bonding. As a rule the larger the value of $\sqrt{(M)}\Theta$ the larger is the cohesive energy. (See also F. H. Herbstein, *Phil. Mag. Supp.* **10**, 313 (1961) for a review article on methods of measuring Θ.)

(*b*) *Electronic Specific Heat*

Table XIX gives a list of values of γ, the electronic specific heat coefficient in eqn. (56). These do not reveal any simple relationship to the electronic structure except that they appear to fall into two categories, the transition

TABLE XVII

VARIOUS RECOGNIZED CONTRIBUTIONS TO THE SPECIFIC HEAT, THEIR TEMPERATURE DEPENDENCES, AND THE INFORMATION GAINED FROM THEIR SEPARATIONS

Contribution	Approximate temperature dependence	Validity	Information gained	Remarks
Lattice	$C = 464T^3/\Theta^3$ $C = C(\Theta/T)[1 + 3\alpha\gamma T]$	$0 < T < \Theta/10$ $0 \rightarrow T_m$	Debye Θ	Lindemann formula helpful $\Theta = 137T_m^{1/2}\rho^{1/3}/M^{5/6}$
Electronic	$C = \gamma T$	$0 \rightarrow T_m$	Approximate density of states at Fermi level	For $\gamma > 20 \times 10^{-4}$ cal/mol/deg^2 the linear relation may not hold at very high temperatures
Magnetic	$C = \frac{0{\cdot}113R}{b}\left(\frac{kT}{2JS}\right)^{3/2}$	$0 < T < T_c/5$	Exchange energy J	$b = 2$ for b.c.c. $b = 4$ for f.c.c. and h.c.p.
	$C =$ lambda type	entire range	Exchange energy J	Total energy $= RzJS^2/k$ (exchange energy J) Total entropy $= R\ln(2J + 1)$ (quantum number J)
Order-disorder	$C =$ lambda type	entire range if $T_c > 500°K$	Energy difference between ordered and disordered state	1. Peak in lambda type curve "washed out" if alloy is not stoichiometric 2. Low temperature specific heat not observable due to small diffusion rates

(Table continued on next page)

TABLE XVII—*continued*

Contribution	Approximate temperature dependence	Validity	Information gained	Remarks
Nuclear spin (magnetic substances)	$C = R\frac{(I+1)}{3I}\frac{(\mu H_{eff})^2}{(kT)^2}$	$kT \gg \mu H_{eff}$	H_{eff} = effective field at nucleus due to unpaired *s* electrons, orbital motion of *d* or *f* electrons, etc.	Particularly suited to alloys where nuclear techniques (Chap. XI) do not work too well
Nuclear quadrupole moment	$C = bR\left(\frac{e^2qQ}{kT}\right)^2$	$kT \gg e^2qQ$	eq = electric field gradient at nucleus	b = 1/16, 7/200, 3/112, 11/480 for I = 3/2, 5/2, 7/2, 9/2 respectively b = 0 for I = 1/2 or 0.
Superconductivity	$C = ae^{-bT_c/T}$	$0 < T < T_c/2$	Energy gap of superconductor	Energy gap $\cong 3{\cdot}5\ kT_c$

α = coefficient of linear expansion
γ = Gruneisen constant
$\overline{H_{eff}}$ = effective magnetic field at nucleus
T_c = critical temperature
z = number at nearest neighbours

γ = electronic specific heat coefficient
μ = magnetic moment of nuclei
I = nuclear spin
S = total m_s quantum number
eQ = electric quadrupole moment
eq = electric field gradient

J = exchange energy or quantum number
M = atomic weight
R = 1·986 cal/mol/deg
Θ = Debye Θ, °K
T_m = melting temperature in °K.
ρ = density in g/cm³

TABLE XVIII

1. APPROXIMATE DEBYE Θ VALUES FOR THE ELEMENTS. 2. VALUES OF $\sqrt{(M)}\Theta$ FOR IDENTICAL CRYSTAL STRUCTURES GROUPED ACCORDING TO THEIR COLUMN IN THE PERIODIC TABLE. (M = ATOMIC WEIGHT). ALL VALUES IN °K

1. Θ (°K)

Element	Θ	Element	Θ	Element	Θ	Element	Θ	Element	Θ	Element	Θ	Element	Θ
Ar	85	Cd	175	Ge	360	α Mn	~380	Pd	290	(white) Sn	170	W	310
Ag	220	Ce	119	Gd	170	β Mn	~380	Pr	141	Sr	140	Y	214
Al	395	h.c.p. Co	390	Hf	200	f.c.c. Mn	~355	Pt	233	Ta	230	Zn	235
As	285	f.c.c. Co	385	Hg	100	b.c.c. Mn	~370	Rb	61	Te	180	h.c.p. Zr	260
Au	177	Cr	490	Ho	162	Mo	380	Re	275	Tb	170	b.c.c. Zr	212
B	1250	Cs	45	In	129	Na	150	Rh	370	Th	140		
Ba	110	Cu	315	Ir	285	Nb	250	Ru	400	Tm	167		
Be	920	Dy	157	K	100	Nd	145	Sb	210	h.c.p. Ti	365		
Bi	120	b.c.c. Fe	432	Li	360	Ne	63	Sc	~400	b.c.c. Ti	300		
C (diamond)	1860	f.c.c. Fe	420	La	132	Ni	390	Si	625	Tl	94		
C (graphite)	~400	Er	167	Lu	166	Os	250	Sm	150	U	160		
Ca	230	Ga	240	Mg	318	Pb	88	(grey) Sn	260	V	399		

2. $\sqrt{(M)}\Theta$ (°K)

Li	948	Be	2760	h.c.p. Ti	2520	Va	2380	Cr	3530	Co	2950	Ni	2990	Cu	2500	Zn	1900	C	6450	b.c.c. Ti	2070
Na	720	Mg	1562	h.c.p. Zr	2480	Nb	2140	Mo	3720	Rh	3740	Pd	2990	Ag	2280	Cd	1855	Si	3300	b.c.c. Zr	2030
K	626	Ca	1450	h.c.p. Hf	2670	Ta	3090	W	4190	Ir	3950	Pt	3250	Au	2480			Ge	3070		
Rb	562	Sr	1310															grey Sn	2840		
Cs	520	Ba	1290																		

TABLE XIX

TABLE OF γ VALUES IN 10^{-4} cal/mol/deg²

Li 4·18 Na 4·3 K 5	Va 21·4 Nb 20·3 Ta 16	Co 12·0 Rh 11·1 Ir 7·6	Cu 1·72 Ag 1·57 Au 1·76	La 24·1 Sm 31 Lu 23 Sc 25 Y 20
Be 0·54 Mg 3·23 Ca 2·91	Cr 3·68 Mo 5·12 W 2·86	Ni 13·0* Pd 22·2 Pt 15·3	Al 3·5 Ga 1·2 In 4·32 Tl 3·5	Th 12·0 U 26·0 αPu ~100
h.c.p. Ti 8·25 h.c.p. Zr 7·1 h.c.p. Hf 6·8 b.c.c. Ti 5·7 b.c.c. Zr 4·4	α Mn 25·3 β Mn 22·5† f.c.c. Mn 11·2 b.c.c. Mn 22·5 Re 5·5	b.c.c. Fe 12·0 f.c.c. Fe 8·0 Ru 8·0 Os 5·6	White Sn 4·35 Pb 7·5	Bi 0·15
			Zn 1·42 Cd 1·70 Hg 4·5	

metals and rare earths which have γ values of ~ 10 cal/mol/deg² and all the rest which have values ~ 1 cal/mol/deg². This is a quite crude division but does support our general ideas that the transition metals and rare earths have higher density of states at the Fermi level due to the outer d electrons. The remaining elements employ s and p electrons for bonding. A γ value of about 10 cal/mol/deg² in the metals corresponds to a density of states of about 1 level per eV per atom.

(c) *Magnetic Specific Heat*

Figure 103 shows the separated magnetic specific heat of gadolinium, a ferromagnetic rare earth metal with seven unpaired $4f$ electrons ($S = 7/2$, $L = 0$, $J = 7/2$). The total entropy agrees with the expected entropy $R \ln(2J + 1) = 4{\cdot}16$ e.u. and the total magnetic energy $W_{ex} = 800$ cal/mol yields a value of the exchange energy $J = 2{\cdot}1 \times 10^{-4}$ eV based on nearest neighbour interactions only ($W_{ex} = RzJS^2/k$). From low temperature magnetization measurements we obtain $J = 1{\cdot}4 \times 10^{-4}$ eV from eqn. (62), a large enough difference from the specific heat determination to indicate that the theory is oversimplified. If we substitute this latter value of J into eqn. (60), the spin wave specific heat, we obtain the dotted curve in Fig. 103, again in rather poor agreement with experiment. If we consider only nearest neighbours in the case of iron metal, we find the value of J obtained from the total magnetic

* Below 20°K Ni has a γ value of $\sim 17 \times 10^{-4}$ cal/mol/deg². The value quoted is the value above the Curie point, 360°C.

† Below ~ 50°K β Mn has a γ value of $\sim 120 \times 10^{-4}$ cal/mol/deg². The value in the table is the value at temperatures above 1000°K.

energy in the specific heat is a little more than half that obtained from low temperature saturation magnetization measurements (0·010 compared to ~0·018 eV). The probable reason for these discrepancies which are of opposite sign in gadolinium and iron is that the second nearest neighbour exchange energy is important. This point has already been raised to explain the presence of spirals in chromium and in the rare earths. The difference in sign between gadolinium and iron may be qualitatively explained by assuming the second nearest neighbour interactions to be antiferromagnetic and ferromagnetic, respectively.

Since the measured entropy agrees with the calculated entropy one can conclude that the value of gJ per atom is independent of temperature since

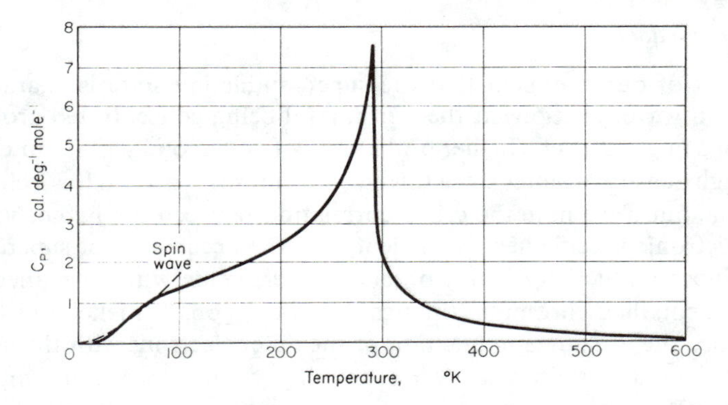

FIG. 103. The separated magnetic specific heat of Gd (solid curve) and the low temperature spin wave specific heat calculated with the exchange energy determined from low temperature saturation magnetization measurements.

the entropy measurement covers a very wide range of temperatures.* This suggests that there is very little change in the outer electron wave functions as one passes through the Curie temperature. This is supported by the fact that other physical properties such as Debye Θ, lattice parameter, elastic constants etc. do not change appreciably through the Curie temperature.

(*d*) *Order-Disorder Specific Heat*

There is very little data on this subject. β brass exhibits a pronounced lambda type anomaly at $T_c = 733$°K. Both the entropy, $R \ln 2$, and energy, $\sim RT_c/2$, agree reasonably well with theoretical calculations based on nearest neighbour interactions. One can only observe the entire order-disorder specific heat for alloys for which $T_c > 300$°C due to the requirement that equilibrium be reached in a reasonable time.

* This does not appear to be so for antiferromagnetic chromium and α manganese.

(*e*) *Nuclear Specific Heat (spin and quadrupole)*

In the iron-cobalt system the room temperature average value of H_{eff} varies smoothly from -333 kgauss in iron to about -217 kgauss in cobalt *with no marked changes at the phase boundaries, b.c.c., f.c.c., h.c.p.* This suggests little change in electronic structure for different crystal structures since H_{eff} is a sensitive measure of the value of the unpaired s wave functions at the nucleus. Of course the energy differences of these phases is so small this may not be surprising.

The quadrupole interaction has not been observed in a sufficient number of cases to offer any comments.

(*f*) *Superconductivity*

In light of our comment that the superconducting state is characterized by electron wave functions at the Fermi level being concentrated around the atoms, one might expect a higher probability of superconductivity in elements with a high density of states since this suggests more d or p and less s character. It is borne out if we note the crude correlation between the higher values of γ, the electronic specific heat coefficient, and the occurrence of superconductivity. Superconductivity is not observed in elements with permanent magnetic moments like chromium, manganese, iron, cobalt, nickel and the rare earths since the exchange interaction of the outer electrons with the unpaired electrons would destroy the perfect pairing of momenta and m_s values (cf. chap. VIII).

2. Phonon Transport

Experimental Measurement of Thermal Conductivity

Specific measurements of the motion of phonons along a sample are divided into a low and a high frequency range solely as a result of the practical problem of introducing the phonons. In the high frequency range $\nu \sim 10^{13}$ cps one end of a bar is heated and the progress of these excited phonons is followed down the bar. Since the mean free path between collisions of high frequency phonons is only a few Å they make many collisions as they move along the sample and we measure their net drift (called *thermal conductivity*) just as we measure the net drift of electrons in electrical conductivity. In the low frequency *ultrasonic* range ($\nu \sim 10^{10}$ cps) the mean free path is of the order of centimeters or longer in pure materials enabling us to follow individual phonons as they traverse the sample. In addition it is possible to produce monochromatic low frequency phonons but we are limited to a broad range of frequencies at high frequency, although the use of "reststrahlen" may alter this in the latter case. We shall discuss these two ranges separately.

(a) Thermal Conductivity

There are several techniques for measuring thermal conductivity but we shall only describe two which appear to be representative of the approach. The first is a dynamical method.[55] A long thin rod of the metal ($\sim$3 mm diameter by 500 mm length) is heated at one end with a sinusoidally varying heat pulse. This is accomplished by heating a wire with a current source that varies sinusoidally. Further along the rod are two thermocouples spaced at some convenient separation like 100 mm and a moving chart records the oscillating temperature versus time curve for the two. The entire specimen and heat source are placed in a furnace so that thermal conductivity can be measured at several temperatures. If one measures the velocity with which the individual heat oscillations pass from the first thermocouple to the second and the amplitude of these temperature oscillations which decrease as we move away from the heat source we have the *thermal diffusivity*, κ, from

$$\kappa = Lv/2 \ln q = \text{thermal diffusivity} \tag{188}$$

where L is the distance between thermocouples, v the velocity of the pulses and q the ratio of the temperature amplitudes. The thermal conductivity K is given by

$$\text{K} = \kappa C \rho A = \text{thermal conductivity} \tag{189}$$

where ρ is the density, C the specific heat and A the atomic weight. K has the units cal/cm sec deg and κ the units cm^2/sec. The advantage of this technique is that one is not bothered by radiation losses and the disadvantage is that one must know C in order to determine K. Typical values are $\kappa = 1$ cm^2/sec and K $= 0{\cdot}98$ cal/sec cm deg for copper at room temperature.

The second method is static in that a steady state is achieved. A sketch of this method is seen in Fig. 104. Heat is introduced at one end of a long rod of uniform cross-sectional area, A, at a rate $\mathrm{d}Q/\mathrm{d}t$ (= power) and the steady state temperature difference ΔT of two thermocouples a distance L apart, is measured. The other end of the rod is attached to a heat sink, a good massive conductor whose temperature rise is negligible with the introduction of the power $\mathrm{d}Q/\mathrm{d}t$. The thermal conductivity is

$$\text{K} = \frac{L}{A}(\mathrm{d}Q/\mathrm{d}t)/\Delta T \tag{190}$$

The major error is due to heat loss to the surroundings, but this can be minimized by using a vacuum furnace, keeping ΔT as small as possible and employing auxiliary heating coils to match the thermal gradient in the sample so as to eliminate radiation losses. In spite of these precautions there are still losses and it is best to measure K for various values of $\mathrm{d}Q/\mathrm{d}t$ and extrapolate to $\mathrm{d}Q/\mathrm{d}t = 0$.

Both dynamic and static techniques show a practical error of about 1% although demands for higher accuracy have not been necessary.

The transport of phonons in a lattice is governed by two mechanisms, collisions with conduction electrons in metals and collisions with other phonons. In the first the phonon is absorbed by the conduction electron which rapidly carries this energy along the crystal and in the second the phonon collides with another phonon. The two phonons destroy each other and create two new phonons. In a metal the electron process is primarily responsible for the transport of energy since it moves so much faster than phonons. Since the transport of electrons is governed by the resistivity of a metal there is a relationship in the free electron theory between the thermal conductivity due to electrons and the electrical resistivity, ρ,

$$K_e = \frac{T}{\rho}\left(\frac{k}{e}\right)^2 = \frac{T}{\rho}(0{\cdot}536 \times 10^{-8}) \tag{191}$$

where K_e is in cal/sec cm deg and ρ is in micro ohm cm. This relationship is called the *Wiedemann–Franz ratio* and $K_e\rho/T$ is called the *Lorenz number*.

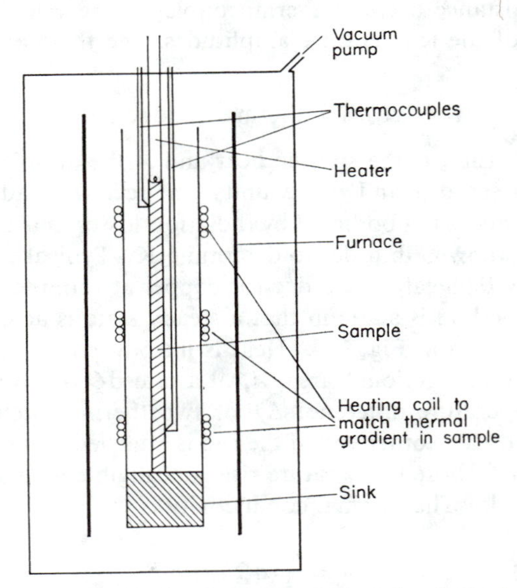

FIG. 104. A schematic sketch of a static method for determining the thermal conductivity of a rod-like sample.

Experimentally the Lorenz number is surprisingly close to $0{\cdot}536 \times 10^{-8}$ but in any event it is reasonably constant with temperature. Thus in a metal the introduction of phonons or heat at one end results in exciting the conduction electrons at that end which move rapidly along the metal exciting phonons. This is why a metal is a good thermal conductor and a piece of wood is not.

Thermal Conductivity of Copper and Nickel. Perhaps, though, a few illustrative examples might be illuminating. Firstly let us look at the thermal conductivity of copper, nickel and a copper-nickel alloy from 0°C to ~650°C. In this range $T \geqslant \Theta$ for pure copper so that ρ is proportional to T and we expect K_e in eqn. (191) to be constant. This can be seen to be the case in Fig. 105. For pure nickel ρ rises faster than T up to the Curie temperature and is then practically constant since the magnetic resistivity is such a large part of the resistivity of nickel (see p. 236). From eqn. (191) we expect K_e to decrease with T up to the Curie temperature at 360°C and then to reverse

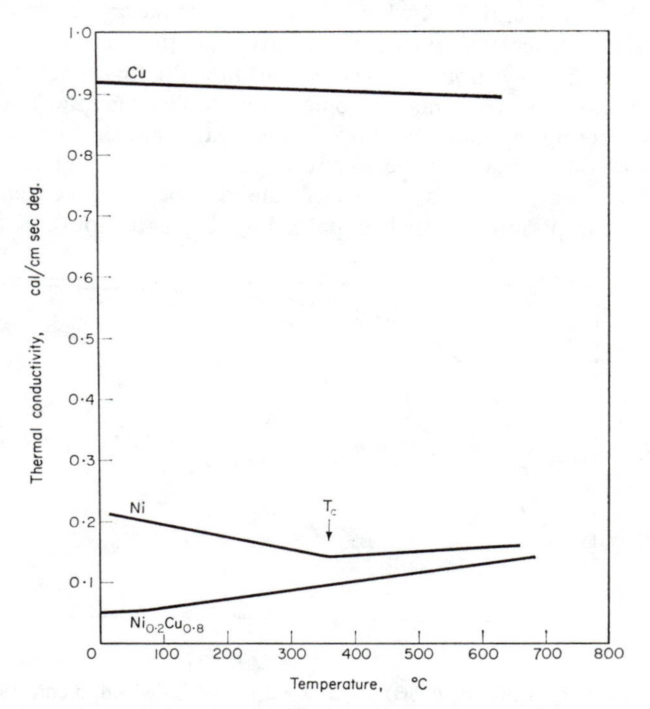

FIG. 105. The thermal conductivity of copper, nickel and $Ni_{0{\cdot}2}Cu_{0{\cdot}8}$ in cal/cm sec deg as a function of temperature in degrees centrigrade.

itself and increase with T beyond! This is also seen in Fig. 105. For the alloy $Ni_{0{\cdot}2}Cu_{0{\cdot}3}$ the resistivity consists of a large alloy contribution, ρ_i, and a term linear in T due to lattice vibrations. The net effect is to expect K_e to increase with temperature, as it does in Fig. 105.

At temperatures below 4°K the temperature independent impurity resistance predominates in most materials and we expect K_e to be proportional to T. Experimentally this is found to be the case and generally includes agreement with the coefficient in the free electron theory ($0{\cdot}536 \times 10^{-8}$).

In the intermediate range $\Theta/3 < T < \Theta$ the theory is poor but it is found empirically that the thermal conductivity in metallic elements is related to the limiting constant value of K_e at high temperatures by

$$K_e(\infty)/K_e(T) = 2(T/\Theta)^2 \int_0^{\Theta/T} \frac{x^3\,dx}{(e^x - 1)(1 - e^{-x})} = J(T/\Theta) \qquad (192)$$

where $J(T/\Theta)$ is plotted in Fig. 106.

Thermal Conductivity at Low Temperatures. In the low temperature region between about 4°K and 20°K where ρ_T begins to make an appreciable contribution to the electrical resistivity the thermal conductivity is extremely sensitive to impurities, imperfections and cold working. In this temperature range phonon-phonon scattering may become comparable or greater than phonon-electron scattering especially in alloys. This means that the heat is carried by the phonons rather than via the electrons. This can occur in alloys because the mean free path of conduction electrons can be shortened considerably more than the phonon mean free path. Fig. 107 is a plot of K for various

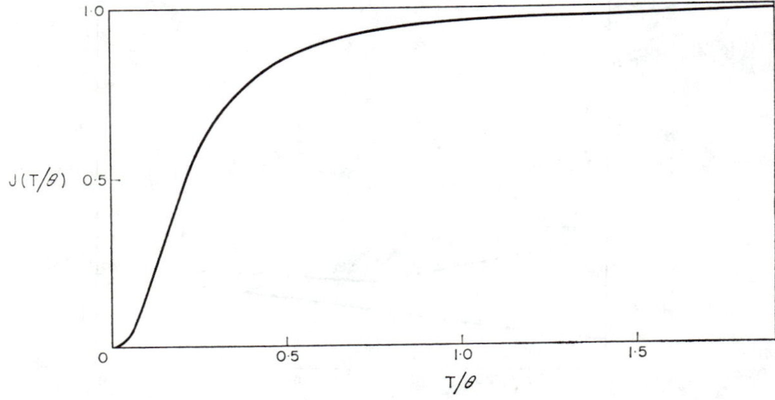

FIG. 106. A plot of $J(T/\Theta)$ versus T/Θ. $J(T/\Theta)$ is defined in eqn. 192.

samples of copper[(50)] and reveals the extreme sensitivity. We cannot assess the various contributions to the thermal conductivity in this region (imperfections, impurities, phonon-phonon etc.) since we do not have an accurate independent measurement of the precise state of these samples. In fact it appears that *thermal conductivity in this temperature range is the most sensitive measurement in revealing subtle differences in sample perfection* and will ultimately be a useful tool. In the interim a concerted experimental and theoretical effort to separate the various contributions to the thermal conductivity is necessary in order to make sense out of the measurements.

Theoretically the problem of phonon-phonon scattering is quite complicated. The process relies on the presence of anharmonic terms in the interaction between atoms. If the forces were harmonic there would be no phonon-phonon scattering. When the experimental problem is solved for producing and following monochromatic high frequency phonons we shall learn a good deal about anharmonic forces. In the interim the theoretical problem is one of the most difficult in solid state physics.

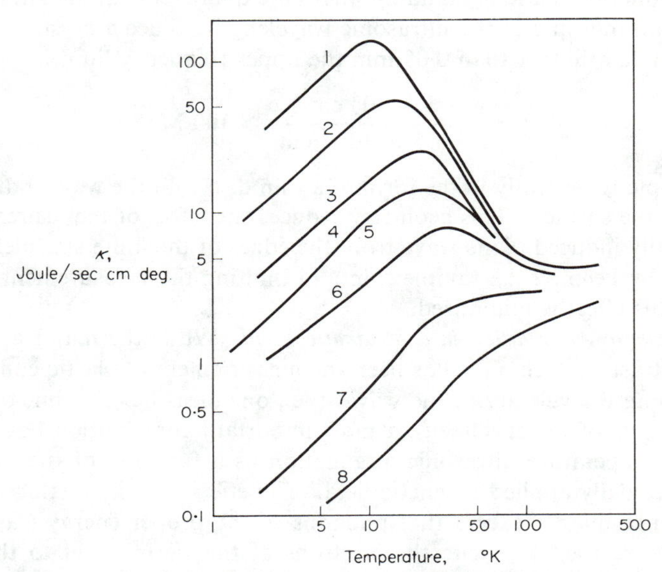

FIG. 107. The thermal conductivity in joules/sec cm deg of various samples of copper metal showing the extreme sensitivity in the temperature range ~Θ/30.

1	Cu	99·999
2	Cu	99·999
3	Cu	99·98
4	Cu (2) Cold Worked	
5	Cu	99_t9
6	Cu + 0·6% Te	
7	Cu + 0·056% Fe	
8	Cu + 0·1% P	

(*b*) *Ultrasonic Attenuation*

Ultrasonic Experimental Technique. In this measurement transverse and/or longitudinal sound waves in frequency ranges up to several hundred megacycles/sec are generated in a piezoelectric crystal-like quartz. An alternating electric field applied to the quartz causes it to expand and contract with the direction and frequency of the applied electric field. This produces forced vibrations in the quartz which can be transmitted to a specimen that is brought into contact with it. A second quartz crystal, converting these vibrations back into electric pulses, can be attached to the other end of the specimen and used as a detector but more often the same quartz crystal is used as source and detector by analyzing the vibrations that are reflected back from the end of the specimen. The contact between the quartz and the specimen is made with a thin fluid layer of special material that does not distort the signal. The problem of high temperature contact fluids has not been adequately solved ($T > 1000°K$). Ultrasonic attenuation is measured from the amplitudes of successive echoes by allowing the wave to make many traverses. In order to create monochromatic or standing waves the quartz crystal thickness is made an integral multiple of the ultrasonic wavelength. Since a crystal cannot be ground much thinner than 0·05 mm the upper frequency limit is

$$\nu = \frac{v}{\lambda} = \frac{5 \times 10^5 \text{ cm/sec}}{5 \times 10^{-3} \text{ cm}} \cong 100 \text{ Mc/sec}$$

The sample is generally in the form of a thin disc with the wave induced normally to the surface. This geometry reduces the effect of the "spreading" of the initially induced plane wave from the edges of the finite size piezoelectric crystal. By keeping the sample thin and limiting the measurement to a few echoes this effect is minimized.

Ultrasonic Attenuation in a Magnetic Field. While the major application of the ultrasonic technique has been the measurement of elastic constants by determining the velocity of the waves (i.e., one measures the time to traverse the thickness of the specimen), a most important contribution has arisen in the low temperature ultrasonic attenuation as a function of the magnitude of an externally applied magnetic field. The effect is quite analogous to the cyclotron resonance except that phonons of the proper energy (rather than photons) are used to excite the electrons at the Fermi level to the higher quantized levels. Both phonons and photons can give up energy $h\nu$, so that both experiments are performed in the range of 10–100 Mc/sec. In the ultrasonic attenuation measurement a single crystal is cut along a principal plane (like 110 in f.c.c.) and cooled to 4°K. The ultrasonic wave is introduced normally to the surface and the external magnetic field is at right angles to the direction of ultrasonic propagation. By varying the magnitude and direction of the magnetic field, H, a periodicity in $1/H$ is found due to the excitation of electrons at the Fermi surface whose direction of propagation is normal to both magnetic field direction and ultrasonic propagation direction. The wave number of these electrons is

$$k_F = \frac{e}{2cP} \tag{193}$$

where P is the period in $1/H\lambda$ and λ is the ultrasonic wave length. Oscillations due to the "neck" have been observed[56] in copper, silver and gold and provide the fifth technique confirming the contact of the Fermi surface with the 111 Brillouin zone faces, Fig. 90.

This measurement has the advantage over cyclotron resonance in that the ultrasonic wave penetrates the entire volume of the specimen. In the cyclotron resonance one is limited to about 100 Å, the depth of penetration of the microwaves.

3. Heat and Entropy of Alloy Formation

Thermodynamics and Cohesion of Alloys

Metallurgists are interested in alloy formation since pure elements rarely find commercial applicability (except electrical conductors and aluminium foil for the kitchen). The physical properties of these alloys have had an enormous effect on our economy and over the years metallurgical empiricism has produced very desirable alloys with no help from the theoretical solid state physicist. Nor has the vast bulk of information concerning physical properties of alloys enabled the physicist to glean any more than a qualitative understanding of the electronic structure of metals.

We have already discussed the cohesive energy of some of the elements in Chapter II. There is little further we can add at this point except to underscore the qualitative observation that the more orbitals available for bonding the higher the cohesive energy. For most elements we cannot calculate the cohesive energy to better than about a factor of two. However, since the metallurgist is more interested in alloys what are some of the qualitative comments we can make about alloys and the thermodynamic measurements made to understand their behaviour?

Firstly, *there are no reliable rules that enable us to predict what shall emerge if we mix various elements together except that elements with similar free atom outer electron configurations tend to favour alloying as compared to ones which are dissimilar*. Thus virtually any two transition elements either form a solid solution or an intermetallic compound, and any two rare earth metals will form a solid solution. On the other hand we do not understand why copper and silver are immiscible while copper and gold, and silver and gold form solid solutions. Even the system copper-nickel, considered a classic example of a continuous solid solution, thermodynamically prefers to segregate into copper rich and nickel rich phases at absolute zero. (Of course it does not do so because the atomic diffusion necessary to accomplish this ceases.)

Secondly, in most cases the energy favouring alloying is only a small fraction of the cohesive energy of the solid. For example the cohesive energy of pure gold or pure copper is about 80 kilocalories per mole (~ 3 eV per atom) while the energy is lowered only an additional 1·1 kilocalories per mole in forming the disordered Cu_3Au alloy and another half per cent in the ordered state. (These small energy differences have also been found in the case of most allotropic transformations.) This gives us a fair idea of the capabilities of predicting or calculating these energy differences since we are generally unable to calculate the cohesive energy of metals to any precision. *The energies of alloy formation are of the order of* 1000 *cal/mol and are one to two orders of magnitude smaller than the errors in any fundamental calculation.*

Thirdly, and in spite of the second point, one can still make *estimates* of differences in energy even though the energies themselves cannot be calculated provided we have some clues as to how the wave functions are altered on alloying. *Even though thermodynamics gives us these clues in measurements of heat formation, entropy of formation, Debye* Θ, *elastic constants and lattice parameters of alloys, very little experimental work has been done in this respect except possibly for lattice parameters.*

Hume-Rothery Rules. We should like at this point to mention the *Hume-Rothery rules.* These rules emerged from an attempt by Hume-Rothery to find some systematic rules for alloy formation. He noticed, for example, that solid solubility was less likely the greater the disparity in atomic volume for the two elements. He also noted that in several alloys of copper, silver and gold the beta brass (CsCl) structure seemed to occur at an average outer electron per atom ratio of 1·5. For example in copper-zinc, copper has one outer electron ($4s$) and zinc has two ($4s^2$) so that the average electron per atom ratio of 1·5 occurs at 50–50. This is approximately where beta brass appears. It was further suggested that this might arise because the Fermi surface of α brass grew larger relative to the Brillouin zone as zinc was added to copper and that the Fermi surface just made contact with the 111 zone faces at the 50–50 composition. It then became energetically more favourable to form beta brass since the Brillouin zone was larger and the Fermi surface did not touch the zone. A few other structures (like γ brass) appeared at other electron-atom ratios.

The Hume-Rothery rule about atomic sizes is a good rule to remember although there are occasional exceptions. The rule about electron per atom ratios, unfortunately has had very limited applicability being confined principally to noble metal alloys. In addition we now know that the Fermi surface makes contact in pure copper so that the model of a Fermi surface just making contact at 50–50 in α brass cannot be so simply applied. Actually the Hume-Rothery rules have been around so long that one might think that if only one knew how to apply the rules one could work out most phase diagrams. Such of course is not the case and phase diagrams are still entirely empirical. Nonetheless, such attempts to systematize our knowledge of alloys are always worthwhile for they provide part of our *modus operandi* toward making progress. For example more recently P. Beck and his co-workers have attempted to systematize the appearance of σ phase in transition metal alloys and considerable effort has been made in identifying laves and other phases. Alas! the metallurgist and crystallographer are creating new phases at a faster rate than the physicist is progressing in solving the Schrödinger equation accurately enough to predict these phases.

Thermodynamics of Copper-Gold and Gold-Platinum Alloys. Let us return to our original discussion by considering what we know or can guess about the specific alloy systems copper-gold and platinum-gold. Neither of these form genuinely new phases although the copper-gold system forms an

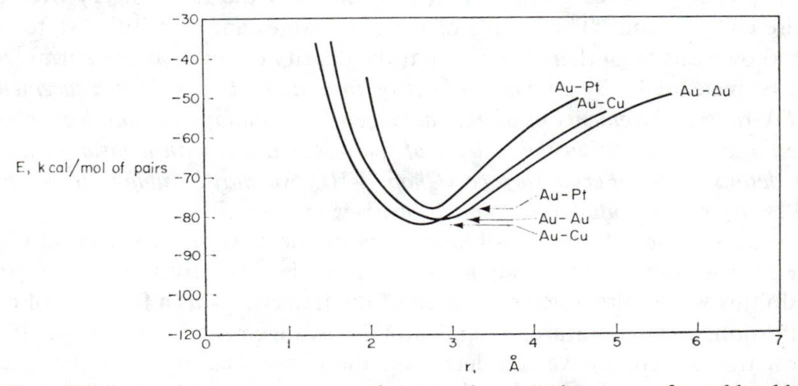

FIG. 108. Approximate energy versus interatomic separation curves for gold-gold, gold-platinum and gold-copper pairs. In both alloys the nearest neighbour distance is shorter than in pure gold. The addition of platinum to gold raises the energy minimum while the addition of copper lowers it. The spring constants, determined by the curvature at the minima, are approximately unaltered with the addition of copper to gold but are increased with the addition of platinum.

ordered f.c.c. alloy and an ordered face-centred tetragonal alloy and gold-platinum forms segregated f.c.c. gold- and platinum-rich phases. At absolute zero in gold-rich alloys the replacement of gold atoms by copper atoms or platinum atoms requires an adjustment of the gold wave functions in the vicinity of the substitutional atom. We do not know how the wave functions are changed but presumably the change is small since the additional heat of formation of the alloy is small compared to the cohesive energy of pure gold. The addition of copper lowers the energy and the addition of platinum raises the energy. Furthermore the shape of the energy versus interatomic separation curve may be altered which may change the spring constants between the atoms. From X-ray measurements of the Debye Θ in Cu_3Au it appears that the spring constant between copper and gold atoms is about the same as in pure copper or pure gold (see values of $\sqrt{(M)}\Theta$ in Table XVIII) while it increases somewhat for gold-platinum. This latter point can be inferred from the smooth variation with composition of the melting point of the alloy and the Lindemann formula for the Debye Θ. From lattice parameter measurements we know there is a negative deviation from Vegard's law so that the minimum of the energy versus interatomic separation occurs at smaller lattice spacings. From the heat of formation of copper-gold and from an estimate obtained from the consolute point* in the gold-platinum phase diagram the energy of the alloy is lowered about $\sim$1000 cal/mol of copper-gold pairs and raised $\sim$3000 cal/mol of gold-platinum pairs. A qualitative sketch of the gold-gold, gold-copper and gold-platinum energy versus interatomic separation curves is shown in Fig. 108. Of course we may also expect similar but smaller effects from second nearest neighbours, etc.

The energy curves given in Fig. 108 represent the total energy over the entire energy band. The details of the eigenvalues and density of states are not known although there is hope that the density of states at the Fermi level can be measured. *By combining heat of formation, Debye Θ measurements, and X-ray measurements of atomic arrangements in alloys we can draw curves like those in Fig. 108 in the vicinity of the minima, and with measurements of the details of the Fermi surface (Chap. VIII) we may be able to understand the changes in eigenfunctions and eigenvalues in alloys.*

Since it is not always possible to measure the heat of formation at absolute zero we rely on thermodynamics to provide this information. In order to do this we require a determination of the free energy as a function of concentration, c, temperature, T, and atomic arrangement in the alloy. If we know the free energy we can determine the phase diagram since the system always attempts to minimize the free energy. Rewriting the free energy, eqn. (63), we have

$$G = H_0 - TS_0 + H(T) - TS(T) = H_0 - TS_0 + \int_0^T C\,\mathrm{d}T - T\int_0^T \frac{C}{T}\,\mathrm{d}T \tag{194}$$

where H_0 is the energy of the alloy at absolute zero, and S_0 is the entropy of mixing and depends on the atomic arrangement. If the alloy is completely ordered S_0 is zero and if completely disordered

$$S_0 = -R[c\ln c + (1-c)\ln(1-c)] \tag{195}$$

It is generally quite difficult to determine the entropy of intermediate atomic arrangements. The energy of the alloy can be determined relative to the pure elements by measuring the heat of formation, and this is generally sufficient since the energy of the pure elements can be treated as an additive constant. Thus if one measures the heat of formation (at some temperature T) and the atomic arrangement both as a function of concentration and then measures the specific heat from $T = 0$°K to T *for each atomic arrangement and concentration* we would be able to determine the phase diagram from eqn. (194). Obviously it is impossible to force an alloy into any atomic arrangement one desires so we must rely on fewer measurements and critical judgement to extract the desired information. *In fact we invariably work backwards by using the phase diagram to tell us about the atomic arrangement.*

* Temperature of the maximum in the miscibility gap.

Experimental Techniques

Before proceeding with a description of the practical methods of extracting our desired information we shall describe several methods which yield helpful information. These measurements are: (*a*) Vapour pressure, (*b*) Electrolytic cell, (*c*) Heat of Mixing.

(*a*) *Vapour Pressure*

There have been several methods employed to measure the vapour pressure of one or all of the constituents of an alloy but we shall limit our discussion to the dew point and effusion methods.

In the dew point method the alloy specimen is maintained at some temperature T_2 at one end of a long tube, the other end being at some variable temperature T_1 (lower than T_2). As the more volatile constituent of the alloy passes down the tube it will condense in its pure state if T_1 is low enough. One adjusts T_1 so that the vapour just condenses. From a separate measurement of vapour pressure versus temperature for the volatile constituent in its pure state we know the vapour pressure at T_1 which is then just equal to the vapour pressure in the alloy at T_2. This technique is generally only useful for low melting and boiling point metals like zinc, cadmium, mercury, tin etc., alloyed with high melting point metals but does yield good results.

The effusion method is more generally applicable but a less tested technique. The vapour pressure, as a function of T, is directly measured by comparing the rate of flow through a small orifice of the vapourized element in the pure metal and in the alloy. The best detector is a mass spectograph and this requires ionizing the vapour as it passes through the orifice. This technique has the advantage that all constituents of an alloy can be measured, since the mass spectrograph separates them. Furthermore one only requires the change in pressure between the alloy and the pure elements, and an absolute calibration of vapour pressure versus temperature is not necessary for the pure elements. (It would be difficult to convert the intensity of the ionized atoms measured by the mass spectrograph to an absolute pressure.) Another advantage of the technique is the low vapour pressure required since the mass spectrograph is very sensitive, but the temperature must be high enough in the alloy so that the vapour is always in equilibrium with the bulk of the specimen not just the surface. This requires reasonable diffusion rates so that an atom can move from the centre of the specimen to the surface in in a time comparable to the time required for the measurement.

The ratio of the vapour pressure of the element in the pure metal to the vapour pressure in the alloy at some temperature T is called the *activity*, a. The activity divided by the concentration, c, is called the *activity coefficient*,

$\gamma = a/c$. The *partial molar free energy* is the change in free energy between a mole of atoms of an element in its pure state and in an alloy. This is given as

$$\Delta G = RT \ln a \tag{196}$$

Thus a knowledge of the free energy as a function of T for the pure elements and a measurement of a as a function of T for all constituents of an alloy enables one to determine the free energy of an alloy from the weighted sum of the partial molar free energies. If one only measures the activity of one of the constituents over a limited range of concentrations one must make an educated guess of the activity of the other in order to determine the free energy of the alloy. From the temperature dependence of ΔG we can determine the partial molar entropy, $\Delta S = \Delta G/\Delta T$.

(*b*) *Electrolytic Cell*

In this method the element in pure metal form and the alloy are made electrodes in a cell with the electrolyte a salt of the pure metal. For example Zn, ZnSn and $ZnCl_2$ permit a flow of Zn^{++} ions with the application of an external d.c. voltage. The cell must be reversible so that a reversal of the external voltage returns the cell to its original state. If one measures the potential, ε, between the electrodes at some temperature T then the partial molar free energy is given as (for example, Zn in ZnSn)

$$\Delta G = 23{,}060\varepsilon v \quad \text{cal/mole} \tag{197}$$

where v is the valence of the element in the salt (two in the case of $ZnCl_2$). This technique has had extremely limited success since it is difficult to find reversible cells. Furthermore we are not certain about the influence of the surface conditions of the electrodes in creating spurious results. More work is required to gauge the reliability of this method.

(*c*) *Heat of Formation (mixing)*

In this technique the alloy and pure elements are separately dissolved in some liquid so that the final state is the same. For elements other than transition metals and rare earths ~400 g of liquid tin[57][58] at ~350°C can be used. About 10–15 g of the pure elements at temperature T are separately dissolved in the tin and their heats of solution measured from the temperature rise or fall of the tin-bath. This procedure is repeated for the alloy and the difference per mole in the heats of solution is the heat of formation per mole. For transition metals and rare earths a room temperature acid solution may be feasible, if the metals dissolve readily, and if the heat of formation is greater than ~1000 cal/mol. In most cases the heats of formation are ~1000 cal/mol.

Gold-Platinum. Let us now attempt to analyze the available data on gold-platinum and gold-copper to see what can be deduced. Fig. 109 is the approximate phase diagram for gold-platinum, the phase boundaries below 500°K being extrapolated from the high temperature data. At all temperatures below about 1550°K there are concentrations which segregate into two f.c.c. structures, one gold-rich and one platinum-rich. For example at 1000°K the

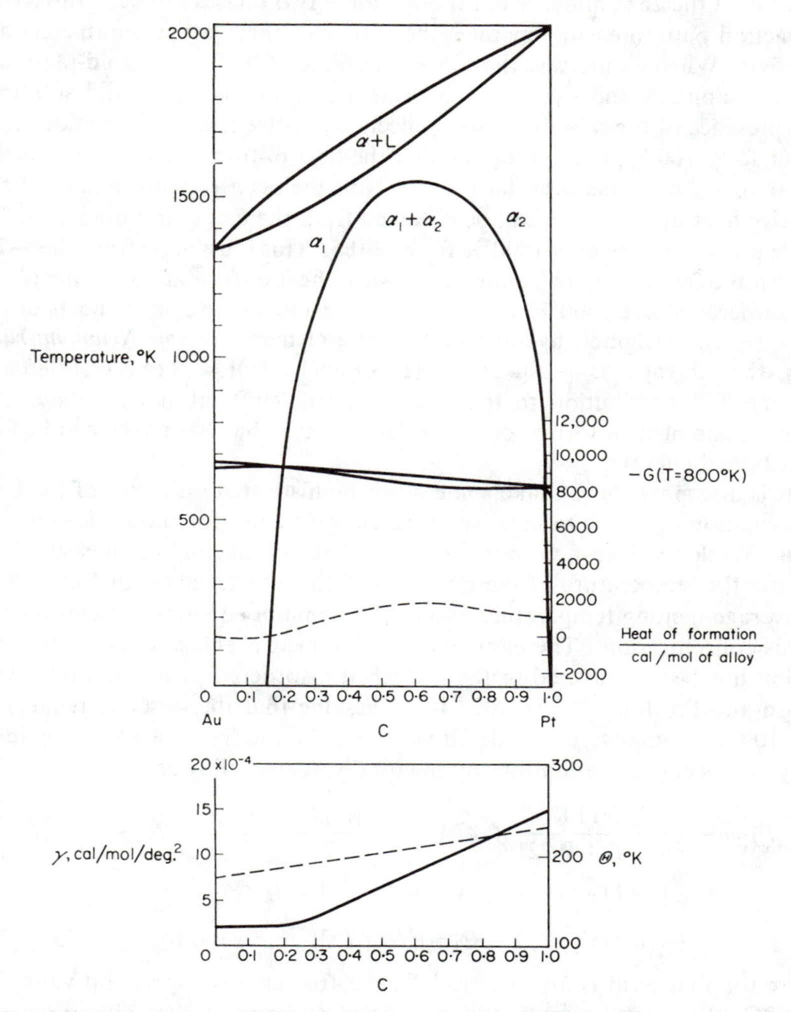

FIG. 109. *Upper curve*—the phase diagram of the gold-platinum system showing the miscibility gap between the two f.c.c. phases α_1 and α_2. The dashed curve is the estimated heat of formation of the alloy at $T = 0$°K. The solid curve with the two maxima is the negative of the free energy curve for the disordered alloy at $T = 800$°K. The common tangent determines the boundaries of the miscibility gap. *Lower curve*—the measured electronic specific heat coefficient, γ, in cal/mol/deg² versus concentration for the disordered alloys (solid curve). The dashed curve is the estimated Debye Θ of the disordered alloy determined from the Lindemann melting temperature formula.

alloy forms a continuous solid solution from $c = 0$ to $c = 0{\cdot}26$ of platinum. Further increases in alloy concentration are reflected in the build-up of a second f.c.c. phase containing 96% platinum and with smaller lattice parameter than the 26% alloy. Even though these two phases are f.c.c. the X-ray diffraction pattern easily separates the two since their lattice parameters are different. When we increase the concentration to 96% the 26% gold-platinum phase disappears and the last 4% forms a continuous f.c.c. solid solution. The presence of the miscibility gap reflects a positive heat of formation from about 15%–100% platinum but neither the heat of formation nor the partial molar free energy has been measured! Nonetheless a rough estimate of the positive heat of formation can be obtained from the fact that a homogeneous single phase is stable at $\sim 1500°K$ for $c = 0{\cdot}5$. This must arise from the $-TS$ term in the free energy balancing the positive heat of formation. If the phase is disordered above 1500°K and the electronic and lattice specific heats of the alloy are the weighted average of the pure elements (*Kopp–Neumann rule*) then the entropy is $-R[c \ln c + (1 - c)\ln(1 - c)] = R \ln 2 = 1{\cdot}386$ e.u. and the TS contribution to the free energy is 2100 cal/mol of alloy. An approximate heat of formation curve for $T = 0°K$ has been sketched in Fig. 109 (dashed curve).

It is also possible to make some more sophisticated estimates of the heat of formation at absolute zero by using Debye theory for the lattice specific heats. While the Debye Θ of the disordered alloys has not been measured we can use the temperature of the midpoint of the $\alpha + L$ curves in Fig. 109 as an average melting temperature and the Lindemann equation to determine Θ versus concentration. The electronic specific heat coefficient versus concentration has been measured in the quenched disordered phase by Budworth, Hoare and Preston,[59] Fig. 109. If we assume that the miscible regions in Fig. 109 are completely disordered we can write the free energy per mole of alloy versus concentration for any binary disordered alloy as

$$G(c, T) = -\frac{5{\cdot}11T}{\Theta(c)/T + 0{\cdot}221} + \tfrac{9}{8}R\Theta(c) - \frac{\gamma(c)T^2}{2} + RT[c \ln c + (1 - c)\ln(1 - c)] + H_0(c) - 6\alpha(c)\{[3T/2 - \Theta(c)]^2 + \tfrac{3}{4}T^2\} \quad \text{valid for } T > 2\Theta \qquad (198)$$

where the first term is an empirical Debye free energy expression valid for $T > 2\Theta$, the second term is the zero point energy of lattice vibrations, the third term is the free energy due to the electronic excitations, the fourth term is the free energy of mixing, the fifth term is the heat of formation at absolute zero and the sixth term is an approximate free energy term due to anharmonicity (α is the coefficient of linear expansion). Thus by estimating $\Theta(c)$ from the melting temperature, and from the measured values of $\gamma(c)$ and $\alpha(c)$ we have all the quantities in eqn. (198) except $G(c, T)$ and $H_0(c)$. However we do have the phase diagram and can vary the temperature-independent $H_0(c)$ until $G(c, T)$ reproduces the phase diagram. At all values of T below 1550°K a plot of G versus c yields two minima which determine the phase boundaries from their common tangents such as that at 800°K given in Fig. 109.

What can we say about the electronic structure of gold-platinum alloys from this analysis? Firstly, both the reversal in sign of H_0 at ~15% platinum and the upturn in the γ values at the same concentration suggests that the 5*d* states are passing through the Fermi level at this concentration. In pure gold as in pure copper the *d* band is several volts below the Fermi level but in platinum there are *d* states at the Fermi level. The high γ value in platinum and the low γ value in gold support this conclusion. It is natural that as one varies the alloy concentration some gradual transition of the top of the *d* band relative to the Fermi level should occur.

Copper-Gold. In the copper-gold system the preference is towards ordering and this occurs at Cu_3Au, Au_3Cu and CuAu although there is partial order at other compositions. The crystal structures of both Cu_3Au and Au_3Cu are identical, Fig. 43. In copper-gold there are sheets of copper and gold along the cube axes and since there is a preference for opposite nearest neighbours the sheets are closer together than the atoms in the sheets. As a result the f.c.c. crystal structure distorts into face-centred tetragonal with $c/a = 0{\cdot}925$.

The heat of formation has been measured[57][60] and this is shown in Fig. 110. The energy difference between the ordered and disordered states is

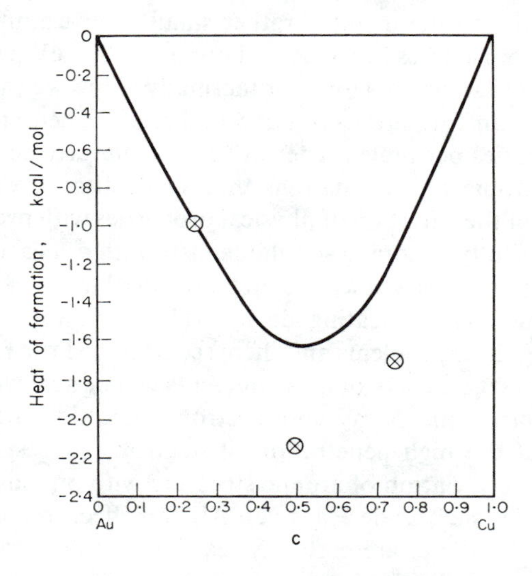

FIG. 110. The measured heat of formation of disordered Cu-Au alloys (solid curve) and ordered Cu-Au, $Cu\text{-}Au_3$, Cu_3Au (crosses). The Au_3Cu alloy was not completely ordered but nevertheless reflects an asymmetry in the ordering energy.

~500 cal/mole for copper-gold ($T_c = 683°K$), and 526 cal/mol for Cu_3Au ($T_c = 633°K$) and quite small for Au_3Cu ($T_c = 472°K$). X-ray measurements of the Debye Θ indicate very little change in the spring constants between ordered and disordered Cu_3Au. The major effect in the alloy seems to be a lowering of energy without a change in curvature (see Fig. 108). The oddity in this system is the asymmetry in the preference for ordering, in that Cu_3Au is more preferred than Au_3Cu. This may be due to strain effects since the gold atom is larger than the copper atom and the strain energy is greater putting one gold atom in a copper lattice than one copper atom in a gold lattice. In the ordered states Cu_3Au and Au_3Cu, the strain energy is reduced but presumably more in Cu_3Au. This system is one of the few with considerable experimental information but even here the thermodynamic data is not complete.*

In summary, qualitative conclusions about the electron structure of alloys can be gained from measurements of heats of formation, Debye Θ values, electronic specific heat coefficients, atomic arrangements, etc., but very little data is available, particularly for transition metals. *In fact the author knows of no alloy system for which all these measurements have been performed.*

4. Pressure Studies

Pressure Measurements

The variation of the physical properties of metals as a function of pressure is a rather young and relatively unexplored field. The range of energies that can be introduced into the metal is rather small. For example compressing iron to 100,000 atmospheres introduces about 5×10^{-2} eV per atom whereas one can introduce 10 times that amount thermally. Thus we can only expect to induce allotropic transformations if the second phase is denser and less than a few thousand calories per mole higher in energy. Nevertheless there are probably many allotropic transformations that await discovery under pressure.

In the matter of the variation of physical properties with pressure the major obstacle is the difficulty of coping simultaneously with a measuring device and a pressure vessel particularly since the pressure vessel tends to be considerably more cumbersome than a heating device. However, there has been little effort in attacking these problems and there seems to be little reason at present to doubt that as great a variety of measurements will be achieved as in the field of temperature variation. X-ray and electron diffraction measurements are quite difficult but the high penetration of neutrons suggests this technique for determining the variation of atomic structure with pressure. Many of the transport properties such as de Haas van Alphen effect, resistivity, magnetoresistance, Hall effect, etc. are certainly feasible since electrical connections

* At 31·6 atomic per cent aluminium, R. E. Scott[61] reports the presence of antiphase domains about 10 unit cells apart. This suggests that more than nearest neighbour interactions must be considered (cf. Cr, p. 171).

can withstand pressure and magnetic fields can be applied externally. Even nuclear resonance and ultrasonic measurements have been done under pressure. In short a whole field of studies awaits investigation.

One problem that has not yet been solved is the matter of an accurate pressure scale since there is no physical property whose pressure variation up to $\sim$100,000 atmosphere can be independently calculated. In the case of a temperature scale such things as the susceptibility of cerium magnesium nitrate, the pressure of a fixed volume of helium gas etc. can be calculated. The common unit of pressure is the kilobar = 10^9 dynes/cm^2. One bar is approximately 15 psi, close to one atmosphere.

Inasmuch as this field is in the midst of a sudden expansion we shall not sketch any of the apparatus but refer the interested reader to the article by Jamieson and Lawson.* We shall select some representative measurements of interest. As these measurements become more abundant they will be treated simultaneously with the variation of physical properties with temperature rather than separately as we are now doing. The field is still "strange" and requires some orientation in thinking about it.

(*a*) *The α-γ Transition in Iron at Room Temperature*

There is some evidence that α iron will transform to γ iron in the vicinity of $\sim$130 kilobars.[62] Since γ iron is more closely packed this change is expected, and we can write the free energy at the transformation pressure as

$$G^\alpha = G^\gamma = H^\alpha - H_0 + PV^\alpha - TS^\alpha = H^\gamma - H_0^\gamma + PV \;- TS^\gamma \quad (199)$$

The free energy difference of α and γ iron at $P \cong 0$, $T = 300°$K can be estimated if the percentage difference in volume of the two phases is independent of pressure (room temperature volume change is 4·2%) and the differences $H^\alpha - H^\gamma$, $H_0^\alpha - H_0^\gamma$, and $S^\alpha - S^\gamma$ are relatively insensitive to pressure. From eqn. (199) this gives 1080 cal/mol as the difference in free energy at $T = 300°$K. Thus we can employ these pressure measurements to aid us in the mystery of the electronic structure of γ iron by measuring the free energy difference of α and γ iron versus temperature. Precise measurements of this transition pressure coupled with precise measurements of the volume change should produce accurate free energy values for γ iron. These can be converted to specific heat by differentiation and should enable one to separate the lattice, magnetic and electronic components.

(*b*) *Variation of Curie Temperature with Pressure*

Figure 111 plots the change in Curie temperature deduced from measurements of Patrick.[63] Iron, cobalt, and a nickel-rich iron-nickel alloy show no

* L. Marton, (1959) *Methods of Experimental Physics*, **6A**, 407, Academic Press, N.Y.

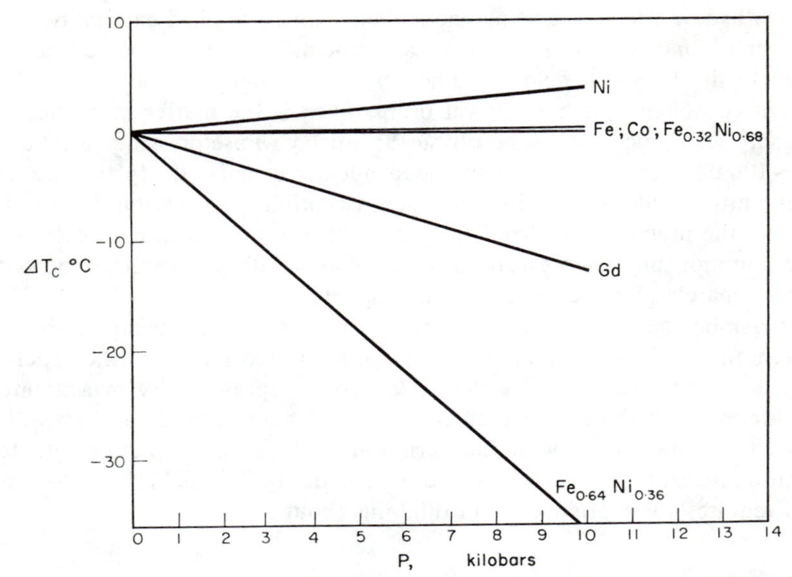

FIG. 111. The measured variation of Curie temperature with pressure (in kilobars) for several ferromagnetic metals.

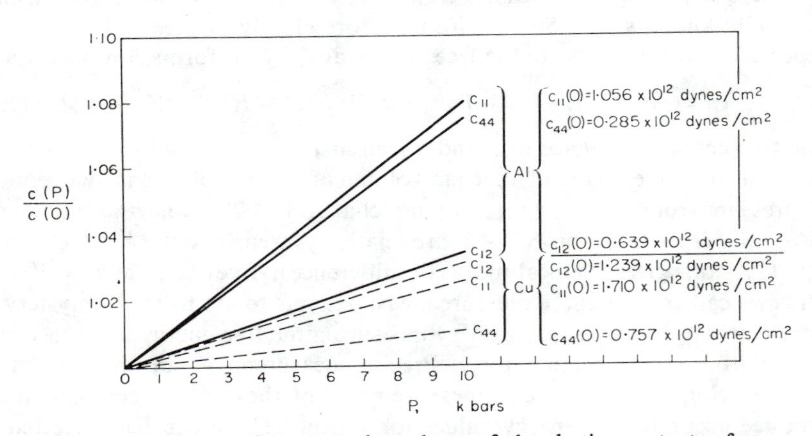

FIG. 112. The measured pressure dependence of the elastic constants of copper and aluminium. The ratio of the elastic constants at pressure divided by the elastic constants at zero pressure, $c(p)/c(0)$, is plotted against pressure in kilobars.

pronounced change; nickel increases with pressure; and gadolinium and an iron-rich iron-nickel alloy decrease with pressure. Since we do not understand the origin of the magnetic exchange energy we cannot interpret these results. One may have to couple such measurements with nuclear resonance,* Hall

* See section on nuclear resonance.

effect etc. in order to gather some clues as to the variation of the outer electron wave functions with pressure. Since one can now reach pressures greater than 100 kilobars it might be interesting to see if gadolinium continues to drop so precipitously.

(c) *Variation of Elastic Constants with Pressure*

Figure 112 shows the variation of the elastic constants of copper and aluminium with pressure[64] determined from the velocity of ultrasonic waves in single crystals. As is expected they increase since the repulsive forces rise faster than the attractive forces (see Fig. 5 for H_2 molecule). As the compressibility (eqn. (65)) is 1·8 times larger for aluminium than for copper it is not surprising that the relative change in the elastic constants is larger. If the shape of the energy versus separation curve were identical for copper and aluminium then the ratio of elastic constants would vary as the ratio of the compressibilities. The fact that this is not observed shows that the energy versus atomic separation curves are quite complicated. It will require rather high precision theoretical solutions of the Schrödinger equation to calculate the curves in Fig. 112, particularly for transition metals with many more overlapping orbitals than copper or aluminium.

(d) *Variation of Resistivity with Pressure*

Figure 113 shows the variation of resistivity of several metals with pressure. Bridgman has measured a considerable number of these which are re-

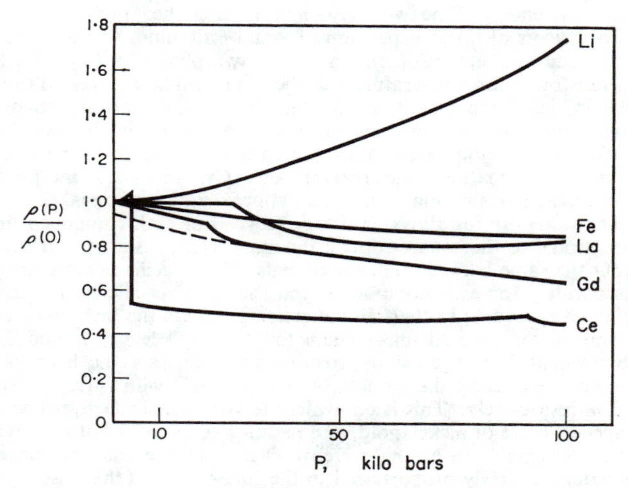

FIG. 113. The pressure dependence of resistivity for lithium, iron, lanthanum, gadolinium and cerium. The ratio of the resistivity at pressure divided by the resistivity at zero pressure, $\rho(p)/\rho(0)$, is plotted against pressure in kilobars.

viewed by Gerritsen.[65] Most metals show a steady decrease in resistivity with increasing pressure probably due to the decreasing amplitude of lattice vibrations with the increasing repulsive forces. Except for phase changes which clearly reveal themselves as discontinuities (cf. cerium, gadolinium and lanthanum in Fig. 113) there is not much more than can be said that is beyond pure conjecture. In the case of iron Drickamer and Balchan (*R.S.I.*, **32**, 308, 1961) have measured the resistivity up to $\sim$400 kbars. The $\alpha - \gamma$ transition at $\sim$130 kbars is accompanied by an increase in resistivity of a factor of 2·2. This suggests a large magnetic contribution in γ Fe at room temperature.

Some nuclear resonance measurements under pressure will be discussed in Chapter XI. The main conclusion is that pressure measurements require a considerable amount of exploration. Measurements like those of Lazarus should be particularly helpful in understanding more of the details of the shape of the energy versus interatomic distance curve.

Problems

1. Describe in detail how you would measure the heat of formation of the iron-cobalt system for completely ordered and completely disordered alloys. Include a description of the preparation of the alloys, the verification of the completeness of order and disorder, the composition of the calorimeter, the solvent, the type of thermocouple, etc. Estimate the cost of such a project and the number of man hours required.

2. Why is aluminium or copper attached to the bottom of stainless steel cooking utensils? How is the copper bonded to the stainless?

How does solder work? What is the purpose of soldering flux?

3. From the tabulated values of Θ and γ for h.c.p. and b.c.c. titanium determine the difference in cohesive energy of the two phases and the latent heat of the α-β transformation. Look up the coefficients of linear expansion of α and β titanium and plot the Morse type energy versus interatomic separation curves for the two phases at $T = 0$°K. Estimate the change in the transformation temperature α-β when titanium is compressed to 100 kilobars.

4. Describe in detail the measurements that you would perform to determine the equilibrium free energy versus concentration and temperature in the vanadium-rhenium system. Include a description of the alloy preparation, X-ray measurements on atomic arrangement, heat of formation measurements, etc. Can you guess the phase diagram?

5. (For the mathematically minded). As an approximation one makes the assumption in a "quasi" Debye theory for alloys, that the Debye Θ of each component in an alloy is inversely proportional to the square root of the mass. This assumes that the spring constants are about the same between the components. The specific heat at any temperature is evaluated separately for each component and the total specific heat determined from their sum weighted according to their abundances. Evaluate the low temperature specific heat and entropy of the ordered linear chain for $M/m = 1{\cdot}1$, $1{\cdot}5$, 2, and 3, and spring constant κ. Show that it is a good approximation to this specific heat to average the specific heats and entropies of the monatomic linear chains with spring constants κ and masses M and m respectively. This is equivalent to the "quasi" Debye theory for alloys. Look up the specific heats of nickel, gold, and nickel-gold from 20°K to room temperature and try to fit each curve with a single Debye Θ and, in the case of nickel-gold, with two Debye Θ values inversely proportional to the square root of the masses of nickel and gold. Does the Kopp–Neumann rule work for this alloy? (Cf. problem 3, Chap. VI.)

6. At temperatures below $\sim$20°K the coefficient of linear expansion of chromium metal is negative. Can you think of any reasons for this?

Summary

Considerable information concerning the electronic structure of metals can be obtained from thermodynamic measurements such as specific heat, heat of formation of alloys and determination of three dimensional phase diagrams (variation of both P and T). Even though we do not have good wave functions and density of states curves for pure elements we can still investigate the changes that occur on alloying. In general, though, studies of alloys have been incomplete and so much of the potential of this field has not been realized.

CHAPTER X

MAGNETIZATION

Introduction

In the absence of any external magnetic field most metals do not create measureable external fields of their own since all the electrons are paired both in respect to their orbital motion and their intrinsic magnetic moments. If any of the nuclei do not have an even number of protons and neutrons they will possess an intrinsic magnetic moment but this is a thousand times smaller than the electron magnetic moment and can generally be neglected in magnetization measurements. If we place a metallic specimen in a magnetic field it becomes a source of a measureable magnetic field and if we could place a magnetic measuring device inside the metal we would measure a magnetic field either greater or smaller than the applied magnetic field. If it is smaller, then the specimen is creating a field opposite to the applied field and is termed diamagnetic while if greater, it is creating a magnetic field parallel to the applied magnetic field and is either ferromagnetic, paramagnetic, ferrimagnetic or antiferromagnetic, depending on the precise arrangement of the electrons causing the magnetic field. If we measure the magnetic moment per cm^3, M, of the sample in an external static field, H, then we express the ratio, M/H, as the *susceptibility* per cm^3, $\pm\chi$, the minus sign signifying diamagnetism. Actually, in any metal there will be at least three separate contributions to the magnetic moment, the sum generally being measured. Table XX lists the various possible contributions to the magnetic moment and we shall discuss each after first describing the techniques for making some of the measurements.

Experimental Techniques

There are many techniques for measuring the magnetic moment of a metal but traditionally they have separated themselves into ferromagnetic measurements on the one hand and non-ferromagnetic measurements on the other, since the latter generally cover ranges of magnetic moments 10^3 to 10^6 times smaller than a ferromagnet.

Of course, all measurements begin with a magnet, preferably an electromagnet which allows us to vary the strength of the magnetic field. In the last few years reliable magnets and controls to maintain constant fields have been available (e.g. Varian Corp.) for fields up to about 20 kilogauss (at an approximately equal cost in dollars). These fields can be obtained in an air gap of about one inch and over a volume of about twelve cubic inches. More

TABLE XX. SUMMARY OF TYPES OF MAGNETIZATION MEASUREMENTS AND INFORMATION TO BE GAINED

STATIC FIELDS

Contributing electrons	Type of contribution	Type of magnetism	Do atoms have magnetic moments before field is applied	Information gained from measuring total magnetic moment	Temperature dependence of magnetism	Field dependence of susceptibility
1. All electrons below Fermi level including core	Orbital motion of charged particles	Dia-magnetism	No	$\langle r^2 \rangle$ of electron probability distribution	Independent	Independent
2. d or f electrons	Unpaired spins or Unpaired orbitals	Ferro-magnetism	Yes	gJ at $T = 0°K$; and Exchange energy $T < T_c/5$	Brillouin Function $T = 0$ to $T = T_c$; $1/T$ above T_c; $T^{3/2}$ spin wave, $T < T_c/5$	$B - H = 4\pi\mathcal{M}$ saturates at high fields
3. d or f electrons	Unpaired spins or Unpaired orbitals	Antiferro-magnetism	Yes	Néel temperature	Peaks at Néel temperature	Relatively Independent
4. d or f electrons	Unpaired spins or Unpaired orbitals	Para-magnetism	Yes	$g\sqrt{[J(J+1)]}$	Curie law $\sim 1/T$	Independent
5. electrons at Fermi level	Unpaired spins	Para-magnetism	No	Density of states at Fermi level	Relatively independent, generally rises or falls slowly	Independent
6. electrons at Fermi level	Orbital motion	(Landau) Diamagnetism	No	Shape of Fermi surface	Relatively independent	Oscillatory at low T, independent at high T
7. electrons in metallic band	Unpaired spins induced by exchange energy with unpaired d or f electrons	Same as unpaired d or f electrons	Yes	s–d or s–f exchange energy	Same as unpaired d or f electrons	Same as unpaired d or f electrons
8. electrons in metallic band	Van Vleck (orbital)	Para-magnetism	No	?	Independent	Independent

(Table continued on next page)

TABLE XX—*continued*

ALTERNATING FIELDS

Contributing electrons	Type of contribution	Type of magnetism	Do atoms have magnetic moments before field is applied	Information gained	Measurement	Frequency range
9. *d* or *f* electrons	Unpaired spins or orbitals	Ferro-magnetism	Yes	*g* factor, extent of *L* and *S* contribution	Ferromagnetic resonance	Microwaves ~20,000 Mc/sec
10. *d* or *f* electrons	Unpaired spins or orbitlas	Antiferro-magnetism	Yes	Exchange energy, *J*	Antiferromagnetic resonance	Microwaves
11. *d* or *f* electrons	Unpaired spins or orbitals	Para-magnetism	Yes	Hyperfine splitting; *g* factor	Paramagnetic resonance	Microwaves
12. electrons at Fermi level	Unpaired spins	Para-magnetism	No	Susceptibility of electrons at Fermi level	Free electron spin resonance	Microwaves
13. *d* or *f* electrons	Unpaired spins or orbitals	Ferro-magnetism	Yes	*g* factor	Gyromagnetic ratio	Acoustic

recently superconducting magnets have become available and these have increased the fields by two or more. For several hundred dollars a device which relies on the magnetic field dependence of the Hall effect of indium-arsenide can be obtained for measuring the field strength to about one per accuracy.* A nuclear resonance probe is about ten times more costly but does provide an accuracy to better than 0·01% since only a frequency measurement is required. The sensing element in the nuclear resonance probe can be made extremely small and this enables one to plot accurately the magnetic field over the volume in the air gap to determine its uniformity.

Even though one may encounter ranges of magnetic moment per unit volume over some eight orders of magnitude and even though the literature is crowded with many different techniques for measuring the magnetic moment, there has not been any technique so outstanding that it has emerged as a favourite among physicists. In general, they are all satisfactory and so we shall limit our description to just one technique that appeals to the writer. This technique[(66)] permits measurements over a wide range of magnetization and relies on standards like nickel and manganous ammonium sulphate to calibrate the device. The sample is placed in a magnetic field gradient and is attached to one end of a long plastic rod, the other end being connected to an acoustic loud speaker which causes the rod to vibrate at some convenient acoustic frequency. The sample vibrates between two series-wound, opposed stationary coils both within the magnetic field. As the sample's magnetic moment cuts the magnetic lines of force of the magnetic field it induces a signal in the coils. The signal is amplified and put through a phase sensitive detector employing the acoustic driving frequency as a phase reference. Generally the electronics will handle only a limited range of signals from the coil so that a range of coils with different numbers of terms are used (the signal is proportional to the number of turns) and calibration is obtained by using a nickel or manganous ammonium sulphate sample. Since unwanted fluctuations in the amplitude of vibrations of the acoustically driven rod will give rise to a fluctuating signal, a small permanent magnet is attached to the end of the plastic rod away from the magnetic field and two series-wound, opposed stationary coils are placed on either side. The signal from these coils is adjusted and fed in opposition to the signal from the sample coils and only the difference signal is used. An accuracy of a per cent or so is obtainable with the Foner apparatus. Considerable effort has gone into the measurement of the magnetic moment of nickel metal and it makes a good standard for samples of high magnetic moment, its room temperature value being 0·948 times its value at 0°K, 0·606 Bohr magnetons per atom. Since one Bohr magneton is $\mu_B \sim 9{\cdot}27 \times 10^{-21}$ erg/gauss we have for the magnetic moment of nickel at 20°C, $M \sim (9{\cdot}27 \times 10^{-21} \times 6{\cdot}025 \times 10^{23} \times 0{\cdot}606 \times 0{\cdot}948) \sim 3{\cdot}21 \times 10^{3}$ erg/gauss/mol $\sim 54{\cdot}65$ erg/gauss/g.

* Bell Inc., 1356 Norton Ave, Columbus 12, Ohio

For samples of lower magnetic moment, manganous ammonium sulphate is quite good at room temperature since its susceptibility follows the Curie law, χ (per g) $= 35\ N_0\mu_B^2/3kT$ where N_0 is the number of manganese atoms per g ($=1{\cdot}54 \times 10^{21}$). Consequently, at 20°C the magnetic moment of manganous ammonium sulphate in a field of 10 kilogauss is $3{\cdot}819 \times 10^{-5}$ erg/gauss/g, or about a million times smaller than nickel.

In measuring very weak susceptibilities, particularly diamagnetic metals, very small traces of ferromagnetic inclusions or isolated atoms of transition elements or rare earths may cause difficulty. Iron is the principal ferromagnetic impurity in metals and, if present in small ferromagnetic inclusions, the susceptibility will show a field dependence. To correct for it, one need only extrapolate the strong magnetic field data ($H >$ ten kilogauss) to $H = 0$. On the other hand isolated atoms of transition elements or rare earths will show no field dependence and one must look for their presence by elemental analysis (X-ray fluorescence). In measuring larger magnetizations or susceptibilities, particularly ferromagnetics around their Curie temperature, there are two major problems. Firstly, the demagnetizing field of the sample causes each atom in the sample to experience a field which is the sum of the external field and the field created by the other atoms in the specimen. If the sample is a sphere the demagnetizing factor is $\frac{1}{3}$ so that the field experienced by the sample is the external field, H_0, minus $\frac{1}{3}(4\pi\mathcal{M})$ where $\mathcal{M}$ is the magnetization of the sample, i.e., the magnetic moment per g times the density. For nickel metal near saturation the demagnetizing field is ~2000 gauss for a sphere, but in spite of its large value the use of a sphere does enable one to make the correction properly. A second problem arises in ferromagnets with large anisotropies or with non-magnetic inclusions so that the sample does not reach saturation* in fields up to ~20 kilogauss obtainable with most magnets. If the anisotropy is high, as it is in most non-cubic metals, it is best to obtain a single crystal. If there are impurities, inclusions, etc., the corrections are difficult and a paper by Danan(67) is a good starting reference. As a rough approximation $\mathcal{M} \cong \mathcal{M}_\infty\ (1 - b/H)$ where the constant, b, is determined from the high field dependence of the measured values of $\mathcal{M}$, and $\mathcal{M}_\infty$ is determined by extrapolating to infinite H. In terms of absolute standardization a considerable effort is required to achieve an accuracy of 1%.

Results of Measurements Listed in Table XX

1. Orbital Diamagnetism

All electrons below the Fermi level give rise to an orbital diamagnetism whose classical analog is the Lenz law for current in a wire loop. Quantum mechanically this diamagnetism is proportional to $\langle r^2 \rangle$. In the case of inert

* All the magnetic moments pointing in the same direction.

gas atoms this orbital diamagnetism is the only contribution to the susceptibility, which is given per mole as

$$\chi = -2{\cdot}832 \times 10^{10}\langle r^2\rangle \qquad \langle r^2\rangle \text{ in cm}^2 \tag{200}$$

where

$$\langle r^2\rangle = \int_0^\infty r^2|R|^2 \mathrm{d}r \qquad \text{(see p. 8)}$$

for inert gas atoms. Table XXI lists some experimental determinations of $\langle r^2\rangle$ for inert gas atoms and compares them to the values calculated by the Hartree–Fock method and, in the case of helium, by an exact many electron wave function of Pekeris.[1]

TABLE XXI
EXPERIMENTAL AND CALCULATED VALUES OF $\langle r^2\rangle$ IN (BOHR UNITS)^2

Element	Experimental	Hartree–Fock calculation	Error	Correlated wave function
He	2·386	2·386	~0	2·387
Ne	8·9	9·40	+6%	—
Ar	24·95	26·03	+4%	—

Thus we see that the H.F. theory is quite good for helium but yields values 4–6% in error for neon and argon. Helium is probably a special case in that it has only *s* electrons. We have already mentioned in Chapter I that the error in the Hartree–Fock eigenvalues is something like 1 eV per electron. Considered with the 4–6% error in $\langle r^2\rangle$ we have an overall estimate of several percent for the errors in the eigenfunctions and eigenvalues due to the one electron approximation.

In many metals, particularly transition metals, the diamagnetic susceptibility is only a small fraction of the total and is outweighed by the positive contributions from items 5 and 8 in Table XX. However, this is not the case for copper, silver, gold, zinc, cadmium, mercury and beryllium, all of which are diamagnetic. In copper, for example, the measured value of χ is -5×10^{-6} erg/(gauss)^2/mol. This is composed approximately of -15×10^{-6} from item 1, $+14 \times 10^{-6}$ from item 5 and -4×10^{-6} from item 6 in Table XX; but this is only a guess since we have been unable to separate these contributions experimentally. Any one of the items can be in error by some 50%.

2. Ferromagnetic Metals

At $T = 0$°K a ferromagnet like pure iron consists of domains, several microns in size, in which all the atomic magnetic moments are rigidly aligned in some specific direction. Different domains are aligned in different directions and the regions between the domains, called *Bloch walls*, consist of a gradual rotation of the atomic magnetic moments smoothly joining the neighbouring domains. The atomic magnetic moments in each domain are aligned along any of the identical principal crystallographic axes like [100], [010], [001], [–100], [0–10], [00–1] in iron. (Even in a single crystal of iron at $T = 0°K$ domains still exist and are aligned in one of the six [100] directions.) While the neighbouring atomic magnetic moments are coupled parallel to each other by exchange forces their preference for some specific crystallographic direction is due to spin orbit coupling and is called *magnetic anisotropy* (see page 306). The unpaired 3*d* electrons in iron have a slight preference for the e_g orbitals which point along the 100 direction (Fig. 17) while in nickel the preference is for the t_{2g} orbitals which point along the 111 directions. Thus the spin orbit coupling makes the *easy direction of magnetization*, [100] in iron, and [111] in nickel. This anistropy energy is about 10^4 times smaller than the exchange energy in the transition metals but can be considerably larger in some rare earths.

By breaking up into domains one can form small closed magnetic circuits within the iron itself and minimize the resistance of the flow of magnetic flux through the air. This is shown schematically in Fig. 114. The thickness of the

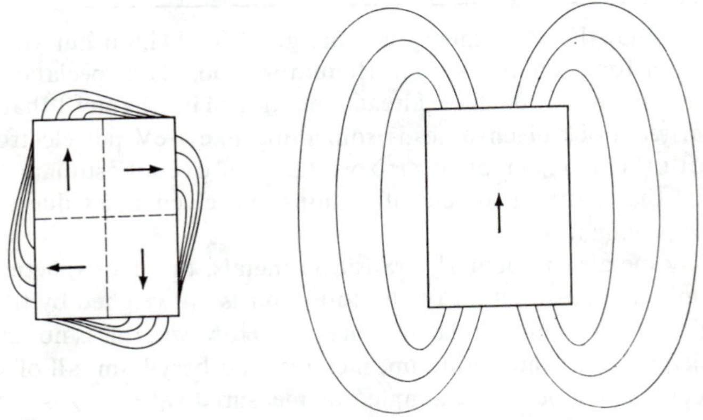

FIG. 114. A schematic sketch of the magnetic lines of force in a multi-domain and a single-domain magnet, demonstrating the preference for domain structures in pure ferromagnets.

Bloch wall becomes a compromise between the magnetic anisotropy which would have the magnetic moments make the transition between two domains 90° apart as rapidly as possible so as to minimize the number of magnetic moments that point in the wrong crystallographic direction, and the exchange energy which would have the transition as slowly as possible so as to keep the neighbouring atoms as parallel as possible. This compromise results in Bloch domain walls of several hundred Å thickness.

If we apply an external magnetic field to iron at $T = 0°K$ the domains rotate to line up in the direction of the field, in order to form the lowest energy state. In iron or nickel single crystals, only a few gauss, are required to align all the domains in the easy direction of magnetization and several hundred gauss for the "hard directions", [111] and [100] respectively. In h.c.p. cobalt the easy direction is along the *c*-axis but this reverts to the *a*-axis above about 300°C. To align the domains along the hard *a*-axis direction in a single crystal of cobalt at room temperature requires several thousand gauss. In some rare earth metals where the spin-orbit coupling is much stronger, several million gauss are required to align the domains in the hard direction (estimated by extrapolation). As yet we do not have sufficient measurements to indicate the easy and hard directions in all the rare earth elements. For gadolinium ($L = 0$ for the half filled $4f$ shell) the spin-orbit coupling is quite small. It is not identically zero since the $5d$ wave functions are different depending on whether their m_s is parallel or antiparallel to the seven $4f$ electrons and this provides a source for weak spin-orbit coupling.

The manner in which the domain walls move as the domains rotate, and eventually disappear as the external field is increased is very sensitive to impurities, cold working etc. We are all probably familiar with the so-called hysteresis loops that develop when we plot the magnetic moment of the sample as we increase and then decrease and reverse the external field. The reader is referred to Bozorth's book[(68)] for details of this process.

If we apply a magnetic field and remove all the domain walls of a ferromagnet at $T = 0°K$ and measure the magnetization, we can determine the average value of the magnetic moment per atom. For example, in iron this leads to $gJeh/4\pi m_0 c = 2{\cdot}2\mu_B$. A separate measurement of the g factor (1·93) tells us that $L = 0{\cdot}08$, $S = 1{\cdot}06$ and $J = L + S = 1{\cdot}14$, so that the magnetic moment per iron atom is primarily due to the intrinsic moment of the unpaired electrons rather than their orbital motion ($L \ll S$). As the temperature is raised the spin waves are excited and the magnetization varies according to eqn. (62) if only nearest neighbours are considered. This enables us to determine the exchange energy J. Table XXII gives the values of the exchange energy, J, for several ferromagnetic elements, deduced from various sources such as the total magnetic energy in the specific heat, neutron diffraction, etc. Theoretical statistical mechanics also enable us to determine J from a knowledge of the Curie temperature and this value is included. To within 10% or so, the statisticle mechanical value of J is given by $J \cong kT_c/S^{3/2}z$ where z is the number of nearest neighbours and S the spin.

In all cases there is a complete lack of agreement by a factor of two or so and this suggests that next nearest neighbour interactions may play a role in the exchange energy. However, none of the theories have yet considered second nearest neighbour interactions.

At temperatures above $T/T_c \cong 1/5$ the spin wave theory is no longer valid

TABLE XXII

VALUES OF THE EXCHANGE ENERGY J IN eV FOR VARIOUS FERROMAGNETIC METALS DERIVED FROM VARIOUS SOURCES (BASED ON NEAREST NEIGHBOUR INTERACTIONS ONLY)

Element	Spin wave magnetization	Magnetic energy in specific heat	Neutron diff.	Theory from Curie temp.
b.c.c. Fe	0·018	0·010	—	0·011
f.c.c. Ni	0·020	0·011	—	0·020
f.c.c. Co	0·02	—	0·016	0·012
h.c.p. Gd	0·00014	0·00021	—	0·00032

but it is found that the classical Brillouin function describes approximately the saturation magnetization versus temperature;[69] and that above the Curie temperature the susceptibility per mole approximately follows the *Curie–Weiss law.*

$$\chi = N_0 J(J + 1)g^2\mu_B^2/3k(T - \theta) \tag{201}$$

where N_0 is the number of magnetic atoms per mole, $gJ\mu_B$ their magnetic moment, and θ a constant approximately equal to the Curie temperature. Both the Brillouin function and Curie–Weiss law are plotted in Fig. 115 for

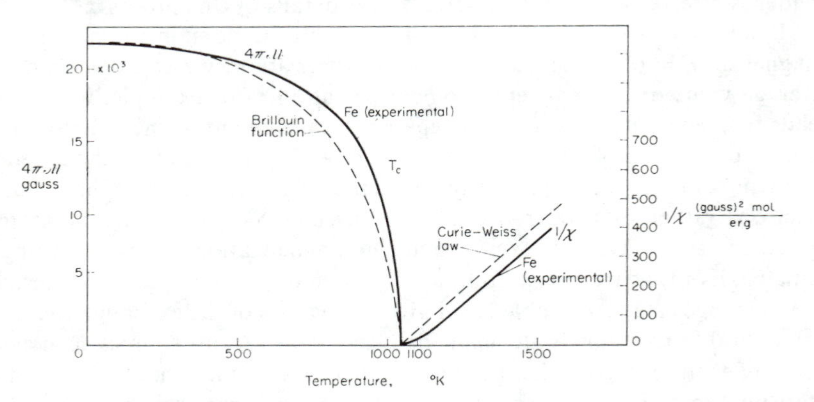

FIG. 115. The observed saturation magnetization (times 4π) versus temperature in °K for α iron (solid curve); and the observed reciprocal of the susceptibility in $(\text{emu})^{-1} = (\text{erg}/(\text{gauss})^2\text{mol})^{-1}$ versus temperature above the Curie temperature at 1043°K (solid curve). The dashed curve for magnetization is calculated from the Brillouin function ($J = 1$) and the dashed curve for susceptibility from the Curie–Weiss law ($g = 1{\cdot}93$, $J = 1{\cdot}15$).

iron as well as the experimental data. The most serious discrepancy is just above the Curie temperature where eqn. (201) fails to account for the short range magnetic order.

All theoretical expressions like eqn. (201) have been derived on the basis of an integral number of unpaired electrons. In the case of nickel we shall continue to use the artifice of assuming nickel to be 60% $gJ = \frac{1}{2}$ and 40% $gJ = 0$ (see p. 62). Until the theory is improved to cover these cases this is all we can do.

Possibly the main conclusion to be drawn from the slope of the susceptibility measurements above the Curie point (eqn. 201) and from the saturation magnetization measurements at $T = 0°$K is that the number of unpaired electrons per atom is independent of temperature. This had already been concluded from the magnetic entropy and magnetic resistivity in iron, gadolinium and nickel.

An interesting example of the spiralling effect in the rare earths occurs in the magnetization of h.c.p. thulium[70] shown in Fig. 116. Thulium has an unpaired electron configuration, $4f^{12}$, which by application of Hunds' rule and spin orbit coupling yields the quantum numbers $J = 6$, $L = 5$, $S = 5/2$ and $g = 7/6$. Due to the large orbital contribution there is an appreciable anisotropy and the curves in Fig. 116 were taken on polycrystalline specimens which could not be saturated in fields up to 20 kilogauss. Apparently J_1/J_2, the ratio of the nearest neighbour and next nearest neighbour exchange energies, is in a critical region. The material is ferromagnetic up to $\sim$20°K but above 20°K spiralling sets in so that the net magnetization is zero (Fig. 59). However, an external magnetic field of $\sim$4 to 12 kilogauss eliminates the spirals so that the material develops a large magnetization. At 60°K the Curie or Néel temperature is reached and the magnetization then falls as $1/T$. The susceptibility above 60°K follows the Curie–Weiss law and its slope gives the expected value, $g^2J(J + 1) = 57{\cdot}1$, confirming the fact that the $4f$ electrons are essentially unaffected in the bonding. However, the anisotropy at lower temperatures shows there is some very weak coupling to the lattice. It has been suggested by Yosida and Miwa (*J. App. Phys.* **32**, 85 (1961)) that this anisotropy is responsible for the transition to the spiral arrangement.

3. Antiferromagnetic Metals

The theory of the magnetization of antiferromagnetic metals is in rather a poor state. However we do know that at $T = 0°$K the metal consists of atoms with magnetic moments coupled antiparallel to their neighbours. There are domains, as in a ferromagnet, although the reason for their existence is not as obvious since the equal numbers of magnetic moments in each direction in a domain produce no net magnetic field. Nevertheless, there is subtle evidence that they exist in metals and optical evidence shows they exist in antiferromagnetic insulating compounds like nickel oxide. The domain structure is probably a result of imperfections, impurities etc.

If an external magnetic field is applied to a domain along the direction of

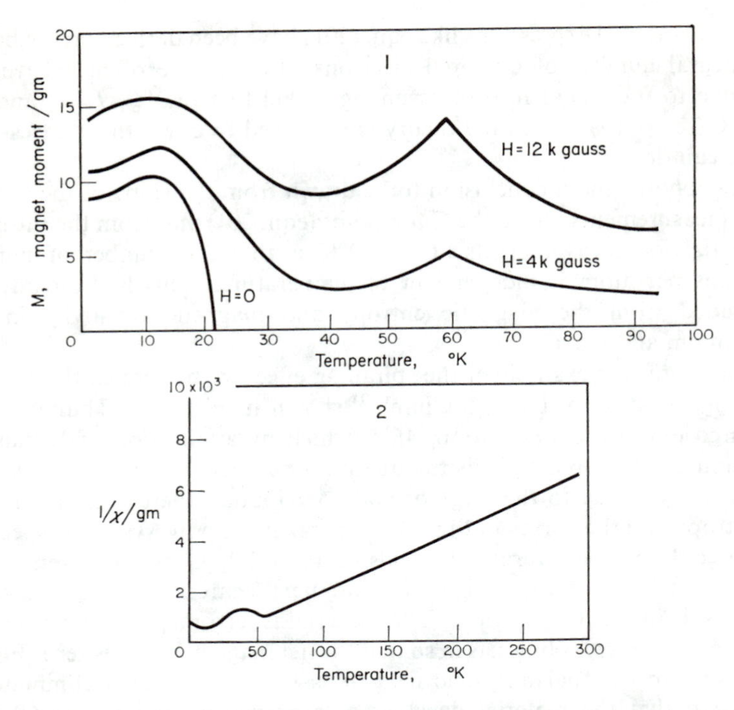

FIG. 116. The magnetic moment per g (erg/gauss/g) of thulium metal as a function of temperature for $H = 0$, 4 kilogauss and 12 kilogauss. In zero field the metal is ferromagnetic up to about 20°K at which temperature spiralling of the magnetic moments sets in. This spiralling can be partially eliminated with the application of a strong field. The Néel temperature in zero field occurs at $\sim$60°K. The reciprocal of the susceptibility $(\text{erg}/(\text{gauss})^2/\text{g})^{-1}$ is plotted in diagram 2. The slope above 60°K confirms the expected value of $g^2J(J+1) = 57{\cdot}1$.

magnetic moments, Fig. 117, diagram 1, nothing happens at $T = 0$°K since the magnetic moments are rigidly locked in an antiparallel array by the exchange energy. On the other hand, if the field is applied at 90° then both spins can rotate a little, Fig. 117, diagram 2, and lower their energy since the exchange energy is raised by the cosine of the angle while the magnetic moments, interacting with the external field, lower their energy by the sine of the angle. The resulting angle of minimum energy is approximately $\theta \sim \mu_B H/2J_1 S$ or about several minutes in a typical case. Thus, antiferromagnetic metals develop a small magnetization in an external field as long as there are domains not parallel to H. As the temperature is increased an external field can now induce some magnetization in the parallel domains since some of the magnetic moments have been thermally uncoupled, but this external field will have relatively little *additional* effect on the domains oriented normal to the external field. A polycrystalline or multidomain specimen will show some average effect between the two orientations. Above the Néel temperature one expects a Curie–Weiss law. These theoretical arguments on temperature and field effects are borne out in experimental results of non-metallic antiferromagnetic substances and, indeed, most antiferromagnets are non-metallic but the antiferromagnetic transition metals chromium, manganese, and f.c.c. iron-manganese alloys, do not behave this way. The susceptibility of chromium, for example, is practically independent of T, even around the Néel temperature.

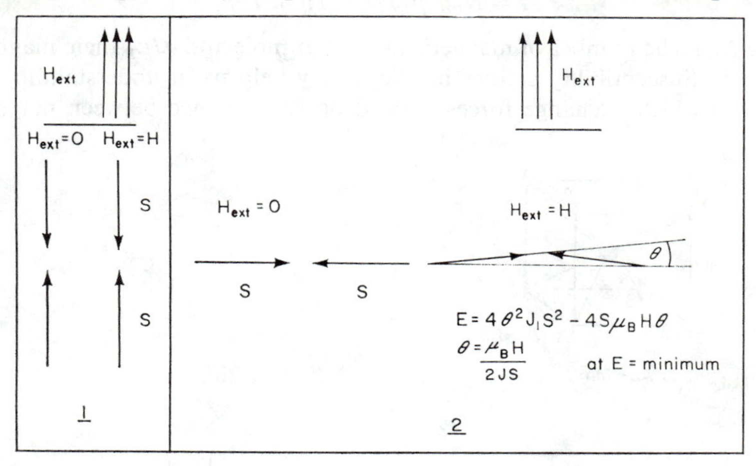

FIG. 117. Two electrons coupled antiparallel in an external magnetic field. In diagram 1 the field is applied parallel to the quantum mechanical component of magnetic moment while in diagram 2 it is applied normally. In the latter case the energy of the system is lowered by a slight deviation from antiparallel alignment in the direction of the applied field.

We can only conclude that we are very ignorant of the electronic structure of these antiferromagnetic transition metals.

On the other hand, gold-manganese forms an antiferromagnetic CsCl structure whose susceptibility does behave as expected.[(71)] However, this alloy is a bit unusual since the gold atoms do not have a significant magnetic moment but provide the exchange mechanism for coupling the manganese atoms anti-parallel along the 111 direction. The structure and susceptibility of gold-manganese are shown in Fig. 118. When a non-magnetic intervening atom provides the magnetic exchange coupling between two magnetic atoms this is called *superexchange*. The outer electron wave functions on the gold atom overlap the manganese atom and interact with the unpaired $3d$ electrons.

4. Paramagnetism

There is no element with unpaired d or f electrons that is only paramagnetic. If atoms with unpaired electrons are nearest neighbours there is always an exchange force that produces ferromagnetic or antiferromagnetic alignment at $T = 0°K$. However, if one has a sufficiently dilute alloy such as one-half per cent iron in f.c.c. aluminium then the substance is paramagnetic. If the atoms with unpaired electrons are further apart than third nearest neighbours then the exchange forces are practically zero. In such cases the susceptibility per mole is given by the *Curie law*,

$$\chi = N_0 g^2 \mu_B^2 J(J+1)/3kT \tag{202}$$

where N_0 is the number of magnetic atoms per mole and $gJ\mu_B$ their magnetic moment. Susceptibility studies in alloys may help us in understanding the extent to which exchange forces depend on the distance between magnetic

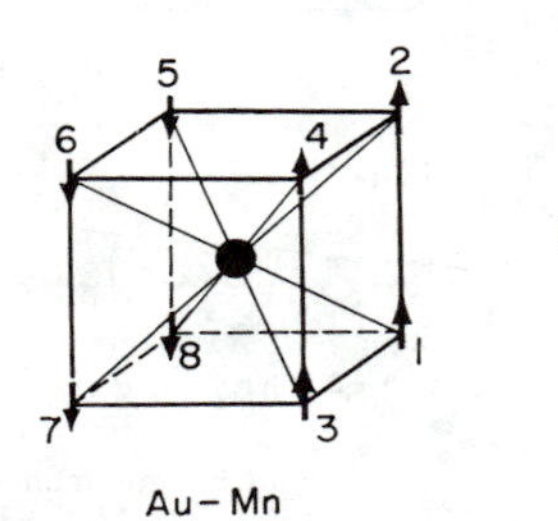

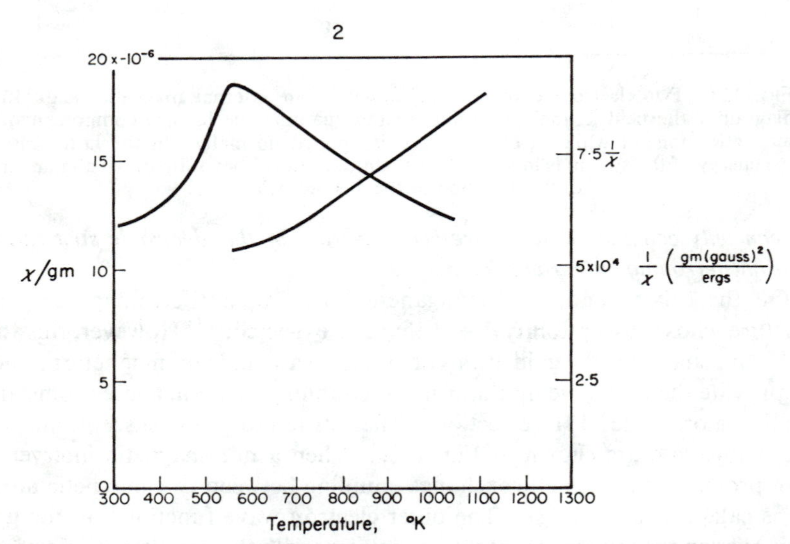

FIG. 118. Diagram 1. A sketch of the CsCl crystal structure of gold-manganese showing the gold atom at the body centre position. There is no appreciable magnetic moment on the gold atom even though the exchange interaction between manganese atoms is via the outer electrons on the gold atom.

Diagram 2. The susceptibility in ergs/(gauss)^2/gm of gold-manganese and the reciprocal of the susceptibility. The slope of the curve of $1/\chi$ versus T yields a value of $g^2J(J+1) = 24$ corresponding to $g \cong 2$, $L = 0$, $S = 2$.

atoms and the type of solvent atom. When combined with paramagnetic resonance which gives the g factor, the value of L, S and J can be determined. As yet there is insufficient data on dilute transition metal alloys to say very much. The cases this author has noted do not seem to fit any recognizable pattern. For example, very dilute solutions of the transition elements vanadium through nickel dissolved in f.c.c. aluminium have no magnetic moment except possibly for iron.

Because the $4f$ electrons are not part of the band in the rare earth atoms dilute solutions of rare earths invariably obey eqn. (202).

5. Pauli Paramagnetism of Electrons at Fermi Surface

In the absence of an external magnetic field a metal like copper has a complete balance of electron spins. In the one electron Bloch wave treatment the electrons near the Fermi level occupy a closely spaced set of energy levels. If we limit our attention to the intrinsic magnetic moments of the electrons, quantum mechanics tell us that a measurement of each magnetic moment in an external magnetic field will reveal it to be either parallel to the field, $m_s = +\frac{1}{2}$, or antiparallel, $m_s = -\frac{1}{2}$. The energy of those electrons with spin parallel to the field is lowered by $\mu_B H$ and the energy is raised $\mu_B H$ for those electrons antiparallel. Since $\mu_B H$ is generally larger than the spacing of the energy levels, those electrons of antiparallel spin within an energy $2\mu_B H$ of the Fermi level can lower their total energy by occupying an unfilled higher energy level and reversing their spin. This is shown schematically in Fig. 119. Diagram 1 shows the energy levels occupied up to the Fermi level in the absence of a magnetic field. At the moment the field is applied, diagram 2, the energy of all the electrons is lowered or raised $\mu_B H$ depending on whether m_s is parallel or antiparallel to the field. This is followed by the electrons of higher energy reversing their m_s and occupying the higher unfilled levels until a balance in energy is reached. This leaves more electrons with spin parallel to the applied field than antiparallel and they create a net magnetization parallel to the field. The closer the spacing of the energy levels at the Fermi level (higher density of states) the more electrons will reverse their m_s, hence the higher the paramagnetic susceptibility.

If we now raise the temperature, two things can happen. Firstly, electrons at the Fermi level are thermally excited to higher energy states so that the application of a magnetic field allows more electrons to reverse their m_s since they can now occupy the levels vacated by the thermally excited electrons. This tends to increase the paramagnetic susceptibility. Secondly, the thermal energy tends to disorient those electrons that have been aligned by the magnetic field and tends to decrease the susceptibility. The two effects approximately cancel in the noble and alkali metals so that the Pauli paramagnetism is approximately temperature-independent. In most of the nonmagnetic transition metals, however, one or the other effect dominates and we find the Pauli paramagnetism increasing or decreasing with temperature.

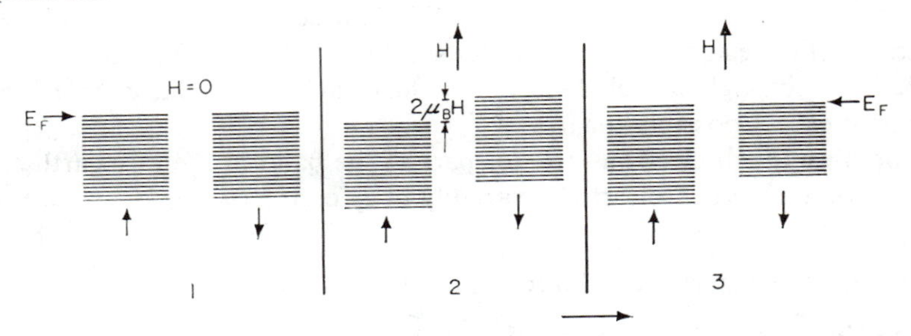

FIG. 119. Diagram 1. The population of the energy levels near the Fermi level in a nonmagnetic metal,

Diagram 2. The population of the energy levels at the instant of application of an external field,

Diagram 3. The population of the energy levels after the electrons have found the lowest energy state, leaving a net paramagnetism.

It is extremely difficult to calculate either the absolute magnitude or temperature dependence of the paramagnetic susceptibility of the electrons at the Fermi level. Nor have the experimental observations yet provided us with the eigenvalues and eigenfunctions near the Fermi level required for such a calculation. In principle one has to solve separately the Schrödinger equation for the metal with and without a magnetic field although an *accurate* solution without the magnetic field might provide the wave functions that can be used with the variational principle to solve for the eigenvalues with the magnetic field applied. In the interim we use the free electron theory and intuition.

Crudely, however, we can say that the Pauli paramagnetic susceptibility, χ, is proportional to the density of states as is the electronic specific heat coefficient, γ, and that χ is high when γ is high. An example of this correspondence is the alloy series rhodium through paladium to silver(59) shown in Fig. 120. This gives rather direct evidence that the density of states at the Fermi level is higher in palladium than in rhodium and higher in rhodium than silver. In addition a comparison of the temperature dependence of χ for palladium, palladium (5% rhodium), rhodium and silver (silver is temperature-independent and slightly diamagnetic) suggests a much more intricate density of states curve for palladium than for rhodium or silver. Quite possibly the *d* band in palladium is making contact with the Brillouin zone. In silver the *d* levels are filled below the Fermi level.

Room temperature susceptibility values of some transition elements are: Ti (+) 161; Zr (+) 121; Hf (+) 68; V (−) 287; Nb (−) 214; Ta (−) 154; Cr (+) 165; Mo (+) 79; W (+) 53; Mn (−) 534; Re (+) 65; Ru (+) 34·5;

Os (+) 10; Rh (+) 105; Ir (+) 27; Pd (−) 570; Pt (−) 187. All the susceptibilities are paramagnetic in units of 10^{-6} emu/mole. The sign (+ or −) refers to increasing or decreasing values with temperature.

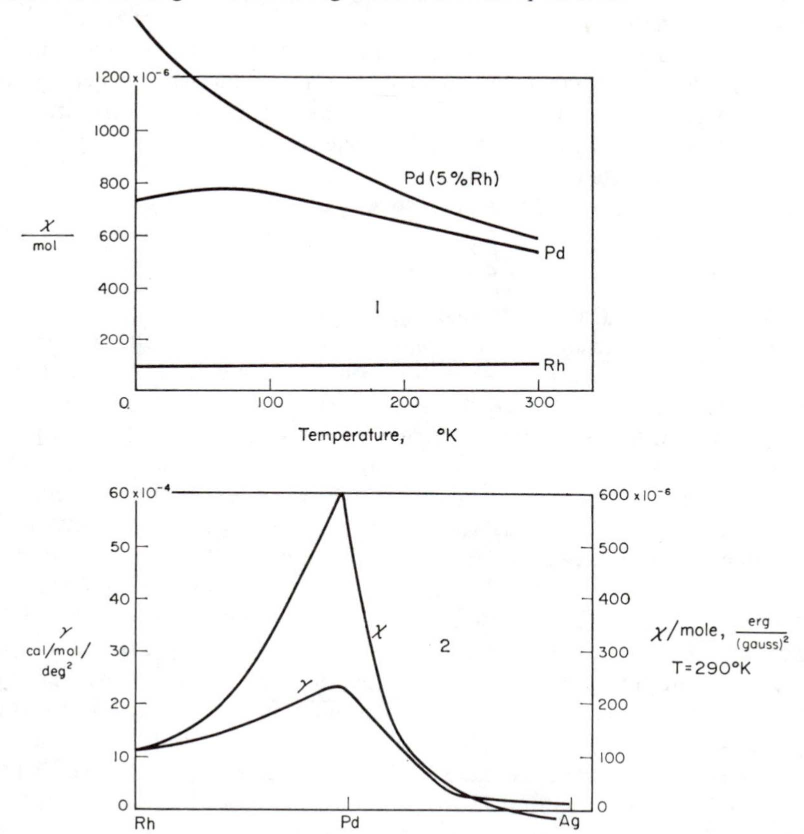

FIG. 120. Diagram 1. The susceptibility in ergs/(gauss)²/mol for palladium, rhodium and palladium (5% rhodium) as a function of temperature.

Diagram 2. The electronic specific heat coefficient, γ, of the alloys rhodium-palladium and palladium-silver and the room temperature susceptibility of the same alloys showing the correlation with the density of states at the Fermi level.

6. Orbital (Landau) Diamagnetism

We have already discussed the orbital diamagnetism of the electrons at the Fermi surface and how the quantization of orbits in a magnetic field leads to the low temperature de Haas van Alphen effect, but we have not discussed the magnitude of the susceptibility. In the free electron theory the magnitude is about $\frac{1}{3}$ the magnitude of the Pauli paramagnetism. We cannot calculate this quantity with any accuracy in real metals nor can we experimentally separate it from the other contributions. It is generally small, however, and until the various contributions to the susceptibility are either calculated or measured more accurately we shall not generally concern ourselves with its magnitude. In item 11, paramagnetic resonance of the electrons at the Fermi surface, the cases of lithium and sodium are discussed (p. 328) and indicate that the orbital diamagnetism is only 10% of the Pauli paramagnetism rather than $\frac{1}{3}$. The de Haas van Alphen effect did not depend on the magnitude of χ, only on the frequency and amplitude of the oscillations with magnetic field. One cannot separate the orbital diamagnetism by virtue of its periodic nature since the minima do not correspond to zero diamagnetism.

7. Exchange Polarization of Electrons in Metallic Bands

If we have unpaired *d* or *f* electrons in a metal they interact with the other electrons of parallel m_s through the exchange forces resulting from antisymmetrization of the total wave function. As a result two things happen. Firstly, all the electrons well below the Fermi level, including the strongly bound ones, will have slightly different wave functions if their m_s is parallel or antiparallel to the unpaired *d* or *f* electrons. Those with m_s parallel will have their radial wave functions more concentrated in the region of the unpaired *d* or *f* electrons as compared to those with their m_s antiparallel. This does not alter the exact balance in the number of electrons with $m_s = +\frac{1}{2}$ and $m_s = -\frac{1}{2}$ so that the net magnetization is zero. Secondly, the outer or bonding electrons will have their wave functions significantly altered by the exchange effects since the exchange energy is an appreciable fraction of their eigenvalue in the metal. Thus the eigenfunction, eigenvalues, density of states, Fermi surface, etc., are significantly different for the outer electrons with m_s parallel and m_s antiparallel to the unpaired *d* or *f* electrons. Thus quite a bit can happen. There may actually be more electrons with m_s parallel than antiparallel to the unpaired electrons so that they contribute to the net magnetization. The evidence suggests this to be about 0·1 unpaired electron in gadolinium. The fraction of *d* to *s* character may vary and the evidence in transition metals is that there is a larger fraction of *d* to *s* character for the electrons that are parallel to the unpaired electrons. Of course both these effects vary with temperature in the same way as the magnetization of the unpaired *d* or *f* electrons do so that the average effect disappears above the Curie temperature. However, the effect remains on each atom at all temperatures since the number of unpaired electrons per atom is approximately independent of temperatures, (chromium and αMn excepted).

8. Van Vleck Paramagnetism

It is possible that the electrons in the entire metallic band can actually change their wave functions slightly upon application of a magnetic field, selecting orbitals which have lower energy in the field. This will always favour paramagnetism. We have no ideas yet as to its magnitude although estimates suggest it to be significant, particularly in transition metals.

9. Ferromagnetic Resonance

The first eight topics covered measurements in static or d.c. magnetic fields. We now proceed to measurements in which an alternating magnetic field is applied to the sample. Neglecting orbital motion, let us consider the quantum mechanical energy levels of a single electron in a magnetic field H. Its eigenvalues are $-\mu_B H$ and $\mu_B H$ for its magnetic moment parallel and antiparallel to the field, respectively. The energy difference is $2\mu_B H$ and the electron in the lower energy level can absorb this energy from the electromagnetic field if it is of the right frequency, $h\nu = 2\mu_B H$. This corresponds to a microwave frequency of $2{\cdot}8 \times 10^{10}$ cycles/sec (wavelength $\lambda \cong 1$ cm) in a field of ten kilogauss. If there are two electrons coupled ferromagnetically to each other, then the quantum mechanical eigenvalues are $-2\mu_B H$, 0 and $+2\mu_B H$. Conservation of angular momentum permits these coupled electrons to absorb a photon equal to the energy difference between two adjacent levels. The energy difference, $2\mu_B H$, is identical to the case of a single electron and we find that no matter how many electrons are coupled together the spacing between the eigenvalues is identical, $2\mu_B H$. The number of energy levels is one greater than the number of electrons. The eigenvalues and eigenfunctions for the cases of one, two, three or four electrons are shown in Fig. 121. If we deal with a ferromagnetic metal the number of electrons is of the order of 10^{23} so that the coupled magnetic moments make a small change in angle with the field between adjacent levels (the energy difference is still $2\mu_B H$) and we have practically a continuous distribution that can be treated classically. The resonant frequency is not exactly $\nu = 2\mu_B H/h$ but must be modified by the orbital contribution to the magnetic moment of each electron, and by the demagnetizing factor that alters the field experienced by the other unpaired electrons. The resonant frequency for a thin disc is

$$\nu \cong \frac{[1-(1/g)]^{-1}(HB)^{1/2}}{4\pi m_0 c} \tag{203}$$

where g is the Landé g factor, eqn. (179). Thus a measurement of the ferromagnetic resonance frequency gives a value for g. Coupled with the saturation magnetization measurement which yields gJ we can determine L and S for each atom, if we assume that they couple according to the rules for spin orbit coupling. The values of g are 1·92, 1·85, and 1·84 for iron, cobalt and nickel respectively and 1·95 for gadolinium.[72] We have already discussed the question of the partially unquenched orbitals in the transition metals (p. 174) and have indicated that there is very little quenching in the rare earth metals.

Gadolinium is of special interest since its seven 4*f* electrons would be expected to give $\Sigma m_l = L = 0, \Sigma m_s = S = 7/2$ and a g factor of 2·00. The measured g factor of 1·95 must reflect the influence of the 4*f* electrons on the outer electrons. If we combine this with a measured value of $gJ \cong 7{\cdot}2$ to 7·3* from

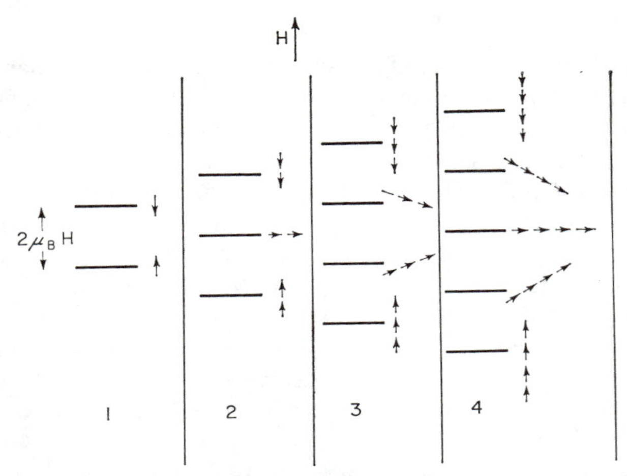

FIG. 121. The eigenvalues and eigenfunctions (represented by the magnetic moment directions) of one, two, three and four ferromagnetically coupled electrons in an external field.

saturation magnetization at low temperatures we must conclude that there is an excess of the outer $5d^2 6s$ electrons with spin parallel to the 4*f* electrons than with spin antiparallel. In fact an excess of 0·05 of a 5*d* electron with $m_l = 2$ would lead to $J = 3{\cdot}74$, $S = 3{\cdot}55$, $L = 0{\cdot}19$, $gJ = 7{\cdot}3$, $g = 1{\cdot}95$. This provides one of the clearest assessments of the effect of unpaired electrons on the outer electron distribution in the metallic bands.

In performing the ferromagnetic resonance experiment the sample, in the form of a thin disc, is made one end of a microwave cavity operating at a fixed frequency. The external magnetic field is varied until resonance is reached at which point there is absorption of energy by the specimen.

10. Antiferromagnetic Resonance

Very little work has been reported in the detection of antiferromagnetic resonance. Its principle is somewhat different from the ferromagnetic resonance in which the entire spin system remains rigidly coupled as it is excited from one energy level to the next. If the magnetic moments are rigidly coupled in an antiferromagnetic array then the total energy is not altered in an external field since there is a complete balancing of energies. However,

* Corrected for impurities.

it is always possible to break the antiferromagnetic coupling by adding sufficient energy to the system. Let us return to our electrons in an external field, H, and couple them antiferromagnetically in a linear chain of ten electrons with nearest neighbour exchange energy $2J_1\boldsymbol{S}\cdot\boldsymbol{S} = J_1/2$ per pair.* Fig. 122 shows what happens if one of the electrons (other than the end ones) reverses its m_s value. In the ground state there is a slight magnetization due to the external field competing with the exchange energy. If the encircled electron reverses its spin two bonds are broken requiring an energy $2J_1$, the resultant array becomes ferromagnetic and its interaction with the external field produces four eigenvalues, $2\mu_B H$ apart. Since J_1 is generally very much greater than $2\mu_B H$ we can neglect the fine separation of the upper levels. In a real crystal the nearest neighbour energy required to reverse one atomic magnetic moment is $2zJS^2$ where z is the number of nearest neighbours. Furthermore, anisotropy energy is always present in real crystals. It turns out that the frequency of electromagnetic radiation required to excite the system by reversing an atomic magnetic moment is independent of H and is given by

$$\nu \cong \left[A\left(4zJS^2 \frac{M}{M_0} + A\right)\right]^{1/2} \Big/ h \tag{204}$$

where A is the average anisotropy energy and M/M_0 is the correction due to the thermal disorder of the long range alignment of the atomic moments. In MnF_2[73] the resonant frequency is $\sim 3 \times 10^{11}$ cycles/sec corresponding to 1 mm wavelength radiation and a value of $zJ \cong 2{\cdot}5 \times 10^{-3}$ eV. The magnetic energy in the specific heat gives $zJ \cong 2{\cdot}4 \times 10^{-3}$ eV. There is no such data available for metals. It is also possible to obtain M/M_0 versus temperature from these measurements and eqn. (204) by estimating the anistropy energy A. It is found that the long range order in non-metallic antiferromagnets varies approximately as a Brillouin function.

11. Paramagnetic Resonance of Transition Metal Atoms and Rare Earth Atoms

In this case the measurement is made above the Curie or Néel temperature. The frequency condition is simply the same as in ferromagnetic resonance except that one need not worry about demagnetizing factors so that

$$\nu \cong \left(1 - \frac{1}{g}\right)^{-1} eH/2m_0c \tag{205}$$

again yielding g directly. Paramagnetic resonance has been observed in gadolinium metal[72] and gives $g = 1{\cdot}95$ agreeing with the value obtained from ferromagnetic resonance, and this suggests no alteration above T_c of the basic exchange mechanism discussed under ferromagnetic resonance. Most of the

* Quantum mechanically $\boldsymbol{S}\cdot\boldsymbol{S} = \Sigma S(\Sigma S + 1) - \frac{3}{2}$ where ΣS is the sum of the m_s for an isolated pair of nearest neighbours. For a long chain we follow the procedure on p. 62.

work in paramagnetic resonance has been done on non-metals and provides a simple and accurate determination of g factors. There are no measurements for iron, nickel or cobalt metals above their Curie temperatures.

12. Paramagnetic Resonance of Electrons at Fermi Level

In lithium and sodium the resonance has been observed for the essentially isolated electrons at the Fermi level. The frequency can be calculated since we know the g factor of the free electron (2·0023)* but the intensity of the resonance absorption line depends on the paramagnetic susceptibility of the

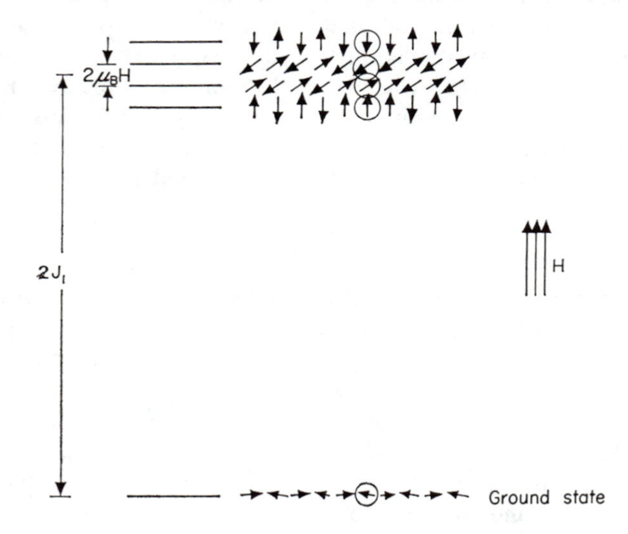

FIG. 122. The approximate eigenvalues and eigenfunctions of an antiferromagnetic linear chain of electrons in the ground state; and in the excited state in which m_s is reversed for one electron.

metal. Schumacher and Slichter[(74)] determined the paramagnetic susceptibility by placing their measurements on an absolute basis. The value is considerably larger than the value given by the free electron theory $\chi/\text{mol} = 3N_0\mu_B^2/2E_F$ where E_F is the energy at the Fermi level and N_0 is the number of free electrons per mole. The following table lists the results and *enables one to deduce the orbital diamagnetism of the electrons at the Fermi level from the measured susceptibility of the metal.*

This suggests that the Landau diamagnetism is considerably smaller than the free electron estimate of $\frac{1}{3}$ the Pauli spin paramagnetism. The most important result of this measurement is to indicate that the Pauli paramagnetism is

* The g factor is not exactly 2·00 due to refinements in the theory of the structure of the electron.

larger by a factor 1·5 to 2 and the orbital diamagnetism smaller by a factor ~10 than anticipated from the free electron theory, particularly in the case of sodium where it would have been expected to give reasonable answers.

TABLE XXIII

VARIOUS CONTRIBUTIONS TO THE SUSCEPTIBILITY OF LITHIUM AND SODIUM METAL (χ per mole $\times 10^6$)

Contribution to susceptibility	Li	Na
Paramagnetic spin Susceptibility (measured)	+27·1	+22·5
Estimated core diamagnetism ($\langle r^2 \rangle$)	−0·07	−4·1
Total measured Susceptibility in metal	+24·6	+16·1
Residue due to Orbital Diamagnetism (Landau)	−2·4	−2·3

13. Einstein de Haas (and Barnett) Effect

One can determine the g factor by measuring directly the ratio of the magnetic moment to the angular momentum of a specimen. The ratio of the intrinsic magnetic moment of an electron to its intrinsic spin angular momentum is $(e\hbar/2m_0c)/\hbar/2 = e/m_0c$ and the ratio of its orbital magnetic moment to its orbital angular momentum is $(e\hbar/2m_0c)/\hbar = e/2m_0c$, the former being twice as great.

If one suspends a long magnetized ferromagnetic rod from one end and oscillates the rod about its long axis at some convenient frequency, then the total angular momentum of the rod is the sum of the angular momentum of its mass rotation and a very small angular momentum due to the unpaired electrons. Of course this latter angular momentum is present even without spinning the rod. The rod is oscillated inside a magnetic coil which keeps it magnetized. Suppose one suddenly reverses the direction of magnetization by reversing the current in the coil, then the angular momentum due to the unpaired electrons will now subtract from the massive angular momentum. Since the massive angular momentum arises from the rotation of the nuclear masses around the axis of the rod and the mass of the nuclei is unchanged when the magnetization direction is reversed, the massive angular momentum must be altered in the field reversing process in order to conserve the *total* angular momentum. An optical device measures the change in frequency of the specimen. The ratio of the change in angular momentum to the change in magnetic moment in units e/m_0c is called the *gyromagnetic ratio*,* ρ. The change in the magnetic moment of the rod when the magnetic field is reversed, can be determined from its saturation magnetization or from an actual measurement. The Landé g factor for a ferromagnet is

$$g = 1 + \frac{J^2 + S^2 - L^2}{2J^2} \cong 2/\rho \tag{206}$$

The values of g obtained by this method agree with those obtained by ferromagnetic resonance,[75] (cf. Sec. 9). A bibliography of g factor measurements is given by S. P. Heims and E. T. Jaynes *Rev. Mod. Phys.* **34**, 154 (1962). It is quite common in the literature to find g in eqn. (206) called g' and $[1 - (1/g)]^{-1}$ in eqn. (203) called g. It is rather confusing!

In the Barnett method one measures the magnetization produced as one alters the rotation of the specimen. Both methods are equivalent.

Problems

1. Look up the room temperature susceptibility per mole of as many non-magnetic metallic elements as you can find and see if there is any correlation with their electronic specific heat coefficient or their position in the periodic table. Can you conclude anything about the electronic structure of a metal from its room temperature susceptibility?

2. Use the free electron theory to calculate the susceptibility of the electrons at the Fermi level in aluminium. Estimate the core diamagnetism of aluminium by using hydrogenic wave functions, determining Z_{eff} from the X-ray absorption edges. Compare this with the measured value.

3. Write a report on the relationship of magnetic permeability to the purity and state of cold work in iron metal.

4. Write a report on the preparation of alnico magnets and the current ideas on why they remain permanent magnets in the absence of an external field as contrasted to a piece or pure iron.

5. Assume the earth's centre to be a solid $Fe_{0 \cdot 8}Ni_{0 \cdot 2}$ mass 2,000 miles in diameter and 2·5 times the density of this alloy at zero pressure. Could this core possibly account for the earth's magnetic field of ~ 1 gauss at the surface? Do the $3p$ electrons have much overlap at these interatomic distances?

6. Write a report on the crystal structure and magnetic properties of magnetite. What are the commerical uses of ferrites? How does a tape recorder work (magnetic aspects only)?

7. Compare the initial costs and fields attainable with superconducting and conventional magnets. Considering operating costs as well as initial costs, which of the two are more economical for fields of 20 kilogauss.

Summary

Measurements of the magnetization of a metal in a static magnetic field do not generally provide clear cut information about the electron eigenfunctions or eigenvalues. In ferromagnetic metals the low temperature

* The reciprocal of twice the gyromagnetic ratio is sometimes called the *magnetomechanical factor*.

magnetization enables an empirical determination of the exchange energy while in non magnetic metals the susceptibility provides a qualitative indication of the density of states. In some cases, particularly the rare earths, the slope of the susceptibility versus temperature curve enables one to determine the magnetic moment per atom.

Measurements in a.c. magnetic fields primarily provide the g values of magnetic atoms.

CHAPTER XI

NUCLEAR MEASUREMENTS

Introduction

In this chapter we shall discuss experiments which rely on the nucleus or high energy particles emitted by the nucleus for gathering information about the solid. It is convenient to group these measurements into four categories:

1. Nuclear magnetic and quadrupole resonance
2. Mössbauer effect
3. Radioactive tracers
4. Positron annihilation

In the first group of measurements one measures the magnitude of the magnetic fields or electric field gradients created by the electrons at the position of the nucleus. Since the nucleus is in effect a point at the centre of the atom the information is limited to the magnitude of the fields at this point. In spite of this restriction it is possible to gain some information about the electron wave functions.

The second group of measurements, the Mössbauer effect, utilizes a rather special property of the excited or isomeric state of certain nuclei like Fe^{57}. The energy widths of the ground state and the excited state are extremely sharp and a fraction, e^{-M}, of the gamma rays, emitted by the excited state, Fe^{57*}, are just the right energy for resonance absorption by Fe^{57} in the ground state. If the absorber is moved relative to the emitter with a velocity of only a few cm/sec, then the "relative" energy of the 14·4 keV γ-ray is altered by about 10^{-7} eV and this is enough to eliminate the resonance absorption! Thus other small energy shifts such as the hyperfine interaction of the nuclear magnetic moment with local magnetic fields is sufficient to alter the resonance conditions. In such a case one determines the relative velocity necessary to re-establish the resonance condition and thus measure the energy shift and effective magnetic field at the nucleus.

The third group of measurements, the use of radioactive tracers, utilizes the high-energy decay products (α, β, γ-rays) to detect individual atoms. Its principal application in solid state physics has been the measurement of diffusion coefficients although radiation damage studies and elemental analysis have been of interest.

In positron annihilation, the fourth group of measurements, one measures the momenta of the γ-rays emitted when a positron captures an outer electron in a metal. In principle one can hope to gain information about the momenta of outer electrons in metals but in practice it has yet to prove its usefulness.

1. Nuclear Magnetic Resonance (NMR), and Nuclear Quadrupole Resonance (NQR)

Theoretical NMR and NQR Background

Commercial resonance equipment is available from Varian Corp., consisting of an electromagnet and a radio frequency source and detector. For a good deal of NMR work the equipment can be used as received but measurements as a function of temperature and certain specialized measurements require home-made adaptation. The field is still new and the apparatus is not as commonplace as X-ray diffraction equipment. We shall first outline the basic ideas in NMR and NQR and then return to some of the questions of types of apparatus.

The eigenvalues of a nucleus of spin I and magnetic moment μ in a magnetic field, H, are

$$-\frac{\mu}{I}IH, \qquad -\frac{\mu}{I}(I-1)H, \qquad -\frac{\mu}{I}(I-2)H, \ldots +\frac{\mu}{I}IH.$$

There are $2I + 1$ eigenvalues and the quantized values $I, I - 1, I - 2 \ldots$ are designated by quantum numbers m_I. Thus a nucleus with spin $I = 5/2$ has the quantum numbers m_I, of $-5/2$, $-3/2$, $-1/2$, $+1/2$, $+3/2$, $+5/2$. The eigenvalues and eigenfunctions (represented by the magnetic moment directions) are shown in Fig. 123 for the two cases $I = 1/2$ and $I = 5/2$. The spacing between levels is $(\mu/I)H$. The energy scale in Fig. 123 is merely representative.

If we add an electric quadrupole moment, eQ, to the nucleus and place it in an axially symmetric electric field gradient, eq, then the eigenvalues for the so-called *first-order quadrupole effect* (interaction e^2Qq small relative to μH; θ the angle between the direction of the field, H, and symmetry axis)

$$E = \frac{e^2Qq}{4I(2I-1)}[3m_I^2 - I(I+1)] \qquad \text{valid for } I \geq 1,\ H = 0 \qquad (207)$$

$$E = \frac{\mu H}{I}m_I + \frac{e^2Qq}{8I(2I-1)}[3m_I^2 - I(I+1)](3\cos^2\theta - 1) \qquad \text{valid for } H = H.$$

The eigenvalues and eigenfunctions for the quadrupole interaction only ($H = 0$) are shown schematically in Fig. 124 diagram 2 for the cases $I = 1/2$ and $I = 5/2$.* Conservation of angular momentum permits energy changes

* There are higher order quadrupole terms whose contribution to the energy levels are[76] $= [2m_I I/\mu H][e^2qQ/4I(2I-1)]^2[\{(\partial^2V/\partial x\partial z) + (i\partial^2V/\partial y\partial z)^2\}\{8m_I^2 - 4I(I+1)/3 + 1\} + \left\{\frac{1}{2}\left(\frac{\partial^2V}{\partial x^2} - \frac{\partial^2V}{\partial y^2}\right) + i\frac{\partial^2V}{\partial x\partial y}\right\}^2\{-2m_I^2 + 2I(I+1) - 1\}]$

which alter m_I by one unit so that the energy separation of the levels in Fig. 124 are $3e^2Qq/20$ and $6e^2Qq/20$. If we now add a magnetic field to the first order quadrupole interaction we have the level spacing of diagram 3, Fig. 124 (heavy lines). The eigenvalues for $eq = 0$ have been drawn in (dashed lines) as a reference. The energy level spacing for a change in m_I of one unit are $\mu H + 6e^2Qq/20$; $\mu H + 3e^2Qq/20$; μH; $\mu H - 3e^2Qq/20$; $\mu H - 6e^2qQ/20$

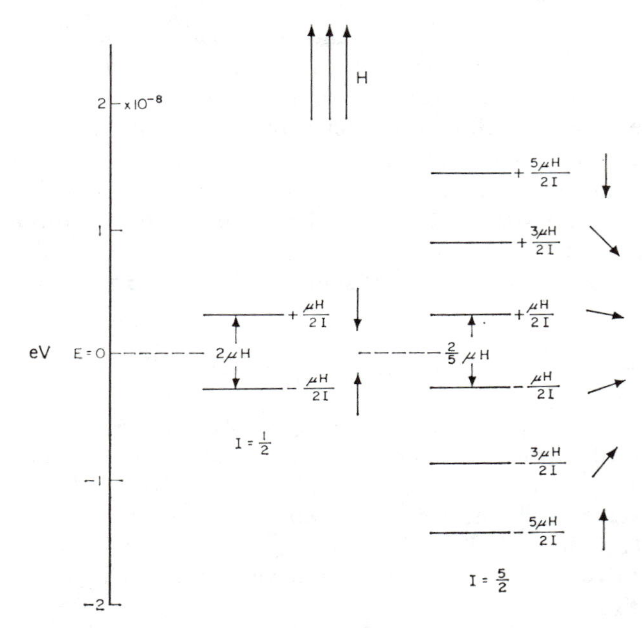

FIG. 123. The energy levels of nuclei of spin 1/2 and 5/2 in an external magnetic field. The energy scale is merely representative.

leading to five lines with the *central line* $m_I - \frac{1}{2} \rightarrow +\frac{1}{2}$ *unaltered in first order.* If the quadrupole interaction is strong enough even this central line is altered as indicated in the footnote.

At $T = 0$°K all the nuclei are in their lowest energy state but at any finite temperature the probability of finding a nucleus in any one of the energy

---

where V is the electric potential. If the electric field gradient in the crystal is axially symmetric then the central line $m_I = \pm 1/2$ has a total energy in second order

$$E = \frac{\mu H}{I} m_I + \frac{e^2Qq}{4I(2I-1)}\left[\frac{3}{4} - I(I+1)\right](3\cos^2\theta - 1) - m_I\left(\frac{3e^2qQ}{2I(2I-1)}\right)^2 \frac{1}{16\mu H}$$
$$\times\ [(I(I+1) - 3/4)(1 - \cos^2\theta)(9\cos^2\theta - 1)]$$

where θ is the angle between H and the axially symmetric axis. The second order contribution can be decreased by increasing H.

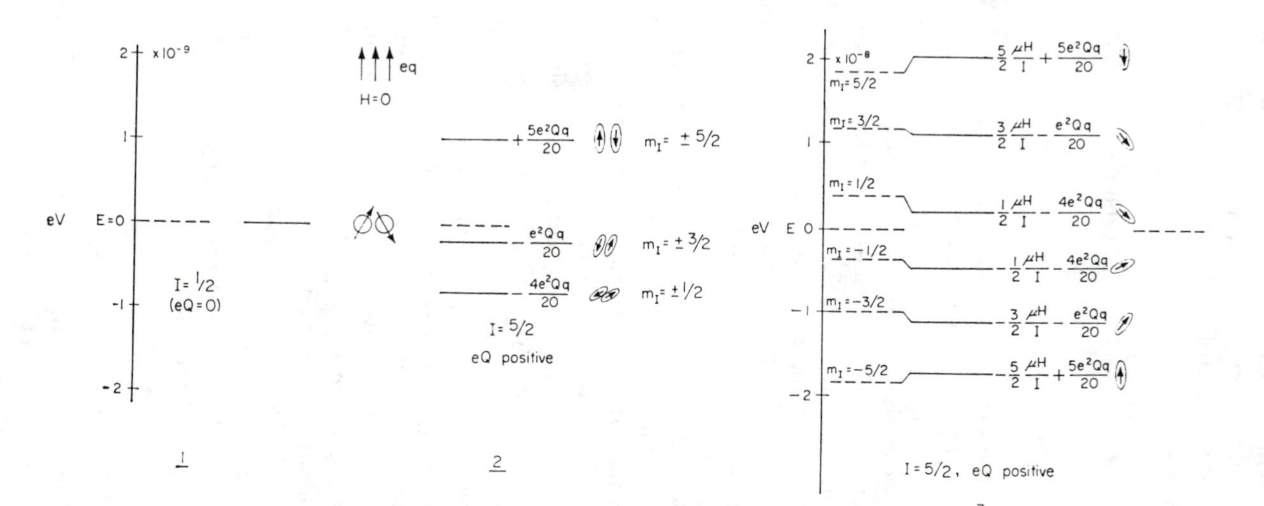

FIG. 124. Diagrams 1 and 2. The energy levels of nuclei of spin 1/2 and 5/2 in an electric field gradient and zero magnetic field. The energy scale is merely representative.
Diagram 3. The energy levels of a nucleus of spin 5/2 in a magnetic field and in an electric field gradient parallel to H. The coupling constant $e^2qQ \ll \mu H$. The levels with $eq = 0$ are given as a reference.

states, E (eqn. (207)), is given by eqn. (41). Since the energies are so small in these nuclear interactions, $kT \gg E$, we can expand the exponentials ($e^{-x} \to 1 - x$) so that the probability of finding a nucleus of spin I in any one of the energy levels is

$$P = \frac{(1 - E/kT)}{2I + 1} \quad \text{valid for } kT \gg E \tag{208}$$

Thus the $(2I + 1)$ energy levels are almost equally populated at ordinary temperatures.

In performing a resonance experiment the sample is placed inside a coil which emits a radio frequency signal and both are placed in a magnetic field. When the energy, $h\nu$, of the electromagnetic radiation equals the energy difference of any two levels whose m_I differs by one unit then nuclei in the lower level can absorb energy and be excited to the upper level. On the other hand a peculiarity of quantum mechanics is a process called *stimulated emission* whereby a nucleus in the upper level is stimulated to occupy the lower level by emitting a photon of frequency $h\nu$. This stimulation comes, peculiarly enough, from a photon of the same frequency which is near the nucleus but does not actually take part in the process (a sort of catalyst). If there was just as much stimulated emission as absorption then the system would not absorb energy and one could not detect the resonance. However there is ordinarily more absorption than stimulated emission because at thermal equilibrium there are more nuclei in the lower level than upper level. If the absorption of radiation of the resonant frequency continued we should eventually have an equal population of the upper and lower level so that an equilibrium between absorption and stimulated emission would exist. This is called *saturation*. However if thermal equilibrium can be quickly established at all times, eqn. (208), then some of the nuclei in the upper level would occupy the lower level by transmitting their energy to the lattice vibrations. Thus the system of nuclei are continuously absorbing energy from the electromagnetic field and transferring this energy to the lattice vibrations. The time required for thermal equilibrium to be established is called the *spin-lattice relaxation* time and is commonly denoted T_1. It is generally in the range of milliseconds for a typical metal like copper and as long as this time is short compared to the rate at which the radio frequency field is exciting half the nuclei to the upper level, saturation will not be reached.

How the Measurements are Made

In the actual measurement there are several modifications of technique in detecting the resonance. They all depend on the fact that at resonance not only are the electromagnetic waves absorbed but they are also strongly scattered in phase with the incident radiation. In the Pound–Purcell technique the scattering is observed in the emitting coil as a back e.m.f. It is a very weak signal and in order to amplify it against a good deal of noise one adds to the d.c. magnetic field a weak audio frequency oscillating magnetic field of several gauss amplitude or an audio frequency variation to the radio frequency of the oscillator. These are called amplitude and frequency modulation respectively. This technique enables one to eliminate noise by using a narrow band phase-sensitive amplifier relying on the audio frequency of the amplitude or frequency modulation as a reference signal.

In the Bloch technique, employed in the Varian equipment, the scattered signal is detected by a second "pick-up" coil near the sample and at right angles to the emitting coil. The right angle prevents the "pick-up" coil from detecting the direct signal of the radio frequency emitting coil. Amplitude modulation is employed to reduce the background noise. It should be borne in mind that both the Pound–Purcell and Bloch techniques depend on the fact that the radio frequency photons produced by the emitting coil are all in phase. If they were independent photons of the same frequency one would not be able to observe the weak signal by these techniques.

It has been found that 300 mesh metal powders are adequately fine for a sample. Powder particles must be used since the depth of penetration of electromagnetic radiation is limited to $\sim$1–50 microns. It may be necessary, especially in alloys, to produce these by filing and subsequent annealing may be necessary to remove the cold working (unless one is studying cold working). Fairly complete annealing of cold work is accomplished by annealing at about half the absolute melting temperature. It should, of course, be done in vacuum or inert atmosphere. In one case though, copper, such annealing drives the oxide surface layer of each powder particle into the particle so that the individual particles are not insulated from each other. In effect this makes the powder specimen a large porous metal specimen and the skin depth effect prevents the radiofrequency signal from penetrating the sample. However exposing the sample to air for a few minutes builds up sufficient oxide layer to again insulate the particles.

If one is investigating an ordered cubic alloy, such as beta brass, it has been found that it is virtually impossible to obtain an alloy well enough ordered and of proper stoichiometry so that the quadrupole effect does not remove all but the central component from the resonance line. The problem of performing measurements as a function of crystal orientation has been solved by Hofmann and Sagalyn in a few cases (such as copper and aluminium) where the signal is so strong that a single crystal rod has sufficient surface area. In general, though, one may have to build up sandwiches of thin crystals between insulating layers to obtain a sufficiently strong signal.

Table XXIV separates nuclear resonance measurements into various types, the phenomena investigated and the information sought. We shall discuss each separately.

TABLE XXIV

THE VARIOUS TYPES OF NUCLEAR RESONANCE MEASUREMENTS AND THE INFORMATION SOUGHT FROM THEM

Measurement	Phenomena investigated	Information sought
a. Frequency of resonance	i Knight shift	$\lvert\psi(0)\rvert^2$, Pauli susceptibility
	ii Ferromagnetic shift	Effective field at nucleus
	iii Pure quadrupole resonance	Electric field gradient in non-cubic elements and non-cubic ordered alloys
	iv Pressure and temp. dependence	Variation of above with inter-atomic distance
b. Intensity of resonance	"Wipe out" numbers	Electron probability distribution around an impurity atom
c. Shape of resonance	i Second moment, Anisotropic Knight shift, exchange broadening	Nearest neighbour configuration in alloys, *p* character of electrons at Fermi level
	ii Line shape analysis	Strains due to cold work, distribution of electric field gradients
	iii Line narrowing	Atomic diffusion rate vs. temp.
d. Relaxation times, spin echoes	T_1, T_2	Spin-lattice coupling (Korringa relation), nuclear-nuclear coupling
e. Electron spin and NMR (simultaneous)	Overhauser effect	Polarization of nuclei

a. Frequency of Resonance

i. Knight Shift

The Knight Shift in Lithium, Sodium and Beryllium. Since, in the absence of quadrupole effects, the frequency of the nuclear magnetic resonance depends only on the product of the nuclear magnetic moment and the magnetic field at the nucleus one might expect that the frequency could be determined from a knowledge of the nuclear magnetic moment and the external magnetic field applied to the sample. That such was not the case was first discovered by W. Knight when he noticed that the copper nuclear resonance occurred at a slightly higher frequency in the metal than in the diamagnetic compound copper chloride. It was suggested by C. H. Townes that the external field created a small additional field at the nucleus due to the Pauli susceptibility of the electrons at the Fermi level. Since the electrons at the Fermi level in copper have considerable 4*s* character and since the 4*s* wave functions are finite at the nucleus, the predominance of 4s electrons with m_s parallel to the external field gives rise to an additional effective field at the nucleus

$$H_{\text{eff}} \cong \frac{8\pi}{3} \mu_B \{\uparrow|\psi(0)|^2 - \downarrow|\psi(0)|^2\}\mathscr{R}$$
$$\mathscr{R} \cong \frac{1}{\sqrt{\left[1 - \left(\frac{Z}{137}\right)^2\right]\left\{1 - \frac{4}{3}\left(\frac{Z}{137}\right)^2\right\}}} \tag{209}$$

where $\uparrow|\psi(0)|$ and $\downarrow|\psi(0)|$ are the values of the s wave functions at the nucleus for electrons with spin, m_s, parallel and antiparallel to the external field and $\mathscr{R}$ is a small relativistic correction to the wave functions (Z is the atomic number). This is called an effective field since it is the average magnetic field experienced by the nucleus and is due to the magnetic moments of the s electrons weighted according to their electron probability distribution at the nucleus. Since $\uparrow|\psi(0)| \cong \downarrow|\psi(0)|$ for all s electrons except those few that have reversed their m_s value at the Fermi level we can rewrite eqn. (209) as the Knight shift, $\Delta\nu/\nu$,

$$\frac{\Delta H}{H} = \frac{8\pi}{3} \chi|\psi(0)|_F^2 = \frac{\Delta\nu}{\nu} \tag{210}$$

where $|\psi(0)|_F$ is the value of the wave function at the nucleus for the s electrons at the Fermi level, χ is the Pauli paramagnetic susceptibility per atom and H is the magnetic field at which resonance would have occurred if the electrons at the Fermi level did not reverse their m_s values.* The Knight shift, $\Delta\nu/\nu$, is of the order of 0·1% to 2·0% and is independent of the external field since the frequency shift, $\Delta\nu$, and the frequency are both proportional to the field. A measurement of the Knight shift thus provides us with the product $\chi|\psi(0)|_F^2$.

Since we are unable to calculate χ or $|\psi(0)|_F$ for a metal, a measurement of the Knight shift still leaves us with the problem of apportioning the observed value between χ and $|\psi(0)|_F$. Lithium, sodium and beryllium are exceptions since χ has been directly measured by electron spin resonance, and this allows us to determine $|\psi(0)|_F$.

* Quite often one approximates H (or ν) from a measurement of a diamagnetic salt such as copper chloride. There are two conditions to be considered in this approximation, the orbital diamagnetism and the chemical shift. The orbital motion of the electrons creates a small diamagnetic field at the nucleus $\Delta H/H \cong 2{\cdot}3 \times 10^{-5} Z^{1{\cdot}38}$ which is about the same for any atom independent of its environment while the chemical shift (~0·01%) arises from slight changes in the outer electron wave functions and their accompanying diamagnetic changes. These effects are negligible if the Knight shift is greater than 0·1%. If not, the entire correction can be determined from the radial wave function.

$$\Delta H/H = -\frac{e}{3m_0c^2}\int_0^a |R|^2 r \, dr$$

where $\int_0^a |R|^2 r^2 \, dr = 1$ and a is the atomic radius.

It is generally expressed as $\xi = |\psi(0)|_F^2/|\psi(0)|_A^2$ which is the ratio of the s electron probability distribution at the nucleus for the s electrons at the Fermi level in the metal and for the outer s electron in the free atom ($2s$ for lithium, $2s$ for beryllium, $3s$ for sodium). These ratios are $\xi = 0{\cdot}43$ for lithium, ~ 0 for beryllium, and 0·72 for sodium. The results suggest that the wave functions at the Fermi level are about 50% $2p$ like in lithium, all $2p$ like in beryllium and 25% $3p$ like in sodium since the p electrons (whose wave functions vanish at the nucleus) do not give any appreciable Knight shift. In the case of lithium and sodium the quantum defect calculations, p. 67, yield fairly good theoretical agreement with the measured values. That beryllium should be devoid of s character at the Fermi level is interesting. Its large positive ordinary Hall coefficient also implies considerable mixture of $2p$ wave functions at the Fermi level.

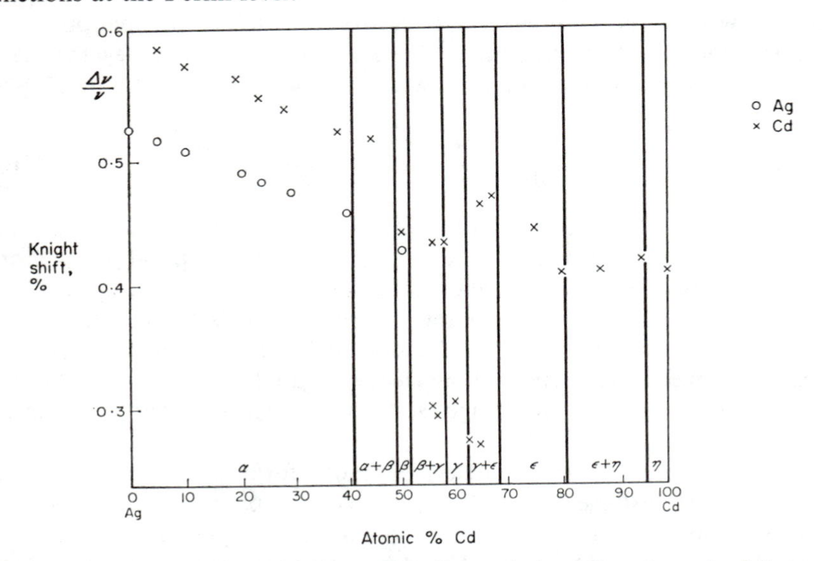

FIG. 125. The Knight shift in per cent for silver-cadmium alloys in each of their various phases. At high cadmium concentrations the silver resonance was too weak to be seen.

The Knight Shift in Silver-Cadmium Alloys. Perhaps the greatest effort in the Knight shift measurements has been the variation of Knight shift with alloy concentration.[77] Even though one is unable to calculate the Knight shift for any but the alkali metals it is felt that *changes* in the Knight shift might suggest the nature of the *changes* of the wave functions in alloys. We shall describe the system silver-cadmium since both elements have spin $\frac{1}{2}$ and quadrupole effects do not affect the resonance. The silver-cadmium system is similar to copper-zinc in its phase diagram and physical properties, and the results are qualitatively applicable to brass. Figure 125 separates the room temperature phases of silver-cadmium; α is f.c.c., β is b.c.c., γ is complex cubic, ε and η are h.c.p. In the α-phase the cadmium dissolves in the silver with a Knight shift about 50% larger than in pure cadmium. This can be due either to the cadmium atoms developing a larger percentage of *s* character when they are part of a f.c.c. silver lattice or else the contribution of the paramagnetic susceptibility at the Fermi level is greater in the vicinity of the cadmium atom. The addition of cadmium to silver, however, decreases the paramagnetism compared to pure silver so that we choose the change in the percentage of *s* character as more likely. The addition of cadmium to silver decreases the Knight shift at the same rate for both silver and cadmium and suggests that the density of states at the Fermi level is decreasing in the α-phase since the paramagnetism is also decreasing. The Knight shift for cadmium is about the same in the β, ε and η phases but is reduced by a factor of two in the complex γ-phase. Since complex crystal structures generally indicate complex orbital wave functions (the functions must show the symmetry of the crystal) the decreased Knight shift in the γ-phase is probably due to a decreased percentage of *s* character. It is not obvious why there is so little change in the Knight shift of cadmium between the h.c.p. ε and η-phases. In the ε-phase $a_0 = 3{\cdot}060$ Å, $c_0 = 4{\cdot}81$ Å, $c_0/a_0 = 1{\cdot}572$ and in the η-phase (pure cadmium) $a_0 = 2{\cdot}9728$ Å, $c_0 = 5{\cdot}6054$ Å, $c_0/a_0 = 1{\cdot}8856$. The η-phase has about a 10% larger atomic volume. The large deviation from close packing in the η-phase suggests a large admixture of 5*p* bonding character particularly in the basal plane but this is not reflected in the Knight shift. Unfortunately the nuclear resonance of silver is rather weak and was not detected in the cadmium-rich alloys. Measurements of the electronic specific heat and susceptibility of the silver-cadmium system might help unscramble the relative χ and $|\psi(0)|^2_F$ contribution to the Knight shift particularly in the ε and η-phases.

There are at least a dozen metallic elements with nuclear spin $I = \frac{1}{2}$, so that the Knight shift could be studied in many alloys without quadrupole interference. Parallel measurements of the electronic specific heat, susceptibility and transport properties might aid in understanding the changes in the density of states and wave functions at the Fermi level.

ii. Ferromagnetic shift

The Ferromagnetic Shift in Iron. The unpaired electrons in a ferromagnet contribute to the effective magnetic field at the nucleus in several ways. The most obvious way, the actual field from the intrinsic magnetic moments of the unpaired *d* or *f* electrons is zero if the *d* or *f* electron probability distribution is spherically symmetric or has cubic symmetry as the t_{2g} or e_g orbitals in Fig. 17. This latter condition follows simply from a direct calculation of the field of a magnetic dipole. The *d* or *f* electrons also contribute from any unquenched orbital motion. This is approximately equal to $H_{eff} \cong 2S\mu_B(2 - g)\langle 1/r^3 \rangle$ where S is the unpaired spin, g is the Landé g factor and $\langle 1/r^3 \rangle$ is the average value of $1/r^3$ for the unpaired electrons. A typical value of $\langle 1/r^3 \rangle$ is about 30/Å^3 for 3*d* electrons and 40/Å^3 for 4*f* electrons so that H_{eff} is about 40 kilogauss for iron. The largest contribution in iron, though, arises from the alteration of the 1*s*, 2*s*, 3*s* and 4*s* wave functions with m_s parallel to the unpaired 3*d* electrons due to antisymmetrization. This has already been mentioned on p. 61. The ferromagnetic shift has been observed in iron to be due to an effective field of −333 kgauss. Hartree–Fock free atom wave functions enable one to calculate the properly antisymmetrized 1*s*, 2*s* and 3*s* wave functions since they should not be altered appreciably in the metal. Only the 4*s* contribution, being part of the metallic band, cannot be readily calculated so that the experimental value of −333 kgauss enables one to determine the 4*s* contribution. The various contributions are given in Table XXV. The alteration of the *s* wave functions at the nucleus by the unpaired electrons is called *exchange polarization*, (see p. 324).

TABLE XXV

THE ESTIMATED ONE ELECTRON CONTRIBUTIONS TO THE EFFECTIVE FIELD, H_{eff}, IN IRON METAL

Contributor	Amount in kilogauss
3*d* dipole contribution	~0
1*s* exchange polarization	−8
2*s* exchange polarization	−700
3*s* exchange polarization	+355
unquenched orbital	+40
total (calculated)	−313
total (observed)	−333
4*s* exchange polarization (difference)	−20

It would be fallacious to conclude from these results that there are more outer electrons with m_s antiparallel to the unpaired 3*d* electrons than parallel since this is contrary to our observations in free atoms. However, as we have already indicated, one can conclude that there is more 4*s* character in the band of eigenfunctions with m_s antiparallel to the unpaired 3*d* electrons than parallel. The total magnetization of the outer electrons, though, *is* parallel and is due to the prevalence of the 3*d* character of the wave functions.*

Recently Benedek and Armstrong[78] have studied the temperature and pressure dependence of the Fe^{57} nuclear resonance in metallic iron. At −77°C, 0°C and 84·2°C the resonant frequency decreases linearly with pressure at a rate of 0·2% per ten kilobars. This represents a very slight

* In the case of β-Mn (see p. 350) there is a negative (Knight) shift of ~0·13% whose origin is similar to that of iron. The difference, though, is that the unpaired *d* electrons in β-Mn are induced by the application of the magnetic field whereas they naturally exist in iron.

change in H_{eff} from about -333 to -332 kilogauss and probably represents the change in the relative $4s$–$3d$ admixture of the eigenfunctions with m_s parallel to the unpaired $3d$ as compared to antiparallel. This change is subtle and well beyond any theoretical calculation.

At one atmosphere of pressure the frequency of the Fe^{57} resonance decreases $\sim 4\%$ from -77°C to 84·2°C which is due partly to the decreasing magnetization with increasing temperature. Equation (209) is the expression for the effective field on each atom but the nuclear resonance measurement averages the effect over all atoms so that the right side of eqn. (209) has to be multiplied by the measured Brillouin function for iron, Fig. 115. However the frequency decreases faster with temperature than the Brillouin function. By using the pressure variation of the resonant frequency to provide the volume dependence, Benedek corrected the temperature dependent data for the volume expansion and change in magnetization with temperature and concluded that in the range 0 to 600°K the effective field, H_{eff} is given: $H_{eff} = -333\ (1 - 7{\cdot}7 \times 10^{-8} T^2)$ kilogauss.

Thus decreasing the volume at constant temperature and increasing the temperature at constant volume both decrease the magnitude of H_{eff}. A decrease in the magnitude of H_{eff} is primarily due to an equalization of the $4s$ character of the band with m_s parallel and antiparallel to the unpaired $3d$ electrons. Beyond this we can say little else except that the Curie temperature, hence magnetization appears to be pressure insensitive (see p. 303).

Negative effective fields have been found in Co* (-217 kgauss, f.c.c.; -228 kgauss, h.c.p.) and nickel (-170 kgauss).

iii. Pure Quadrupole Resonance

Pure Quadrupole Resonance in Gallium. In the absence of a magnetic field the eigenvalues of a nucleus with quadrupole moment, eQ, in an electric field gradient, eq are given in eqn. (207). If the electron probability distribution has spherical and/or cubic symmetry around the nculeus the electric field gradient is zero at the nucleus. This is not true for a disordered alloy even if it is cubic since the nearest neighbour distances are not all identical (size effect) nor is the electron probability distribution cubic since it may overlap one type of neighbour more than another.

In non cubic gallium ($I = \frac{3}{2}$) the quadrupole interaction produces two energy levels and a single quadrupole resonance line which excites the system between these levels.[79] The frequency of the line is 10·908 Mc/sec (Ga^{69}) corresponding to a separation of levels of $\sim 4{\cdot}5 \times 10^{-8}$ eV. In indium ($I = \frac{9}{2}$)

* The gamma rays emitted from radioactive cobalt nuclei have known anisotropic distributions relative to the direction of their magnetic moments. A study of the anisotropy at low temperatures indicates the number of cobalt nuclei which are aligned parallel to the magnetization direction of the ferromagnetic state and from this H_{eff} can be determined just as in specific heat (p. 278).

Knight finds the *coupling constant* $e^2qQ/h = 45$ Mc/sec, about twice that of gallium. In indium there are 4 eigenvalues (see eqn. (207)).

Even though the coupling constants have been measured for gallium and indium there still exist two problems in determining the electric field gradients due to the outer electrons, and from this an indication of the outer electron probability distribution. The first problem is that of the so-called *anti-shielding factor*, γ. This arises from the readjustment of the inner electron probability distribution in the presence of an external electric field gradient. For low Z elements the effect is small and negative but for elements with $Z > 10$ the effect is to multiply the external electric field gradient at the nucleus by factors as high as 100. This arises because the electric field gradient at the nucleus due to the inner electrons is zero only because of perfect spherical symmetry. These inner electrons are so close to the nucleus that any slight distortion from spherical symmetry creates large electric field gradients at the nucleus. This magnification or anti-shielding factor has not been measured but has been calculated. Thus the value of eq determined from the coupling constant must be divided by a large theoretical number in order to determine the outer electron probability distribution. The second problem is that we are not quite sure about the errors in eQ since the nuclear quadrupole moments are always determined on atoms with electrons surrounding the nuclei. Thus an anti-shielding factor correction has to be made! This problem does not arise in NMR since the external magnetic field is only altered slightly at the nucleus.

Nonetheless these problems will some day be worked out and values of eq will certainly help in deciding the nature of the outer electron wave functions. While the quadrupole interactions in silver, indium and beryllium are probably too small to be observed in specific heat measurements and the low temperature specific heat effects observed in rhenium and zinc are too large to be seen in the radiofrequency range nevertheless there is room for considerable effort in both fields.

iv. Pressure and Temperature Dependence

Pressure and Temperature Dependence of Knight Shift. We have already mentioned the pressure and temperature dependence of the Fe^{57} resonance. In the case of the Knight shift there is little pressure or temperature dependence, except for the alkalis, since the paramagnetic susceptibility and s wave function character are generally insensitive to the pressures attained in the laboratory.

Benedek and Kushida[80] have studied the variation of the Knight shift with pressure for lithium, sodium, rubidium, caesium and copper. The Knight shifts are 0·0249%, 0·113%, 0·635%, 1·49% and 0·232%, respectively, at zero pressure and room temperature. At 10 kilobars pressure the Knight shifts are changed by −1·1% for lithium, −1·3% for sodium, 7% for rubidium, 48% for caesium and −0·6% for copper. Using the quantum defect method, Brooks has calculated the values of $\xi = |\psi(0)|_F^2/|\psi(0)|_A^2$ as a function of volume and these compare favourably with the pressure dependence of the Knight shift (corrected to volume dependence) for lithium, sodium and rubidium but only fair agreement is obtained for caesium. This assumes that χ, the paramagnetic susceptibility, is independent of volume. No such calculation is available for copper. This underscores the success of the quantum defect method for the alkali metals.

No measurements are yet available as to the pressure dependence of the quadrupole resonance but the temperature dependence of the NQR in indium[(81)] is found to vary by 23% between 4°K and 222°K. In the case of gallium there is an increase in the coupling constant of 3% in cooling from 300°K to 77°K. These pieces of information are not very helpful on their own but then there are very few other measurements on indium and gallium from which to make any deductions about the electron probability distribution.

b. Intensity of Resonance

Intensity of Cold Worked Copper. If we take a well annealed sample of pure 400 mesh copper powder, then every Cu^{63} nucleus ($I = \frac{3}{2}$) will resonate at one frequency and every Cu^{65} nucleus at a second frequency. The line from either isotope is not perfectly sharp for reasons we shall discuss below but the integrated intensity of the line over a frequency range $\sim \Delta\nu/\nu = 0{\cdot}2\%$ includes all the nuclei of that isotope in the sample. The quadrupole effects have cancelled out due to cubic symmetry.

If we now compare the *peak* intensity in this sample with that in extremely cold worked copper filings we should find the *peak* intensity reduced by a factor of 2·5. Cold work introduces dislocations which distort the cubic symmetry over a wide range depending on the proximity of the particular nucleus to the dislocation line. This produces an electric field gradient at the nuclei and reintroduces the quadrupolar shift in eigenvalues. Figure 126 shows the effect in copper to second order and averaged over all directions of the strain.* If $\frac{1}{2}e^2qQ$ is large compared to the width of the resonance ($\frac{1}{2}e^2qQ \gg H/1000$) then in first order one expects two of the three lines to move to some other frequency and reduce the intensity by a factor of about 3 (from quantum mechanical considerations it turns out to be 2·5). If $\frac{1}{2}e^2qQ$ is of the order of the line width then one expects the outer two transitions $\mu H/I - \frac{1}{2}e^2qQ$ and $\mu H/I + \frac{1}{2}e^2qQ$ to move out to the "wings" of the central line which is at $\mu H/I$. Indeed the factor 2·5 found in cold worked copper respresents the reduction of the central maximum but a measurement of the integrated intensity including the wings of the now broadened line reveals 90–95% of

* We have used $\overline{3\cos^2\theta - 1} = 1$ as an approximate average. The lines in Fig. 126 represent the average of the energy levels, now somewhat broadened due to the distribution in θ values.

the nuclei still contributing to the resonance. Thus severe cold working of copper yields a small value of *eq* and the first order quadrupole approximation is adequate. The actual value of *eq* for each copper nucleus depends on its distance from the dislocation line.

Intensity of the Resonance in Dilute Copper Alloys. If we substitute some other atoms in annealed copper (like zinc) the cubic symmetry of the electron probability distribution is altered in the vicinity of the zinc atoms. The same energy level diagram as in Fig. 126 applies in this case, although the actual value of *eq* depends on the proximity of the copper nucleus to the zinc atom. Experimentally the effects are much larger than cold working since only 2% zinc addition reduces the central maximum by as much as extreme cold working. The term "wipe out" number is used and indicates the number of solvent atoms whose removal would decrease the *peak* intensity by the same

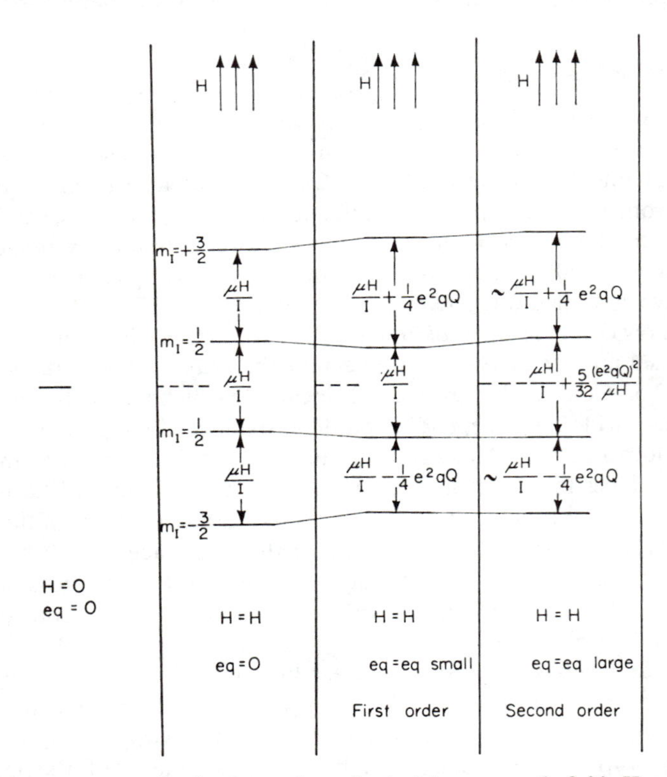

FIG. 126. The energy levels of a nucleus of spin $\frac{3}{2}$ in a magnetic field, *H*, and in an electric field gradient, *eq*, in first order and second order. In both of these cases an approximate average has been taken for various directions of *eq* relative to *H*. While the energy levels are indicated as sharp they merely represent the centre of gravity of a broadened distribution. The levels for $eq = 0$ and $H = 0$, $H = H$ are given as a reference.

amount as the addition of one solute atom. Thus a "wipe out" number of 18 for zinc added to copper indicates that the peak intensity is reduced as much as would occur if the 12 nearest and 6 next nearest copper neighbours to a zinc atom had their resonance shifted so that it no longer contributed to the *peak*. This does not reduce the integrated intensity by that amount.

The very small effect due to cold working has not been too surprising especially since estimates of the strains and atomic arrangement in cold worked powders can be obtained from X-ray line shape analysis. Such estimates enable one to determine the local deviations from cubic symmetry. Considerably more challenging has been the calculation of the large "wipe out" numbers in alloys. One assumes, for example, that a zinc atom ($3d^{10}4s^2$, free atom configuration) introduces an extra outer electron in the region of overlap with the copper atoms ($3d^{10}4s$, free atom configuration). While the theories concerning the change in the outer electron wave functions of those copper atoms near the zinc atom must be necessarily simple it is believed that this extra zinc electron introduces an oscillatory electron probability distribution diminishing in amplitude as we move away from the zinc. These oscillations in electron probability distribution induce oscillations in the electric field gradient at the copper nuclei surrounding the zinc impurity and extend as far as seventh and eighth nearest neighbours. In addition the larger zinc atom causes strains due to the size effect. These strains do not simply move the copper atoms away radially from the zinc but cause them to move in a rather more complicated way since the copper atoms are connected to each other. The precise value of the electric field gradient at the copper nuclei surrounding a zinc atom due to both these effects is impossible to calculate. Only crude calculations are possible and they are based generally on the free electron approximation. Nevertheless Table XXVI gives a list of "wipe out" numbers measured by Rowland(82) for various impurities in copper and a theoretical estimate by Harrison, Paskin and Sagalyn(83) of the effect of strain and oscillations on the electron probability distribution is given. The principal conclusions are: 1. The effect of the oscillations extends to at least seventh nearest neighbours, 2. Both the strain effect and oscillations contribute to the "wipe out" number. 3. The greater the number of outer electrons in the impurity the larger the effect of the oscillations. The importance of the effect of charge oscillations is due to Kohn and Vosko,(84) while the suggestion of strains comes from Harrison, Paskin and Sagalyn.

This field of endeavour appears most promising particularly if one analyzes the shape of the line to determine the distribution in *eq*. The above work merely reports the number of copper nuclei whose resonance is shifted enough to remove its contribution from the peak. The wipe out numbers are all a measure of the second order quadrupole effect since the peak intensity is relative to cold worked copper for which the $(\mu/I)H \pm \frac{1}{4}e^2qQ$ lines (Fig. 126) have disappeared from the peak (not from the integrated intensity).

TABLE XXVI

THEORETICAL AND EXPERIMENTAL "WIPE OUT" NUMBERS FOR VARIOUS IMPURITIES IN COPPER. THE THEORETICAL "WIPE OUT" NUMBERS ARE CALCULATED WITH AND WITHOUT THE STRAIN EFFECT, IN THAT ORDER

Solute	No. of outer electrons	Experimental wipe out number	Theoretical wipe out numbers	
			With strain	*No strain*
Silver	1	25	20	14
Gold	1	44	29	24
Zinc	2	18	24	24
Magnesium	2	23	32	24
Cadmium	2	32	37	24
Aluminium	3	27	30	31
Gallium	3	38	33	34
Indium	3	48	44	29
Silicon	4	61	61	72
Germanium	4	63	64	71
Tin	4	67	69	63
Phosphorous	5	75	73	87
Arsenic	5	80	82	87
Antimony	5	87	86	76

c. Shape of Resonance

i. Second Moment, Anisotropic Knight Shift, Exchange Broadening

Second Moment. Even in pure annealed copper the resonance is not infinitely sharp because the magnetic moments of the neighbouring copper nuclei create weak magnetic fields of the order of a few gauss at their neighbouring nuclei. In spite of the cubic symmetry of the crystal this effect does not average to zero since the neighbouring nuclear magnetic moments are not ordinarily aligned. The average field they create at each nucleus is zero but the square is not and is given in terms of the *second moment*.

$$\langle(\Delta H)^2\rangle = \frac{3}{5}\frac{I(I+1)}{I^2}\mu^2 \sum_i \frac{P_i}{r_i^6} + \frac{4}{15}\sum_f \sum_i P_f \frac{\mu_f^2}{r_i^6}\frac{I(I+1)_f}{I^2} \qquad (211)$$

where P_i is the number of nuclei of the same isotope at a distance r_i, μ and I are the magnetic moment and spin of the isotope in question, P_f is the number of nuclei of other isotopes with spin I and magnetic moment μ_f at a distance r_i. Thus each atom of the same isotope is weighted 2·25 times more than other isotopes (3/5 against 4/15). Since the effect falls off rapidly with distance only the nearest neighbours need be considered in most cases. In many annealed substances eqn. (211) is a reasonable approximation to the width of the line although some cubic metals exhibit broader lines for unknown reasons.

Anisotropic Knight Shift in Tin, Thallium, Cadmium and Mercury. Bloembergen and Rowland[85] observed that the nuclear resonance lines in powders of tin, thallium, cadmium and mercury (all non-cubic) were considerably broader than given by eqn. (211). For example the line was seven times broader in tin. Since these all have spin $I = \frac{1}{2}$, the quadrupole effect is absent and this led the authors to suggest an anisotropic Knight shift as the cause. If the electrons at the Fermi level have cubic symmetry or are entirely spherically symmetric (pure s character) then the effective field at the nucleus due to the paramagnetism of these electrons is independent of the crystal orientation relative to the field. However any p character at the Fermi level in a non-cubic crystal will give a finite contribution at the nucleus when these electrons give rise to paramagnetism in an external field. It is not surprising that tetragonal tin has p character at the Fermi level since tin would not be tetragonal without such p character in its outer electrons. Bloembergen and Rowland made a crude estimate of the amount of p character at the Fermi level in tin from the line width and estimated $\sim 50\%$. The anisotropic Knight shift thus offers a guide to the amount of "non-s" character in the paramagnetic electrons at the Fermi level, but its quantitative determination involves assumptions about the specific p orbitals involved. By employing single crystals and relying on the fact that the anisotropic Knight shift depends on the magnitude of the field one should be able to learn something of the wave functions at the Fermi level.

Exchange Broadening in Silver. The third contribution to line broadening is entitled *exchange broadening* since it involves a type of exchange interaction between nuclei that is similar to the one which couples the magnetic moments in the rare earth metals. The difference is that the exchange interaction between the outer electrons and the $4f$ electrons in the rare earth metals is electrostatic and involves the antisymmetrization properties of electron wave functions whereas the nuclear interaction with the outer s electrons is purely magnetic. For example, in silver metal the wave functions of the $5s$ electrons have a slightly different value at the nucleus depending on whether their m_s is parallel or antiparallel to the nuclear magnetic moment. This difference is actually induced by the nuclear magnetic moment. Since the $5s$ electrons overlap neighbouring atoms the net effect is to lower the energy of the crystal for either parallel or antiparallel alignment of the nuclei depending on the precise nature of the $5s$ wave functions. Thus nuclear ferromagnetism or antiferromagnetism could be observed at $T = 0°K$ but the energy is so low that the nuclear magnetic moments in silver are completely disordered at any temperature attainable in the laboratory. Nevertheless at any one instant this interaction between neighbouring nuclei does not average to zero even in a cubic crystal. Peculiarly enough, the second moment, $\langle(\Delta H)^2\rangle$, does average to zero for like isotopes but not for unlike isotopes. In addition all higher moments like $\langle(\Delta H)^4\rangle$ are not zero for both like and unlike isotopes.

The interaction is written in the form of the vector model for magnetism, $A_{ij}\, \mathbf{s}_i \cdot \mathbf{s}_j$, where A_{ij} is the exchange energy between unlike isotopes (comparable to J the magnetic exchange energy) and $\mathbf{s}_i$ and $\mathbf{s}_j$ are the nuclear spins. The second moment is

$$\langle (\Delta H)^2 \rangle = \frac{1}{3} \frac{I(I+1)\mu^2}{I^2} \sum_i A_{ij}^2 \tag{212}$$

where μ and I are the magnetic moment and spin quantum numbers of the neighbouring *unlike* nuclei. In silver the measured second moment is about twice that calculated from eqn. (211) and if the difference is attributable to exchange broadening this leads to a value of $A_{ij} \sim 10^{-11}$ eV. This is $\sim 10^7$ times smaller than typical values of J but is not surprising since it is magnetic not electrostatic in origin.

If one introduces an atom with unpaired electrons into a non-magnetic metal like manganese in silver there should be additional magnetic fields at the silver nuclei but this has not been done for this alloy. It has been for manganese in copper but quadrupolar effects obscure the magnetic effects.

Nuclear Resonance in Manganese. In the case of α and β-manganese (Jaccarino, Peter and Wernick, *Phys. Rev. Letters*, (1960) **5**, 53) a broad resonance $\sim$250 gauss was found in α-manganese and a reasonably sharp resonance $\sim$10 gauss was found in β-manganese. This is rather surprising since the large quadrupole moment of manganese and the fact that the nearest neighbours do not have cubic symmetry presumably should have "washed out" the resonance. It may indicate the electronic probability distribution is approximately spherically symmetric.

ii. Line Shape Analysis

Line Shape in Cold Worked Copper. In a pure annealed elemental metal the quadrupole effects are either zero in a cubic material or else some specific value for all equivalent nuclei, i.e. all nuclei with identical non-cubic surroundings. In this latter case the line is generally not broadened excessively but is split sufficiently to create new lines. Sources of broadening due to the electric field gradients created by neighbouring nuclei ($I > \frac{1}{2}$) appear to be small although other sources are present.[(81)]

We have already mentioned the broadening in cold worked copper. Averbuch *et al.*[(86)] used copper filings and compared their results to X-ray line broadening measurements while Faulkner used thinly rolled copper sheets for which stored energy measurements were available (energy stored due to cold work). In both cases the integrated intensity was within a few per cent of an annealed sample. In X-ray measurements the integrated intensities of annealed and cold worked samples are identical. Both the nuclear resonance and X-ray measurements indicate that local strains are never very large in cold working.

It is possible to analyze the broadened X-ray line into a distribution of strains in the lattice but a formidable problem still remains in calculating the distribution in electric field gradients which would then tell us the shape of the nuclear resonance line. The problem involves a determination of the distortion of the outer electron probability distributions from cubic symmetry with a given distribution of strains, and an estimate of the shielding factor (~ 18 for copper). Such a calculation is rather too difficult at present but if one assumes one can calculate the shielding factor then one should be able to combine the X-ray and nuclear resonance line shapes to determine the distortion in the electron probability distributions. Such an analysis has not yet been attempted.

iii. Line Narrowing

Diffusion in Sodium. It is possible to eliminate all sources of broadening by holding a nucleus fixed for a time comparable to radio-frequency times ($\sim 10^{-6}$sec) and spinning its entire world around in a spherically symmetric fashion so that all sources of magnetic fields and electric field gradients are spherically symmetric. In effect this is what happens in a liquid or a cubic solid at high temperature, for the atomic arrangements are changing rapidly and randomly every 10^{-6} sec. However this does not eliminate the ordinary Knight shift which depends on the unbalance in the atoms outer s electrons with m_s parallel and antiparallel to the magnetic field.

Figure 127 is a study of the width of the Na^{23} nuclear resonance as a function of temperature[87] and shows the sudden narrowing of the line at $\sim 180°K$ due to the rapid diffusion rate of the sodium atoms through the lattice. The width of the resonance is related to the diffusion coefficient, $D = D_0 e^{-E/RT}$ (where E is the activation energy for diffusion) by

$$(\delta \nu)^2 \cong (\delta \nu_0)^2 \frac{2}{\pi} \arctan(4{\cdot}2 r^2 \delta \nu / 12 D) \tag{213}$$

where $\delta \nu_0$ is the half width in kc/sec in the temperature range before motional narrowing occurs, and r is the nearest neighbour distance. The formula is only approximate as some assumptions are made about the line shape. It gives fair results when compared to radioactive tracer methods. A more accurate method, involving relaxation times, will be mentioned below.

The nuclear resonance technique has an advantage over the radioactive tracer method since the former follows actual atomic motion whereas the latter measures the bulk motion of material. In the tracer method diffusion along grain boundaries and dislocation lines which provide short-circuiting paths are not easily separated from the true volume diffusion through the lattice. While the nuclear resonance is also affected by the short circuiting paths its contribution is only proportional to the fraction of atoms in these paths at any time and this is small.

Theory of Diffusion. It would be appropriate now to indicate the problems involved in relating diffusion studies to electron structure. The theory of diffusion has not primarily concerned itself with this problem but rather with the geometry of diffusion, i.e. how do atoms move through a crystal lattice with the energies available from thermal motion? The simplest mechanism is that of vacancy diffusion since every metal has at least 1 in 10^4 or 10^5 vacant lattice sites. Even though it requires a certain amount of energy ($E \sim 1$ eV per atom) to create a vacant lattice site the entropy, S, of the crystal is increased so that the total free energy is lowered

$$\Delta G = cE - RTc \ln c \tag{214}$$

where c is the fraction of vacancies. If one minimizes the free energy, eqn. (214), we determine $c = e^{-E/RT}$ which is between 10^{-4} and 10^{-5} for most metals near their melting point. With such concentrations of vacancies and typical atomic vibrational frequencies one obtains typical diffusion rates ($D_0 \cong \nu\delta^2$ where $\nu \cong 10^{13}$/sec and $\delta \cong$ 2–3 Å). Experimentally then, one determines D_0 and E.

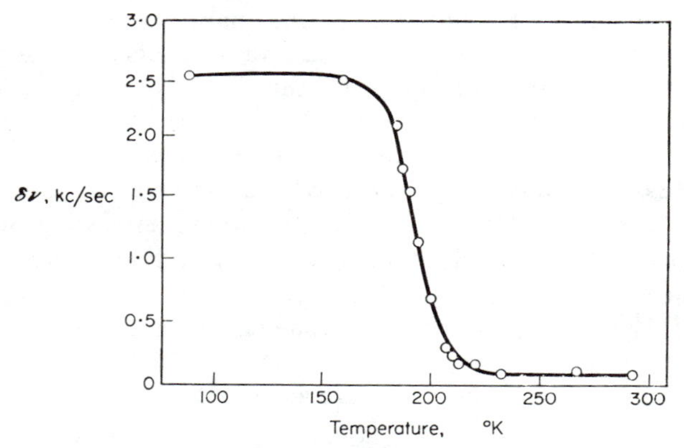

FIG. 127. The width of the sodium resonance in sodium metal as a function cf temperature, showing the motional narrowing due to the rapid diffusion of sodium at temperatures above 175°K.

The geometrical problems involve the hopping of atoms from one lattice site to an adjacent vacant site, followed by a second atom hopping to the new vacant site. Sometimes the original atom which hops into an adjacent vacant site hops back and we are left as we began, or else the vacant lattice site is shifted several atomic distances by the co-operative motion of several atoms. The geometrical problem then involves all possible hopping arrangements so that one ultimately mixes the atoms together. Each of these specific hopping arrangements involves overcoming an energy barrier E and one must calculate how long an atom or group of atoms must wait before they have sufficient energy and momentum to make this transition. If one knows E for the various configurations then thermodynamics enables us to make this calculation but a calculation of E involves details of the wave functions. The actual quantum mechanical calculation of these wave functions (and eigenvalues) is rather prodigious for it involves determining the wave functions for a nonperiodic crystal. The Bloch wave solutions are no longer applicable. In the main, semi-classical methods are employed and the field is beginning to evoke theoretical interest. The experimental results are not complete nor accurate enough to provide the clues as to the proper quantum mechanical approach to the wave function around a vacancy.

d. Relaxation Times

Relaxation Times in Aluminium and Copper. One generally speaks of two relaxation times in nuclear resonance, T_1 and T_2. T_1 is the time required for the nuclear magnetic moments to return to the thermal equilibrium distribution given by eqn. (208), since the input of radiofrequency energy at resonance disturbs the thermal equilibrium. T_1 is called the *spin-lattice relaxation time* since it is the time required for the nuclear magnetic moments to come into equilibrium with the lattice. The sequence of events is as follows: 1. The outer or bonding electrons determine the vibrational motion of the atoms due to their thermal energy, kT. 2. These outer electrons interact with the nuclear magnetic moments via their s character which has a finite probability distribution at the nucleus. 3. Through this interaction the phonon distribution is continuously altered until the distribution of nuclear magnetic moments in an external field is given by eqn. (208). After this there is a continual excitation and de-excitation of nuclear magnetic moments but if T_1 is very short then the eigenvalues of Fig. 123 are broadened by an excessive amount ($\Delta E \sim h/T_1$) and may noticeably broaden the line.

T_2 is called the *spin-spin relaxation time* since it depends on the magnetic interaction between nuclear magnetic moments. It is this same interaction that leads to the dipolar broadening of the resonance line, eqn. (211). Specifically T_2 is the time required for the nuclear spin system to come into thermal equilibrium with itself, independent of whether it has reached thermal equilibrium with the lattice. Here again the equilibrium has generally been destroyed by the radiofrequency energy at resonance. It is almost obvious that T_1 is the upper limit to T_2 for if T_1 is shorter than T_2 the spin-lattice relaxation produces thermal equilibrium and the mechanism leading to T_2 is ineffective. In aluminium, for example, T_2 is 0·1 millisecond and T_1 is 3 milliseconds at room temperature. Thus, any energy absorbed by the nuclear magnetic moments from the radiofrequency radiation at resonance requires about 0·1 milliseconds to be distributed equally amongst the nuclei via their magnetic dipolar interaction. In this state the nuclei may still be effectively at a higher temperature according to the population of the eigenvalues in the external magnetic field, eqn. (208). In ~3 milliseconds, though, the nuclear magnetic moments have lowered their energy via the spin-lattice relaxation mechanism, thereby reducing their temperature to that of the lattice. If T_1 was shorter than T_2 then the nuclear magnetic moments would reach thermal equilibrium with the lattice and any further interaction between the nuclear magnetic dipoles would not alter their distribution.*

There are, essentially, two methods employed in the measurement of T_1 and T_2, the spin echo method[88] and the saturation method.[89] Since these techniques are rather complicated and since the latter technique has seen only limited use the reader is referred to the references for experimental details. The spin echo method relies on the coherence of the excited nuclei, since the exciting r.f. field is always in phase rather than consisting of a group of uncorrelated photons. This coherence of the excited nuclei leads to a net magnetization of these nuclei and this magnetization precesses at the resonant frequency around the external magnetic field, H, with a component normal to H; but due to the magnetic field contributions of the neighbouring dipoles the coherence of this normal component is eventually lost. This happens in a time T_2 or T_1 whichever is shorter. If a strong r.f. pulse, of duration $\ll T_2$ and at the resonant frequency, is placed on the system at right angles to the external field, H, followed by a second pulse at time interval, τ, then the precessional component of the magnetization normal to H develops certain phase relationships which at a later time, τ, add constructively to emit a pulse of its own called a spin echo. The amplitude of the spin echo is related to τ, T_1 and T_2 and by appropriate choices of τ, both T_1 and T_2 can be separated. While the method is quite good it is not easily adapted to the Varian unit.

Fig. 128 shows the data of Spokas and Slichter for aluminium[88] separated into T_1 and T_2 components. At temperatures below 360°K T_2 is constant but at higher temperatures the diffusional and motional narrowing of aluminium decreases the interaction between the nuclear dipoles (since the magnetic field from neighbouring dipoles is more likely to average to zero) and thus increases the relaxation time due to this mechanism. On the other hand the increased thermal motion of the atoms decreases the spin-lattice relaxation time, T_1, approximately as $1/T$. There is a well-known relationship between the spin-lattice relaxation time and the Knight shift, since the same interaction with the outer s electrons causes both. This is called the *Korringa relation*

$$(\Delta H/H)^2 = \frac{\hbar I \mu_B}{\pi k T T_1 \mu} = (\text{Knight shift})^2 \tag{215}$$

* If, $T_1 \gg T_2$ a bizarre situation can be created by suddenly reversing the direction of the magnetic field after equilibrium has been established. For times short compared to T_1 the upper levels in eqn. (208) are more populated than the lower levels. In this non-equilibrium state continued use of eqn. (208) would require T to be negative. As equilibrium is again established T first goes through $\mp\infty$ before resuming positive values.

where μ_B is the Bohr magnetron, μ is the nuclear magnetic moment and I its spin. This relationship is approximately satisfied in copper and aluminium. In the molten state of aluminium $T_1 = T_2 = 2{\cdot}1$ milliseconds.

The second method is limited to T_1 and involves the measurement of the resonant absorption as a function of the amplitude of the r.f. signal $H(= H \cos \omega t)$ for as it saturates the absorption signal decreases according to the ratio

$$R = \left[1 + \frac{\mu^2 H^2 T_1}{2\hbar^2 I^2}\, g(\nu_0)\right]^{-1} \tag{216}$$

where $g(\nu)$ is the functional shape of the resonance and $g(\nu_0)$ is the value at the resonant frequency ν_0, ($\int g(\nu)\, d\nu = 1$). Both spin echo and saturation methods[89] yield comparable results for T_1 in copper and aluminium.

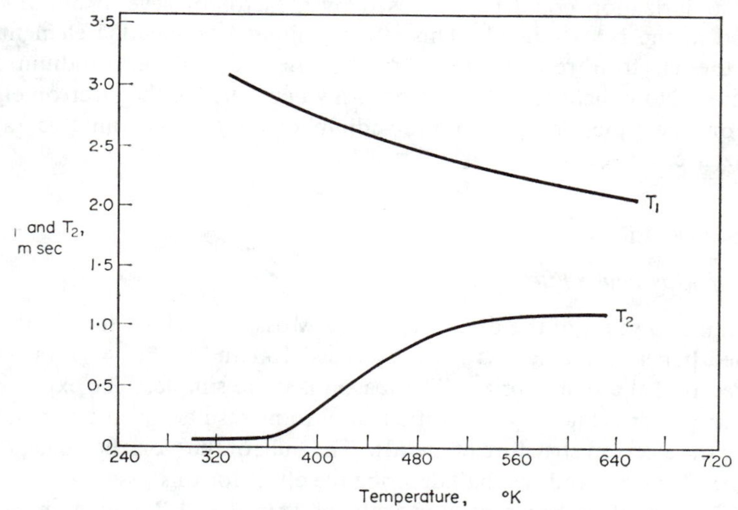

FIG. 128. The relaxation times T_1 and T_2 of aluminium in milliseconds, versus temperature as determined from spin echo measurements. In the melt $T_1 = T_2 = 2{\cdot}1$ milliseconds.

There is a general sparsity of relaxation time data so there is little we can say about it.

e. Electron Spin and NMR

Overhauser Effect

By combining nuclear resonance and paramagnetic resonance of the electrons at the Fermi level Overhauser pointed out that it is possible to magnetize (polarize) the nuclei. The effect relies on the strong interaction between the s character of the electrons at the Fermi level and the nuclear magnetic moments. This causes a nuclear spin to reverse its m_s when an electron at the Fermi level reverses its m_s.

Suppose we place a sample of lithium metal in a magnetic field. Some of the electrons at the Fermi level will reverse their m_s value due to the paramagnetic susceptibility, and the nuclear magnetic moments ($I = 3/2$) will occupy the eigenvalues of Fig. 126 ($eq = 0$) according to eqn. (208). Let us now introduce a strong microwave signal at the free electron resonance frequency (p. 328) so that we saturate the electron resonance. This produces equal population of the two m_s values of the electrons and the paramagnetic susceptibility and Knight shift disappear. However due to the strong interaction with the nuclei this new distribution of the electrons at the Fermi level leaves a net polarization of the nuclear magnetic moments.* In the case of Li^7 in lithium metal the nuclear polarization could be increased by a factor of several hundred as observed in the NMR signal. This effect is limited to metallic elements in which the electron resonance has been observed (lithium, sodium and beryllium). No conclusions have been drawn concerning the electron eigenvalues or eigenfunctions in lithium, sodium or beryllium from this rather specialized effect.

2. Mössbauer Effect

Theory of Mössbauer Effect

Within two years of the discovery of the Mössbauer effect the number of published papers utilizing this effect exceeded 100 and within 4 years Mössbauer received the Nobel prize. The reason is quite simple. The experiment is easy to perform and it yields rather important results toward our understanding the electron structure of metals. The bulk of the work has been done on the Fe^{57} nucleus and we shall describe the effect for this case.

The Fe^{57} nucleus has a ground state of spin $I = 1/2$ and a magnetic moment $\mu = +0{\cdot}0903$ nuclear magnetons. It has an excited isomeric state (Fe^{57*}) 14·4 keV above this ground state with spin $I = 3/2$, magnetic moment $\mu = -0{\cdot}153$ nuclear magnetons, and a small quadrupole moment. The excited state decays to the ground state by emitting a 14·4 keV gamma ray with a half life of 10^{-7} sec. From the half life the uncertainty principle tells us that the uncertainty in energy of the excited state is $\Delta E = \hbar/\Delta t = 6{\cdot}62 \times 10^{-27}$ erg sec$/10^{-7}$ sec $\cong 4 \times 10^{-8}$ eV. Since the excited state has an uncertainty in energy of 4×10^{-8} eV and since the ground state has zero uncertainty since it has an infinite half life, the 14·4 keV gamma ray has an uncertainty of 4×10^{-8} eV or only about one part in 3×10^{11}! If

* For nuclei of spin $\frac{1}{2}$ the ratio of nuclei with $m_I = \frac{1}{2}$ to $m_I = -\frac{1}{2}$ at saturation becomes $\exp(2\mu_B H/kT)$ where μ_B is the Bohr magneton.

some Fe^{57} nuclei in the their ground state are bombarded with gamma rays it requires a gamma ray of exactly the right energy within about 4×10^{-8} eV to excite the Fe^{57} to its excited state Fe^{57*}. *It is the sharpness of this resonance absorption that is the basis for the Mössbauer effect.* If we have a sample of pure iron metal containing some Fe^{57*} nuclei the gamma rays, emmitted as these nuclei decay to their ground state, can be absorbed by the Fe^{57} nuclei in another piece of pure iron. The absorption cross-section is quite high for gamma rays $\sim 1{\cdot}5 \times 10^6$ barns (10^{-24} cm^2). However if we move the absorbing piece of iron metal relative to the emitting piece of iron metal by only a velocity $v \cong 1$ cm/sec then the Fe^{57} nuclei in the moving absorber will see a gamma ray which has undergone a Doppler shift so that the apparent energy of the gamma ray is altered by a factor $= (1 \pm v/c) = 1 \pm 1/3 \times 10^{10} = 1 \pm 0{\cdot}33 \times 10^{-10}$ or about 4×10^{-7} eV. This is greater than the half width of the resonance and the cross-section per atom has dropped to $\sim 6{\cdot}3 \times 10^2$ barns, (principally electron fluorescence) so that the gamma rays are no longer strongly absorbed by the moving foil. This velocity effect is experimentally demonstrated in Fig. 129[(90)] by the sharpness of

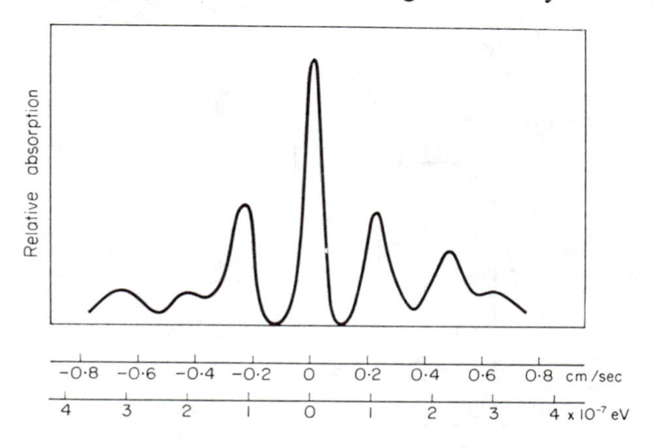

FIG. 129. The Mössbauer effect in iron metal (relative absorption versus relative velocity of absorber). The width of the central peak is a measure of the sharpness of the resonance.

the central component (half width $\sim 0{\cdot}1$ cm/sec). The additional lines in Fig. 129 arises from structure in the energy levels due to magnetic effects and actually provide the greatest source of interest in the measurement.

Mössbauer Effect in Iron

In addition to the velocity effect altering the apparent energy of the γ-ray, it is also possible to alter the γ-ray by placing the atom in different surroundings so that there is a different interaction with the atomic electrons. In iron metal the largest observable effect arises from the effective magnetic field at the nucleus ($H_{eff} = -333$ kilogauss) due to the interaction of the unpaired d electrons with the 1s, 2s, 3s and 4s wave functions (see p. 342). This leads to the nuclear energy levels of the ground state and excited state as shown in Fig. 130. The excited state is split into 4 levels $m_I = \pm 1/2$, $m_I = \pm 3/2$ and the ground state into two levels $m_I = \pm 1/2$. Quantum mechanics permits excitations in which m_I changes by zero or ± 1 so that we expect the six lines shown in Fig. 130

$$\left(E_0 \pm \left(\frac{\mu^* \pm 3\mu}{3}\right) H_{eff}; \quad E_0 \pm (\mu - \mu^*) H_{eff}\right).$$

If the Fe^{57*} emitting metal sample is stationary relative to the Fe^{57} absorbing metal sample all six lines will be absorbed as shown in Fig. 130, diagram 1, and this accounts for the central sharp line in Fig. 129. If the absorber is moved at a velocity of ~0·1 cm/sec then the resonance disappears as in diagram 2

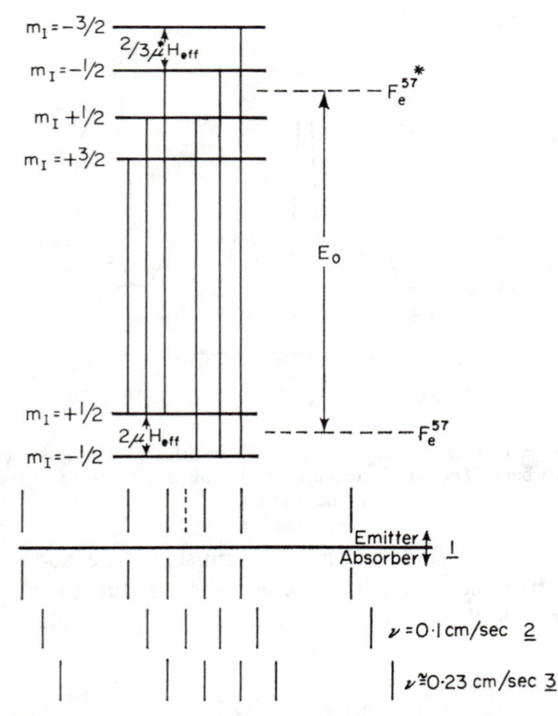

Fig. 130. The energy levels of the ground state and excited isomeric state of Fe^{57} employed in the Mössbauer measurement (upper curve). In the lower curve the relative shift of the energy levels is indicated as the absorber is moved relative to the emitter.

of Fig. 130 but if the velocity is increased to about 0·23 cm/sec, as in diagram 3 of Fig. 130, then two levels again coincide and we have a resonance again, but only weaker. This movement of the absorber energy levels relative to the emitter accounts for the pattern in Fig. 129. From a knowledge of the spin and magnetic moment of the ground state and of the excited state in Fe^{57} one can determine H_{eff} from the pattern in Fig. 129.

The emitting sample is prepared by plating some radioactive Co^{57} onto an iron foil and heating it at ~800°C for an hour so that it diffuses into the iron about 1000 Å but not so deeply that the 14·4 keV gamma ray will be self-absorbed before leaving the emitting foil.* The radioactive Co^{57} nucleus decays into our Fe^{57*} nucleus by emitting a 250 keV positron with a half life of 270 days and this is instantaneously followed by a 1 MeV γ-ray. The Fe^{57*} nucleus then decays in 10^{-7} sec, emitting the 14·4 keV Mössbauer gamma ray. The 1 MeV γ-ray and the positron do not bother us since scintillating crystals can discriminate against them. The thin absorber (~1 mil) is then placed on a rotating turntable and slits are arranged so that we concern ourselves with the tangential velocity of the sample which is arranged to be parallel to the line from the emitter to the counter. Since the resonance is very sharp it is imperative that the turntable moves at a constant speed without oscillation or flutter.

A problem that must be considered in the Mössbauer experiment is the recoil of the Fe^{57} nucleus as the gamma ray is emitted. Since the Fe^{57} nucleus is part of a crystal the recoil can be elastic by the entire crystal recoiling or inelastic by absorbing or exciting a phonon. The fraction of times the emission or absorption of the Mössbauer gamma ray is elastic is given by the Debye–Waller factor, e^{-M} with $\sin\theta = 1$ (eqn. (111)).† The wavelength of the 14·4 keV Mössbauer gamma ray is 0·856 Å so that e^{-M} at room temperature is 0·63 for iron. If the gamma ray absorbs or excites a phonon the energy change is ~0·025 eV and completely removes it from the resonance region.

The number of nuclei available for Mössbauer experiments is reasonably large and includes Ni^{61}, Zn^{67}, Pd^{105}, Sn^{119}, Nd^{145}, Gd^{155}, Dy^{161}, Hf^{176}, W^{182}, W^{184}, Ir^{193}, Pt^{193}, Au^{197}, Hg^{201}, and Th^{232}.These can also be used to gather data about electric field gradients if the nuclei have a quadrupole moment in either the ground or excited states. *In fact any electronic effect that changes the nuclear energy levels can be investigated and the potentialities*

* The effect of self-absorption can be minimized by simultaneously co-plating Fe^{56} and Co^{57}. The Fe^{56} is a relatively inexpensive stable isotope of iron and does not absorb the 14·4 keV Mössbauer gamma ray. The Co^{57} is produced by cyclotron bombardment of natural iron and is chemically separated.

† The factor is e^{-M} not e^{-2M} as in Bragg scattering of X-rays for in the latter case there is a factor e^{-M} both for the absorption and re-emission of the X-ray ($e^{-M} \times e^{-M} = e^{-2M}$). Of course there is also a factor e^{-2M} in the Mössbauer experiment when we consider both the emission and absorption process.

are only just being realized. For example, the *isomer shift*[91] depends on a slightly different electrostatic interaction between *all* the s electrons and the distribution of protons which differs in the Fe^{57} and Fe^{57*} nuclei. This alters E_0 slightly in Fig. 130 and can be related to the outer electron configuration by comparing with the isomer shift in an inorganic compound where we believe we know the electron configuration. This has been determined to be $\sim 3d^7 4s$ in b.c.c. iron.

3. Radioactive Tracers

In this section we shall review some of the solid state and metallurgical uses of the high energy radiations emitted by excited states of nuclei. The high energy of these particles either permit easy detection and identification of its parent or else act as a source of energy to damage crystalline lattices. For convenience, we shall discuss these topics under the following headings:

a. Diffusion Studies
b. Micrographic Studies
c. Radiation Damage
d. Elemental Analysis

a. Diffusion Studies

Diffusion Studies with Radioactive Tracers

The use of a radioactive isotope of a metal to measure self-diffusion has been utilized to advantage. The technique consists of plating the radioactive material on one end of a cylindrical bar of the stable material, and heating the bar for a time t and to a temperature T sufficiently high so that reasonable diffusion occurs. The bar is then sectioned by machining, say, a thousandth of an inch at a time and measuring the radioactivity of the machinings. Each subsequent thousandth of an inch will indicate a smaller radioactivity per unit weight. If the radioactivity is plotted per thousandth of an inch we expect a curve of the form

$$c = \frac{S}{(\pi D t)^{1/2}} e^{-x^2/4Dt} \tag{217}$$

where c is the concentration of radioactivity, x the perpendicular distance from the surface, D the diffusion coefficient and S the amount of radioactivity per cm^2 originally plated on the surface. Figure 131 is a plot of the concentration of radioactive silver in silver. A silver disk, 7/8 inch in diameter and 1/4 inch thick, was plated with radioactive silver and a second similar disk was welded to the first so that there was diffusion in both directions. In this case eqn. (217) would have an additional factor of ½ on the right side. The diffusion coefficient D is obtained by fitting eqn. (217) to the curve in Fig. 131.

Other than the preparation of the radioactive tracer, the principal difficulty is separating the true volume diffusion from the diffusion along short circuiting paths like grain boundaries. The reason grain boundaries provide an easy path for diffusion is that there are generally open spaces along grain boundaries since the atoms can never match up as perfectly as in the crystal lattice. Even if one uses metal single crystals there are always very small angle boundaries called lineage which provide short-circuiting paths. It may be possible to start with a strain anneal metal single crystal that has a very narrow X-ray rocking curve, cold rolling it, measuring the diffusion rate versus rocking curve width, and extrapolating to zero width. By using a technique called auto-radiography it has been possible to show clearly the rapid diffusion along grain boundaries and to separate the two effects.

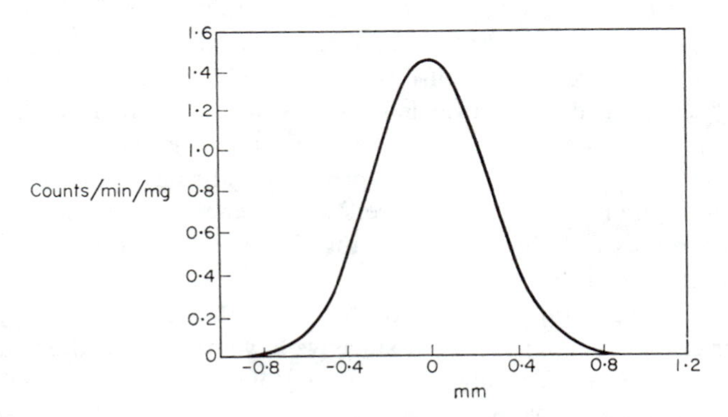

FIG. 131. The amount of radioactive silver diffused into a silver sandwich by heating at 876°C for 4·78 days.[92]

In selecting a radioactive isotope of an element one must find one with a reasonable half life, since it will decay and disappear if the half life is too short or it may be too weak if the half life is too long. In addition, any technique for producing radioactive isotopes may produce extraneous radioactivities not desired and separation may be difficult.

An attractive variation of the radioactive tracer method is the use of an enriched isotope, and a mass spectrograph to determine the isotopic ratios. This has the advantage that it is suitable to more elements than the radioactive tracer technique. In some cases one can even use nuclear resonance to ascertain the isotopic ratios in the sectioning method.

The use of radioactive or stable isotopes is of course mandatory in self-diffusion studies, but in the case of alloys conventional chemical methods or even a nondestructive X-ray fluorescence can be employed.

The comments concerning interpretation of diffusion coefficients, made on p. 352, apply equally well to this section.

In an interesting study, J. A. Davies (*J. Electrochem. Soc.* 999, Oct. (1962)) bombards aluminium with inert gas ions that decay via internal conversion. The ions remain on the surface and subsequent oxidation will increase the absorption of the conversion electrons. In this way he can measure the thickness of the oxide as well as make some deductions about the oxidizing mechanism.

b. Micrographic Studies—Auto-radiography

Auto-radiography

It has been possible to determine microstructural details by placing a sample containing a radioactive material, whose disposition throughout the sample is desired, in contact with a fine grained film. Figure 132 is an auto-radiograph taken in Prof. Lacombe's laboratory in Paris and clearly demonstrates the enhanced self-diffusion of aluminium along sub-boundaries.[93][94] Considerable knowledge exists of fine grained film techniques, particularly through the efforts of the nuclear physicists who use them to study high energy particles. Under proper conditions a resolution better than one micron (10^{-4} cm) is possible.[95] In general, pure beta ray emitters are best since they only have a short range in the metal and good resolution can be maintained.

Such microstructural studies of the imperfect lattice must be approached in a rather crude way, theoretically, since we have not solved the Schrödinger equation for the imperfect lattice. We can still make semi-classical models, using thermodymanics whenever possible, and still hope to glean some understanding of the behaviour of the imperfect crystal.

c. Radiation Damage

Radiation Damage in Copper

With the advent of nuclear reactors the field of radiation damage became quite popular and at about the same time interest was aroused in naturally occurring imperfections in crystals, such as vacancies, dislocations, interstitials, etc. Since many of the physical properties of metals depended on these imperfections it was natural to suppose that the artificial introduction of imperfections by radiation damage would help in the understanding of the more naturally occurring imperfections. After some fifteen years' effort and at least a thousand published papers it has become apparent that it is extremely difficult to relate imperfections, due to reactor radiation damage, to the naturally occurring imperfections.

FIG. 132. An auto-radiograph of radioactive aluminium in aluminium metal showing preferential diffusion along grain boundaries and sub-boundaries.

Facing page 362

If a piece of copper metal, for example, is placed in a reactor it will be bombarded by neutrons, β-rays and γ-rays, but the principal damage comes from the very high energy neutrons. A one MeV neutron colliding head-on with a copper nucleus, transfers $4/A$ or less of its energy to the copper nucleus (A is the atomic weight of copper, 63·5). This collision sends the copper nucleus hurtling through the lattice at 60 keV energy leaving many of its weakly bound electrons behind. The neutron recoils with 0·94 MeV and goes on to hit another nucleus possibly dislodging it. This continues until the neutron energy is about 400 eV below which it cannot impart the 25 eV to the copper nucleus necessary to dislodge it permanently from the lattice. Each of the copper nuclei, which have been dislodged, can dislodge other copper nuclei, can excite electrons, etc. The whole cataclysmic process* caused by a 1 MeV neutron leaves the vicinity of each collision in somewhat of a mess. A theoretical estimate is that $\sim$1000 copper atoms are displaced per 1 MeV neutron but the experimental evidence indicates about 100. The difference between the two suggests that the copper nucleus is not dislodged in each high energy collision, but transfers its energy to the lattice vibrations by "shaking" a whole group of copper atoms together.

On the other hand if we bombard our copper with one MeV monoenergetic electrons, which can be produced by a Van de Graaff accelerator, then each electron can only transfer about 30 eV to a copper atom since its mass is $\sim$1/2000 of a neutron mass. This is just barely enough to dislodge one copper atom from its lattice site and is the only damage created. Thus electron bombardment creates relatively simple and isolated damage and by controlling the initial energy of the electron the threshold energy for permanent damage can be determined.

While it is also possible to perform radiation damage with cyclotrons (protons or deuterons) this creates as much of a mess as neutrons.

One of the most sensitive measurements to the state of damage in a metal is resisitivity and Fig. 133, prepared by Dr. J. Antal, gives the relative change in resistivity for a copper sample, irradiated with electrons at 4°K and allowed to anneal at various temperatures. It appears that 90% of the damage anneals at temperatures less than 77°K. The various steps in Fig. 133, indicating annealing of various types of imperfections, are identified from A to I and the suggested types of defects are:

i. *Vacancy-interstitial Anneal* (A, B, C)

In this case an atom is dislodged from its lattice position moving into some nearby interstitial position and leaving a vacancy behind. The annealing steps A, B, C refer to the recombination of this pair by thermal energy, the three distinct groups indicating that the vacancy and interstitial are different

* Sometimes called a "*thermal spike*".

distances apart and require different energies for annealing. This type of anneal returns the lattice to its undamaged state.

ii. *Crowdion or Split Interstitial Anneal* (*D*, *E*)

A *crowdion* consists of adding an extra atom in a f.c.c. lattice along the 110 close packed direction. A line of atoms, some 10 atomic distances long, adjusts itself to make room for the extra atom. A split interstitial is shown in Fig. 133. Ordinarily one might envisage an interstitial atom at the body centre position of the cube but in the split interstitial the extra atom pushes one of the atoms aside so that the two atoms are opposed on either side of the former position of the atom that has been pushed.

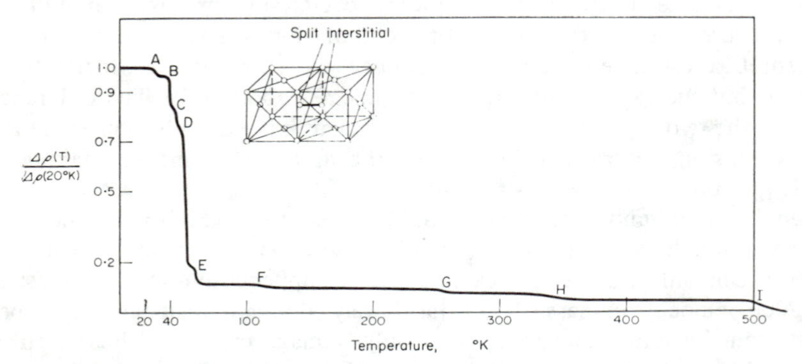

FIG. 133. Relative additional resistivity, $\Delta\rho$, of copper irradiated at 4°K as a function of annealing temperature. $\Delta\rho(20°K)$ is the initial amount of additional resistivity before annealing. The inset is a sketch of a split interstitial.

iii. (*F*)

At this writing people are still thinking about this one.

iv. *Vacancy Anneal* (*G*)

The vacancy, left behind by an interstitial that does not recombine with this vacancy, moves to the surface of the metal or to a grain boundary both of which are stable configurations for it.

v. *Tri-vacancy Anneal* (*H*)

Three adjoining vacancies break up into a divacancy (vacancy pair), which instantly moves to the surface, and a vacancy which moves to the surface or to a grain boundary. The divacancy is reported to be very mobile and anneals even at $T = 0°K$.

vi. *Dislocation* (*I*)

In this case a small dislocation loop moves to the surface and disappears.

There is virtually no clear and independent measurement of the imperfections suggested above. These suggestions are based on crude calculations and other estimates, but are the best we can do. The annealing curve for reactor-irradiated copper evidences the same sort of overall structure as in Fig. 134 although not so clear in its details. In cold worked copper there is some evidence for annealing at the points *G*, *H*, and *I* but it is not yet clear that the animals are the same and not merely first cousins.

The main conclusion is that radiation damage artificially produces lattice imperfections and a study of these imperfections and their energy can lead to an understanding of the nature of the wave functions around these imperfections. But, before we can do that, it is necessary to identify these imperfections with more certainty. Diffraction measurements have not revealed any effects in metals since the amount of damage is not sufficient. One requires at least 0·1% displaced atoms to do X-ray diffraction. While diffraction measurements in non-metals do reveal line broadening effects this is only because heavily irradiated non-metals retain more of their damage than metals.

d. Elemental Analysis

Elemental Analysis by Activation

Since the radiation from radioactive substances is capable of identifying an individual atom, we are approaching the ultimate in sensitivity for elemental analysis. In some cases, trace quantities of less than one part in 10^9 can be identified by induced radioactivity. The element can be identified by the type of radiation, the energy of the radiation, the half life of the radiation and a knowledge of the type of particle used for irradiation. In general this type of analysis is not straightforward and one should not attempt this approach if more straightforward techniques such as chemical analysis or X-ray fluorescence will suffice.

The two sources of irradiation commonly used are the reactor for slow neutron bombardment and the Van de Graaff accelerator for proton, deuteron and alpha particle bombardment. Electrons and γ-rays do not interact appreciably with nuclei and are not suited to this problem. Of the two sources, the slow neutron bombardment is the most fruitful. This technique works best when the trace element sought has a high *activation cross-section** (>5 barns), a reasonable half life (hours to years), and emits radiation of reasonable energy (β-rays $>$ 1 MeV, γ-rays $>$ 10 keV). The matrix metal containing the sought for impurities should have just the opposite properties so that it provides little or no background to the measurement. Table XXVII lists

* The activation cross-section is that fraction of the absorption cross-section that leads to a radioactive isotope.

those elements with suitable isotopes for trace identification and those elements which are particularly good matrix elements.

In employing Table XXVII a certain amount of metallurgical intuition is

TABLE XXVII

SLOW NEUTRON ACTIVATION

TRACE ELEMENTS PARTICULARLY SUITED FOR NEUTRON ACTIVATION AND MATRIX ELEMENTS SUITABLE FOR IMPURITY DETERMINATION BY NEUTRON ACTIVATION

Element	Abundance (%)	σ_{act} (barns)	Half-life	Radiation	Energy (MeV)
Cl^{35}	75·4	30	3×10^5 y	β^-	0·7
Ar^{40}	99·6	0·5	109 m	β^-, β^-, γ	1·25, 2·55, 1·3
K^{39}	93·1	3	$1{\cdot}3 \times 10^9$ y	β^-, γ	1·4, 1·5
Sc^{45}	100	12	85 d	β^-, β^-, γ	0·36, 1·2, 0·89
Mn^{55}	100	13·4	2·6 hr	β^-, γ	wide range up to 3 MeV
Co^{59}	100	20	5·3 y	β^-, γ	0·31, 1·2
Cu^{63}	69	4	12·8 hr	β^-, β^+, γ	0·57, 0·65, 1·34
Zn^{64}	49	0·5	250 d	β^+, γ, γ	0·32, 1·14, 0·2
Ga^{71}	40	3·4	14·2 hr	β^-, γ	wide range to 3 MeV
As^{75}	100	4·2	27 hr	β^-, γ	wide range to 3 MeV
Br^{81}	49	3·5	35·9 hr	β^-, γ	0·47, up to 2 MeV
Rb^{85}	72	0·7	19·5 d	β^-, β^-, γ	1·8, 0·7, 1·1
Y^{89}	100	1·2	63 hr	β^-	2·2
Ru^{102}	31	1·2	41 d	β^-, γ	0·7, 0·5
Pd^{108}	27	12	13·6 hr	β^-	0·95
Ag^{109}	49	2·8	270 d	β^-, γ	0·53; many to 1·5 MeV
Cd^{114}	29	1·1	53 hr	β^-, β^-, γ	0·46, 1·1, 0·52
In^{113}	4	56	49 d	β^-, γ, γ	2, 0·7, 0·55
Sb^{121}	57	6·8	2·8 d	$\beta^-, \beta^-, \gamma, \gamma, \gamma$	1·36, 1·94, 0·6, 1·2, 0·7
Te^{124}	5	5	58 d	e^-, e^-	0·1, 0·035
Cs^{133}	100	26	2·3 y	$\beta^-, \beta^-, \gamma, \gamma, \gamma$	0·65, 0·09, 0·8, 0·6, 0·57
La^{139}	100	8·4	40 hr	β^-, γ	many to 2 MeV
Pr^{141}	100	10	19·2 hr	$\beta^-, \beta^-, \gamma, \gamma$	2·2, 0·64, 1·6, 0·14
Nd^{146}	17	1·8	11·3 d	$\beta^-, \beta^-, \gamma, \gamma$	0·78, 0·35, 0·09, 0·52
Sm^{152}	27	140	47 hr	$\beta^-, \gamma, \gamma, \gamma$	0·82, 0·07, 0·1, 0·6
Eu^{151}	48	1400	9·2 hr	β^-, γ	many to 1·8 MeV
Gd^{158}	25	4	18 hr	β^-, γ, γ	0·85, 0·06, 0·35
Tb^{159}	100	25	73 d	β^-, γ	many to 1 MeV
Ho^{165}	100	60	27·3 hr	$\beta^-, \beta^-, \gamma, \gamma$	1·84, 0·55, 0·08, 1·36
Er^{168}	27	2	9·4 d	β^-	0·33
Tm^{169}	100	130	129 d	β^-, β^-, γ	1, 0·89, 0·09
Yb^{174}	32	60	101 hr	β^-, β^-, γ	0·13, 0·5, 0·35
Lu^{176}	25	4000	6·8 d	β^-, γ	many to 0·5 MeV
Hf^{180}	35	10	46 d	β^-, γ	0·41, many to 0·5 MeV
Ta^{181}	100	19	111 d	β^-, γ	many to 1 MeV
W^{186}	29	34	24 hr	β^-, γ	many to 1·3 MeV
Re^{185}	37	100	92 hr	β^-, γ	many to 1 MeV
Os^{190}	26	8	16 d	β^-, γ, γ	0·14, 0·13, 0·04
Ir^{191}	39	700	74 d	β^-, γ	many to 0·65 MeV
Au^{197}	100	96	2·7 d	β^-, γ	many to 1·4 MeV
Hg^{202}	30	3·8	47 d	β^-, γ	0·21, 0·28
Tl^{203}	30	8	2·7 y	β^-	0·76

TABLE XXVII—*continued*

Element	Abundance (%)	σ_{act} (barns)	Half life	Radiation	Energy (MeV)
Good Matrix Elements (isotope with highest activity)					
H^2	0·015	0·57 mb	12·4 y	β^-	0·018
Li^7	92	0·33 mb	0·85 s	β^-	12
Be^9	100	9 mb	2·7 × 10^6 y	β^-	0·56
B^{11}	81	50 mb	0·03 s	β^-	13·4
C^{13}	1·1	0·9 mb	5570 y	β^-	0·155
N^{15}	0·37	0·024 mb	7·4 s	β^-, β^-, γ	4, 10, 6·5
O^{18}	0·2	0·21 mb	29 s	β^-, β^-, γ	2·9, 4·5, 1·6
F^{19}	100	9 mb	11 s	β^-, γ	5·1, 1·6
Ne^{22}	8·8	36 mb	40 s	β^-	4·2
Mg^{26}	11·3	50 mb	9·5 m	β^-, γ	several to 1·8 MeV
Al^{27}	100	0·21	2·27 m	β^-, γ	3, 1·8
Si^{30}	3	0·11	2·6 h	β^-	1·49
Ti^{50}	5·3	0·14	5·8 m	β^-, γ	1·6, 0·32
Va^{51}	99·8	4·5	3·76 m	γ	0·25
Fe^{58}	0·31	0·9	46 d	β, γ, γ	46, 1·1, 1·3
Ni^{64}	1·1	2·6	2·57 hr	β^-, γ	many to 2 MeV
Nb^{93}	100	1	6·6 m	β^-, γ	1·3, 0·9
Rh^{103}	100	12	4·5 m	e^-	0·05
I^{127}	100	5·5	25 m	β^-, β^-, γ	2, 1·6, 0·43
Pb^{208}	52	0·6 mb	3·2 hr	β^-	0.68
Others Important to Metallurgists					
Pt^{192}	0·78	90	4·3 d	γ, γ	0·126, 1·5
Cr^{50}	4·3	11	27·8 d	γ, γ	0·32, 0·267
Zr^{94}	17·4	0·1	65 d	β^-, γ	many to 1 MeV
Mo^{98}	23·8	0·13	67 hr	β^-, γ	many to 1·2 MeV

called for. For example, if nickel is used as a matrix metal one must realize that cobalt is a common impurity and even so called cobalt-free nickel may contain sufficient cobalt to cause serious radioactivity. Let us consider a specific case as an example, say copper in silicon. If an element with an isotope of abundance, c, a neutron activation cross-section, σ, and half life, τ, is irradiated in a flux, G_0, for a time, T, removed from the reactor and allowed to stand for a time, t, the number of atoms decaying per unit time is

$$\frac{dN}{dT} = c\sigma N G_0 \left[1 - \exp\left(\frac{-0{\cdot}693T}{\tau}\right)\right][\exp(-0{\cdot}693t/\tau)] \tag{218}$$

where N is the number of atoms of the element in the original sample. Thus one microgram of copper irradiated in a flux of 10^{12} neutrons/cm^2/sec for 12 hours and allowed to stand for two hours after removal from the reactor has a decay rate of $\sim 10^4$/sec. On the other hand, ten grams of silicon would decay at a rate of $4{\cdot}2 \times 10^8$/sec. However, if the sample were allowed to "cool off" for two days the respective rates would be 10^3/sec for copper and $2{\cdot}4 \times 10^3$/sec for silicon. If, in addition, the copper is separated from the silicon chemically, an enhancement of the copper activity relative to the silicon activity of at least 100 could be achieved. Furthermore, the silicon does not emit a γ-ray as the copper does and absorbers can easily discriminate against the silicon β-ray. Thus, under proper conditions one part per hundred million of copper can be detected in silicon if nothing else is present that introduces a high background.

Radioactivity and its Health Hazard

The question of the level of radioactivity which presents a health hazard frequently arises in dealing with irradiated samples. The Curie is defined as a sample emitting $3{\cdot}7 \times 10^{10}$ particles per sec but the danger lies in the amount of energy per particle and the ability to shield it. α- and β-rays can be shielded quite easily but not γ-rays. The unit employed for health measurements is the *Roentgen* and a crude estimate is as follows: 1 millicurie of 1·2 MeV γ-rays ($3{\cdot}7 \times 10^7$ per sec) yields 1·31 milliroentgens per hour at 1 metre distance. The rate falls off inversely as the square of the distance but increases roughly as the energy. Each inch of lead reduces this by $\sim$90%. 50 milliroentgens per week is a typical allowable dosage.

Van de Graaff Accelerators

The use of the Van de Graaff accelerator for elemental analysis has certain promising potentialities although they have not yet been proven. Its principal advantage may arise in the non-destructive analysis of O, N and C in metals such as iron.

In performing the experiment a beam of deuterons is accelerated to several MeV at currents of 1–100 microamperes ($\sim 10^{13}$–10^5 deuterons/cm^2/sec) and allowed to impinge on the sample which we shall assume is iron containing C, N, and O. The deuterons are absorbed by the nuclei (cross-section $\sim$10 millibarns) which then emit higher energy protons in energy groups indicative of the different nuclei species. Silicon *p-n* junctions make excellent proportional detectors with energy resolution of about 1%. Since the principal reaction with the iron nuclei is the elastic scattering of the deuterons, the silicon detectors can discriminate against these lower energies.

In principle one should be able to obtain sensitivites of parts per million although this has not been realized. There are practical problems. The proton energies are not sharp since the deuteron loses energy as it enters the iron and the proton loses energy as it emerges. The depth of penetration of a 2 MeV deuteron beam is about five microns in iron so that a polished surface is desirable. Furthermore, all the cross-sections are not known and still await measurement. But in spite of these problems the accurate analysis of C, N, and O in metals is so important that such a technique should be pursued.

4. Positron Annihilation in Solids

If a 1 MeV positron enters a chamber containing some gas at temperature T it will make a series of collisions with the gas atoms, eventually reducing its energy to $\sim kT$. It will then either attach itself to a gas atom or pick up a free electron and form *positronium*. The eigenfunctions and eigenvalues of positronium are similar to the hydrogen atom except that the centre of mass of the system (which is the centre of the co-ordinate system) is at a point half way between the electron and positron. This is due to the relatively light mass of the positron which is equal to the electron mass, m_0. The hydrogenic wave functions about the centre of mass as origin are identical to those given in eqns. (3), (4) and (5) except that the mass m_0 is replaced by $m_0/2$ or one half the rest mass of the electron. The hyperfine structure of the hydrogen atom involving the magnetic interaction between the electron and the nuclear magnetic moment is now considerably greater since the magnetic moment of the positron is some three orders of magnitude larger than that of the proton. Since the positron magnetic moment is equal in magnitude but opposite in sign to that of the electron the magnetic moments are antiparallel in the lowest energy state of positronium and it is called parapositronium. The eigenvalue of the lowest energy state is half a Rydberg and the "Bohr radius" is twice as large. Both positron and electron are in $1s$ states in parapositronium as well as in orthopositronium, the next higher state in which the spins are parallel.

After about 10^{-10} sec parapositronium annihilates itself by converting into two γ-rays each of energy m_0c^2 ($\sim\frac{1}{2}$ MeV) and each with equal and opposite momenta (180° apart). But because of the conservation of intrinsic angular momentum in orthopositronium (the electron and positron m_s^t values are parallel) it must decay via three γ-rays and this takes much longer, $\sim 10^{-7}$ sec. The existence and properties of positronium have been well established in gases.

If we now allow a positron to enter a metal it will make many collisions and end up in thermal equilibrium with the lattice in about 10^{-11} sec. Its final position is not known but some logical possibilities can be presented. It probably does not penetrate inside the region of the core electrons in spite of its attractive Coulomb potential since the nuclear repulsion is greater. In fact any wave function that one contrives for the positron on an atom will have a positive energy about equal to the negative energy of an electron with a similar wave function. Thus one might guess that the positron will choose positions between the atoms where the nuclear repulsion on opposite sides approximately cancels so that a larger attractive potential with the outer electrons is left.

In about 1·5 to 2·5 × 10^{-10} sec from the time the positron was created it will decay in the metal by annihilation with a nearby electron (with opposite m_s) but now any momentum of the positron and electron pair will have to be conserved in the process. As such the γ-rays do not emerge exactly 180° apart but may deviate from this by several degrees. The additional momentum is seen in Fig. 134, diagram 1 and is equal to $2m_0c \sin\theta = 2m_0v$ where θ is one half the angular deviation from 180°. Fig. 134, diagram 2 shows the results for the 3 cases sodium, aluminium and copper plotting the relative number of annihilations in the metals as a function of the angle θ in radians. An angle of 10^{-2} radians corresponds to $v/c \cong 10^{-2}$ for the electron-positron pair or an energy of about 12 eV for the pair. Per electron, this is of the order of the Fermi energy in a metal and indicates that some information about the metal is evident in the results.

Na^{22} is generally used as the positron source (half life, 2·6 y) and is placed near the metal which is in the form of a strip. Two γ-ray scintillation counters are placed on opposite sides of the strip with slits in front of them.

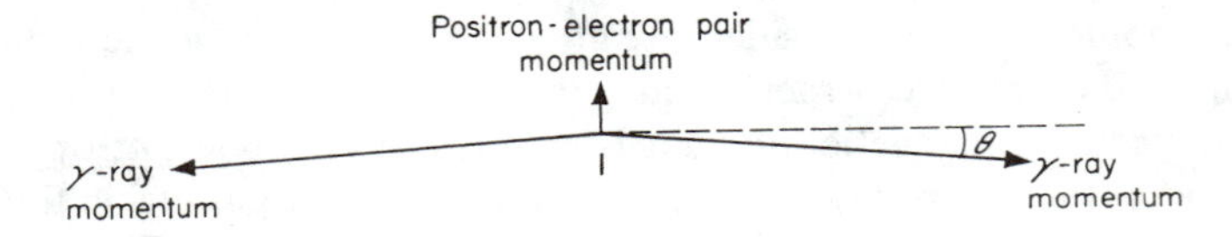

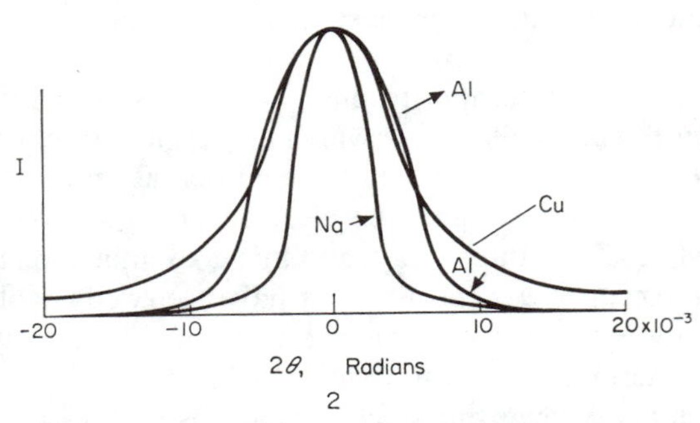

FIG. 134. The angular coincident rate, I (in arbitrary units), of gamma rays emerging from copper, aluminium and sodium following positron annihilation, versus angle in radians.

The Na^{22} source is shielded from the counters with lead and the counters are wired in coincidence so that only photons arriving simultaneously at both counters are recorded. This minimizes all other sources of background counts. In a typical case the slit widths give a resolution of $\sim 10^{-3}$ radians ($\sim$0·1 eV).

The results of many measurements are given in the two references[97][98] and the theory is reviewed by Ferrell.[99] While there are significant differences between various metals, many such as iron, cobalt, and nickel are similar and the interpretation of the results have not yielded significant details of the momentum distribution of the electrons in the band. The annihilation process depends on the positron finding an electron near it and so depends on the overlap of the positron and electron wave functions. One of the difficulties is that the wave functions of all electrons near the positron are seriously altered by the positrons attractive potential. Thus our measuring device greatly disturbs the system we are measuring.

Perhaps one of the most discouraging results are those of Lang and de Benedetti[(100)] who measured the angular distribution in pure palladium and palladium with sufficient hydrogen (0·7 per palladium atom) so that most of the interstitial sites are filled. Since the positron and proton exert an identical attractive Coulomb potential it is expected that their positions in the lattice would be similar even though their masses are different and that they would affect the local electron probability distribution in a similar manner. Within the limit of the experimental error no difference was observed, both curves being similar to the one for copper in Fig. 134.

If one applies the free electron theory to the volume of the crystal outside the closed shell core electrons (the core volume is called the *excluded volume*) and assumes the positron is at rest before annihilation, the *cut off angle* is the largest angular deviation from 180° and corresponds to the momentum of the electrons at the Fermi level. This angle is

$$\theta = 10^{-3}\left[\frac{135}{r_s^2} + \frac{185}{a_s^2}\right] \tag{219}$$

where r_s is the dimensionless radius (in units of Bohr radii) of the sphere whose volume is equal to the volume occupied by each free electron and a_s is the dimensionless radius of the sphere equal in volume to the closed shell core volume which is excluded from the positron. Equation (219) yields answers agreeing with the experiment to within a factor of two. Another expression derived from the free electron theory is the life time τ, of the positron in the metal and depends on the amount of electron overlap. This is

$$\tau \cong r_s^3 \frac{10^{-10}}{1{\cdot}2} \text{ sec} \tag{220}$$

Measurements of lifetime have not been too abundant, but a few examples in units of 10^{-10} sec are: aluminium, 2·4; lead, 3·5; copper, 1·2; silver, 1·5; lithium, 1·5; beryllium, 1·7; and sodium, 1·5. The correlation between the expressions for cut off angle and life time is poor.

We must conclude that positron annihilation has not as yet produced significant enough differences in angular correlation or life time to indicate that it is sensitive to the momentum distribution of the outer electrons in metals. There are a sufficient number of results on various metals to make the field rather discouraging for solid state purposes. However, there is some indication that single crystals are showing variations for different crystal directions but this awaits more extensive investigations.

Problems

1. Evaluate various methods for determining the ratio of austenite to martensite in an iron alloy with 8% nickel by utilizing nuclear properties. Compare these methods with an X-ray method and a magnetic method and estimate the relative sensitivity, particularly in martensite rich alloys. Can you devise other practical methods for making such a determination?

2. Look up the diffusion coefficients (at about $\frac{2}{3}$ the melting temperature) and activation energies for diffusion for as many elemental metals as you can find. See if there is any relationship to the periodic table. Can you construct a general self-diffusion equation for all elements merely as a function of their melting temperature by expressing D_0 and E in terms of melting temperature? What might this indicate?

3. Calculate the second moment in (gauss)^2 for both isotopes of copper in pure copper and in ordered β brass, and compare these values with experiment.

4. (For the mathematically minded) Show that the electric field gradient at the centre of a cube is zero if the surface of the cube is uniformly charged negatively. Evaluate the electric field gradient at the centre if the cube is distorted to a c/a ratio $= 0{\cdot}95$ ($a = 1$ Å) and one electronic charge is placed at the centre of each face. Plot the eigenvalues for a nucleus at the centre with $eQ = 0{\cdot}5 \times 10^{-24}$ cm^2.

5. Cobalt has a neutron resonance at 135 eV with a peak cross-section of 7000 barns. Assume this to be due to either one of the two spin states (neutron m_s parallel or antiparallel to nuclear spin) calculate the absorption cross-section of a magnetized sample of cobalt metal for 1 Å polarized neutrons as a function of temperature for 1·2°K to 4°K.

6. By using one or more of the techniques discussed in this chapter how would you attempt to solve the following problems:
   a. The solubility of vanadium in copper.
   b. The mean square displacement (due to thermal motion) of iron atoms dissolved in a vanadium matrix.
   c. The "authenticity" of a prehistoric piece of steel attributed to a highly developed civilization from the lost city of Atlantis.
   d. The amount of cobalt in a piece of Hadfield steel.

Summary

Nuclear magnetic resonance, nuclear quadrupole resonance and Mössbauer measurements in metals are all rather sensitive to the local electron wave functions. While these measurements have been most successful in giving precise information about the electron wave functions close to the nuclei they unfortunately yield less precise details of the outer (or bonding) electron wave functions. To a great extent the field has been dominated by physicists but it is apparent that metallurgical applications are most promising.

Radioactivity measurements have been useful in self diffusion studies and in elemental analysis of trace quantities. While positron annihilation appeared rather promising at one time, little has been learned of the properties of the outer electrons in metals.

CHAPTER XII

THE JIG-SAW PUZZLE; A PROBLEM IN SYNTHESIS

Introduction

We have endeavoured to present the various experimental measurements made by physicists in their study of the electronic structure of metals and, whenever possible, to relate the results of these measurements to the eigenfunctions and eigenvalues of the electrons. We have pointed out the immense theoretical problem in solving the Schrödinger equation for a solid and that it is rare when the solutions predict a measurable property as accurately as we can measure it. In the main we must rely on intelligent measurements of physical properties to guide us in seeking the theoretical solutions of the Schrödinger equation.

We think it appropriate, however, to itemize the measured physical properties of copper and iron to illustrate how the measurements might solve the jig-saw puzzle and synthesize the outer electron eigenfunctions and eigenvalues. In these two cases we have as many pieces of the puzzle as we have for any metal. These measurements are listed in Table XXVIII. A blank in the deduction column means that we cannot say anything even though the measurement has been made. (Please advise the author when you can fill it in.) No attempt has been made to include those measurements which depend on surface properties such as photoelectric effect or thermionic emission.

Electronic Structure of Copper and Iron. Our picture of copper deduced from the entries in Table XXVIII is perhaps best summarized by the qualitative E versus k curves of Fig. 135, diagram 1. The total band width is ~6 eV with appreciable $3d$ contribution from the bottom of the band to about 2eV below the Fermi level (point D). The remainder of the band is principally $4s$ like with a little $4p$ contribution. The wave functions have considerable overlap with the neighbours at the bottom of the band and this accounts for the large cohesive energy but at energies corresponding to point C in Fig. 135 the wave functions are compressed in the [111] direction (see Fig. 10) and the overlap is very small. At the Fermi level the wave functions overlap in the nearest neighbour direction [110] and next nearest neighbour direction [100] but not [111]. This is approximately what one gets by packing overlapping spheres together. A very crude density of states curve is sketched in Fig. 135, diagram 2. It is interesting that recent band calculations

Table XXVIIIa
Pure Copper

Measurement	Information	Deduction
X-ray diffraction	Crystal structure	Close packing suggests overlapping electrons are approximately spherically symmetric
	Lattice parameter	Significant d contribution to bonding
	Debye Θ and melting point	Significant d contribution to bonding
	Variation of Θ with T	
	Electron probability distribution	
	Frequency spectrum	
Neutron diffraction	Crystal structure	Same as X-ray diffraction
	Lattice parameter	Same as X-ray diffraction
Electron diffraction	Crystal structure	Same as X-ray diffraction
	Lattice parameter	Same as X-ray diffraction
Optical spectroscopy	K, L, M emission bands	Width of $3d$ band $\sim$4 eV, total width of $4s$ band $\sim$6 eV
	Optical absorption	Top of $3d$ band $\sim$ 2·2 eV below Fermi level (red colour)
Transport properties	Resistivity	Fermi level $4s$ like
	Magnetoresistance	Fermi surface contacts 111 Brillouin zone face
	de Haas van Alphen effect	Fermi surface contacts 111 Brillouin zone face, and low density of states at Fermi level
	Anomalous skin effect	Fermi surface contacts 111 Brillouin zone face
	Cyclotron resonance	Fermi surface contacts 111 Brillouin zone face, and low density of states at Fermi level
	Hall effect	Normal current carriers
	Thermoelectric power	
	Lack of superconductivity	

TABLE XXVIIIa. PURE COPPER—*continued*

Measurement	Information	Deduction
Thermodynamics	Elastic constants	Forces not central $c_{12}/c_{44} \neq 1{\cdot}0$
	Debye Θ	Same as X-ray diffraction
	Cohesive energy	3*d* electrons contribute appreciably to bonding
	Coefficient of expansion	
	Electronic specific heat	Low density of states at Fermi level
	Ultrasonic attenuation in magnetic field	Fermi surface contacts 111 Brillouin zone face
	Thermal conductivity	
Pressure	Pressure dependence of elastic constants	Forces between atoms not simple springs
Magnetization	Diamagnetic susceptibility	Small paramagnetic susceptibility —low density of states at Fermi level
Nuclear measurements	Knight shift	
	Positron annihilation	
	Pressure dependence of the Knight shift	
	Relaxation time, T_1	

TABLE XXVIIIb

PURE α IRON

Measurement	Information	Deduction
X-ray diffraction	Crystal structure	Mostly 3*d* electrons used in bonding—t_{2g} preferred slightly
	Lattice parameter	3*d* electrons used in bonding
	Debye Θ and melting point	Large 3*d* contribution to bonding
	Variation of Θ with *T*	
	Electron probability distribution	

TABLE XXVIIIb. PURE α IRON—*continued*

Measurement	Information	Deduction
Neutron diffraction	Crystal structure	Same as X-ray diffraction
	Magnetic structure	
	Unpaired electron Probability distribution	Unpaired electrons have slight preference for e_g orbitals and have negligible overlap with neighbouring atoms
	Magnetic scattering at high temperature	Unpaired electrons remain localized on atom above the Curie temperature
	Frequency spectrum	First and second nearest neighbour forces most dominant
Electron diffraction	Crystal structure	Same as X-ray diffraction
	Lattice parameter	Same as X-ray diffraction
Optical spectroscopy	*K, L, M* emission bands	Band width ~6 eV wide
	Optical absorption	
Transport properties	Magnetic resistivity	Strong exchange interaction between current carriers and unpaired 3*d* electrons
	Ordinary resistivity	Current carriers have large effective mass—3*d* like
	Hall effect (ordinary)	Current carriers—holes, probably 3*d* like
	Hall effect (anomalous)	Current carriers—holes, 3*d* like due to spin-orbit coupling
	Magnetoresistance	
	Orientation effect	
	Thermoelectric power	
Thermodynamics	Elastic constants	Forces not central, $c_{12}/c_{44} \neq 1{\cdot}0$
	Cohesive energy	3*d* electrons used in bonding
	Debye Θ and melting point	Same as X-ray diffraction
	Coefficient of expansion	Small change at Curie point—magnetic electrons not involved in bonding

TABLE XXVIIIb. PURE α IRON—*continued*

Measurement	Information	Deduction
	Electronic specific heat	High density of states at Fermi level—probably 3*d* like
	Magnetic specific heat	Entropy—unpaired 3*d* electrons remain localized above Curie point; energy—$J = 0{\cdot}011$ eV
	Thermal conductivity	
Pressure	Pressure dependence of resistivity	Curie temperature independent of pressure up to 100 kilobars (6% volume decrease)
Magnetization	Saturation magnetization, $T = 0°$K	$gJ = 2{\cdot}2$, non-integral number of unpaired electrons
	Low temperature spin wave	J (= exchange energy) does not agree with specific heat determination. Probably indicates second nearest neighbour interaction
	g factor $= 1{\cdot}93$	
	Magnetostriction	
	Variation of Curie temperature with pressure	
Nuclear measurements	Ferromagnetic shift in nuclear resonance	$H_{eff} = -333$ kgauss indicates more 4*s* contribution antiparallel to 3*d* rather than parallel
	Mössbauer effect	Same as ferromagnetic shift
	Ferromagnetic shift with pressure	3*d*–4*s* admixture varies slightly with atomic distance
	Isomer shift	~ seven 3*d* electrons in metal

have been able to predict the shape of the Fermi surface with remarkable accuracy (B. Segall, *Phys. Rev.* **125**, 109 (1961). In this calculation Segall tried two different starting potentials and found little difference. He also determined the energy gap from the top of the 3*d* band to the Fermi level to be ~2·1 to 2·7 eV in good agreement with optical data. The insensitivity to the choice of starting potential is probably due to the predominantly 4*s* character of the wave functions at the Fermi level and is yet to be demonstrated for transition metals. Nonetheless, Segall's calculation is one of those rare cases where a fundamental calculation predicts an experimental result.

Our deductions for iron are that the band is about 6 eV wide and about seven of the eight outer electrons are $3d$ like. The overlapping electrons at the bottom of the band tend to prefer the t_{2g} orbitals, Fig. 17, and they account for the bonding with nearest neighbours, while the unpaired $3d$ electrons near the top of the band have virtually no overlap and slightly prefer the e_g orbitals. The $4s$ contributions extends throughout the band although there is more $4s$ contribution with the m_s antiparallel to the unpaired $3d$ electrons than parallel. The current carriers near the top of the band are $3d$ like having a relatively large effective mass. We cannot guess the E versus k or density of states curves but there seems to be qualitative agreement with theoretical band calculations like those of John Wood (see Fig. 21 for the $3d$ wave functions).

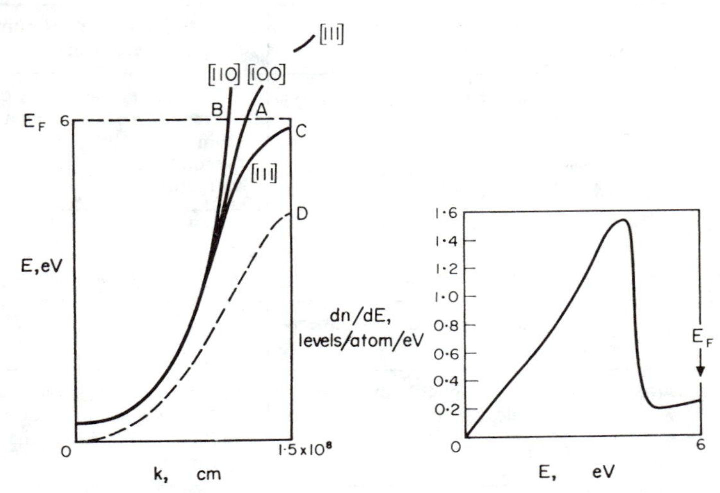

FIG. 135. Diagram 1. Approximate energy versus wave number curves for the $4s$-like orbitals in copper (solid lines) and the $3d$-like orbitals (dashed line). The directions [111], [100], [110] are indicated and show the contact of the $4s$ orbitals with the 111 Brillouin zone face.

Diagram 2. Approximate density of states curve for copper showing the peak in the $3d$ band about 2 eV below the Fermi level.

Solutions of the Schrödinger Equation for the Imperfect Crystal. We shall conclude this chapter by pointing out that the solution of Schrödinger's equaion for imperfect crystals has begun to evoke theoretical effort. Since this will be one of the biggest hurdles in ultimately providing the metallurgist with theoretical calculations for the "impure materials" he deals with, we should like to show an early attempt at this, Fig. 136. Makinson and Roberts[101] have considered the one dimensional square well periodic potential consisting of 2000 atoms in their chain and through laborious calculations have determined the E versus k curves near the energy gap at $k = \frac{1}{2}a$ (a is the repeat distance). For the perfectly periodic case the solution is given by the dashed curve, $\sigma = 0$. They then vary the spacing between the atoms in a random fashion simulating the effect in a liquid. σ is the statistical standard deviation of the spacing between atoms and as long as σ is small the energy gap still exists although it is not as sharp. However, for $\sigma \geqslant 0{\cdot}1$ the energy gap disappears. This corresponds to all ordering disappearing in a distance of $\sim$25 atoms! It is too early to speculate on this result but it does represent the beginnings of a fundamental approach to the imperfect lattice, so dear to the heart of the metallurgist.

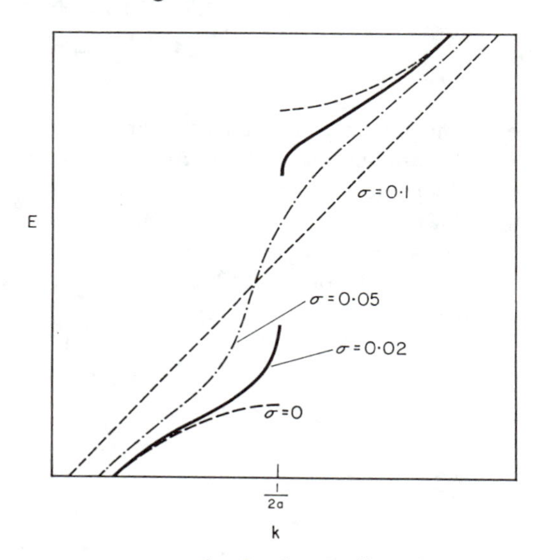

FIG. 136. Energy versus wave number for the one dimensional chain of 2000 square well potentials. $\sigma = 0$ is the ordered case showing the energy gap at $k = \frac{1}{2}a$ (a is the repeat distance). The increased values of σ refer to increased disorder in the spacing, a, between potentials. As the disorder approaches the liquid like structure, the energy gap narrows and eventually disappears.

Problems

1. Try to make an educated guess as to why invar has such a low coefficient of expansion. Collect as much experimental evidence as you can to support this guess.

2. Look up the melting points of the elements and group them according to their column in the periodic table. Are there any trends? What might you conclude about the electron structures from these trends? What other evidence can you gather to support these conclusions?

3. What evidence is there to suggest that there might be a metallic form of germanium? How might it be obtained and what properties might it have?

4. Look up the compressibility of as many metallic elements as you can find and see if there are any trends relating to their positions in the periodic table. Try this also for Poisson's ratio.

5. Look up the hardness of as many pure elements as you can find and see if there is any correlation to their crystal structure and/or their position in the periodic table. Does this suggest that electronic structure plays a role in this property?

6. (For science fiction fans). What would the world be like if the one electron energy levels of the atom strictly followed the hydrogenic sequence $1s$, $2s$, $2p$, $3s$, $3p$, $3d$, $4s$, $4p$, $4d$, $4f$, $5s$, $5p$, etc., with at least 10 eV between each one electron group. For example, Ni would be $1s^2$, $2s^2$, $2p^6$, $3s^2$, $3p^6$, $3d^{10}$ and would suffer the dreary fate of an inert gas. What would our technology be like?

7. Write the atomic numbers 1 to 94 (omitting 2, 10, 18, 36, 43, 54, 84, 85, 86, 87, 88 89, 91) on individual pieces of paper and select random pairs. See if you can predict the phase diagram, density, electrical, magnetic and thermal properties etc. for each alloy system. Make a literature search to check your predictions. What fraction of the 3240 binary systems are virtually unexplored experimentally?

8. (As a class project.) Try to complete Table XXVIII for α Mn, Ni, Al, Gd and Pb.

Summary

The various experimental measurements on copper and iron are enumerated. A semi-quantitative indication of the electronic structure of these metals is given as deduced from the measurements. While these two metals have been subjected to an abundance of measurements we still know very little about their density of states and wave functions. For metals of more practical metallurgical interest (alloys etc.) the experimental and theoretical approaches have been much less fruitful.

CHAPTER XIII

THE PHOTON REVOLUTION

Introduction

Just after WW II the transistor made its revolutionary impact on technology. Based on silicon (diamond structure, Fig. 16), this abundant element is reduced from sand and is characterized by brittle mechanical behavior and semiconducting electronic properties. In its pure form, thermal or photon excitation can raise one of the bonding electrons into a state that permits free movement through the lattice, as in a metal. By clever addition of dopants the energy required for this excitation can be reduced in the vicinity of the impurity atom and one can create a local trap for this electron. The trap has the further property of being addressed through appropriate application of either a negative or positive potential. This trap becomes a piece of information i.e. either plus or minus, or more commonly a binary code 0 or 1.

Through clever miniaturization this storage and reading of the binary code becomes the basis for information processing. We are all aware of the impact silicon technology has had on device design. There is even a valley in California named after the element.

In 1960 the laser was first demonstrated, based on a process called stimulated emission, predicted by Einstein in the early part of the century. Einstein died before the discovery of the laser. If an atom is excited to a metastable higher energy state and a photon of the same energy passes in the vicinity of the atom, the atom is prodded into returning to its ground state with the emission of a photon. The second photon becomes a clone of the first, identical in direction, phase and energy. If many atoms are excited to metastable states, then the stimulated emission initiated by one photon can create a chain reaction—all emitted photons producing a coherent pulse of light. Such lasers are marketed in many shapes, sizes and forms. Unlike an incandescent bulb or a fluorescent lamp the photons are sharply collimated.

Following this development, high purity glass fibers were developed to transmit this well-collimated beam over long distances and replace copper wires in telephonic transmission. Why the new technology?

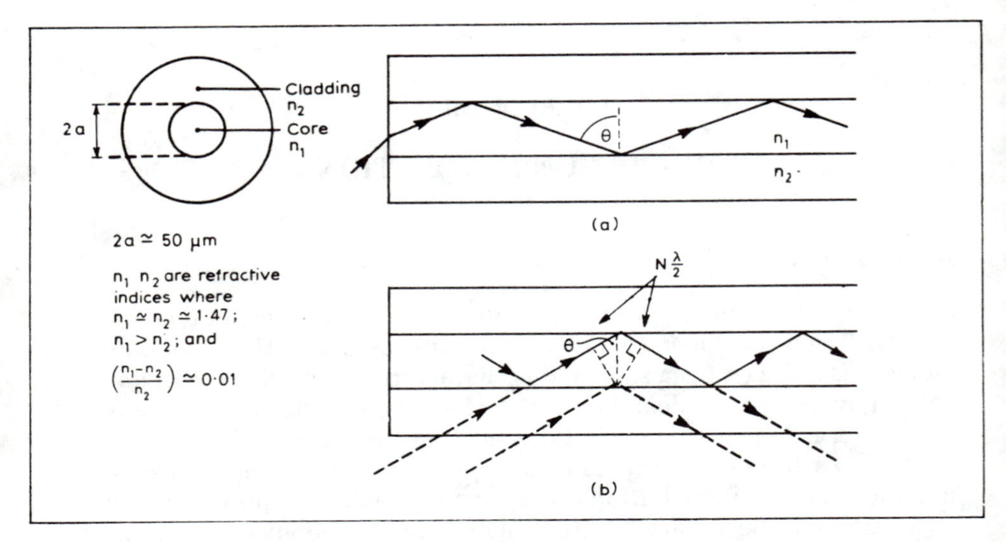

FIG. 137. A typical optic-fibre structure consists of a central core of glass or silica, surrounded by a cladding of slightly lower refractive index (n). If $\sin\theta > n_2/n_1$ the ray is totally internally reflected at the core/cladding boundary so that the ray bounces down the core and is trapped within it. The light is thus 'guided' by the core. Only if the path differences $N\lambda/2$ shown correspond to an integral number of half-wavelengths of the light will the angle lead to a self-reproducing wave interference pattern and thus to a stable 'propagation mode' of the waveguide.

Silica Fibres

A typical fibre structure is shown in Fig. 137. Here we see a central core, normally of glass or silica, surrounded by cladding, also of glass or silica, but which has a slightly lower refractive index (about 1 per cent less). Consider a ray of light travelling within the core. When it strikes the core/cladding boundary it will normally pass into the cladding, travelling then at a slightly reduced angle to the boundary, in accordance with Snell's well-known law of refraction. But, if the incident angle is very shallow, Snell's law cannot be satisfied and the result is a 'total internal reflection' at the boundary; the boundary now acts as a perfect mirror. Under this condition the ray will bounce from side to side along the fibre until it emerges at the far end. The light is thus trapped in the fibre core, which acts as an optical waveguide.

However, not all reflection angles are possible, even when the 'total internal reflection' condition is satisfied; for light is a wave motion and the waves represented by the various rays will interfere with each other to cancel or reinforce, depending on the relative phases of the waves. The result is that a refraction angle is allowed only if it satisfies the condition shown in Fig. 137. The condition requires that, for two parallel rays striking the boundary at opposite ends of a diameter, the difference in effective path length to the boundary points is a whole

TABLE XXIX

	Photons	Electrons
1. Low resistance	Yes	No
2. Polarization	Yes	No
3. Coherent source	Yes	No
4. Frequency separation	Yes	No
5. Two-directional flow	Yes	No
6. No electromagnetic interference	Yes	No
7. Very broad band	Yes	No
8. Vibration sensitive	Yes	No
9. No short circuit hazard	Yes	No
10. Compatible with silicon	No	Yes
11. Injection problems	No	Yes
12. Superconducting	No	Yes
13. Non-linearity	Yes	No

number of half-wavelengths of the light in the core. The difference in phase is thus a 'whole number times π' radians. For each of these allowable angles an interference pattern is formed which is stable and self-reproducing along the fibre.

Photons Versus Electrons

Optical fibre use in communication is now well known to our entire society due to the PR of the telephone companies. The direction of our technology is a replacement of electrons in wires by photons in fibres for conveying information.

Table XXIX provides a short visual guide to the advantages and disadvantages of photons over electrons and these items are explained below.

1. *Low resistance* — Optical fibre is now readily produced with very little attenuation for distances of miles. Repeater stations are necessary for transmission across oceans but this is also true for signals sent by electricity. Photons and electrons are possibly on a par in this respect.

2. *Polarization* — The elecromagnetic character of photons enable one to inject a fibre with photons that have a variety of states of polarization, i.e. the direction of the electric vector relative to the propagation direction. If this is uniaxial, it is plane polarized. If it spirals regularly as it propagates it is circularly polarized, while a combination of plane and circularly polarized light leads to elliptical polarization. The important thing to remember is that the state of polarization can be recognized in an appropriate detection system. The interest lies in the effect certain parameters like stress, pressure, temperature etc. have on the polarization, leading to sensor application.

While electrons have a sense of direction (north and south poles) there is no technique known to control this as they flow in a wire.

3. *Coherent source* — The laser was first demonstrated by T. Maiman in 1960 and in 25 years has led to a multi-billion dollar technology. Funnily enough it was Albert Einstein's theory of stimulated emission that accounts for the laser, i.e. that an excited state of an atom can be more rapidly de-excited with photons of the same energy. In the atom's return to the ground state it emits a photon whose phase and polarity are identical to the photon producing the stimulation. The latter photon is in no way altered in the process and continues on its merry way.

Neither Maiman nor Einstein received the Nobel Prize for this revolution in physics. Maiman was pre-empted by a few years with the work of Charles Townes and others on the Maser, while Einstein passed away before the laser's discovery.

The coherence of the laser source permits interferometric techniques to be applied and this is currently employed in sensors.

4. *Frequency separation* — Photons traversing an optical fibre are easily identified as to their frequency. Most photons maintain this frequency although a small fraction absorb or give up energy to (thermal) vibrational or rotational energy levels and are scattered with a measurable frequency shift.

Electrons in conductors cannot be identified as to their energy. Furthermore most of the resistance to flow comes from collisions with vibrating atoms and this typically occurs over distances of atomic dimensions. We measure the total flow of electrons, little else. At room temperature electrons in metals are scattered more than three orders of magnitude more effectively than photons in optical fibre.

5. *Two-dimensional flow* — Photons generally do not interfere with each other. There are some interference effects when photons come rather close but for practical purposes we can think of photons traversing a fibre as non-interacting. Thus we can inject and detect light at both ends of a fibre.

In electrical conductors we can only measure total electrical flow. There is no way to establish two way traffic at the same time. Alternating current goes only one way at any instant of time.

6. *No electromagnetic interference* — Photons are not deviated when traversing a magnetic or electric field in free space (funnily enough they are slightly deflected by gravity and this has been measured in a high resolution experiment employing X-rays).

But the immunity for electromagnetic noise, such as is created by lightning and arcing, is a major attraction in optical fibre communication development.

7. *Broad band* — Multimode optical fibres are capable of transmitting a broad range of frequencies so that considerable information can be sent. In addition, optical signalling can be fast, i.e. essentially with the speed of light divided by the index of refraction.

Electrical signals transmitted over conductors are limited in their information content and are slower than optical signals.

8. *Vibration sensitive* – One disadvantage with optical fibres is that stresses due to vibration lead to strains in the fibres and to scattering of the photons. This can be advantageous when one uses a fibre to measure strain.

Electrical conductors are insensitive to small strains except in specialty devices such as pressure transducers where differences in electric potential are measured between the arms of an otherwise balanced bridge.

9. *Short-circuit hazard* – Short circuits can create difficulties in electrical systems, especially in hazardous environments where sparks can cause fires or explosions. This is a major consideration favouring optical circuitry.

10. *Compatible with silicon chips* – Where silicon chips are employed in devices the electrical nature of their signals is compatible with electrical circuitry.

The employment of optical fibres to carry signals requires an optical interface and its attendant losses. The industry is trying to overcome these problems. An effort is being made to replace silicon chips with optical read, write and erase devices. Another effort is being made to use materials like doped GaAs which can produce a laser signal from an electrical input.

It is difficult to predict what approach will make the most significant inroad into the industry.

11. *Injection problems* – Common sources of optical signals are LED devices but these are not lasers and only provide a simple and inexpensive optical source. There is a significant loss on injection of the signal into the fibre.

There is no similar problem in electrical circuits where connections can be made simply and with little loss.

Some technique for exciting a laser by special doping within a fibre is an obvious solution to reduce injection problems.

12. *Superconductivity* – Superconductivity is a phenomenon present in certain materials at low temperatures, generally at less than 120°K. In such materials all electrical resistance ceases due to a special coupling of the current-carrying electrons with the vibrational modes of the atoms. It is not understood too well and new superconductors are discovered only by empiricism.

While there is no analogous phenomenon in optical circuitry, optical fibre has low enough resistance to make such a requirement unnecessary at this time.

Superconductivity has found limited use (magnets for example). Only a vacuum shows zero resistance to optical flow.

13. *Non-linearity* – Non-linearity has become a fashionable part of the optical properties of matter. The concept is simple.

As photons enter certain materials they excite electrons to higher energy states. This alters the optical properties of the material. Changes in attenuation, dispersion etc. take place. Thus the properties depend on the light intensity.

In the case of electrons in metals, Ohm's law strictly represents the linear nature of the flow process. In semiconductors temperature changes alter the current flow as more electrons are exponentially excited into the conduction process. This can be viewed as a non-linear phenomenon. The greater the current flow, the hotter the semiconductor and the less the apparent resistance (more electrons contribute to the current).

Non-linear phenomena provide an added dimension to optical circuit strategy. Some of the subtler functions of the human brain are probably due to non-linear effects.

To date, non-linearity is a laboratory study, but it surely looms as a potential in device design.

The advantage of photons over electrons lies in versatility, i.e. changes occur in several optical properties at the same time.

Sensors

In the course of developing optical fibres for long-range communication, researchers became aware of the effects of external parameters such as temperature, stress etc. on the optical signal. This has now been turned around and optical-fibre sensor technology is dedicated to converting a desired parameter into an optical signal.

The possibilities are virtually limitless. We are embarking on a new technology that appears capable of adding new dimensions to the ease, efficiency and potential in measuring the world around us.

Consider this list of possibilities:

1. Temperature
2. Pressure
3. Acoustic phenomena
4. Radioactivity
5. Magnetic fields
6. Electric fields
7. Electromagnetic radiation
8. Chemical species
9. Acceleration (gyroscope)
10. Biological functions
11. Non-destructive evaluation
12. Image analysis

Not only can we prepare special fibres sensitive to items listed above, but the same fibre can be made sensitive to several parameters.

It is difficult to imagine a property not amenable to fibre sensor technology (we make no claims about ESP). The bottleneck at the moment is in materials development.

Inasmuch as sensor technology relaxes the optical fibre requirement for long-range communication (i.e. hundreds of miles) specialty fibres can tolerate higher attenuation coefficients since the observer is typically less than 100 feet away from the system he is measuring. But even if one wished to transmit such a signal over longer distances, one only requires an interface into a long range communications network.

Another exciting prospect is the distributed sensor, i.e. the ability to measure an external parameter at all points along a sensor fibre stretched through a large system of interest. This possibility arises from the ability to time the return signal and so pinpoint each spot along the fibre.

Some elaboration now about the items in the list:

1. *Temperature* — Ordinary optical fibre used for communication is temperature sensitive. Any local temperature increase expands the fibre reducing the electron density (hence index of refraction) as well as setting up stresses that alter the index of refraction and scatter light. Thus intensity changes occur. A second phenomenon is the inelastic Raman scattering, i.e. the excitation of the vibrational or rotational energy levels in the SiO_2 chains in glass. These inelastic changes are clearly separated from the input frequency and by comparing the relative strengths of these lines a temperature scale can be established.

But this is only the beginning. By adding specific elements to the glass that fluoresce (or to other fibres), one can examine the spectrum of fluorescent lines which usually follows an exponential excitation function ($\exp -E/kT$), where E is the excitation energy and T the temperature.

Laboratory work is currently examining the optical properties of glass with various additions and with other transparent materials appropriately doped.

Another phenomenon is the depolarization of light as it passes through regions of varying electron density set up by temperature induced strains.

It is possible to examine the black-body radiation emitted by the fibre. At ambient temperatures most of the emission is in the infra-red and one requires special fibres with good transmission at these frequencies. One can also imbed a liquid crystal into the end of a fibre. This changes colour with temperature.

For high temperatures the black-body spectrum of a sapphire or quartz fibre can provide a temperature measurement.

The possibilities are many. In spite of the ease of use and popularity of thermocouples, they are likely to be superseded by optical thermometers.

2. *Pressure* — The problems are a bit different, depending on

whether we deal with modest hydrostatic pressures (<1 kilobar) or high pressures (up to megabars).

One promising sensor for high pressure is the ruby fluorescence line, whose frequency varies almost linearly up to the highest pressures attainable in the laboratory (~megabars). Such a sensor would consist of a very small ruby chip imbedded in the top of an optical fibre. The fibre tip is placed at the point of interest and the ruby is excited by an external optical source through the fibre. The fluorescent line is examined after it emerges from the same fibre. Accurate calibration of this device is still required, but a 1% accuracy should be feasible.

At lower pressures other dopants must be explored, such as the Na yellow line (sodium is easily contained in glass fibres). Such sensors are still in the laboratory stage.

One cannot easily predict the pressure dependence of the fluorescent lines. The energy levels are altered with pressure since the electron arrangements change and these are too difficult to evaluate theoretically to an accuracy comparable to a good measurement.

One advantage of a fluorescent line is that its width is an indication of the departure from true hydrostatic pressure, i.e. when the sample experiences a pressure gradient.

At low pressures the Raman lines are influenced by pressure as well as temperature, inasmuch as the electron arrangement affects the vibrational and rotational levels. The technique used for temperature measurement can be employed for pressure.

3. *Acoustic phenomena* — The ear is a remarkably sensitive organ. If all the people in the world shouted for one second, the total acoustical energy could just about boil a pot of water. An acoustic wave disturbs an optical fibre sufficiently to affect the propagation of light, both in an air or underwater medium.

Acousto-optic phenomena were reported as long ago as the nineteenth century by Alexander Graham Bell (not with optical fibres).

At the present time magnetic tape is still the mainstay of the recording industry, but optical technology is making inroads in reading, writing and even erasing. This is a transition period—the next 20 years will probably witness a total technology transfer to optical systems.

This does not necessarily mean that magnetic phenomena will be eliminated. Rather, one can envisage fibre sensors imbedded with colloidal magnetic particles whose magnetic sense affects the local stresses, hence the optical properties. Magnetic particles still remain an effective technique for easy writing and erasing.

4. *Radioactivity* — Radioactive decay products represent a potential hazard. There are now on the market plastic optical fibres imbedded with a scintillating compound. This represents a crude start for a device that parallels germanium solid state detectors.

The germanium single crystal detector converts essentially all the radioactive particle's energy into an electron current. One not only detects the particle, but determines its energy as well, to at least 1% accuracy.

There is no reason in principle why optical materials cannot be found that yield such accuracy. NaI crystals with thallium impurities do just this, but unfortunately with an accuracy at least an order of magnitude poorer.

The second stumbling block in developing such fibres is that they must not suffer excessively from radiation damage. Sapphire might be a good candidate.

5. *Magnetic fields* — For many years colloidal ferrite particles have been employed in producing magnetic tape and ferromagnetic fluids. A ferromagnetic fluid consists of a dispersion of ferrite particles with a polymeric coating that inhibits them from clustering. The fluid carrier into which the particles are dispersed is able to carry the particles along even in the presence of high magnetic fields. That is to say that the viscosity of the ferromagnetic fluid is independent of magnetic field strength.

Such colloidal particles can be introduced into most liquids such as molten glass, and they give rise to a fibre with good magnetic field sensitivity.

A polymeric fibre with colloidal magnetite offers another possibility. In the presence of a magnetic field the fibre will alter its shape to conform to the lowest magnetic energy. This will alter the magnetic path.

It is too early to predict the outcome of this technology, but it does have the potential of not only detecting magnetic fields but of encoding magnetic information as is done on magnetic tape.

Actually, any fibre, whether paramagnetic or diamagnetic, will respond to the presence of a field and have its optical properties altered. The most common magnetic field probe currently in use is the Hall probe consisting of a strip of bismuth metal. As an electric current flows along the strip in the presence of a magnetic field a potential develops normal to this current flow. This potential is proportional to the strength of the magnetic field.

Unfortunately, the device is bulky. The optical fibre has the potential of significant size reduction.

6. *Electric fields* — All materials are affected by electric fields. The electrons in solids have distributions that depend on Coulomb interactions with the other electrons and nuclei. Quantum mechanics provides a guide to understanding such distributions.

An external electric field can perturb these distributions. Some materials like PVF_2 (polyvynilidine fluoride) are very sensitive to external electric fields because of the lop-sided nature of its structure (light hydrogens on one side of a carbon backbone, fluorines on the other).

Metals shield out external electric fields while certain ceramics like sapphire and diamond respond weakly.

Quartz crystals, on the other hand, are one of a class of substances that are crystallographically lop-sided, i.e. the atomic arrangement produces a mirror image of itself so that one can define a top and a bottom to a crystal. Crystallographers use the term non-centrosymmetric. In such cases electric fields can have a marked effect and this is why quartz is a piezoelectric crystal, i.e. its dimensions are altered by an electric field. An A.C. field turns quartz into a mechanical oscillator.

A single crystal quartz fibre would provide us with a good optical sensor. Alas, quartz is presently unavailable as a single crystal fibre.

Nonetheless, such problems are not insurmountable, and with appropriate technology FO (fibre optic) sensors should be available that are quite sensitive to electrical fields. If the atoms move (as in quartz) the optical properties will change. It's now a materials problem.

7. *Electromagnetic radiation* — Electromagnetic radiation at virtually all frequencies is detectable. In most cases the radiation detector converts the absorbed photons into a measurable electrical signal. FO sensor interest lies in conversion of the incoming radiation into an optical signal.

One can argue that once you produce an electric signal you can use a device like an LED (light emitting diode) to produce the optical signal, but for FO sensors we want to be cleverer than that.

Basically, we might consider four ranges of electromagnetic energy; I. Hard, which includes X-rays and γ-rays; II. Medium, which includes UV and visible; III. Soft, which includes infra red and microwaves and IV. Very Soft, which includes radiowaves.

(I) *Hard.* X-rays and γ-rays can be detected with Ge or Si solid state devices which, as mentioned above, convert the entire energy into an electron pulse. GaAs (appropriately doped) has a similar diamond-like atomic arrangement to Ge, but has the added advantage in that it can be used as a laser. Unfortunately it emits in the infra red, which presents some problems in fibre optic transmission.

The direction for development is to find a detector (like ZnS) that emits efficiently in a more convenient frequency range. Total conversion of a one MeV γ-ray into visible light would produce about 200,000 photons, easily seen by the naked eye.

(II) *Medium.* UV and visible light can be turned into monochromatic light of a lesser frequency through fluorescence. Plastic optical fibres are commercially available which have been embedded with fluorescent materials.

With some ingenuity, one can employ the incoming radiation to excite a laser so that the outgoing radiation is coherent and more efficiently directed along the fibre.

(III) *Soft.* Infra red and microwave detection is more difficult because the radiation is generally too soft to cause fluorescence. However, it is capable of exciting photons (heat energy) in solids or liquids.

This heat can alter the local dimensions of an optical fibre and cause changes in the optical properties like polarization or attenuation.

Hence such low energy signals can only be employed in sensor applications to *modulate* the optical signal, but the modulation can be directed to the very point of detection.

(IV) *Very soft.* Radio waves are so weak and of such long wave length that one must rely on some clever modulation method for detecting them and converting them into an optical signal.

Unfortunately, we do not have a material that can produce a high order frequency enhancement from radio wave frequencies (megahertz) to optical frequencies (10^{15} hertz), i.e. over ten orders of magnitude.

No doubt special circuit designs using radio detection technology will be used, but an imaginative physicist might find a technique for more direct detection.

The problem is that photons do not interact with each other so that electrons must act as an intermediary in coupling radio waves to visible light.

One can, however, consider an electron gas within a glass tube. When struck by radio waves, this might alter their spatial distribution and modulate an optical signal traversing the same gas. Some of the coherent character of the radio waves can be retained.

Of course, this is hand waving, but this sort of thinking represents the early process in the development of fibre optical sensors for direct radio detection.

8. *Chemical species* — Consider a polymer like polyethylene, a chain of carbon atoms with two hydrogens appended to each carbon. The chains are linked to each other by weak electrostatic forces and the distances between chains is rather large.

If polythene is immersed in a liquid or gas some molecules are expected to enter the space between the polymer chains. This alters the optical density depending on the species absorbed.

This is one principle in developing fibre optics as chemical sensors. Considering the large variety of polymers available and the chemicals of interest, the possibilities are manifold.

9. *Acceleration* — For a number of years the fibre optic gyroscope has been in the process of development. The basic principle lies in the fact that the optical properties of a moving system of atoms differs from one at rest (a variation of the Doppler shift).

Its sensitivity depends on the availability of sharp laser lines. Compared to mechanical rotating gyroscopes, the FO gyroscope is simple and light weight.

10. *Biological functions* — While the endoscope enables doctors to observe bodily functions by insertion of optical fibres, one can imagine a wide variety of biological functions that can modulate a photon beam in an appropriate FO sensor: I. Blood flow and blood condition; II.

Retinal response to visual signals; III. Muscle tone; IV. Bone healing; V. Stomach, intestinal and bowel functions; VI. Kidney function, liver function; VII. Brain responses; VIII. Food metabolism.

Since optical fibres can be very small in diameter, they can be unobtrusively inserted into the body and the recording apparatus can be small enough so that a patient can remain ambulatory. The signal can even be continuously transmitted to the doctor.

The current requirements for such sensor development requires the cooperative talents of medical, optical and materials research teams.

A few comments on the VIII items might help the reader in assessing this promising field of Fibre Optic Biosensors.

Endoscopy combined with directed laser beam surgery has had a dramatic impact on medical practice. Major surgery with its attendant risks has been eliminated in many cases.

The surgeon can burn off carcinoma and cauterize at the same time. The work can be done in the surgeon's office and reduces considerably the risk and cost. Insurance companies will no doubt benefit considerably from the reduction in malpractice suits.

The ability to embed optical fibres in the body (like acupuncture) gives one access to the direct observation of most bodily functions. By adding sensor tips to the fibres one can record considerable information on chemistry and physical functions for selected organs.

X-ray radiology and its attendant risks may ultimately be eliminated. NMR and ultrasound have made useful impacts in the study of bodily functions and will continue to do so, but the ability to imbed a fibre into any point in the body must open up new vistas of medical practice.

(I) *Blood flow and blood condition.* Laser anemometry is commonly employed to determine flow patterns in liquids and gases. Light scattered from a moving particle is Doppler shifted. If the liquid or gas is transparent, small particles can be introduced to scatter the light. The red (or blue) corpuscles can serve to scatter the light from the sensor end of an optical fibre.

When blockages occur in the heart and arterial system, turbulence can occur and the flow pattern is altered considerably.

Within the same sensor system, one can determine the blood viscosity by measuring flow through a small orifice in the fibre. Blood colour can reflect the important red to white cell ratio. Foreign bodies may have specific fluorescent characteristics.

The question to be raised is—''Doc, what do you want to measure?'' Since the sensor tip need only be of dimensions 0.1 mm it should not unduly interfere with the normal function of the organ to be studied.

(II) *Retinal response to light.* For many years the response of frogs' and cats' retinas to visual stimulae have been productive in establishing the way their brains respond to images.

We are still far short of developing a totally electronic device that will recognize your grandmother, but the industry is moving in that direction.

The work on frogs' eyes indicates that they are sensitive to very simple shapes like a dark-light edge or a simple orifice transmitting light.

Image analysis commands significant attention, but the work on frogs and cats has relied on examining the electrical stimulae resulting from an optic signal. It is expected that optical fibres will enhance these measurements since they are both small and can provide direct analysis of the optic signal as well as correlate it with the electric brain response.

(III) *Muscle tone.* Considerable electrochemistry takes place in muscle tissue as well as changes in dynamic stress and temperature.

All of these responses can affect appropriately designed fibres.

Athletes are continuously subjecting their muscles to overstressing. The insurance industry is interested in reduction of sports injuries.

This technology can also be extended to animals used in sports as well as other functions requiring considerable muscle exertion.

(IV) *Bone healing.* Bone healing and hip transplants are common medical concerns. An optical-fibre biosensor can examine the state of the bone as it hardens, for it then transmits traverse acoustical waves unlike soft or liquid-like materials.

An acoustic wave can be transmitted and reflected along a fibre. While ultrasonic scanning can perform such functions, its resolution is poor. The fibre can be embedded at the precise point (or points) of interest.

Hip transplants are a major problem, since they often require replacement after a few years. Fibre sensors might be capable of monitoring degradation of the joint.

(V) *Stomach, intestinal and bowel functions.* The chemistry of these parts of the food and waste flow can be monitored with sensors that record pH, temperature, gas/liquid ratios etc., as well as specific chemical products.

The attraction of FO biosensors is the ability to provide continuous monitoring while the patient proceeds with normal activities.

Potentially, an unobtrusive, light-weight package can be carried by the patient to yield pertinent data for long periods of observation.

(VI) *Kidney and liver function.* Problems associated with diseases or malfunction of these organs are common. Diabetics face a particular problem concerning food and insulin intake. Some fibre optic sensors could record sugar levels and avert disasters.

The onset of kidney stones could be monitored in cases where a patient is susceptible to this problem. Such stones might be observed if the urine flow becomes turbulent.

The chemistry and pH of the liver offers a number of sensor opportunities. Liver functions are sometimes detected by the introduction of certain compounds which can be observed with X-ray radiology.

(VII) *Brain responses.* Brain responses are now being examined magnetically. A continuous monitoring of the alertness of airforce pilots is being measured in order to provide a signal if drowsiness develops.

The size of the magnetic field is perhaps a small fraction of a guess. These measurements are non-intrusive since the magnetic signal easily penetrates human bone and tissue.

At this point the reader may wonder whether we are opening the door to mind reading; one cannot say. The author prefers to keep his thoughts to himself.

(VIII) *Food metabolism.* This might appeal to all those heavyweights out there. At this present time, the calorie value of food is measured by burning it in a calorimeter.

No doubt carbohydrates are oxidized by the body—but how efficiently, and what happens to the fats etc.?

Specific metabolic patterns for individuals might be established by examining localized temperature rises in muscles etc. employing FO sensors.

Each individual can determine his best dietary allowance through heat evolution. By the time the excess food is stored as fat, it is a bit too late to evaluate one's metabolism.

11. *Non-destructive evaluation* — At the present time air frames are tested for cracking and flaws by removing them from service and employing a costly ultrasonic or X-ray scan.

An obvious alternative is to incorporate optical fibres into structural members so that stresses or fractures can be continuously monitored. Thus an on-board microprocessor can provide a detailed survey of the air frame stress distribution.

Optical fibres cast into concrete also provide a technique for continuously monitoring the integrity of the structure. Such techniques can be applied to oil rigs, roads, etc. The idea is already receiving attention.

Optical sensors can be employed in hazardous environments, i.e. high temperatures, high pressures, radioactivity, toxic liquids and gases etc.

Again—tell us your problem and we'll design a sensor.

12. *Image analysis* — If one takes a bundle of fibres that maintain their neighbour configuration one can transmit an image over long distances with no other electronic system or power source.

Such an arrangement of optical fibres is called a coherent bundle and is not easily produced. However, they are available commercially.

If the matrix of say 100 × 100 fibres is digitized as to colour and position, an image is easily stored. The next step in the process is the ability of a system to recognize the image.

If each fibre is identified as to intensity and colour (say 10 colours and 10 ranges of intensity), we have perhaps to deal with 10^6 bits of information.

This outlines one technology for employing fibres in image analysis. The ultimate goal is a robotic system that can identify shapes.

It will not be easy to build robots with a visual system comparable to the human eye and brain. However, the next decade or so will witness an effort to automate pattern recognition through image analysis. Artificial intelligence is the keyword.

13. *Laser Positioning and Machining* — Robotics is a key word describing anything that can sense and play an active part in a process of interest. It does not imply an anthropomorphic device nor a variation of R2D2. A simple example is a sensor that positions a piece in a lathe by measuring the phase of a reflected laser signal. It is possible in this way to turn out extremely flat aluminum mirrors with a single point diamond tool. In addition to employing a low power laser as a measuring device, it is possible to increase the power and perform cutting and drilling operations. Under very high powers the laser beam can become a weapon, shades of Buck Rogers.

The positioning capability relies on the coherency of the laser beam, while the machining and weapon functions depend on the ability to pump atoms to a metastable state before the stimulated chain reaction launches all the laser energy in the same direction. Lasers can be pulsed or, at lower powers, operated in a CW (continuous wave) mode.

Synchrotron Sources

By electrostatically accelerating an electron beam in a magnetic field, the current of particles can be confined to a circular orbit. When an electron undergoes such bending in its path, it conserves momentum by radiating photons along a tangent to its path. This yields a most copious controlled source of photons covering the entire spectrum from infra-red to X-rays. These "photon factories" as they are sometimes called, have been producing research reports on the properties of matter unachievable by other means.

One popular measurement goes under the acronym XAFS (X-ray absorption fine structure). The ejected electron on the high energy side of the absorption edge is coherently scattered by the atomic configuration in its immediate vicinity, a process already alluded to in Fig. 81. This produces a structure that can be analyzed in terms of the distances and configuration of atoms surrounding the absorbing atom.

Non-linear Materials

For certain optically transparent materials, photons in a specific energy range can excite electronic transitions that alter the local electron distribution and change the index of refraction of the material. Subsequent photos entering the substance are subjected to different optical properties. The more photons entering the material, the greater the change in optical transmission. This field is called non-linear optics and bodes importance in optical computing. One photon can write a binary coded message, a second can read it. The message is erased when the excited atom returns to its ground state.

The potential advantage of such non-linear materials is to increase speed compared to silicon chip technology. It appears likely that optical computing will replace transistor technology.

Photon-Electron Interactions

The manipulation of photons in the manner described above is based on the availability of special materials in which the electron distributions and eigenvalues provide the desired interactions. We don't employ photons for practical purposes without appropriate materials that transmit, absorb, scatter, or emit light. The search for new materials in constantly underway. With the recent discovery of non-metallic superconductors the interaction between photons and electrons may soon be re-examined. To date virtually all the interesting optical materials do not depend on the mass transport of electrons, as in metals or semiconductors. Perhaps the next field of development is with optically interesting superconductors, both electrons and photons engaged simultaneously in transmitting information.

Conclusion

It no longer is prudent to produce scientists who specialize in metals only. The photon has become as important as the electron, and structural materials need no longer be metallic. Ceramics such as sialon and diamond are now employed as cutting tools and there is no reason to believe that one must turn to metals to build something. For this reason universities are abandoning the term metallurgist and replacing it with materials scientist.

It is also apparent that the Schrödinger equation is not useful in predicting the properties of complex materials. An approach outlined in chap XII is necessary to develop those new materials that will advance the synergism of photon and electron behavior. In the near future the optical, electronic and structural properties of a material will all be vital in its selection for a particular end item, even a bridge.

APPENDIX I

THERMODYNAMIC TABLES

The following tables express the entropy, $S(x)$, energy, $U(x)$, and specific heat, $C(x)$, as a function of $x = \Theta/T$ where Θ is the Debye temperature in°K. The tables are based on harmonic forces only and the following approximate anharmonic corrections give the total entropy, energy and specific heat from a knowledge of the coefficient of linear expansion, α. The energy does not include the zero point energy, $(9/8)R\Theta$, nor the cohesive energy, H_0. At zero pressure the enthalpy, H, equals the energy, U.

Lattice

Total Entropy	$S \cong S(x) + 6\alpha U(x)$
Total Energy	$U \cong H \cong U(x) + (15/2)R\alpha T^2$
Total Specific Heat	$C \cong C(x)\,[1 + 6\alpha T]$

In a metal one must also add the electronic contribution, based on the electronic specific heat coefficient, γ.

Electronic

Total Entropy	$S = \gamma T$
Total Energy	$U = H = \gamma T^2/2$
Total Specific Heat	$C = \gamma T.$

$x = \frac{\Theta}{T}$	$S(x)$ cal/mol/deg
0·05	25·7783
0·06	24·6923
0·07	23·7743
0·08	22·9793
0·09	22·2750
0·10	21·6510
0·11	21·0838
0·12	20·5661
0·13	20·0900
0·14	19·6492
0·15	19·2387
0·16	18·8551
0·17	18·4950
0·18	18·1551
0·19	17·8336
0·20	17·5293
0·21	17·2399
0·22	16·9635
0·23	16·6996
0·24	16·4473
0·25	16·2051
0·26	15·9721
0·27	15·7484
0·28	15·5325
0·29	15·3245
0·30	15·1233
0·31	14·9276
0·32	14·7400
0·33	14·5580
0·34	14·3818
0·35	14·2106
0·36	14·0438
0·37	13·8821
0·38	13·7242
0·39	13·5706
0·40	13·4213
0·41	13·2755
0·42	13·1335
0·43	12·9946
0·44	12·8590
0·45	12·7263
0·46	12·5970
0·47	12·4697
0·48	12·3460
0·49	12·2242
0·50	12·1051
0·51	11·9888
0·52	11·8753
0·53	11·7627
0·54	11·6528
0·55	11·5451
0·56	11·4393
0·57	11·3256
0·58	11·2340
0·59	11·1335
0·60	11·0354
0·61	10·9388
0·62	10·8437
0·63	10·7497
0·64	10·6585

$x = \frac{\Theta}{T}$	$\frac{C}{3R}$	$\frac{U(x)}{3RT}$
0·0	1·000	1·000
0·1	1·000	0·963
0·2	0·998	0·927
0·3	0·996	0·892
0·4	0·992	0·858
0·5	0·988	0·825
0·6	0·982	0·793
0·7	0·976	0·762
0·8	0·969	0·732
0·9	0·961	0·703
1·0	0·952	0·674
1·1	0·942	0·647
1·2	0·932	0·621
1·3	0·920	0·595
1·4	0·908	0·571
1·5	0·896	0·547
1·6	0·883	0·524
1·7	0·869	0·502
1·8	0·855	0·481
1·9	0·840	0·461
2·0	0·825	0·441
2·1	0·810	0·422
2·2	0·794	0·404
2·3	0·778	0·387
2·4	0·762	0·370
2·5	0·746	0·354
2·6	0·729	0·339
2·7	0·713	0·325
2·8	0·696	0·311
2·9	0·679	0·297
3·0	0·663	0·284
3·1	0·646	0·271
3·2	0·630	0·259
3·3	0·613	0·248
3·4	0·579	0·237
3·5	0·581	0·227
3·6	0·565	0·217
3·7	0·549	0·208
3·8	0·533	0·199
3·9	0·518	0·190
4·0	0·503	0·182
4·1	0·488	0·174
4·2	0·474	0·166
4·3	0·460	0·159
4·4	0·446	0·152
4·5	0·432	0·146
4·6	0·419	0·140
4·7	0·406	0·134
4·8	0·393	0·128
4·9	0·381	0·123
5·0	0·369	0·118
5·1	0·357	0·113
5·2	0·346	0·108
5·3	0·334	0·104
5·4	0·324	0·0993
5·5	0·313	0·0952
5·6	0·303	0·0914
5·7	0·293	0·0877
5·8	0·284	0·0842
5·9	0·275	0·0808

$x = \frac{\Theta}{T}$	$S(x)$ cal/mol/deg
0·65	10·5676
0·66	10·4786
0·67	10·3910
0·68	10·3044
0·69	10·2195
0·70	10·1357
0·71	10·0533
0·72	9·9720
0·73	9·8920
0·74	9·8131
0·75	9·7355
0·76	9·6590
0·77	9·5837
0·78	9·5095
0·79	9·4363
0·80	9·3643
0·81	9·2933
0·82	9·2229
0·83	9·1533
0·84	9·0845
0·85	9·0165
0·86	8·9493
0·87	8·8829
0·88	8·8173
0·89	8·7526
0·90	8·6892
0·91	8·6265
0·92	8·5642
0·93	8·5031
0·94	8·4426
0·95	8·3827
0·96	8·3234
0·97	8·2647
0·98	8·2066
0·99	8·1491
1·00	8·0934
1·10	7·5484
1·20	7·0601
1·30	6·6206
1·40	6·2183
1·50	5·8491
1·60	5·5068
1·70	5·1906
1·80	4·8947
1·90	4·6176
2·00	4·3680

$x = \frac{\Theta}{T}$	$\frac{C}{3R}$	$\frac{U(x)}{3RT}$
6·0	0·266	0·0766
6·1	0·257	0·0745
6·2	0·249	0·0716
6·3	0·240	0·0688
6·4	0·233	0·0661
6·5	0·225	0·0636
6·6	0·218	0·0612
6·7	0·211	0·0589
6·8	0·204	0·0566
6·9	0·197	0·0545
7·0	0·191	0·0525
7·1	0·185	0·0506
7·2	0·179	0·0487
7·3	0·173	0·0470
7·4	0·168	0·0453
7·5	0·162	0·0437
7·6	0·157	0·0421
7·7	0·152	0·0406
7·8	0·147	0·0392
7·9	0·143	0·0379
8·0	0·138	0·0366
8·1	0·134	0·0353
8·2	0·130	0·0341
8·3	0·126	0·0330
8·4	0·122	0·0319
8·5	0·118	0·0308
8·6	0·115	0·0298
8·7	0·111	0·0289
8·8	0·108	0·0279
8·9	0·105	0·0271
9·0	0·101	0·0262
9·1	0·098	0·0254
9·2	0·096	0·0246
9·3	0·093	0·0238
9·4	0·090	0·0231
9·5	0·088	0·0224
9·6	0·085	0·0217
9·7	0·083	0·0211
9·8	0·080	0·0205
9·9	0·078	0·0199
10·0	0·076	0·0193
10·1	0·074	0·0187
10·2	0·072	0·0182
10·3	0·070	0·0177
10·4	0·068	0·0172
10·5	0·066	0·0167
10·6	0·064	0·0163
10·7	0·063	0·0158
10·8	0·061	0·0154
10·9	0·059	0·0150
11·0	0·058	0·0146
11·1	0·056	0·0142
11·2	0·055	0·0138
11·3	0·053	0·0135
11·4	0·052	0·0131
11·5	0·051	0·0128
11·6	0·049	0·0124
11·7	0·048	0·0121
11·8	0·047	0·0118
11·9	0·046	0·0115
$x > 12$	$\frac{77·93}{x^3}$	$\frac{19·48}{x^3}$

APPENDIX II

NUCLEAR TABLES

Tables of nuclear magnetic moments in nuclear magnetons, nuclear quadrupole moments $(e)Q$, and nuclear spin I.

The neutron absorption cross-sections, σ_{abs}, are for neutrons of 1·28 Å. The coherent scattering cross-section is related to the coherent scattering amplitude, b, by

$$\sigma_{\text{COH}} = 4\pi b^2$$

and the sign of b is given in parenthesis. $\bar{\sigma}_s$ is the average total scattering cross-section for thermal neutrons.

TABLE OF NUCLEAR MAGNETIC MOMENTS, NUCLEAR QUADRUPOLE MOMENTS, AND NUCLEAR SPINS

Element $_{Z}^{A}$	μ (n.m)	Q 10^{-24} cm^2	I	Element $_{Z}^{A}$	μ (n.m)	Q 10^{-24} cm^2	I
${}_{0}n^{1}$	−1·91315	0	1/2	${}_{20}Ca^{43}$	−1·31534		7/2
${}_{1}H^{1}$	+2·79267	0	1/2	${}_{21}Sc^{45}$	4·7492		7/2
${}_{1}H^{2}$	+0·85738	+0·00282	1	${}_{22}Ti^{47}$	−0·78686		5/2
${}_{1}H^{3*}$	2·97876	0	1/2	${}_{22}Ti^{49}$	−1·1020		7/2
${}_{2}He^{3}$	−2·12742	0	1/2	${}_{23}Va^{49}$	+4·68		7/2
${}_{3}Li^{6}$	+0·82192	−0·00046	1	${}_{23}Va^{50}$	+3·3413		6
${}_{3}Li^{7}$	3·25598	−0·042	3/2	${}_{23}Va^{51}$	+5·1392	0·30	7/2
${}_{4}Be^{9}$	−1·1774	+0·02	3/2	${}_{24}Cr^{53}$	−0·47353		3/2
${}_{5}B^{10}$	+1·8004	+0·111	3	${}_{25}Mn^{54}$	5·050		2
${}_{5}B^{11}$	2·68816	+0·0355	3/2	${}_{25}Mn^{55}$	+3·4610	+0·5	5/2
${}_{6}C^{13}$	0·70220	0	1/2	${}_{26}Fe^{57}$	+0·0903	0	1/2
${}_{7}N^{14}$	+0·40358	+0·0071	1	${}_{27}Co^{56*}$	3·855		4
${}_{7}N^{15}$	−0·2830	0	1/2	${}_{27}Co^{57*}$	+4·64		7/2
${}_{8}O^{17}$	−1·89295	−0·004	5/2	${}_{27}Co^{58*}$	+4·052		2
${}_{9}F^{19}$	2·62728	0	1/2	${}_{27}Co^{59}$	+4·6389	+0·5	7/2
${}_{11}Na^{22}$	+1·746		3	${}_{27}Co^{60*}$	+3·800		5
${}_{11}Na^{23}$	2·21624	+0·10	3/2	${}_{28}Ni^{61}$	0·05	0	1/2
${}_{11}Na^{24*}$	+1·687		4	${}_{29}Cu^{63}$	+2·2213	−0·16	3/2
${}_{12}Mg^{25}$	−0·8547		5/2	${}_{29}Cu^{64*}$	0·4		1
${}_{13}Al^{27}$	3·6385	+0·149	5/2	${}_{29}Cu^{65}$	+2·3791	−0·15	3/2
${}_{14}Si^{29}$	−0·5548	0	1/2	${}_{30}Zn^{67}$	+0·8735	+0·18	5/2
${}_{15}P^{31}$	1·1305	0	1/2	${}_{31}Ga^{69}$	2·0108	+0·2318	3/2
${}_{16}S^{33}$	0·6428	−0·064	3/2	${}_{31}Ga^{71}$	2·5546	+0·1461	3/2
${}_{16}S^{35*}$	1·00	+0·045	3/2	${}_{32}Ge^{73}$	−0·8767	−0·2	9/2
${}_{17}Cl^{35*}$	0·82086	−0·0789	3/2	${}_{33}As^{75}$	+1·4349	+0·3	3/2
${}_{17}Cl^{36*}$	+1·2839	−0·0168	2	${}_{34}Se^{77}$	+0·53249	0	1/2
${}_{17}Cl^{37}$	0·68331	−0·0621	3/2	${}_{34}Se^{79*}$	−1·01	+0·7	7/2
${}_{19}K^{39}$	0·39096	+0·14	3/2	${}_{35}Br^{79}$	+2·0991	+0·335	3/2
${}_{19}K^{40*}$	−1·2964		4	${}_{35}Br^{81}$	+2·2626	+0·280	3/2
${}_{19}K^{41}$	0·21488		3/2	${}_{36}Kr^{83}$	−0·96706	+0·15	9/2
${}_{19}K^{42*}$	−1·135		2	${}_{36}Kr^{85*}$	−1·002	+2·5	9/2

* Radioactive ; n.m.= nuclear magneton

Table of Nuclear Magnetic Moments, Nuclear Quadrupole Moments, and Nuclear Spins—*continued*

Element ($_Z$Element^A)	μ (n.m)	Q 10^{-24} cm^2	I	Element ($_Z$Element^A)	μ (n.m)	Q 10^{-24} cm^2	I
${}_{37}Rb^{81*}$	2·00		3/2	${}_{53}I^{127}$	+2·7937	−0·69	5/2
${}_{37}Rb^{85}$	+1·3482	0·31	5/2	${}_{53}I^{129*}$	+2·6030	−0·43	7/2
${}_{37}Rb^{86*}$	−1·68		2	${}_{53}I^{131*}$	−0·412		3/2
${}_{37}Rb^{87*}$	2·74138	0·15	3/2	${}_{54}Xe^{129}$	−0·7725		1/2
${}_{38}Sr^{87}$	−1·0893		9/2	${}_{54}Xe^{131}$	+0·68677	−0·12	3/2
${}_{39}Y^{89}$	−0·13682	0	1/2	${}_{55}Cs^{131*}$	+3·46		5/2
${}_{40}Zr^{91}$	−1·29802		5/2	${}_{55}Cs^{133}$	+2·56422	−0·003	7/2
${}_{41}Nb^{93}$	+6·1435	−0·16	9/2	${}_{55}Cs^{134}$	+2·9729		4
${}_{42}Mo^{95}$	−0·9099		5/2	${}_{55}Cs^{134m*}$	+1·10		8
${}_{42}Mo^{97}$	−0·9290		5/2	${}_{55}Cs^{135*}$	+2·7134		7/2
${}_{43}Tc^{99*}$	+5·6571	0·3	9/2	${}_{55}Cs^{137*}$	+2·8219		7/2
${}_{44}Ru^{99}$	−0·63		5/2	${}_{56}Ba^{135}$	+0·83229		3/2
${}_{44}Ru^{101}$	−0·69		5/2	${}_{56}Ba^{137}$	+0·93107		3/2
${}_{45}Rh^{103}$	−0·0879	0	1/2	${}_{57}La^{135*}$	+3·684	+2·7	5
${}_{46}Pd^{105}$	−0·57		5/2	${}_{57}La^{139}$	+2·7613	0·6	7/2
${}_{47}Ag^{107}$	−0·11303		1/2	${}_{58}Ce^{141*}$	0·16		7/2
${}_{47}Ag^{109}$	−0·12992	0	1/2	${}_{59}Pr^{141}$	+3·92	−0·054	5/2
${}_{47}Ag^{111*}$	−0·145	0	1/2	${}_{60}Nd^{143}$	−1·03	1	1/2
${}_{48}Cd^{111}$	−0·059215	0	1/2	${}_{60}Nd^{145}$	−0·64	1	7/2
${}_{48}Cd^{113}$	−0·61946	0	1/2	${}_{62}Sm^{147}$	−0·83	0·72	7/2
${}_{49}In^{113}$	+5·4961	1·144	9/2	${}_{62}Sm^{149}$	−0·68	0·72	7/2
${}_{49}In^{114*}$	4·7		5	${}_{63}Eu^{151}$	+3·6	+1·2	5/2
${}_{49}In^{115}$	+5·5073	1·161	9/2	${}_{63}Eu^{153}$	+1·6	+2·5	5/2
${}_{49}In^{116*}$	4·2		5	${}_{64}Gd^{155}$	−0·24	1·1	3/2
${}_{50}Sn^{115}$	−0·91318	0	1/2	${}_{64}Gd^{157}$	−0·32	1·0	3/2
${}_{50}Sn^{117}$	−0·9948	0	1/2	${}_{65}Tb^{159}$	1·5		3/2
${}_{50}Sn^{119}$	−1·04085	0	1/2	${}_{66}Dy^{161}$	−0·38		5/2
${}_{51}Sb^{121}$	+3·3415	−0·53	5/2	${}_{66}Dy^{163}$	−0·53		5/2
${}_{51}Sb^{123}$	+2·53336	−0·68	7/2	${}_{67}Ho^{165}$	3·29	+2	7/2
${}_{52}Te^{123}$	−0·7319	0	1/2	${}_{68}Er^{167}$	−0·48	+9·4	7/2
${}_{52}Te^{125}$	−0·8824	0	1·2	${}_{69}Tm^{169}$	−0·20	0	1/2

Table of Nuclear Magnetic Moments, Nuclear Quadrupole Moments, and Nuclear Spins—*continued*

Element $_Z^A$	μ (n.m)	Q 10^{-24} cm^2	I	Element $_Z^A$	μ (n.m)	Q 10^{-24} cm^2	I
$_{70}Yb^{171}$	+0·43		1/2	$_{79}Au^{198}$	0·50		3/2
$_{70}Yb^{173}$	−0·60	3·9	5/2	$_{79}Au^{199*}$	+0·24		2
$_{71}Lu^{175}$	+2·9	+5·5	7/2	$_{80}Hg^{199}$	+0·49926		1/2
$_{71}Lu^{176*}$	+4·2	6	7	$_{80}Hg^{201}$	−0·607	+0·45	3/2
$_{72}Hf^{177}$	+0·61	3	9/2	$_{81}Tl^{203}$	+1·5958	0	1/2
$_{72}Hf^{179}$	−0·47	3	9/2	$_{81}Tl^{205}$	+1·6114	0	1/2
$_{73}Ta^{181}$	+2·340	+4·0	7/2	$_{82}Pb^{207}$	+0·5835	0	1/2
$_{74}W^{183}$	+0·115	0	1/2	$_{83}Bi^{209}$	+4·03771	−0·4	9/2
$_{75}Re^{185}$	+3·1435	+2·8	5/2	$_{89}Ac^{227*}$	+1·1	−1·7	3/2
$_{75}Re^{187}$	+3·1760	+2·6	5/2	$_{92}U^{233*}$	0·54	3·4	5/2
$_{76}Os^{187}$	+0·12		1/2	$_{92}U^{235}$	0·35	3·8	7/2
$_{76}Os^{189}$	+0·65000	+0·8	3/2	$_{94}Pu^{239}$	20·4		1/2
$_{77}Ir^{191}$	+0·16	+1·5	3/2	$_{94}Pu^{241}$	+1·4		5/2
$_{77}Ir^{193}$	+0·17	+1·5	3/2	$_{95}Am^{241*}$	+1·4	+4·9	5/2
$_{78}Pt^{195}$	+0·6003	0	1/2	$_{95}Am^{243*}$	+1·4	+4·9	5/2
$_{79}Au^{197}$	+0·1439	+0·6	3/2				

TABLE OF NEUTRON ABSORPTION CROSS-SECTIONS IN BARNS (10^{-24} cm²); NEUTRON COHERENT CROSS-SECTIONS IN BARNS; AND TOTAL SCATTERING CROSS-SECTIONS IN BARNS. $\sigma_{coh} = 4\pi b^2$ WHERE b IS THE SCATTERING AMPLITUDE WHOSE SIGN IS GIVEN IN PARENTHESIS UNDER σ_{coh}. THE UNCERTAINTIES IN THE VALUES ARE IN THE LAST SIGNIFICANT FIGURE

Element	σ_{abs} (10^{-24} cm²)	σ_{coh} (sign) (10^{-24} cm²)	$\bar{\sigma}_s$ (10^{-24} cm²)
$_1$H	0·33	1·79 (−)	38
$_2$He	~0	1·1 (+)	0·8
$_3$Li	71·0	0·4 (−)	1·4
$_4$Be	0·010	7·53 (+)	7
$_5$B	755		4
$_6$C	0·003	5·50 (+)	4·8
$_7$N	1·88	11·0 (+)	10
$_8$O	<0·0002	4·2 (+)	4·2
$_9$F	<0·01	3·8 (+)	3·9
$_{10}$Ne	2·8		2·4
$_{11}$Na	0·505	1·55 (+)	4·0
$_{12}$Mg	0·063	3·60 (+)	3·6
$_{13}$Al	0·230	1·5 (+)	1·4
$_{14}$Si	0·13	2·0 (+)	1·7
$_{15}$P	0·19	3·1 (+)	5
$_{16}$S	0·49	1·20 (+)	1·1
$_{17}$Cl	31·6	12·1 (+)	16
$_{18}$Ar	0·62	0·5 (+)	1·5
$_{19}$K	1·97	1·5 (+)	1·5
$_{20}$Ca	0·43	3·0 (+)	
$_{21}$Sc	24·0	17·5 (+)	24
$_{22}$Ti	5·6	1·4 (−)	4
$_{23}$Va	5·1	0·032 (−)	5
$_{24}$Cr	2·9	1·56 (+)	3·0
$_{25}$Mn	13·2	1·7 (−)	2·3
$_{26}$Fe	2·53	11·37 (+)	11
$_{27}$Co	37·0	1·00 (+)	7
$_{28}$Ni	4·6	13·2 (+)	17·5
$_{29}$Cu	3·69	7·0 (+)	7·2
$_{30}$Zn	1·06	4·3 (+)	3·6
$_{31}$Ga	2·77		4
$_{32}$Ge	2·35	8·8 (+)	3
$_{33}$As	4·1	5·0 (+)	6
$_{34}$Se	11·8	10·0 (+)	11
$_{35}$Br	6·6	5·7 (+)	6
$_{36}$Kr	28		7·2
$_{37}$Rb	0·7	3·8 (+)	12
$_{38}$Sr	1·16	4·1 (+)	10
$_{39}$Y	1·38		3
$_{40}$Zr	0·180	5·0 (+)	8
$_{41}$Nb	1·1	6·0 (+)	5
$_{42}$Mo	2·5	5·6 (+)	7
$_{43}$Tc	100		
$_{44}$Ru	2·46		6
$_{45}$Rh	150	4·5 (+)	5
$_{46}$Pd	8·0	5·0 (+)	3·6

Table of Neutron Cross-Sections—*continued*

Element	$\sigma_{abs}(10^{-24}\ cm^2)$	σ_{coh} (sign) $(10^{-24}\ cm^2)$	$\bar{\sigma}_s\ (10^{-24}\ cm^2)$
$_{47}Ag$	6·2	4·6 (+)	6
$_{48}Cd$	2550		7
$_{49}In$	190	~1 (+)	2·2
$_{50}Sn$	0·60	4·6 (+)	4
$_{51}Sb$	5·5	3·7 (+)	4·3
$_{52}Te$	4·5	4·0 (+)	5
$_{53}I$	6·7	3·4 (+)	3·6
$_{54}Xe$	35		4·3
$_{55}Cs$	29·0	3·0 (+)	20
$_{56}Ba$	1·17	3·5 (+)	8
$_{57}La$	8·9	8·7 (+)	15
$_{58}Ce$	0·70	2·7 (+)	9
$_{59}Pr$	11·2	2·4 (+)	
$_{60}Nd$	46	6·5 (+)	
$_{62}Sm$	5500		
$_{63}Eu$	4600		8
$_{64}Gd$	46000		
$_{65}Tb$	44		
$_{66}Dy$	1100		100
$_{67}Ho$	64		
$_{68}Er$	166	7·8 (+)	15
$_{69}Tm$	118		
$_{70}Yb$	36		12
$_{71}Lu$	108		
$_{72}Hf$	105		8·2
$_{73}Ta$	21·3	6·1 (+)	5
$_{74}W$	19·2	2·74 (+)	5
$_{75}Re$	84		14
$_{76}Os$	14·7		11
$_{77}Ir$	430		
$_{78}Pt$	8·1	11·2 (+)	10
$_{79}Au$	98·0	7·3 (+)	9·3
$_{80}Hg$	380	22 (+)	20
$_{81}Tl$	3·3	9 (+)	14
$_{82}Pb$	0·170	11·5 (+)	11
$_{83}Bi$	0·032	9·35 (+)	9
$_{90}Th$	7·0	12·8 (+)	12·5
$_{92}U$	2·7	9·0 (+)	8·3

Note—Some cross-sections σ_{coh} for separated isotopes are:
H^2 5·4(+); Li^6 6(+); C^{13} 4·5(+); Fe^{54} 2·2(+); Fe^{56} 12·8(+); Fe^{57} 0·64(+); Ni^{58} 25·9(+); Ni^{60} 1·1(+); Ni^{62} 9·5(−); Ag^{107} 8·7(+); Ag^{109} 2·3(+).

APPENDIX III

REFERENCES, FUNDAMENTAL CONSTANTS, LIST OF SYMBOLS

Suggested Reference Volumes for Metallurgists
(For further detailed reading and derivation of equations used in this text)

Chapters I, II

Introduction to Atomic Spectra. H. E. White, McGraw-Hill, N.Y., (1934)
Elementary Wave Mechanics. W. Heitler, Clarendon Press, Oxford, (1956)
Introduction to Quantum Mechanics. Pauling and Wilson, McGraw-Hill, N.Y., (1935)
The Wave Mechanics of Electrons in Metals. S. Raimes, North-Holland, Amsterdam, (1961)
Theory of Cohesion. M. A. Jaswon, Pergamon Press Ltd., London, (1954)
Quantum Theory of Matter. J. C. Slater, McGraw-Hill, N.Y., (1951)

Chapters III, IX

Physical Chemistry of Metals. L. S. Darken and R. W. Gurry, McGraw-Hill, N.Y., (1953)
Thermodynamics of Alloys. J. Lumsden, Institute of Metals, London, (1952)
Experimental Techniques in Low Temperature Physics. G. K. White, Clarendon Press, Oxford, (1959)
C. A. Swenson, *Solid State Physics*, **11**, 41, (1960) (pressure)

Chapter IV

Radioactivity and Nuclear Physics. J. M. Cork, van Nostrand, N.Y., (1947)
Introduction to Atomic and Nuclear Physics, H. Semat, Holt, Rinehart & Wilson, N.Y., 4th ed., (1962)

Chapter V

Methods of Experimental Physics. K. Lark Horovitz, V. Johnson, **6**, Academic Press, N.Y., (1959)
Modern Research Techniques in Physical Metallurgy. A.S.M., Cleveland, Ohio, (1953)
Symposium on Determination of Gases in Metals. A.S.T.M., Philadelphia, Pa., (1957)

Chapter VI

Neutron Diffraction. G. E. Bacon, Clarendon Press, Oxford, 2nd ed., (1962).
Techniques for Electron Microscopy. D. Kay, Blackwell Scientific Publications, Oxford, (1961)
Electron Diffraction. Z. G. Pinsker, Butterworths, London, (1953)
Handbuch der Physik XXX. Springer-Verlag, Berlin, (1957)
X-ray Diffraction Procedures. H. P. Klug and L. E. Alexander, John Wiley, N.Y., (1954)
X-rays in Theory and Experiment. A. H. Compton and S. K. Allison, van Nostrand, N.Y., (1935)
Optical Principles of the Diffraction of X-rays. R. W. James, G. Bell, London, (1954)
A Handbook of Lattice Spacings and Structures of Metals and Alloys. W. B. Pearson, Pergamon Press, London, (1958)

Chapter VII

Advances in Physics, **6**, 101, (1957)
Handbuch der Physik XXX. Springer-Verlag, Berlin, (1957)

Chapter VIII

The Fermi Surface. John Wiley, N.Y., (1960)
Handbuch der Physik, XIV. Springer-Verlag, Berlin, (1956)

	Handbuch der Physik, XIX. Springer-Verlag, Berlin, (1956)
	The Hall Effect and Related Phenomena. E. H. Putley, Butterworths, London, (1960)
Chapter X	*Ferromagnetism.* R. M. Bozorth, van Nostrand, N.Y., (1951)
	Ferromagnetic Properties of Metals and Alloys. K. Hoselitz, Clarendon Press, Oxford, (1952)
	Magneto Chemistry. P. W. Selwood, Interscience, N.Y., (1956)
Chapter XI	*Nuclear Magnetic Resonance.* E. R. Andrew, University Press, Cambridge, (1955)
	Radiation Effects in Solids. G. J. Dienes and G. H. Vineyard, Interscience, N.Y., (1957)
	Radiation Damage in Solids. D. S. Billington and S. H. Crawford, Princeton University Press, N.J., (1961)
	P. R. Wallace, *Solid State Physics*, **10**, 1, (1960) (positrons)
	D. Lazarus, *Solid State Physics*, **10**, 71, (1960) (diffusion)
	W. M. Lomer, *Progress in Metal Physics*, **8**, 255, (1959) Pergamon Press, London (defects in metals)
General	*Introduction to Solid State Physics.* C. Kittel, John Wiley, N.Y., (1956)
	Solid State Physics. A. J. Decker, Prentice Hall Inc., New Jersey, (1957)
	Properties of Metals and Alloys. N. F. Mott and H. Jones, Oxford University Press, (1936)
	Modern Theory of Solids. F. Seitz, McGraw-Hill, N.Y., (1940)
	Solid State Physics. F. Seitz and D. Turnbull, Academic Press, N.Y., (1956) **1-12**
	The Structure of Metals and Alloys. W. Hume-Rothery, Institute of Metals, London, (1954)
	Theoretical Structural Metallurgy. A. H. Cottrell, Edward Arnold, London, (1948)
	American Institute of Physics Handbook. McGraw-Hill, N.Y., (1957)
	Rare Earth Alloys. K. A. Gschneidner Jr., van Nostrand, Princeton, N.J., 1961

Constants and Conversion Factors

Rydberg $= me^4/2\hbar^2 = 13{\cdot}598$ eV (Hydrogen)
Bohr radius $= a_1 = \hbar^2/me^4 = 0{\cdot}5292 \times 10^{-8}$ cm
1 eV = 8066·0 wave numbers $= 1{\cdot}6021 \times 10^{-12}$ ergs = 23060 cal/mol
R = gas constant = 1·98647 cal/mol/deg; $3R$ = 5·9594 cal/mol/deg
Bohr magneton $= e\hbar/2m_0c = 0{\cdot}9273 \times 10^{-20}$ erg/gauss
Nuclear magneton $= e\hbar/2M_pc = 0{\cdot}5050 \times 10^{-23}$ erg/gauss
1 calorie = 4·1855 joules
1 amu = 931·4 MeV $= 1{\cdot}49 \times 10^{-3}$ ergs $= 1{\cdot}66 \times 10^{-24}$ gm
m_0 = mass of the electron $= 9{\cdot}1083 \times 10^{-28}$ g $= 5{\cdot}488 \times 10^{-4}$ amu
$m_0c^2 = 0{\cdot}511$ MeV
e = charge on the electron $= 4{\cdot}8029 \times 10^{-10}$ esu $= 1{\cdot}6 \times 10^{-20}$ emu
N_0 = Avogadro's number $= 6{\cdot}0248 \times 10^{23}$/mol
c = velocity of light $= 2{\cdot}99793 \times 10^{10}$ cm/sec
k = Boltzman's constant $= 1{\cdot}3804 \times 10^{-16}$ ergs/deg $= 0{\cdot}8616 \times 10^{-4}$ eV/deg
h = Planck's constant $= 6{\cdot}625 \times 10^{-27}$ erg sec $= 4{\cdot}135 \times 10^{-15}$ eV-sec
M_n = mass of the neutron = 1·008982 amu
M_p = mass of the proton $= 1{\cdot}67248 \times 10^{-24}$ g = 1·008142 amu
M_H = mass of the hydrogen atom $= 1{\cdot}67339 \times 10^{-24}$ g
g = earths gravitational acceleration $\cong$ 980·665 cm/sec^2
1 kilobar $= 1{\cdot}0197 \times 10^3$ kilograms/cm^2 $= 0{\cdot}98692 \times 10^3$ atmospheres $= 10^9$ dynes/cm^2
$e^2/m_0c^2 = 2{\cdot}8179 \times 10^{-13}$ cm

F = Farad = 96,489 Coulombs/mol/per unit valence
0°K = Absolute zero = −273·18°C
1 Coulomb = $6{\cdot}281 \times 10^{18}$ electronic charges
1 inch = 2·540005 cm.
1 Å = 10^{-8} cm = 1 X.U./1·002

Magnetic field due to a magnetic dipole $= \mathbf{H} = \dfrac{\boldsymbol{\mu}}{r^3} - \dfrac{3(\boldsymbol{\mu}\cdot\mathbf{r})\,\mathbf{r}}{r^5}$ $\left\{\begin{array}{l}\mu \text{ in erg/gauss}\\ r \text{ in cm}\\ H \text{ in gauss}\end{array}\right.$

Wavelength of neutron λ (Å) $= \dfrac{0{\cdot}287}{\sqrt{E(\text{eV})}}$

General References (page numbers refer to this text)

1. C. L. Pekeris, *Phys. Rev.*, **115**, 1216, (1959) p. 12
2. C. Herring and A. G. Hill, *Phys. Rev.*, **58**, 138, (1940) pp. 50; 218
3. R. Stuart and W. Marshall, *Phys. Rev.*, **120**, 353, (1960) p. 60
4. G. Liebfried, *Handbuch Der Physik*, **7/1**, 104, (1955) p. 76
5. K. Lark Horovitz and V. Johnson, *Methods of Experimental Physics*, **6B**, 39, (1959) p. 116
6. *International Tables for X-ray Crystallography*, Kynoch Press, Birmingham, England, (1952) p. 121
7. A. J. Freeman, *Acta. Cryst.*, **12**, 261, (1959) p. 121
8. L. H. Thomas and K. Umeda, *J. Chem. Phys.*, **26**, 295, (1957) p. 121
9. W. B. Pearson, *A Handbook of Lattice Spacings and Structures of Metals and Alloys*, Pergamon Press, London, (1958) p. 122
10. B. Borie, *Acta Cryst.*, **10**, 89, (1957) p. 126
11. T. Muto and Y. Takagi, *Solid State Physics*, **1**, 193, Academic Press Inc., New York, (1955). p. 133
12. C. H. Dauben and D. Templeton, *Acta Cryst.*, **8**, 841, (1955) p. 147
13. C. B. Walker, *Phys. Rev.*, **103**, 558, (1956) p. 157
14. D. R. Chipman and A. Paskin, *J. App. Phys.*, **30**, 1998, (1959) p. 160
15. D. R. Chipman, *J. App. Phys.*, **31**, 2012, (1960) p. 161
16. J. J. DeMarco and D. R. Chipman, *M.R.L. Report* No. 30, Ordnance Materials Res. Lab., Watertown, Mass, U.S.A., (1959) p. 161
17. C. B. Walker, *Phys. Rev.*, **103**, 547, (1956) p. 162
18. Batterman, Chipman and DeMarco, *Phys. Rev.*, **122**, 68, (1961) p. 151
19. Bensch, Witte and Wölfel, *Zeit. für Phys. Chemie*, **4**, 65, (1955) p. 151
20. A. J. Bradley, *Proc. Phys. Soc. London*, **47**, 879, (1935) p. 168
21. J. S. Kasper and R. M. Waterstrat, *Phys. Rev.*, **109**, 1551, (1958) p. 174
22. C. G. Shull and M. K. Wilkinson, *Phys. Rev.*, **97**, 304, (1955) pp. 176; 182
23. Gersch, Shull and Wilkinson, *Phys. Rev.*, **103**, 525, (1956) p. 178
24. N. F. Ramsey *Molecular Beams*, Clarendon Press, Oxford, 195, (1956) p. 180
25. Nathans, Shull, Shirane and Anderson, *J. Phys. Chem. Solids*, **10**, 135, (1959) p. 181
26. C. G. Shull and Y. Yamada, *International Conference on Magnetism*, Kyoto, (1961) paper 306. p. 184
27. R. J. Weiss and A. J. Freeman, *J. Phys. Chem. Solids*, **10**, 147, (1959) p. 184
28. B. N. Brockhouse and N. T. Stewart, *Rev. Mod. Phys.*, **30**, 236, (1958) p. 187
29. Palevsky, Hughes, Kley and Turkelo, *Phys. Rev. Letters*, **2**, 258, (1959) p. 189
30. Domb, Maradudin, Montroll and Weiss, *Phys. Rev.*, **115**, 24, (1959) p. 190
31. A. T. Stewart and B. N. Brockhouse, *Rev. Mod. Phys.*, **30**, 236, (1958) pp. 192; 193
32. R. N. Sinclair and B. N. Brockhouse, *Phys. Rev.*, **120**, 1638, (1960) p. 193
33. L. O. Brockway and L. S. Bartell, *R.S.I.*, **25**, 569, (1954) p. 196
34. L. S. Bartell and L. O. Brockway, *Phys. Rev.*, **90**, 833, (1953) p. 200
35. Hirsch, Horne and Whelan, *Phil. Mag.*, *Eighth Series*, **1**, 677, (1956) p. 203
36. L. G. Parratt, *Phys. Rev.*, **56**, 295, (1939) p. 210

37. R. E. Watson, *Technical Rept.* No. 12 Solid State and Molecular Theory Group, M.I.T., Cambridge, Mass. p. 212
38. W. W. Beeman and H. Friedman, *Phys. Rev.*, **56**, 392, (1939) p. 215
39. A. Sandström, *Handbuch Der Physik*, *XXX*, 230, (1957) Springer-Verlag, Berlin p. 216
40. C. H. Shaw, *Theory of Alloy Phases*, p. 13, A.S.M., Cleveland, Ohio, (1956) p. 217
41. Skinner, Bullen and Johnston, *Phil. Mag.*, **45**, 1070, (1954) p. 217
42. L. G. Schulz, *J.O.S.A.*, **44**, 357, (1954) p. 224
43. *Methods of Exp. Physics*, **6**, 283, Academic Press N.Y. (1959) p. 226
44. R. M. Bozorth, *Ferromagnetism*, D. van Nostrand Co. N.Y., 748, (1951) p. 240
45. Y. P. Galdukov, *J.E.T.P.*, **37**, 913, (U.S.S.R.) (1960) p. 244
46. D. Shoenberg, *Progress in Low Temp. Physics*, II, 245, Interscience Publishers, N.Y., (1957) p. 246
47. D. N. Langenberg and T. W. Moore, *Phys. Rev. Letters*, **3**, 328, (1959) p. 248
48. B. Pippard, *Phil. Trans.*, **A250**, 325, (1957) p. 249
49. Corak, Goodman, Satterwaite and Wexler, *Phys. Rev.*, **96**, 1442, (1954) p. 262
50. G. K. White, *Experimental Techniques in Low Temp. Physics*, Oxford Press, Chap. IV (1959) p. 269
51. Wallace, Sidles and Danielson, *J. Appl. Phys.*, **31**, 168, (1960) p. 273
52. C. V. Heer and R. A. Erikson, *Phys. Rev.*, **108**, 896, (1957) p. 278
53. P. H. Keesom and C. A. Bryant, *Phys. Rev. Letters*, **2**, 260, (1959) p. 279
54. G. Seidel and P. H. Keesom, *Phys. Rev. Letters*, **2**, 261, (1959) p. 279
55. P. H. Sidles and G. C. Danielson, *J. App. Phys.*, **25**, 58, (1954) p. 287
56. R. W. Morse and J. D. Gavenda, *Phys. Rev. Letters*, **2**, 250, (1959) p. 292
57. R. L. Orr, *Acta Met.*, **8**, 489, (1960) p. 298
58. Orr, Goldberg and Hultgren, *R.S.I.*, **28**, 767, (1957) p. 298
59. Budworth, Hoare and Preston, *Proc. Roy. Soc.*, **A251**, 250, (1960) pp. 300; 323
60. Rubin, Leach, and Bever, *J. Metals*, **7**, 421, (1955) p. 301
61. R. E. Scott, *J. App. Phys.*, **31**, 2122, (1960) p. 302
62. Kaufman, Clougherty and Weiss, *Acta Met.* (1963) p. 303
63. L. Patrick, *Phys. Rev.*, **93**, 384, (1954) p. 303
64. D. Lazarus, *Phys. Rev.*, **76**, 545, (1949) p. 305
65. *Handbuch Der Physik*, *XIV*, Springer Verlag, Berlin, (1956) p. 306
66. S. Foner, *R.S.I.*, **30**, 557, (1959) p. 311
67. H. Danan, *J. de Phys. et Rad.*, **20**, 203, (1959) p. 312
68. R. M. Bozorth, *Ferromagnetism*, D. van Nostrand Co. N.Y. Chap. II, (1951) p. 315
69. R. M. Bozorth, *Ferromagnetism*, D. van Nostrand Co. N.Y., 431, (1951) p. 316
70. D. D. Davis and R. M. Bozorth, *Phys. Rev.*, **118**, 1543, (1960) p. 317
71. A. J. P. Meyer, *J. de Phys. et Rad.*, **20**, 430, (1959) p. 319
72. A. F. Kip, *Rev. Mod. Phys.*, **25**, 229, (1953) p. 325
73. F. M. Johnson and A. H. Nethercot Jr., *Phys. Rev.*, **114**, 705, (1960) p. 327
74. R. T. Schumacher and C. P. Slichter, *Phys. Rev.*, **101**, 18, (1956) p. 328
75. A. J. P. Meyer and S. Brown, *J. de Phys. et Rad.*, **18**, 161, (1957) p. 330
76. M. H. Cohen and F. Reif, *Solid State Physics*, **5**, 332, Academic Press, N.Y. (1957) p. 333
77. L. Drain, *Phil. Mag.*, **4**, 484, (1959) p. 340
78. G. B. Benedek and J. Armstrong, *Tech. Report* No. 332, Cruft. Laboratory, Harvard p. 342
79. Knight, Hewitt and Pomerantz, *Phys. Rev.*, **104**, 271, (1956) p. 343
80. G. B. Benedek and T. Kushida, *J. Phys. Chem. of Solids*, **5**, 241, (1958) p. 344
81. W. W. Simmons and C. P. Slichter, *Phys. Rev.*, **121**, 1580, (1961) p. 345
82. T. J. Rowland, *Phys. Rev.*, **119**, 900, (1960) p. 347
83. Sagalyn, Paskin and Harrison, *O.M.R.L. Report* No. 96, Watertown Mass., (1961) p. 347
84. W. Kohn and S. H. Vosko, *Phys. Rev.*, **119**, 912, (1960) p. 347
85. N. Bloembergen and T. Rowland, *Acta Met.*, **1**, 731, (1953) p. 349
86. Averbuch, Bergevin and Muller-Warmuth, *Comptes Rendus*, **22**, 2315, (1959) p. 350
87. H. S. Gutowsky, *Phys. Rev.*, **83**, 1074, (1951) p. 351
88. J. J. Spokas and C. P. Slichter, *Phys. Rev.*, **113**, 1462, (1959) p. 354

89. A. G. Redfield, *Phys. Rev.*, **98**, 1787, (1955) p. 354
90. De Pasquali, Frauenfelder, Margulies and Peacock, *Phys. Rev. Letters*, **4**, 71, (1960) p. 357
91. Walker, Wertheim and Jaccarino, *Phys. Rev. Letters*, **6**, 98, (1961) p. 360
92. W. Jost, *Diffusion in Solids, Liquids and Gases*, Academic Press, N.Y., (1952) p. 361
93. Berghézan, Lacombe and Chaudron, *Comptes Rendus*, **231**, 576, (1950) p. 362
94. C. Leymome and P. Lacombe, *J. App. Rad. and Isotopes*, **5**, 175, (1959) p. 362
95. H. J. Gomberg, *Nucleonics*, **9**, 28, (1951) p. 362
96. Michael, Leavitt, Bever and Spedden, *J. App. Phys.*, **22**, 1403, (1952) p. 362
97. Lang, de Benedetti and Smolochowski, *Phys. Rev.*, **99**, 596, (1955) p. 370
98. A. T. Stewart, *Phys. Rev.*, **99**, 54, (1955) p. 370
99. R. A. Ferrell, *Rev. Mod. Phys.*, **28**, 308, (1956) p. 370
100. G. Lang and S. Benedetti, *Phys. Rev.*, **108**, 920, (1957) p. 371
101. R. E. B. Makinson and A. P. Roberts, *Aust. J. of Phys.*, **13**, 437, (1959) p. 378

List of Symbols According to Chapter

a	acceleration	I
a_1	Bohr radius	I, II, VI
a_0	lattice parameter	I, II, III, IV, V, VI, VII, XI
a	lattice constant, unit cell	VI, XII
$\mathscr{A}_0^*$	extremal area of Fermi surface	VIII
Å	angstrom	I, II, III, VI, VII, VIII, IX, X
A	atomic number, $(Z + N)$	IV, VI
A	extinction parameter (p. 138)	VI
A	area	VI, VIII, IX
a	activity	IX
A	anisotropy energy	X
A	exchange energy	XI
a_s	excluded volume (radius) p. 371	XI
b.c.c.	body centered cubic	I, II, V, VI, IX, XI
b	lattice constant, unit cell	VI
b_0	lattice parameter	VI
b	nuclear scattering amplitude	VI
B	binding energy	VII
B	thermal resistivity constant, ρ_T	VIII
B	magnetic induction	X
c	velocity of light	I, IV, VI, VII, VIII, IX, X, XI
°C	degrees centigrade	III

C	specific heat	III, VIII, IX
c_{11}, c_{12}, c_{44}	elastic constants	III, XII
c	lattice constant, unit cell	VI
c_0	lattice parameter	VI, XI
C	integrated intensity	VI
c	fraction, atomic	IX, XI
c	concentration	XI
dia.	diamond	I
d	interplanar spacing	II
D, d	orbital quantum number	I, II, IX, X, XI, XII
D	minimum energy in Morse potential	III, VI
D	size of crystalline grain	VI
D	spacing between rows of atoms	VI
D, D_0	diffusion coefficient	XI
e_g	d orbital in cubic crystal	II, VI, IX, X, XI, XII
E	total energy	I, II, III, IV, VI, VII, VIII, XI, XII
e	charge on the electron	I, II, IV, VI, VII, VIII, IX, X
eV	electron volt	I, II, III, IV, VII, VIII, IX, X, XI, XII
E_F	Fermi energy	X
E	electric field	VIII
e.u.	entropy units	IX
$\mathscr{E}$	electromotive potential	IX
f.c.c.	face centered cubic	I, II, VI, IX, XI
F	orbital quantum number	I
F	force	III
F	free energy, $U - TS$	III, VIII
f	partition function	III
F	structure factor	VI
f, f_s, f_L	atomic scattering factor	VI, VII
$\mathscr{F}$	Compton scattering function	VI
f	orbital quantum number	I, II, X, XI

G	free energy, $H - TS = U + PV - TS$	III, IX
g_0, g_1	degeneracy	III
g	statistical factor	IV, VI
G	shear modulus	VI
G_0	flux	VI, XI
g_λ	Maxwellian distribution	VI
g	Landẹ g factor	VI, VIII, IX, X, XII
g	oscillator strength	VII
$g(\nu)$	shape of resonance	XI
h	Miller index	I, II, VI, VII
h.c.p.	hexagonal close packed	I, II, XI
h	Planck's constant	I, II, III, IV, VII, X, XI
$\hbar$	Planck's constant divided by $2h$	I, II, III, IV, VI, VIII
H	hamiltonian, total energy operator	I
H	enthalpy	III, VIII, IX
H	magnetic field	VIII, IX, X, XI
H_c	critical field	VIII
H_{eff}	effective magnetic field at nucleus	IX, XI, XII
I	nuclear spin quantum number	IV, IX, XI
I, I_0	intensity	IV, VI, VII
I	current	IX
J	exchange energy	II, VIII, IX, X, XI, XII
J	total orbital quantum number	I, VI, VIII, IX, X, XII
j	index number	VI
j	multiplicity of planes	VI
j	spherical Bessel function	VI
J	mechanical equivalent of heat	IX
$\mathscr{J}$	Debye resistivity function	VIII
k	Miller index	I, II, VI, VII
$\mathbf{k}$	wave number	II
K.E.	kinetic energy	I
k	Boltzman constant	III, VI, VIII, IX, X, XI

K	polarization factor	VI
K	compressibility	III
°K	degrees Kelvin	III, V, VIII, IX, X, XI, XII
K	$1s$ electron	IV, V
k	wave number = $2\pi/\lambda$	II, IV, VIII, IX, XII
K_α, K_β	X-ray lines	VI, VII, XII
k	imaginary part of index of refraction	VII
l	Miller index	I, II, VI, VII
L, l	length	II, III, VIII, IX
l, L	quantum number	I, II, VI, IX, XI
L	$2s$ or $2p$ electron	V
L	X-ray lines	VII
m_0	rest mass of electron	I, VII, VIII, XI
m	mass	I, II, III, IV, VI, VIII, IX, X
m	quantum number	III
m_l, m_s	quantum numbers, orbit and spin	I, II, III, IV, VI, VIII, X, XI
M_x, M_y, M_z	angular momentum	I
M	mass	II, III, IV, VI
m_I	nuclear quantum number	XI
M	atomic weight	IX
M	magnetic moment per unit volume	IX, X
m^*	effective mass	VII, VIII
$\mathcal{M}$	magnetization	III, VIII, X
$2M$	Debye-Waller factor	VI, XI
$2M'$	size-effect factor	VI
m_A, m_B	atomic faction	VI
n	index of refraction	VI, VII
n	quantum number	I, II, III, VIII
n	neutron	IV
N	number of atoms	III, XI
N	number of neutrons in nucleus	IV
N	non-umklapp process	VIII
n_0	neutrino	IV
$n.m.$	nuclear magneton	VI
n_0	number of conduction electrons per atom	VIII

N_0	number of atoms/cm^3	IV, VI, IX
N_0	number of atoms per g	X
N_0	Avogadro's number	III
P	probability	I, III, VIII, XI
p	momentum	I, II, III, IV
P, p	orbital quantum number	I, II, VI, XI
P	pressure	III
p	proton	IV
p	probability	VI
p	magnetic scattering amplitude	VI
p_F	momentum at Fermi level	VIII
P	period in $1/H$	IX
Q_{op}	quantum mechanical operator	I
Q	an observable quantity	I
$q(\nu)$	frequency distributions of normal modes	III, VI
$(e)Q$	electric quadrupole moment	IV, IX, XI
Q	X-ray intensity parameter (p. 139)	VI
Q	thermoelectric power	VIII
Q	thermal energy	IX
eq	electric field gradient	IX, XI
q	temperature ratio	IX
R	wave function	I, II, VI, X, XI
R	length	II
r	coordinate, spherical system	I, II, IV, VI, VIII, X, XI
r_e	equilibrium distance in Morse potential	III
R	gas constant	III, IX, XI
R	radius of nucleus	IV
R	radius of atom	VI
R	reflecting power	VII
R	resistance	VIII, IX

R_0, R_1	Hall coefficients	VIII
$\mathscr{R}$	size of atom	VI
$\mathscr{R}$	relativistic correction, p. 209	XI
S, s	orbital quantum number	I, II, IV, VI, VIII, IX, X, XI, XII
S	entropy	III, VIII, IX
$\mathbf{s}$, s	momentum change on scattering	VI
S	long-range order parameter	VI
S	amount of radio-activity	XI
t_{2g}	*d* orbital in cubic crystal	I, II, VI, IX, X, XI, XII
t	time	I, VIII, IX, XI
T	absolute temperature	III, VI, VIII, IX, X, XI
T_m	melting temperature	III, IX
T_c	Curie temperature, critical temperature	III, VI, VIII, IX, X
t_0	mosaic block size	VI
T, t	sample thickness	VI, VIII
T_1, T_2	relaxation times	XI, XII
T	time	XI
U	internal energy	III
$\mathbf{u}$	displacement atomic	VI
U	umklapp process	VIII
v	velocity	I, III, IX, XI
V	potential energy	I, II, III, VI, XI
V	volume	III
V	volume of unit cell	VI
v	volume	VI
V	electric potential	VIII
v	valence	IX
W_{ex}	exchange energy	II, IX
W	strain energy	VI
x	coordinate cubic system	I, II, III, IV, VI, VII, VIII, XI
$\mathbf{x}$	displacement	III
x	Θ/T	VI, VIII
x	atomic per cent	VIII
y	coordinate, cubic system	I, II, III, VI, VII, VIII, XI

Symbol	Meaning	Chapters
z	coordinate, cubic system	I, II, III, VI, VII, VIII, XI
Z	number of protons in nucleus	I, II, IV, V, VI, VII, XI
z	number of neighbours	III, VI, X
Z_{eff}	effective value of Z	I, II, X
α	electron spin	I
α	coefficient of linear expansion	III, VI, IX
α	alpha particle	IV, XI
α	short-range order parameter	VI
α	Bragg angle of monochromating crystal	VI
α'	angle	VI
β	electron spin	I
β^+	positron	IV
β, β^-	beta ray (electron)	IV, XI
β	size effect parameter	VI
$\underline{\gamma}$	Gruneisen constant	III, VI, IX
γ	electronic specific heat coefficient	III, VIII, IX, X
γ	gamma ray	IV, XI
Γ_n, Γ_γ	Breit-Wigner resonance widths	IV
γ	E/m_0c^2	VI
γ	magnetic moment of neutron	VI
Γ	resonance width	VII
γ	relaxation frequency	VII
γ	activity coefficient	IX
γ	anti-shielding factor	XI
δ	phase factor	III
$\underline{\delta}$	transmission factor (p. 144)	VI
δ	roughness correction (p. 146)	VI
δ	atomic distance, diffusion	XI
ε	dielectric constant	VII
η	normal coordinate	III

η	average angle between mosaic blocks	VI
Θ	wave function	I, II
θ	angle	I, VI, VII, IX, XI
θ	coordinate, spherical system	I, IV
Θ	Debye temperature	III, VI, VII, XII
θ_B	Bragg angle	VI, VII
Θ	Curie constant	X
κ	bulk modulus	VI
κ	spring constant	III, VI
κ	index of absorption	VI
κ	thermal diffusivity	IX
K_e, K	thermal conductivity	IX
λ	wavelength	I, II, III, IV, VI, VII, IX
λ	decay constant	IV
Λ	mean-free path of electron	VIII
μ_l, μ_s	magnetic moment	I
μ_B	Bohr magneton	I, IX, X, XI
μ_N	nuclear magneton	IV
μ	absorption coefficient	VI, VII
μ	nuclear magnetic moment	IX, XI
μ^*	magnetic moment of isomeric state	XI
ν	frequency	III, VI, VII, VIII, IX, X, XI
ξ	normal coordinate	III
ξ	ratio of s wave function at nucleus, metal to atom	XI
ρ	gyromagnetic ratio	X
ρ	density	III, VI, IX
ρ	probability distribution	IV
ρ	phase factor (p. 199)	VI
ρ	resistivity	VIII, IX
σ	wave number	III, VI
σ	cross-section	IV, VI, XI
σ	diffuse scattering correction (p. 160)	VI
σ_0	conductivity	VII
σ	standard deviation	XII

τ	half life	IV, XI
τ	relativistic correction for electrons (p. 196)	VI
τ	relaxation time	VIII
τ	spin echo time (p. 354)	XI
τ	life time	XI
Φ	wave function	I, II
ϕ	coordinate, spherical system	I, IV
φ	wave function	II
ϕ, φ	angle	II, VI, VIII
ϕ	inner potential for electrons	VI
χ	susceptibility	IX, X, XI
Ψ	total wave function	I, II, VI
ψ	one electron wave function	I, II, VI, VII, VIII, XI
Ω	solid angle	VI
ω	angular velocity	VI
$\mathfrak{S}$	surface area of Fermi surface	VIII

INDEX